Landscape and Land Capacity

The Handbook of Natural Resources, Second Edition

Series Editor:
Yeqiao Wang

Volume 1
Terrestrial Ecosystems and Biodiversity

Volume 2
Landscape and Land Capacity

Volume 3
Wetlands and Habitats

Volume 4
Fresh Water and Watersheds

Volume 5
Coastal and Marine Environments

Volume 6
Atmosphere and Climate

Landscape and Land Capacity

Edited by
Yeqiao Wang

CRC Press is an imprint of the
Taylor & Francis Group, an **informa** business

CRC Press
Taylor & Francis Group
6000 Broken Sound Parkway NW, Suite 300
Boca Raton, FL 33487-2742

Printed on acid-free paper

International Standard Book Number-13: 978-1-138-33408-3 (Hardback)

Library of Congress Cataloging-in-Publication Data

Names: Wang, Yeqiao, editor.
Title: Handbook of natural resources / edited by Yeqiao Wang.
Other titles: Encyclopedia of natural resources.
Description: Second edition. | Boca Raton: CRC Press, [2020] | Revised
edition of: Encyclopedia of natural resources. [2014]. | Includes
bibliographical references and index. | Contents: volume 1. Ecosystems
and biodiversity — volume 2. Landscape and land capacity — volume 3.
Wetland and habitats — volume 4. Fresh water and watersheds — volume 5.
Coastal and marine environments — volume 6. Atmosphere and climate. |
Summary: "This volume covers topical areas of terrestrial ecosystems, their biodiversity, services, and ecosystem management. Organized for ease of reference, the handbook provides fundamental information on terrestrial systems and a complete overview on the impacts of climate change on natural vegetation and forests. New to this edition are discussions on decision support systems, biodiversity conservation, gross and net primary production, soil microbiology, and land surface phenology. The book demonstrates the key processes, methods, and models used through several practical case studies from around the world" — Provided by publisher.
Identifiers: LCCN 2019051202 | ISBN 9781138333918 (volume 1 ; hardback) |
ISBN 9780429445651 (volume 1 ; ebook)
Subjects: LCSH: Natural resources. | Land use. | Climatic changes.
Classification: LCC HC85 .E493 2020 | DDC 333.95—dc23
LC record available at https://lccn.loc.gov/2019051202

Visit the Taylor & Francis Web site at
http://www.taylorandfrancis.com

and the CRC Press Web site at
http://www.crcpress.com

Contents

SECTION II Genetic Resource and Land Capability

SECTION III Soil

SECTION IV Landscape Change and Ecological Security

Preface

Landscape and Land Capacity is the second volume of *The Handbook of Natural Resources, Second edition (HNR)*. This volume consists of 49 chapters authored by 73 contributors from 11 countries. The contents are organized in four sections: *Landscape Composition, Configuration and Change* (10 chapters); *Genetic Resource and Land Capability* (13 chapters); *Soil* (15 chapters); and *Landscape Change and Ecological Security* (11 chapters).

The amount of land available for production of food, fiber, and other necessary products for human sustainability is limited. As the Earth's human population is expected to nine or ten billion by the mid-21st century, the proper use of land becomes a critical component in human sustainability. Land capability implies that the choice of land for a particular use contributes to the success or failure of that use. It further implies that the choice of land for a particular use will determine the potential impact of that use on surrounding resources such as air and water. Among essential land capacities is soil. All soils are not the same and they are not of the same capability for every use. Land capability is a broader concept than soil quality, which is considered as the degree of fitness of a soil for a specific use. Contemporary soil resource exhaustion has resulted in severe land degradation in the world. Combatting land degradation is humanity's major priority for global security.

A landscape is a place where natural and human forces interact. Both forces lead to landscape change over large spatial extents and long temporal spans. Landscape processes are spatially continuous and directly related to landscape position, spatial heterogeneity, and patch geometrics and adjacencies. Landscape connectivity facilitates or impedes movement among resource patches, which is crucial for a number of different ecological processes including migration, dispersal, and colonization of locally extinct habitat patches. Metrics have been developed to study the effect of landscape connectivity on a variety of organisms.

Land change, or land-use and land-cover change, is an interdisciplinary scientific theme that includes to perform repeated inventories of landscape change; to develop scientific understanding and models necessary to simulate the processes taking place; to evaluate consequences of observed and predicted changes; and to understand consequences on environmental goods and services and management. Land change is important for biogeochemical cycles such as nutrient cycling, for biodiversity through its impacts on habitats, and for climate by changing sources and sinks of greenhouse gases and land-surface properties. Land change has important implications for the provision of ecosystem services on which societies depend. Remote sensing and geospatial modeling are among critical approaches to bridge driving forces and the consequences of land changes so that impacts on environments and ecosystems can be effectively modeled.

Genetic diversity is the raw material used by natural selection to allow a population to adapt to a changing environment. It is an integral component of biodiversity that must be conserved. Ecological security is an essential cornerstone for the sustainability of any human and nature system. Ecological security depends on the balance between human demands and actions in consumption and alteration of resource base and the sustainability, vulnerability and resilience of environmental systems that provide ecosystem services. This is a particular concern for regions with sensitive and fragile ecosystems and ecotones, with trans-region and cross-boundary movements such as transportation of pollutants through water and air systems and migration of human populations and other species, and with potential intra- and inter-state conflicts in demographic, environmental, political, and resource issues.

With the challenges and concerns, the 49 chapters in this volume cover topics in *Landscape Composition, Configuration and Change*, including fragmentation and isolation, connectivity and ecological effects, dynamics, disturbance and succession, wildland fires, land use and land cover change, edge effects on wildlife, protected area management, remote sensing of protected areas, remote sensing of land surface temperature, and stewardship in conserved lands; in *Genetic Resource and Land Capability*, including genetic diversity in natural resources management, *in situ* and *ex situ* genetic resource conservations, farmer conservation and crop management for genetic resources, herbicides in environment, herbicide-resistant crops and weeds, economic impacts of insets, insects in flower and fruit feedings, integrated pest management, and land capability classification and analysis; in *Soil*, including soil carbon and nitrogen cycling, soil degradation and food security, global assessment of soil degradation, assessment of soil erosion, soil fertility and nutrient management, soli organic matter, soil organic matter and water capacity, spatial variability of soil, soil taxonomy, soil microbial ecology, soil fauna, soil invertebrates and response to forest types, spatial and temporal distribution of soil macrofauna communities; and in *Landscape Change and Ecological Security*, including sustainability and sustainable development, social aspects of sustainable agriculture, effects of volcanic eruptions on forests, ecological factors influencing landscape epidemiology of tickborne disease, simulation of post-fire vegetation recovery, landscape pattern and change with stonewall feature identification, remote sensing of urban environment and dynamics, and LANDIS Pro for forest landscape modeling.

The chapters provide updated knowledge and information in general environmental and natural science education and serve as a value-added collection of references for scientific research and for management practices.

Yeqiao Wang
University of Rhode Island

About The Handbook of Natural Resources

With unprecedented attentions to the changing environment on the planet Earth, one of the central focuses is about the availability and sustainability of natural resources and the native biodiversity. It is critical to gain a full understanding about the consequences of the changing natural resources to the degradation of ecological integrity and the sustainability of life. Natural resources represent such a broad scope of complex and challenging topics.

The Handbook of Natural Resources, Second Edition (HNR), is a restructured and retitled book series based on the 2014 publication of the *Encyclopedia of Natural Resources (ENR)*. The *ENR* was reviewed favorably in February 2015 by CHOICE and commented as *highly recommended for lower-division undergraduates through professionals and practitioners*. This *HNR* is a continuation of the theme reference with restructured sectional design and extended topical coverage. The chapters included in the *HNR* provide authoritative references under the systematic relevance to the subject of the volumes. The case studies presented in the chapters cover diversified examples from local to global scales, and from addressing fundamental science questions to the needs in management practices.

The Handbook of Natural Resources consists of six volumes with 241 chapters organized by topical sections as summarized below.

Volume 1. Terrestrial Ecosystems and Biodiversity
Section I. Biodiversity and Conservation (15 Chapters)
Section II. Ecosystem Type, Function and Service (13 Chapters)
Section III. Ecological Processes (12 Chapters)
Section IV. Ecosystem Monitoring (6 Chapters)

Volume 2. Landscape and Land Capacity
Section I. Landscape Composition, Configuration and Change (10 Chapters)
Section II. Genetic Resource and Land Capability (13 Chapters)
Section III. Soil (15 Chapters)
Section IV. Landscape Change and Ecological Security (11 Chapters)

Volume 3. Wetlands and Habitats
Section I. Riparian Zone and Management (13 Chapters)
Section II. Wetland Ecosystem (8 Chapters)
Section III. Wetland Assessment and Monitoring (9 Chapters)

Volume 4. Fresh Water and Watersheds
Section I. Fresh Water and Hydrology (16 Chapters)
Section II. Water Management (16 Chapters)
Section III. Water and Watershed Monitoring (8 Chapters)

Volume 5. Coastal and Marine Environments
Section I. Terrestrial Coastal Environment (14 Chapters)
Section II. Marine Environment (13 Chapters)
Section III. Coastal Change and Monitoring (9 Chapters)

Volume 6. Atmosphere and Climate
Section I. Atmosphere (16 Chapters)
Section II. Weather and Climate (16 Chapters)
Section III. Climate Change (8 Chapters)

With the challenges and uncertainties ahead, I hope that the collective wisdom, the improved science, technology and awareness and willingness of the people could lead us toward the right direction and decision in governance of natural resources and make responsible collaborative efforts in balancing the equilibrium between societal demands and the capacity of natural resources base. I hope that this *HNR* series can help facilitate the understanding about the consequences of changing resource base to the ecological integrity and the sustainability of life on the planet Earth.

Yeqiao Wang
University of Rhode Island

Acknowledgments

I am honored to have this opportunity and privilege to work on *The Handbook of Natural Resources, Second Edition (HNR)*. It would be impossible to complete such a task without the tremendous amount of support from so many individuals and groups during the process. First and foremost, I thank the 342 contributors from 28 countries around the world, namely Australia, Austria, Brazil, China, Cameroon, Canada, Czech Republic, Finland, France, Germany, Hungary, India, Israel, Japan, Nepal, New Zealand, Norway, Puerto Rico, Spain, Sweden, Switzerland, Syria, Turkey, Uganda, the United Kingdom, the United States, Uzbekistan, and Venezuela. Their expertise, insights, dedication, hard work, and professionalism ensure the quality of this important publication. I wish to express my gratitude in particular to those contributors who authored chapters for this HNR and those who provided revisions from their original articles published in the *Encyclopedia of Natural Resources*.

The preparation for the development of this HNR started in 2017. I appreciate the visionary initiation of the restructure idea and the guidance throughout the preparation of this HNR from Irma Shagla Britton, Senior Editor for Environmental Sciences and Engineering of the Taylor & Francis Group/CRC Press. I appreciate the professional assistance and support from Claudia Kisielewicz and Rebecca Pringle of the Taylor & Francis Group/CRC Press, which are vital toward the completion of this important reference series.

The inspiration for working on this reference series came from my over 30 years of research and teaching experiences in different stages of my professional career. I am grateful for the opportunities to work with many top-notch scholars, colleagues, staff members, administrators, and enthusiastic students, domestic and international, throughout the time. Many of my former graduate students are among and/or becoming world-class scholars, scientists, educators, resource managers, and administrators, and they are playing leadership roles in scientific exploration and in management practice. I appreciate their dedication toward the advancement of science and technology for governing the precious natural resources. I am thankful for their contributions in HNR chapters.

As always, the most special appreciation is due to my wife and daughters for their love, patience, understanding, and encouragement during the preparation of this publication. I wish my late parents, who were past professors of soil ecology and of climatology from the School of Geographical Sciences, Northeast Normal University, could see this set of publications.

Yeqiao Wang
University of Rhode Island

Acknowledgments

Aims and Scope

Land, water, and air are the most precious natural resources that sustain life and civilization. Maintenance of clean air and water and preservation of land resources and native biological diversity are among the challenges that we are facing for the sustainability and well-being of all on the planet Earth. Natural and anthropogenic forces have affected constantly land, water, and air resources through interactive processes such as shifting climate patterns, disturbing hydrological regimes, and alternating landscape configurations and compositions. Improvements in understanding of the complexity of land, water, and air systems and their interactions with human activities and disturbances represent priorities in scientific research, technology development, education programs, and administrative actions for conservation and management of natural resources.

The chapters of *The Handbook of Natural Resources, Second Edition (HNR)*, are authored by world-class scientists and scholars. The theme topics of the chapters reflect the state-of-the-art science and technology, and management practices and understanding. The chapters are written at the level that allows a broad scope of audience to understand. The graphical and photographic support and list of references provide the helpful information for extended understanding.

Public and private libraries, educational and research institutions, scientists, scholars, resource managers, and graduate and undergraduate students will be the primary audience of this set of reference series. The full set of the HNR and individual volumes and chapters can be used as the references in general environmental science and natural science courses at different levels and disciplines, such as biology, geography, Earth system science, environmental and life sciences, ecology, and natural resources science. The chapters can be a value-added collection of references for scientific research and management practices.

Editor

Yeqiao Wang, PhD, is a professor at the Department of Natural Resources Science, College of the Environment and Life Sciences, University of Rhode Island. He earned his BS from the Northeast Normal University in 1982 and his MS degree from the Chinese Academy of Sciences in 1987. He earned the MS and PhD degrees in natural resources management and engineering from the University of Connecticut in 1992 and 1995, respectively. From 1995 to 1999, he held the position of assistant professor in the Department of Geography and the Department of Anthropology, University of Illinois at Chicago. He has been on the faculty of the University of Rhode Island since 1999. Among his awards and recognitions, Dr. Wang was a recipient of the prestigious Presidential Early Career Award for Scientists and Engineers (PECASE) in 2000 by former U.S. President William J. Clinton, for his outstanding research in the area of land cover and land use in the Greater Chicago area in connection with the Chicago Wilderness Program.

Dr. Wang's specialties and research interests are in terrestrial remote sensing and the applications in natural resources analysis and mapping. One of his primary interests is the land change science, which includes performing repeated inventories of landscape dynamics and land-use and land-cover change from space, developing scientific understanding and models necessary to simulate the processes taking place, evaluating consequences of observed and predicted changes, and understanding the consequences of change on environmental goods and services and management of natural resources. His research and scholarships are aimed to provide scientific foundations in understanding of the sustainability, vulnerability and resilience of land and water systems, and the management and governance of their uses. His study areas include various regions in the United States, East and West Africa, and China.

Dr. Wang published over 170 refereed articles, edited *Remote Sensing of Coastal Environments* and *Remote Sensing of Protected Lands*, published by CRC Press in 2009 and 2011, respectively. He served as the editor-in-chief for the *Encyclopedia of Natural Resources* published by CRC Press in 2014, which was the first edition of *The Handbook of Natural Resources.*

Contributors

Erhan Akça
Adiyaman University
Adiyaman, Turkey

Jose A. Amador
Laboratory of Soil Ecology and Microbiology
University of Rhode Island
Kingston, Rhode Island

Kim A. Anderson
Environmental and Molecular Toxicology
Oregon State University
Corvallis, Oregon

Peter V. August
Department of Natural Resources Science
University of Rhode Island
Kingston, Rhode Island

John Boardman
Environmental Change Institute
University of Oxford
Oxford, United Kingdom

and

Department of Geography
University of the Free State
Bloemfontein, South Africa

Sylvie M. Brouder
Department of Agronomy
Purdue University
West Lafayette, Indiana

Mary T. Burke
UC Davis Arboreturm
University of California
Davis, California

Frederick H. Buttel
Department of Rural Sociology
University of Wisconsin, Madison
Madison, Wisconsin

Wenting Cao
Department of Geological and Atmospheric Sciences
Iowa State University
Ames, Iowa

David A. Cleveland
Environmental Studies Program
University of California
Santa Barbara, California

Janet Coit
Rhode Island Department of Environmental Management
Providence, Rhode Island

Ray Correll
Commonwealth Scientific and Industrial Research Organisation (CSIRO)
Adelaide, South Australia, Australia

Marinés de la Peña-Domene
Department of Biological Sciences
University of Illinois at Chicago
Chicago, Illinois

Eddy De Pauw
International Center for Agricultural Research in the Dry Areas (ICARDA)
Aleppo, Syria

Peter Dillon
Commonwealth Scientific and Industrial Research Organisation (CSIRO)
Adelaide, South Australia, Australia

Florent Engelmann
Institute of Research for Development (IRD)
Montpellier, France

Daniel B. Fagre
Northern Rocky Mountain Science Center
U.S. Geological Survey (USGS)
West Glacier, Montana

Thomas E. Fenton
Soil Morphology and Genesis
Agronomy Department
Iowa State University
Ames, Iowa

Joseph Fiksel
Emeritus Professor of Integrated Systems Engineering
Executive Director (retired), Sustainability Institute
The Ohio State University
Columbus, Ohio

L. R. Gopinath
M.S. Swaminathan Research Foundation
Chennai, India

Josef H. Görres
Department of Plant and Soil Science
University of Vermont
Burlington, Vermont

David Gregg
Rhode Island Natural History Survey
Kingston, Rhode Island

Burak Güneralp
Department of Geography
Texas A&M University
College Station, Texas

John L. Havlin
Department of Soil Science
North Carolina State University
Raleigh, North Carolina

Hong S. He
School of Natural Resources
University of Missouri
Columbia, Missouri

Alan D. Hecht
Office of Research and Development
U.S. Environmental Protection Agency (EPA)
Washington, District of Columbia

T. G. Huntington
U.S. Geological Survey (USGS)
Augusta, Maine

Selim Kapur
University of Çukurova
Adana, Turkey

Marcos Kogan
Integrated Plant Protection Center
Oregon State University
Corvallis, Oregon

Rai Kookana
Commonwealth Scientific and Industrial Research Organisation (CSIRO)
Adelaide, South Australia, Australia

Heather L. Kopsco
Department of Natural Resources Science
University of Rhode Island
Kingston, Rhode Island

Anne Kuhn
Atlantic Ecology Division
U.S. Environmental Protection Agency (EPA)
Narragansett, Rhode Island

William P. Kustas
Hydrology and Remote Sensing Lab
Agricultural Research Service
U.S. Department of Agriculture (USDA-ARS)
Beltsville, Maryland

R. Lal
Carbon Management and Sequestration Center
The Ohio State University
Columbus, Ohio

Yuling Liu
Cooperative Institute for Climate and Satellites
University of Maryland
College Park, Maryland

Chen Ma
School of Geographical Sciences
Northeast Normal University
Changchun, China

Carol Mallory-Smith
Oregon State University
Corvallis, Oregon

Jeffrey A. Markert
Biology Department
Providence College
Providence, Rhode Island

Thomas Mather
Department of Plant Science
University of Rhode Island
Kingston, Rhode Island

Thomas Joseph McGreevy, Jr.
Biology Department
Boston University
Boston, Massachusetts

Mallavarapu Megharaj
Commonwealth Scientific and Industrial Research Organisation (CSIRO)
Adelaide, South Australia, Australia

Emily S. Minor
Department of Biological Sciences and Institute for Environmental Science and Policy
University of Illinois at Chicago
Chicago, Illinois

Ravendra Naidu
Commonwealth Scientific and Industrial Research Organisation (CSIRO)
Adelaide, South Australia, Australia

V. Arivudai Nambi
M.S. Swaminathan Research Foundation
Chennai, India

Micheal D. K. Owen
Department of Agronomy
Iowa State University
Ames, Iowa

Antônio R. Panizzi
Embrapa Trigo
Passo Fundo, Brazil

Peter W. C. Paton
Department of Natural Resources Science
University of Rhode Island
Kingston, Rhode Island

David Pimentel
Department of Entomology
Cornell University
Ithaca, New York

Tony Prato
Department of Agricultural and Applied Economics
University of Missouri
Columbia, Missouri

Jeffrey P. Privette
National Climate Data Center
National Environmental Satellite Data Information Service
National Oceanic and Atmospheric Administration (NOAA)
Asheville, North Carolina

Mary C. Savin
Department of Crop, Soil, and Environmental Sciences
University of Arkansas
Fayetteville, Arkansas

Jennifer L. Schaeffer
Food Safety and Environmental Stewardship Program
Oregon State University
Corvallis, Oregon

Xun Shi
Department of Geography
Dartmouth College
Hanover, New Hampshire

Michael J. Singer
Department of Land, Air, and Water Resources
University of California
Davis, California

Daniela Soleri
Geography Department
University of California
Santa Barbara, California

Mark H. Stolt
Department of Natural Resources Science
University of Rhode Island
Kingston, Rhode Island

Donglian Sun
Department of Geography and Geoinformation Science
George Mason University
Fairfax, Virginia

Rebecca Trueman
Department of Natural Resources Science
University of Rhode Island
Kingston, Rhode Island

Ronald F. Turco
Department of Agronomy
Purdue University
West Lafayette, Indiana

Wen J. Wang
Northeast Institute of Geography and Agroecology
Chinese Academy of Sciences
Changchun, China

Shengzhong Wang
Key Laboratory of Geographical Processes and Ecological Security in Changbai Mountains
Ministry of Education
School of Geographical Sciences
Northeast Normal University
Changchun, China

Shusheng Wang
Key Laboratory of Geographical Processes and Ecological Security in Changbai Mountains
Ministry of Education
School of Geographical Sciences
Northeast Normal University
Changchun, China

Yeqiao Wang
Department of Natural Resources Science
University of Rhode Island
Kingston, Rhode Island

W. W. Wenzel
Institute of Soil Research
University of Natural Resources and Life Sciences
Vienna, Austria

Sara Wigginton
Department of Natural Resources Science
University of Rhode Island
Kingston, Rhode Island

Zhengfang Wu
Key Laboratory of Geographical Processes and Ecological Security in Changbai Mountains
Ministry of Education
School of Geographical Sciences
Northeast Normal University
Changchun, China

Jiawei Xu
School of Geographical Sciences
Northeast Normal University
Changchun, China

Jian Yang
Department of Forestry and Natural Resources
University of Kentucky
Lexington, Kentucky

Qingsheng Yang
Department of Resources and Environment
Guangdong University of Business Studies
Guangzhou, China

Xiaojun Yang
Department of Geography
Florida State University
Tallahassee, Florida

Xiuqin Yin
School of Geographical Sciences
Northeast Normal University
Changchun, China

Yunyue Yu
Center for Satellite Applications and Research
National Environmental Satellite Data
Information Service
National Oceanic and Atmospheric
Administration (NOAA)
College Park, Maryland

Hongyan Zhang
Key Laboratory of Geographical Processes and
Ecological Security in Changbai Mountains
Ministry of Education
School of Geographical Sciences
and
Urban Remote Sensing Application Innovation
Center
School of Geographical Sciences
Northeast Normal University
Changchun, China

Yuyu Zhou
Department of Geological and Atmospheric
Sciences
Iowa State University
Ames, Iowa

Michael A. Zoebisch
German Agency for International Cooperation
Tashkent, Uzbekistan

I

Landscape Composition, Configuration, and Change

1

Conserved Lands: Stewardship

Peter V. August
University of Rhode Island

Janet Coit
Rhode Island Department of Environmental Management

David Gregg
Rhode Island Natural History Survey

Introduction

Acquiring ownership of, or development rights to land is one of the most effective ways of conserving natural and cultural resources at local, regional, and national scales. Once acquired, protected land requires constant stewardship. Land protection is done for many reasons: to protect the fauna and flora of a site, to protect cultural resources, to ensure that ecosystem services are maintained, to allow sustainable use of natural resources of the site, and to afford public access for recreational purposes.[1,2] Land conservation occurs at multiple scales, in all ecosystems (terrestrial, aquatic, and marine), and is accomplished by many kinds of institutions and public agencies. National governments protect watersheds, forests, and rangelands for resource protection and in many cases, for sustainable resource harvesting. National governments also conserve large national park systems for the benefit of biota and present and future generations of citizens. Conservation occurs at smaller jurisdictions as well; for example, states, counties, and towns. Large nonprofit organizations, such as The Nature Conservancy and the Audubon Society in the United States, are very effective in protecting land, as are small nonprofit organizations such as local land trusts and conservancies. Regardless of the size of the conservation organization or the property preserved, all protected lands require ongoing management and stewardship.

The consequences of not stewarding protected lands can jeopardize the very resources that are meant to be preserved. Developing and implementing a management plan for protected land is one good way of cataloging and prioritizing stewardship responsibilities and we review steps in management planning here. Threats to protected lands can rapidly change; therefore, monitoring protected lands is an important part of the process. Finally, the technical knowledge within a conservation organization, the availability of staff and equipment, and funding for protected land management are frequently inadequate for the magnitude of the task, thus, innovative collaborations and efficient implementation of management plans are necessary.

Protected land stewardship is a challenging endeavor and has many components which will be reviewed here. Large government agencies or other organizations that own and manage large areas of protected lands typically have dedicated programs for stewardship activities. For example, the National Park Service (NPS) oversees a large and complex network of protected areas in the United States. Every park in the NPS system has a General Management Plan, which articulates the management needs of a park and how the park will meet those needs.[3] Furthermore, the NPS has a sophisticated program—the NPS Inventory and Monitoring Program—to systematically monitor environmental conditions in parks to know if management and stewardship activities are having the desired results and to be vigilant to new or unforeseen threats to the ecological integrity of a park.[4] Other conservation agencies that oversee large areas of land have similar programs such as the U.S. Fish and Wildlife Services,[5] U.S. Bureau of Land Management,[6] and U.S. Forest Service.[7] Small conservation organizations, such as local conservancies and land trusts, typically do not have dedicated staff or program resources for protected land stewardship, yet they own significant areas of land. In the United States, for example, there are 1700 different land trusts that control over 150,000 km^2 of land.[5] The focus of this entry is to describe the conservation land stewardship process that would be followed by a small conservation organization.

Land Protection and Stewardship

There are a variety of reasons to protect land and many stewardship strategies that can be used to meet conservation goals (Table 1.1). Conservation lands are typically protected by securing fee simple ownership, control over future development rights, or establishment of permitted and restricted uses of the land through zoning controls.[2] However a property is controlled, the development and implementation of an ongoing management plan for the protected property are essential. There are many threats to the ecological integrity of protected lands (Table 1.2, Figure 1.1). Pests, pathogens, and invasive species can impact native fauna and flora of a site. Illegal poaching or harvesting natural resources can diminish the biota. Soil compaction, erosion, and vegetation disturbance in fragile ecosystems caused by motor vehicle riding (e.g., all terrain vehicles, motorcycles) can result in serious environmental damage. Malicious acts, such as garbage dumping, littering, theft of cultural resources, and vandalism, can diminish the esthetic value of a site. A carefully designed management plan will be attentive to these threats. Protected land management should happen in a systematic manner (Figure 1.2). The basic steps are discussed in detail in the coming sections.

TABLE 1.1 Examples of Common Conservation Goals and Management Activities That Achieve Them

Goal	Management and Stewardship Actions
Preserve biodiversity	Protect large tracts of land. Connect separate refuges with corridors. Increase the size of refuges. Enforce policies against poaching wildlife or harvesting plants. Ensure adjacent land uses are not a source of invasive species, pests, or pathogens. Monitor for invasive species, pests, and pathogens. Monitor population levels of key or indicator species of plants and animals.
Preserve cultural resources	Protect sites or regions of interest; limit public access to sensitive sites.
Preserve ecosystem services	Ensure that conservation land boundaries encompass whole watersheds, maintain diversity of habitats, and allow public access and nondestructive forms of recreation so that patrons can benefit from esthetic values.
Preserve aesthetic or recreational values	Provide trails and interpretive services for public access; encourage hiking, hunting, and fishing if appropriate for the site; manage viewscapes and soundscapes to preserve natural conditions.
Ensure sustainable resource use	If a site permits, allow controlled harvest of sustainable resources, such as wood products, game and fish, plants, fruits, berries, and fungi.
Agricultural preservation	Obtain development rights and easements to working lands and waters to ensure they will remain in agricultural land use.

TABLE 1.2 Examples of Common Threats to Resources on Conserved Lands

Resource	Threat
Biodiversity	Habitat destruction, poaching; illegal harvest of plants or animals; spread of invasive species, pests, and pathogens; loss of fitness due to small population sizes; illegal motor vehicle access in sensitive habitats.
Cultural resources	Vandalism, theft, and alteration of landscape context (e.g., viewscapes, soundscapes).
Ecosystem services	Groundwater withdrawal outside refuge, habitat destruction on borders of refuge, and high-impact land uses outside refuge in watershed.
Aesthetic or recreational values	Trespassing, illegal motor vehicle access, dumping, degradation of natural viewscapes or soundscapes.

FIGURE 1.1 **(See color insert.)** Examples of threats to protected lands: (**a**) Vandalism (destruction of signage); (**b**) illegal dumping of refuse; (**c**) all-terrain vehicle track damage to wetland habitat.
Source: Photo courtesy, The Rhode Island Chapter of The Nature Conservancy.

Site Assessment and Baseline Inventory

An essential first step in conservation land management is a site (the parcel) and landscape (what is around it) survey to inventory current conditions and catalog important habitats. The purpose of the site assessment is to evaluate the presence and condition of important natural resources and to identify threats to the focal resources and the ecosystem as a whole. Target resources can be species, habitats,

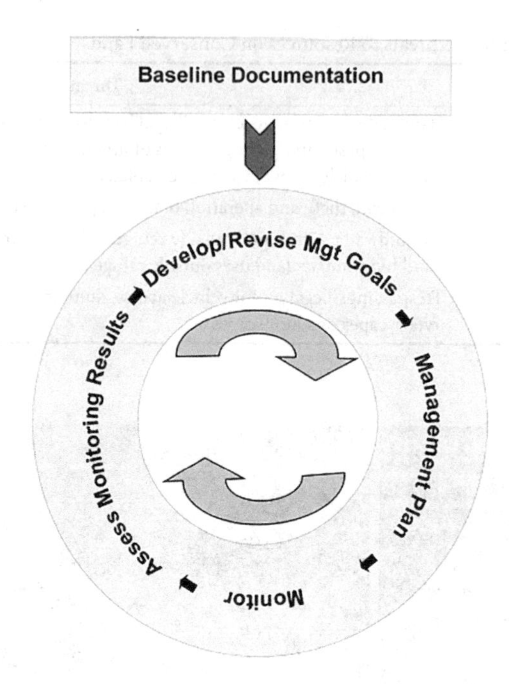

FIGURE 1.2 Management cycle of protected lands: Steps involved in managing protected lands.

rivers, landforms, viewscapes, farms, cultural resources, groundwater, or ecosystem services and ecological processes.[8,9] Site assessment can be a complex activity and require personnel familiar with local ecological conditions. Assessment involves field survey and consolidation of relevant geospatial information such as data from Geographic Information Systems (wetlands, rare species occurrences, land use, soils, landform, etc.), digital imagery, and the results of previous reconnaissance of the area if available. The initial site assessment establishes a baseline condition for easement monitoring and tracking changes in the condition of the property. Landscape analysis provides a regional context for conservation and provides insight in gains and losses of dispersal corridors, up-watershed threats that would jeopardize site-level habitats or species, and land use changes on nearby properties that might enhance or diminish the conservation value of a particular parcel.[10] There is a growing number of protocols that have been advanced to perform a baseline inventory.[11,12]

Development of Management Goals and Plan

Development of management goals is an essential step and the goals of the management plan will reflect the values of the conservation institution and the purpose for acquiring the conservation land.[13] Common management goals include stewarding the land for water (ground and/or surface) protection, conservation of biodiversity, forestry, farming, esthetic values, fish and wildlife management, and recreation (hiking, paddling, fishing, hunting, etc.). Management goals of one property can affect the viability of nearby conservation properties owned by others; hence there is a need for coordinated management when protected lands are in a mosaic of small parcels managed by different institutions. It is important that a management plan establish a basis or rationale for prioritizing resources and stewardship actions, so that scarce resources are applied over time can still achieve large, long-term goals. Once goals for a property are established, a management plan can be developed and clear measures of success must be identified. These measures are the basis for ongoing monitoring and are the indicators of success or failure of the management plan.

Implementation of Management Plan

Management plans can vary in complexity; simple management plans for conservation lands may require little activity beyond periodic monitoring. Other stewardship activities, however, could require considerable work, expertise, and investment in personnel, supplies, equipment, and time in the field. Examples of expensive and complex management tasks include habitat restoration, creating and maintaining trail systems, forest management (selective cutting to maintain the age and target species composition of the forest), and removal or control of invasive species.[14,15] The implementation of a management plan can be a challenge for conservation organizations. Large land owners do not always have the staff or resources to manage extensive properties. Similarly, small, local conservation organizations, such as land trusts, rarely have the technical wherewithal or financial ability to take on complex management activities. Partnerships and collaborative projects are one way to perform complex stewardship tasks.

Monitoring

Vigilant monitoring is required of conservation properties and the land surrounding them. Easements must be monitored on a regular basis to ensure that fee owners are managing properties in a manner that is consistent with the easement that is held by the conservation organization. Violations in easements will have legal and policy ramifications which will need to be addressed.

All properties must be visited on a regular basis to protect against adverse possession claims and to identify inappropriate human activities occurring on or near them. Vandalism, dumping, erosion caused by motor vehicles, illegal hunting, and wood cutting are, unfortunately, common on conservation lands (Figure 1.1). Monitoring for disturbances such as these is relatively straightforward for small properties but requires constant attention. Monitoring over large, expansive conservation lands can be logistically difficult, especially if access is difficult.[16] Ecological monitoring must be done on and around conservation lands to ensure that invasive plants and animals have not become established, pests or pathogens are not present, stewardship activities are yielding desired outcomes, target species are still present, and ecosystem health remains high. Regional changes in the creation of impervious surfaces, which increase storm water runoff, water withdrawal in the watershed, land use conversion, and habitat fragmentation can degrade the condition and viability inside conservation lands.

Synthesis, Reflection, and Adaptive Stewardship

Monitoring provides the data from which decisions about the efficacy of the management plan can be made. A careful analysis of monitoring data determines if the management plan is working and the ecological condition of the property is changing. If the desired results are not occurring, the management plan should be modified to meet the conservation goals of the property. This is a critical step and follows the logic of the adaptive management paradigm.[17] New, unanticipated stewardship challenges can emerge rapidly. Management plans must be dynamic documents and capable of changing as knowledge or needs require.

Conclusions

Acquiring land is an effective way to protect natural resources and careful stewardship of conserved lands ensures that the condition of the natural resources is preserved. Stewardship is an ongoing process and a long-term commitment. Many factors can diminish the value of protected land and these must be monitored, and when present, mitigated. The process of protected land management has a number of steps and begins with a careful resource inventory of the protected property and establishment of a suite of management goals that will direct stewardship activities. Ongoing monitoring to ensure that management goals are being met is an essential component of the process.

Large land owners, such as U.S. NPS or U.S. Forest Service, have very complex management programs to ensure that the property in their care retains the environmental and cultural resources the lands were obtained to protect. Small land owners, such as land trusts and local conservancies, frequently do not have dedicated staff resources, knowledge, or budgets to steward their land, but innovative partnerships are one way the resources of many institutions can be leveraged to achieve effective land stewardship even by small land owners. One model is the Rhode Island Conservation Stewardship Collaborative (RICSC),[18] an alliance of federal, state, municipal, and nonprofit conservation organizations who partner to (from its mission statement) "… *advance long-term protection and stewardship of terrestrial, aquatic, coastal, estuarine, and marine areas in Rhode Island that have been conserved by fee, easement, or other means.*" The RICSC tackles systemic, state-wide, impediments to good conservation land stewardship and provides training materials, protocols, and technical capacity to assist conservation land owners in their stewardship challenges.

Acknowledgments

The following individuals have provided us thoughtful guidance and support for our work in conservation land management and stewardship: Julie Sharpe, Peggy Sharpe, Rupert Friday, Cathy Sparks, Scott Ruhren, Sharon Marino, Larry Taft, Jennifer Pereira, Charles Vandemoer, and Kathleen Wainwright. The Rhode Island Conservation Stewardship Collaborative has challenged us to develop procedures, protocols, and guidance for conservation agencies, especially local land trusts, in supporting their protected land management. The University of Rhode Island Cooperative Extension program, URI Department of Natural Resources Science, and USDA Renewable Resources Extension Act have steadily supported our work in conservation land management and stewardship.

References

1. Daily, G.C.; Ehrlich, P.R.; Goulder, L.H.; Alexander, S.; Lubchenco, J.; Matson, P.A.; Mooney, H.A.; Postel, S.; Schneider, S.H.; Tilman, D.; Woodwell, G.M. Ecosystem services: benefits supplied to human societies by natural ecosystems. Issues Ecol. **1997**, *2* (1), 1–16.
2. Duerksen, C.; Snyder, C. *Nature Friendly Communities: Habitat Protection and Land Use Planning*; Island Press: Washington D.C., 2005; 421 pp.
3. National Park Service. General Management Plan Dynamic Sourcebook. http://planning.nps.gov/GMPSourcebook/GMPHome.htm (accessed February 2012).
4. Oakley, K.L.; Thomas, L.P.; Fancy, S.G. Guidelines for long-term monitoring protocols. Wildlife Soc. Bull. **2003**, *31* (4), 1000–1003.
5. United States Fish and Wildlife Service. Comprehensive Conservation Planning Process. http://www.fws.gov/policy/602fw3.html (accessed February 2012).
6. United States Bureau of Land Management. Land Use Planning. http://www.blm.gov/planning/index.html (accessed February 2012).
7. United States Forest Service. Strategic Planning and Resource Assessment. http://www.fs.fed.us/plan (accessed February 2012).
8. Turner, R.K.; Daily, G.C. The ecosystem services framework and natural capital conservation. Environ. Resour. Econ. **2008**, *39* (1), 25–35.
9. Lokocz, E.; Ryan, R.L.; Sadler, A.J. Motivations for land protection and stewardship: Exploring place attachment and rural landscape character in Massachusetts. Landscape Urban Plan. **2011**, *99* (2), 65–76.
10. Theobald, D.M. Targeting conservation action through assessment of protection and exurban threats. Conserv. Biol. **2003**, *17* (6), 1624–1637.
11. Land Trust Standards and Practice, 2004. Land Trust Alliance. http://www.lta.org. (accessed October 2011).

12. Ruhren, S. Baseline documentation and inventory protocol. Rhode Island Conservation Stewardship Collaborative, 2011. http://www.ricsc.org/Projects/Docs_2009/CSC_BDIP.pdf (accessed October 2011).
13. Carwardine, J.; Wilson, K.A.; Watts, M.; Etter, A.; Klein, C.J.; Possingham, H.P. Avoiding costly conservation mistakes: The importance of defining actions and costs in spatial priority setting. PLoS ONE **2008**, *3* (7), e2586. doi:10.1371/journal.pone.0002586 (accessed October 2011).
14. Anderson, P. Ecological restoration and creation: a review, Biol. J. Linnean Soc. **1995**, *56* (S1), 187–211.
15. Mack, R.N.; Simberloff, D.; Lonsdale, W.M.; Evans, H.; Clout, M.; Bazzaz, F.A. Biotic invasions: causes, epidemiology, global consequences and control. Ecol. Appl. **2000**, *10* (3), 689–710.
16. Upgren, A.; Bernard, C.; Clay, R.P.; de Silva, N.; Foster, M.N.; James, R.; Kasecker, T.; Knox, D.; Rial, A.; Roxburgh, L.; Storey, R.J.; Williams, K.J. Key biodiversity areas in wilderness. International Journal of Wilderness, **2009**, *15* (2), 14–17.
17. Walters, C. Adaptive Management of Renewable Resources; MacMillan Publishers: New York, 1986; 374 pp.
18. RICSC. Rhode Island Conservation Stewardship Collaborative. http://www.ricsc.org (accessed October 2011).

2

Edge Effects on Wildlife

Peter W. C. Paton
University of Rhode Island

Introduction

Ecologists refer to the boundary between adjacent habitat types as the edge or ecotone. At the juxtaposition of adjacent habitat types, there is a tendency for increased species diversity and increased densities of certain game species. In 1933, Aldo Leopold[1] coined the phrase *"edge effect"* to refer to this relationship in his classic book on game management, "Carrying capacity in species of high type requirements and low radius varies directly with the interspersion of the types, which is proportional to the sum of the type peripheries. Such game is an *edge effect*." Leopold, who was the founder of the science of wildlife management, developed the "law of interspersion," which hypothesizes that increases in the amount of edge habitat result in higher population densities of some game species.[1] There are a number of hypotheses to explain this increase in density of game species at edges including proximity to two different habitat types preferred for different uses (e.g., roosting, feeding, or nesting) microclimatic changes in light, temperature, and humidity; changes in plant composition due to increased competition from shade-intolerant species; increased colonization rates by invasive plant species; increased colonization by insects; and predation, parasitism, and competition by "weedy" birds and mammals.[2]

As one of the United States' first spatial ecologists, Leopold believed that the composition and interspersion of essential habitat types in relationship to the daily dispersal capabilities of target species were vital to long-term population persistence.[3] Leopold's ideas led many wildlife biologists to create edges to enhance habitat for game species, which in some instances meant fragmenting existing contiguous habitats.[4] In the 1970s and 1980s, wildlife biologists began to realize that Leopold's hypothesis pertained only to game species of low mobility and specific habitat requirements and challenged the idea that edges benefited most species of wildlife.[5,6] Yet, Leopold knew that the relationship between edge and density was complex and that the habitat requirements of forest-interior specialists were the inverse of game species that preferred early-succession habitats, such as old fields.[3]

In 1978, the *"ecological trap hypothesis"* postulated that avian nest predation rates were density dependent, and that the greater density of nests near edges leads to a concomitant increase in depredation rates of avian nests near edges.[7] Much of the interest in the negative influence of edge effects in the 1980s was due to population declines in neotropical migratory birds that nest in North America.[8] Nest success is vital to long-term avian population dynamics,[9] therefore determining which factors influence nest predation is essential for biologists hoping to successfully manage avian populations. Reviews of avian nest predation studies in the mid-to late-1990s found that most, but not all, studies found evidence of increased nest predation near edges.[10,11] However, evidence of an edge effect on nest predation rates is not restricted to birds, as painted turtles (*Chrysemys picta*) can have greater nest predation rates near water edges.[12]

Although there is strong evidence that avian (and other taxa) nest success can be lower for some species near edges, there appear to be many factors that affect nest success,[10] which can include nest type,[13] plot age,[14] predator densities,[15] plot size,[16] and among the most important factors-landscape context.[17] One of the major challenges for biologists hoping to understand the impact of edge effects on wildlife populations is disentangling the effects of habitat fragmentation on avian nesting success at multiple spatial scales, including edge, patch, and landscape.[17,18] There is considerable confusion or disagreement about whether or how different stand types or ages of forest (e.g., recently harvested patches surrounded by mature forest) constitute a "fragmentation" of that larger habitat patch or landscape. In forested landscapes, some biologists feel that forest harvesting does not necessarily "fragment" larger forest patches or forested landscapes, but rather represents a patch with a different disturbance history and resulting structure.

Most research that has examined the effects of habitat fragmentation on avian nesting success used either real or artificial nests at the edge scale, whereas fewer studies have investigated nest success at either patch or landscape scales.[18] Research in the mid-western United States found that avian nest predation and Brown-headed Cowbird (*Molothrus ater*) nest parasitism rates were much higher in fragmented landscapes than unfragmented landscapes, and nest predation rates were much higher in edge habitats than core areas far from edges.[17,19] More importantly, this research clearly showed that the evidence of edge effects on nest predation and cowbird parasitism rates differed among landscapes, with greater depredation rates near edges in highly and moderately fragmented landscapes, but not in unfragmented landscapes.[17,19] At a landscape scale, nest predation rates of many species of forest birds have a negative relationship with the amount of forest cover; that is as forest cover declines, nest predation rates tend to increase.[19] Therefore, it can be challenging to separate the effects of habitat loss, habitat fragmentation, and edge effects on nest predation rates.

Recently, there is considerable evidence that early succession patches within larger forests may be extensively used by forest-interior birds during the post-fledging period for juvenile and postnesting period by adults. Many researchers now believe that the value of these clearcut patches for increasing juvenile and adult survival rates may outweigh the possible negative "edge effects" on nest predation rates, especially in areas with high renesting rates and moderate-to-high production overall.[20]

Conclusion

Much is known about the potential impacts of edge effects on wildlife populations, with most research focusing on the potential negative impact of edges on avian nest success.[10,11] However, research on other taxa has shown that small mammal populations [e.g., red-backed voles (*Clethrionomys californicus*)],[21] and turtles[12] also can be sensitive to habitat edges. Research conducted to date suggests that avian nest predation rates tend to be significantly higher nearer edges, with the most conclusive studies showing that elevated depredation rates generally occur within 50 m of an edge.[10] However, edge effects are most likely to be detected in forests surrounded by farmland and less likely to be detected in landscapes that are mostly forested mosaics.[22] There is strong evidence that landscape context is critical when trying to understand the mechanisms affecting depredation rates.[17–19] In particular, edge effects on nest predation rates are much more likely to be detected in fragmented landscapes; however, the mechanisms leading to increased depredation rates near edges still remain uncertain.[23]

References

1. Leopold, A. *Game management*; University of Wisconsin Press: Madison, WI, 1933; 135.
2. Alverson, W.S.; Waller, D.M.; Solheim, S.L. Forests too deer: Edge effects in Northern Wisconsin. Conserv. Biol. **1998**, *2* (4), 3458–358.
3. Silbernagel, J. Spatial theory in early conservation design: examples from Aldo Leopold's work. Landscape Ecol. **2003**, *18* (2), 635–646.

4. Yoakum, J.; Dasmann, W.P.; Sanderson, H.R.; Nixon, C.M.; H. S. Crawford, H. S. Habitat improvement techniques. In *Wildlife Management Techniques Manual.* Schemnitz, S. D. Ed.; The Wildlife Society: Washington, D.C., 1980; 329–404.
5. Yahner, R.H. Changes in wildlife communities near edges. Conserv. Biol. **1988**, *2* (4), 333–339.
6. Reese, K.P.; Ratti, J.T. Edge effect: A concept under scrutiny. 53rd Transactions of the North American Wildlife Resources Conference, 1988; 127–136.
7. Gates, J. E.; Gysel, L.W. Avian nest dispersion and fledging success in field-forest ecotones. Ecology **1978**, *59* (5), 871–883.
8. Askins, R.A.; Lynch, J.F.; Greenburg, R. Population declines in migratory birds in eastern North American. Curr. Ornithol. **1990**, *7* (1), 1–57.
9. Lack, D.L. *The natural regulation of animal numbers;* Clarendon Press: Oxford, England, 1954; 343 p.
10. Paton, P.W.C. The effect of edge on avian nest success: How strong is the evidence? Conserv. Biol. **1994**, *8* (1), 17–26.
11. Hartley, M.J.; Hunter, M.L. Jr. A meta-analysis of forest cover, edge effects, and artificial nest predation rates. Conserv. Biol. **1998**, *12* (2), 465–469.
12. Kolbe, J.J.; Janzen, F.J. Spatial and temporal dynamics of turtle nest predation: edge effects. Oikos **2002**, *99* (3), 538–444.
13. Møller, A.P. Nest site selection across field-woodland ecotones: the effect of nest predation. Oikos **1989**, *56* (2), 240–246.
14. Yahner, R.H.; Wright, A.L. Depredation of artificial ground nests: effects of edge and plot age. J. Wildlife Manag. **1985**, *49* (3), 508–513.
15. Angelstam, P. Predation on ground-nesting birds' nests in relation to predator densities and habitat edge. Oikos **1986**, *47* (3), 365–373.
16. Wilcove, D.S. Nest predation in forest tracts and the decline of migratory songbirds. Ecology **1985**, *66* (4), 1211–1214.
17. Donovan, T.E.; Jones, P.W.; Annand, E.M.; Thompson, F.R. Variation in local-scale edge effects: mechanisms and landscape context. Ecology **1997**, *78* (7), 2064–2075.
18. Stephens, S.E.; Koons, D.N.; Rotella, J.J.; Willey, D.W. Effects of habitat fragmentation on avian nesting success: a review of the evidence at multiple spatial scales. Biol. Conserv. **2003**, *115* (1), 101–110.
19. Robinson, S.K.; Thompson, F.R.; Donovan, T.M.; Whitehead, D.R.; Faaborg, J. Regional forest fragmentation and the nesting success of migratory birds. Science **1995**, *267* (5206), 1987–1990.
20. Vitz, A.C., Rodewald, A.D. Can regenerating clearcuts benefit mature-forest songbirds? An examination of postbreeding ecology. Biol. Conserv. **2006**, *127* (4), 477–486.
21. Mills, L.S. Edge effects and isolation: red-backed voles on forest remnants. Conserv. Biol. **1995**, *9* (2), 395–403.
22. Andrén, H. Effects of landscape composition on predation rates at habitat edges. In *Mosaic landscapes and ecological processes.* Hansson, L.; Fahrig, L.; Merriam, G., Eds.; Chapman and Hall: London. 1995; 225–242.
23. Driscoll, M.J.L.; Donovan, T.M. Landscape context moderates edge effects: Nesting success of Wood Thrushes in Central New York. Conserv. Biol. **2004**, *18* (5), 1330–1338.

3

Fires: Wildland

Jian Yang
University of Kentucky

Introduction

Wildland fire is any fire burning vegetation that occurs in wildland areas.[1] It is a worldwide phenomenon since the appearance of terrestrial plants about 400 million years ago.[2] Wildland fires co-evolved with vegetation and have become an integral component of the Earth systems,[3] helping to shape global biome distribution[4] and plant traits.[5] It also influences climate systems via feedback to land–atmosphere energy exchange.[6] Wildland fires are widely distributed in many terrestrial ecosystems (Figure 3.1), particularly in regions where the climate is sufficiently moist to allow for growth

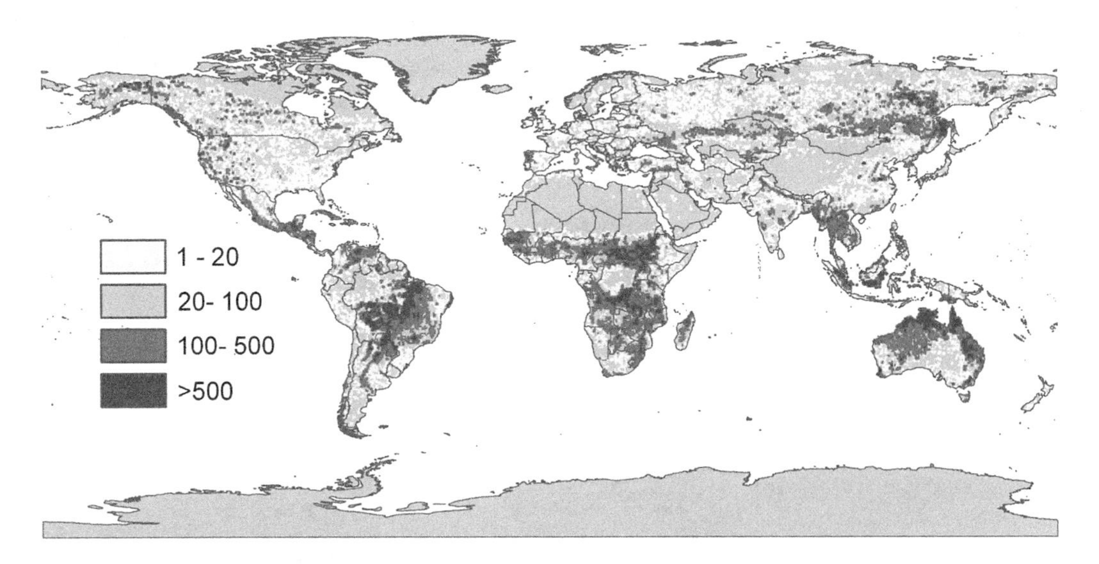

FIGURE 3.1 (See color insert.) Spatial distribution of fire occurrence density (the number of active fire counts detected from the satellite using the Along Track Scanning Radiometers (ATSR) instrument between 1996 and 2006 within a 0.5° × 0.5° grid cell). (Data adapted from http://due.esrin.esa.int/.)

of vegetation but also characterized by extended dry, hot periods (e.g., Mediterranean and tropical ecosystems).[7] It has been estimated that about 2% of terrestrial land is burned annually in the modern period.[8] Due to climate warming and fuel buildup resulting from fire exclusion, many places in the world have experienced more frequent and larger-size wildland fires over the past few decades.[9]

Wildland fires are regulated by a wide range of controls that vary significantly across the gradients of spatial and temporal scales. Fire triangles are often used to depict dominant controls at different scales. The characteristics of a single fire event are described by fire behavior parameters, while the characteristics of recurring fire events in a vast landscape over an extended period are described by fire regime parameters. Wildland fire is an essential natural disturbance, which greatly affects ecosystem productivity, biodiversity, forest landscape dynamics, and climate systems.[10] Humans have traditionally perceived fire as a damaging agent to nature, human lives, and economy. Recent advances in fire ecology have raised public awareness that fire is also an integral component of the Earth systems and has helped foster a shift in fire management policy from fire exclusion to prescribed burning, mechanical fuel treatment, and increased utilization of the existing wildland fire use policy.[11] This entry discusses causes (and controls), characteristics, consequences, and management of wildland fire.

Causes and Controls

Wildland fire occurs as a function of a suitable combination of biomass resources (fuel quantity and fuel quality such as fuel moisture and fuel continuity), conditions conducive to combustion and spread (e.g., fire weather and climate), and ignition agents.[12] Causes of wildland fire starts can be either natural (e.g., lightning, volcanic eruptions, sparks from rockfalls, and spontaneous combustion) or anthropogenic (e.g., arson, discarded cigarettes, sparks from equipment and power line arcs, and out-of-control prescribed burns).[13] There are usually more human-caused than naturally caused wildland fires in many regions during the modern periods. For example, lightning-caused fires contributed only 35% of the fire occurrences in the boreal forest of Canada[14] and <1% of the fires in the Central Hardwood Forest of Missouri, United States, reported in the late 20th century.[15]

Although ignition is a necessary precursor to wildland fire, wildland fire is not a necessary consequence of an ignition source. The factors that regulate wildland fire activity are more complex; subsequent fire growth is mainly controlled by weather conditions, topography, fuel amount, and fuel continuity. Dominant controls of wildland fire process vary at different spatial and temporal scales, often conceptualized as *fire triangles* (Figure 3.2). For example, at the finest microsite scale, the three dominant controls that determine a combustion process are fuel, heat, and oxygen. For a single wildland

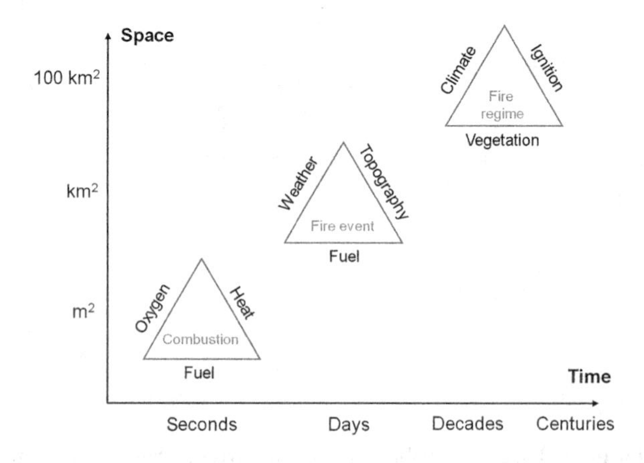

FIGURE 3.2 Dominant controls of wildland fire at multiple temporal and spatial scales conceptualized as fire triangles. (Modified from Moritz et al.[16] and Whitlock et al.[17].)

fire event, the three legs of the fire triangle are fuel, weather, and topography. At the landscape scale, the three dominant factors are vegetation, climate, and ignition.[16,17]

Characteristics of Wildland Fire

Characteristics of a single wildland fire event are often described by fire behavior parameters such as duration, flame length and intensity (i.e., rate of released energy), rate of spread, and fire type (e.g., ground fire, surface fire, and crown fire). Ground fires consume the organic material beneath the surface litter ground, while surface fires burn loose debris (e.g., dead branches, leaves, and low vegetation) on the surface, and crown fires advance from top to top of trees or shrubs independent of a surface fire (www.nwcg.gov/pms/pubs/glossary/index.htm). In contrast, a *fire regime* is used to describe characteristics of cumulative wildland fire events over a broad spatiotemporal domain. Parameters to describe a fire regime include, but are not limited to, frequency, size, seasonality, and severity. *Fire frequency* is the number of fires per unit time at a specific point. *Fire return interval* is the average number of years between fires at a given location. *Fire rotation,* called by some authors the *fire cycle,* is the number of years necessary for the cumulative burned area to equal the entire area of interest. *Fire severity* is defined broadly as the immediate effects of fire on various components of an ecosystem (e.g., vegetation and soil), which is open to individual interpretation. Empirical studies have often defined fire severity operationally as the loss of or change in organic matter aboveground and belowground, although the precise metric varies with management needs.[18] Fire severity differs from *fire intensity* in that the latter term describes only the energy released during various phases of a fire. Fire severity describes, in large part, the mortality of dominant vegetation and changes in the aboveground structure of the plant community. Severe fires in forests are often called "stand-replacement" fires (Figure 3.3).

Fire regimes often vary significantly in different ecosystems and climatic conditions. Wildland fire interacts with ecosystem structures to create fire regimes. For example, low- to mid-elevation ponderosa pine forests in the western United States that were formerly open woodlands with abundant, contiguous fine fuels in the understory historically experienced low- and mixed-severity fire regimes with average fire return intervals of 5–30 years.[19] High-elevation lodgepole pine, Douglas fir, and subalpine fir forests may exhibit dense stand structures and historically have experienced mixed and stand-replacement fire regimes. Fire regime can be relatively constant if climate and vegetation are in an equilibrium state. However, due to anthropogenic activities and climate changes, fire regimes are being modified rapidly in many ecosystems.[10]

FIGURE 3.3 **(See color insert.)** Stand-replacement wildfire in a Chinese boreal forest.

Ecological Effects of Wildland Fire

Wildland fires can exert strong effects on the abiotic and biotic components of an ecosystem, particularly the soil and vegetation. Many physical, chemical, mineralogical, and biological soil properties can be affected by wildland fires.[20] The extended residence time of burning, which is often a result of high levels of coarse fuels, may raise soil temperature substantially. After a burning, wildland fire can still lead to increased diurnal soil surface temperature by removing the shading effect of vegetation and the insulating effect of litter, as well as by reducing the albedo of the soil surface. It may also lead to greater cooling at night through the loss of radiative heat. Soil moisture may be increased in a burned environment where aboveground vegetation is largely consumed due to reduced rain interception and transpiration by plants. However, exposure to sunlight, wind, and evaporation could cause the soil to dry as well. Wildland fire can increase soil bulk density, reduce water holding capacity, and create a discreet or continuous water-repellent crust at the soil surface, all of which will lead to surface run-off and soil erosion.[1]

High-severity wildland fires generally cause significant removal of soil organic matter and considerable loss of nutrients through volatilization, ash entrapment in smoke columns, leaching, and erosion. Soil pH is often increased after the fire due to acid combustion. Wildland fires can also cause a marked alteration of both quantity and composition of microbial and soil-dwelling invertebrate communities. However, the prefire level of most soil properties can be recovered and even enhanced within a relatively short time after the fire if plants succeed in promptly recolonizing the burnt area.[20]

Wildland fire is a global "herbivore" in that it feeds on vegetation (as do herbivores) and converts them to organic and mineral products.[4] As a consumer that has broad dietary preferences, wildland fire interacts with resources (i.e., climate and soil) to determine plant distribution and abundance in flammable terrestrial ecosystems. Many plants exhibit functional traits (e.g., flammability and persistence) that are adaptive to a particular fire regime. For example, Ponderosa pine forests in the southwest United States subject to frequent low-severity surface fires have thick bark and insulated buds, which function to protect living tissues from heat damage; they also self-prune their lower dead branches, which ensure a gap in fuels between the dead surface litter and live canopy to discourage torching. In contrast, lodgepole pine and jack pine forests subject to a crown fire regime have thin bark, retention of lower dead branches, and serotinous cones that synchronously release seeds following high-severity stand-replacing fires. These traits are adaptive in fire-prone environments and are the key to providing resilience to specific fire regimes.[5] The continued existence of particular species and communities often requires wildland fires, but too few or too many fires may both drive the landscape characteristics outside their historical range and variability[21] and lead to degradation of natural communities.[22]

Wildland Fire Policy

Humans and their ancestors have exerted myriad influences on wildland fire. Modern humans can increase or decrease levels of wildland fire activity by forest-fire policymaking. In the United States, federal forest-fire management began in the late 19th century and focused on suppressing all fires on national forests to protect timber resources and rural communities. Decades of fire exclusion have produced uncharacteristically dense forests in many areas. Some forests, which previously burned frequently in low-severity surface fires, are now experiencing fires of much higher intensity than historical levels. The prevailing policy began to be questioned in the 1940s when many foresters recognized the value of controlled burning to clear out understory vegetation and reduce fire hazards. U.S. fire policy has recognized fire as a critical ecosystem process that must be reintroduced to restore forested ecosystems, particularly in regions that historically experienced more frequent low-intensity fires.[23] Improved utilization of the existing wildland fire use policy can provide for a careful and gradual reintroduction of fire into landscapes, with special consideration to the diversity of fire regimes and necessary spatial scale and arrangement of fuel treatment.[11,24]

Conclusions

Wildland fires have been a great natural force in shaping global biome distribution, plant traits, biodiversity, and ecosystem functions. Humans have greatly affected natural fire disturbances both directly by igniting and suppressing fires, and indirectly by altering vegetation and climate. Fire behavior is a major aspect of a single wildland fire event, while fire regime describes characteristics of recurring wildland fire events over a longer period. Many plant species are adapted to specific fire regimes. However, modern fire regimes are currently undergoing rapid change due to climate warming and anthropogenic influences. Changing fire regimes may, in turn, lead to degradation of ecosystem properties and services, novel trajectories in landscape dynamics, and feedbacks to global drivers.

Acknowledgment

Funding support for this work is provided by the National Key R&D Program of China (2017YFA0604403) and the USDA National Institute of Food and Agriculture, McIntire-Stennis Project 1014537.

References

1. Whelan, R.J. *The Ecology of Fire*. Cambridge University Press: Cambridge, **1995**.
2. Scott, A.C. The pre-quaternary history of fire. *Palaeogeogr. Palaeoclimatol. Palaeoecol.* **2000**, *164* (1), 281–329.
3. Bowman, D.M.J.S.; Balch, J.K.; Artaxo, P.; Bond, W.J.; Carlson, J.M.; Cochrane, M.A.; D'Antonio, C.M.; DeFries, R.S.; Doyle, J.C.; Harrison, S.P. Fire in the Earth system. *Science* **2009**, *324* (5926), 481–484.
4. Bond, W.J.; Keeley, J.E. Fire as a global "herbivore": The ecology and evolution of flammable ecosystems. *Trends Ecol. Evol.* **2005**, *20* (7), 387–394.
5. Keeley, J.E.; Pausas, J.G.; Rundel, P.W.; Bond, W.J.; Bradstock, R.A. Fire as an evolutionary pressure shaping plant traits. *Trends Plant Sci.* **2011**, *16* (8), 406–411.
6. Running, S.W. Ecosystem disturbance, carbon, and climate. *Science* **2008**, *321* (5889), 652–653.
7. Pyne, S.J. *Fire in America: A Cultural History of Wildland and Rural Fire.* Princeton University Press: Princeton, NJ, **1982**.
8. Giglio, L.; van der Werf, G.R.; Randerson, J.T.; Collatz, G.J.; Kasibhatla, P. Global estimation of burned area using MODIS active fire observations. *Atmos. Chem. Phys. Discuss.* **2005**, *5* (6), 11091–11141.
9. Westerling, A.L.; Hidalgo, H.G.; Cayan, D.R.; Swetnam, T.W. Warming and earlier spring increase western U.S. forest wildfire activity. *Science* **2006**, *313* (5789), 940–943.
10. Turner, M.G. Disturbance and landscape dynamics in a changing. *World Ecol.* **2010**, *91* (10), 2833–2849.
11. Stephens, S.L.; Ruth, L.W. Federal forest-fire policy in the United States. *Ecol. Appl.* **2005**, *15* (2), 532–542.
12. Krawchuk, M.A.; Moritz, M.A. Constraints on global fire activity vary across a resource gradient. *Ecology* **2011**, *92* (1), 121–132.
13. NWCG [National Wildfire Coordinating Group]. *Wildfire Prevention Strategies.* National Interagency Fire Center: Boise, ID, **1998**.
14. Weber, M.G.; Stocks, B.J. Forest fires and sustainability in the boreal forests of Canada. *Ambio* **1998**, *27* (7), 545–550.
15. Yang, J.; He, H.S.; Shifley, S.R.; Gustafson, E.J. Spatial patterns of modern period human-caused fire occurrence in the Missouri Ozark Highlands. *Forest Sci.* **2007**, *53* (1), 1–15.
16. Moritz, M.A.; Morais, M.E.; Summerell, L.A.; Carlson, J.M.; Doyle, J. Wildfires, complexity, and highly optimized tolerance. *Proc. Nat. Acad. Sci. U. S .A.* **2005**, *102* (50), 17912–17917.

17. Whitlock, C.; Higuera, P.E.; McWethy, D.B.; Briles, C.E. Paleoecological perspectives on fire ecology: Revisiting the fire-regime concept. *Open Ecol. J.* **2010**, *3*, 6–23.
18. Keeley, J.E. Fire intensity, fire severity and burn severity: A brief review and suggested usage. *Int. J. Wildland Fire* **2009**, *18* (1), 116–126.
19. Schoennagel, T.; Veblen, T.T.; Romme, W.H. The interaction of fire, fuels, and climate across Rocky Mountain forests. *BioScience* **2004**, *54* (7), 661–676.
20. Certini, G. Effects of fire on properties of forest soils: A review. *Oecologia* **2005**, *143* (1), 1–10.
21. Keane, R.E.; Hessburg, P.F.; Landres, P.B.; Swanson, F.J. The use of historical range and variability (HRV) in landscape management. *Forest Ecol. Manag.* **2009**, *258* (7), 1025–1037.
22. Hobbs, R.J.; Huenneke, L.F. Disturbance, diversity, and invasion: Implications for conservation. *Conserv. Biol.* **1992**, *6* (3), 324–337.
23. NWCG [National Wildfire Coordinating Group]. *Review and Update of the 1995 Federal Wildland Fire Management Policy.* National Interagency Fire Center: Boise, ID, **2001**.
24. Finney, M.A. Design of regular landscape fuel treatment patterns for modifying fire growth and behavior. *Forest Sci.* **2001**, *47* (2), 219–228.

4

Fragmentation and Isolation

Anne Kuhn
U.S. Environmental Protection Agency (EPA)

Introduction

Habitat fragmentation is the landscape-level process by which large contiguous native habitat is segmented or "fractured" into smaller and more isolated remnants.[1–4] This fracturing of native habitat can be caused by natural forces such as fire, hydrologic and geological processes, or as a result of direct and indirect anthropogenic activities including land transformation. An ecological consequence of human-induced fragmentation is that the remaining smaller habitat fragments or patches become isolated in space and time from the original larger intact habitat, disrupting species interactions and distribution patterns. This ecological or "effective isolation" is species dependent with a range of responses linked closely to the dispersal traits of a species.[5,6] Habitat loss per se is another ecological consequence of fragmentation and is an interrelated process associated with fragmentation and isolation. The five primary aspects of fragmentation are 1) reduction of area or habitat loss; 2) increase in boundary or edge habitat; 3) increased isolation of fragmented habitat patches; 4) altered patch shape; and 5) altered matrix structure. The relative intensity of each of these factors is mediated by the shape and spatial arrangement of the remnant habitats and the structure and quality of the surrounding area or matrix habitat.[6] Matrix habitat is described as the most predominant, most connected, or most influential landscape element of an area.[7]

The fragmentation and isolation of natural habitat have been identified as two of the leading threats to biodiversity and have contributed to population decline in many species. Fragmentation along with habitat loss has also led to significant alterations in species interaction and composition within communities, disrupting ecological flows and ecosystem functioning. Habitat fragmentation is not an isolated phenomenon and co-occurs and interacts synergistically with many other anthropogenic threats to biodiversity and species persistence. Some of these threats may be non-mechanical disturbances such as chemicals, pesticides, herbicides, hydrological changes, livestock grazing, emerging infectious diseases, and invasive species all of which can further disrupt and degrade already fragmented habitats.[8]

The global scientific community, largely ecologists and especially conservation biologists and landscape ecologists, have been studying and measuring the effects of human-induced habitat fragmentation and isolation since the 1960s. This focus on the effects of fragmentation and isolated habitats coincided with the "environmental awakening" during the 1960s and awareness that human activities were inducing ecological deterioration of the global environment.[9] Ecologists began to apply one of the most important and influential contemporary ecological concepts, Island Biogeography Theory (IBT)[10] (described in the next section) to fragmented habitats, using an explicit analogy between habitat fragments and islands, first noted by Preston (1960).[11] From this foundational theory, fragmentation research evolved and diversified into many interrelated subdisciplines ranging from landscape ecology and conservation reserve design to metapopulation dynamics, and from conservation genetics to population viability analysis (PVA).[8]

This entry provides a broad overview of the theoretical and conceptual foundation and progression of the science involved in measuring the effects of habitat fragmentation and isolation, as well as the major landscape issues confronting researchers, practitioners, and conservationists. Future research directions along with advanced innovative methods currently being used to measure the effects of human-induced fragmentation and isolation are presented with the aim of providing a more pragmatic understanding of the ecological consequences of fragmented terrestrial and aquatic ecosystems.

Theoretical Foundation for Fragmentation and Isolation

Island Biogeography Theory

IBT[10] provides the theoretical foundation for understanding habitat fragmentation. The conceptual basis of IBT describes the relationship between species abundance and the size and isolation of available habitat or island.[12] IBT predicts an expected number of species when the opposing forces of immigration and local extinction achieve a dynamic balance or equilibrium as a function of the island's size and degree of isolation. The theory predicts that islands located farther from the mainland (more isolated) will have lower immigration rates than those close to the mainland, and dynamic equilibrium of species will occur with fewer species on distant, more isolated islands. Islands located closer to the mainland will have higher immigration rates and are predicted to support more species. IBT also logically postulates that larger islands will have more species than smaller islands (which could also be attributed to a greater variety of habits on larger islands).

In the late 1960s, IBT captured the imagination of ecologists and population biologists who began applying and testing the general concepts with "habitat islands" or terrestrial patches of habitat within landscapes surrounded by an inhospitable or resistant matrix. These fragmented and isolated habitat patches could be metaphorically considered islands. One of the most important aspects of IBT was the notion that species composition could and would fluctuate over time, but overall the number of species or species richness would remain relatively stable.[12] This dynamic equilibrium concept has since been applied to a range of environments from "sky islands" or isolated mountain tops (Figure 4.1) to fragmented forest patches, to freshwater lakes and aquatic streams disconnected by dams, to conservation reserves to molecular levels, such as genetically isolated populations. Another overarching concept of IBT that has proved to be most pragmatic is the relationship between habitat area and species persistence. This one crucial concept spawned decades of empirical studies measuring the effects of habitat size on a wide range of plant, terrestrial, and aquatic species and their ability to sustain viable population sizes. The conservation community adopted the basic principles of IBT and applied the theory to the design of protected areas and reserves, advancing the notion that single large reserves were better for preserving population-level persistence than several small reserves of comparable area.

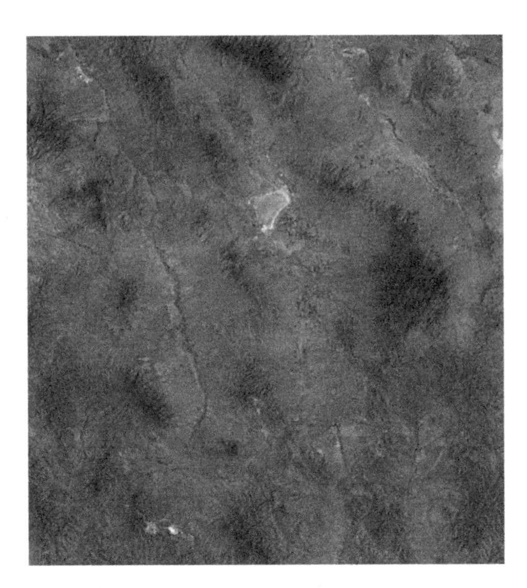

FIGURE 4.1 Madrean Sky Islands: These mountain "islands," forested ranges separated by vast expanses of desert and grassland plains, are among the most diverse ecosystems in the world because of their great topographic complexity and unique location at the meeting point of several major desert and forest biological provinces, southwestern United States and northwestern Mexico.

This idea became known as single large or several small reserves ("SLOSS") and ignited more research and an enduring debate within the conservation design community.[13–17]

Transcending IBT

By the mid-1980s the habitat fragmentation theoretical foundation shifted as some of the basic tenets of IBT were challenged. One of the main criticisms of IBT was with the "closed community" concept where community equilibrium is controlled by inter-species competition alone.[18] Evidence from empirical field studies suggested that species composition and diversity of local assemblages were regulated by local or regional processes. These findings led to a broadened perspective of species distributions and a focus on issues of scale specifically: design of sampling relative to ecological processes and the hierarchical organization of ecological processes in space and time.[9] These issues of scale were being directly addressed in the emerging research field of "Landscape Ecology." From a landscape perspective, new hypotheses were suggesting that non-equilibrium processes were determining landscape dynamics and that fragmentation of landscapes must be assessed at multiple scales and be defined from an organismal perspective.[19] A landscape perspective also shifted the focus from species equilibrium to population level studies at landscape scales in heterogeneous environments or metapopulation dynamics[20,21] (Figure 4.2).

Field-based evidence also suggested that using the IBT to equate habitat fragments with islands was problematic. Field studies demonstrated that "matrix matters" and that not all fragmented patches were surrounded by inhospitable and resistant matrix and sharp boundaries, and that connectivity among fragmented habitats and the ability to traverse the surrounding matrix was determined by species-specific dispersal traits, habitat requirements, and general adaptability (e.g., habitat generalist vs. specialist). This species-specific "effective isolation" has important conservation implications relative to the genetic and demographic risk of isolation.[5] Another observed effect of fragmentation that was not considered by IBT was the "edge effect," a non-discrete boundary line around a patch, which can have negative or positive effects and is species and context dependent. Despite the discrepancies found

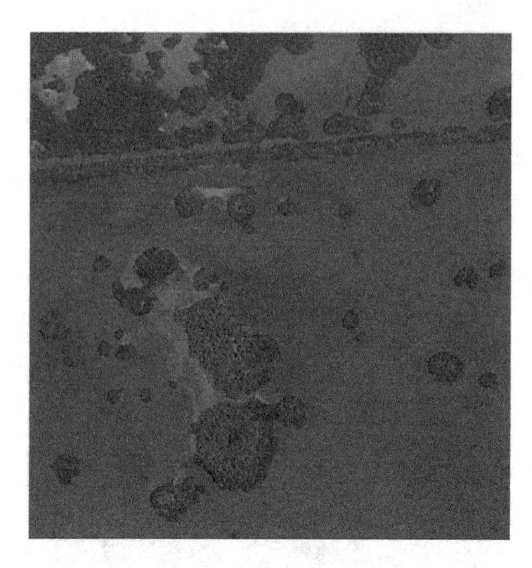

FIGURE 4.2 (See color insert.) Example of metapopulation patch dynamics analogy using mangrove islands in Everglades National Park, Florida, USA. These island patches illustrate how focal populations occur within a network of local populations and may interact via migration within a landscape or region.

between IBT and empirical studies, IBT has had an enormous impact on the study of human-altered landscapes and still provides a heuristic and conceptual, albeit a simplified framework for understanding habitat fragmentation.

Fragmentation research today has advanced far beyond the original intended scope of IBT[8] and approaches habitat fragmentation as a landscape-level phenomenon incorporating more complex processes to elucidate the direct and indirect causal relationships of species responses to landscape change.[22] Currently, a conceptual framework of "interdependence" in both landscape effects on species and species responses to landscape change is being proposed as an integrative perspective to advance the mechanistic understanding of fragmentation and isolation processes.[23] This conceptual framework incorporates the "integrated community" concept[24] recognizing a range of interdependence in natural communities' responses to habitat fragmentation from highly individualistic to highly interdependent. This context-dependence approach recognizes that the patterns, processes, and consequences of habitat fragmentation and isolation are interrelated. Using this interdependence framework as a basis, Didham et al. (2012)[23] developed a hierarchical conceptual model of potential direct and indirect causal paths by which the amount and spatial arrangement of habitat can affect a measured response variable (Figure 4.3). These most recent advances in fragmentation research acknowledge context dependence in ecosystem responses at multiple spatial and temporal scales, along with trait-dependent species responses and synergistic interactions between fragmentation and other components of global environmental change.[22]

Measuring the Effects of Fragmentation and Isolation

Scale of Assessment

The importance of scale is widely recognized and considered throughout the broad field of ecology and is of specific relevance to the field of Landscape Ecology. Landscape Ecology specifically addresses broad and fine-scale spatial patterns and ecological processes and the hierarchical nature of these ecological processes.[25–28] Fragmentation and the isolation of remnant habitat can occur on different spatial scales

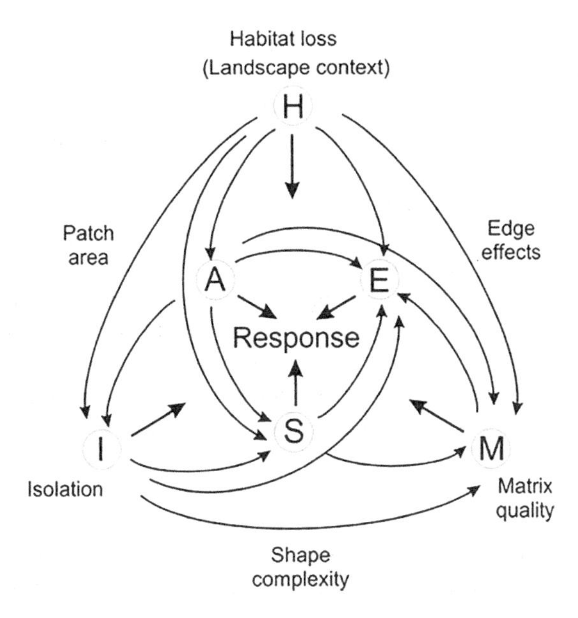

FIGURE 4.3 A hierarchical conceptual model of potential direct and indirect causal paths by which the amount and spatial arrangement of habitat can affect a measured response variable (termed "Response").
Source: © Oikos 2012 as included in Rethinking the conceptual foundations of habitat fragmentation research, R. K. Didham; V. Kapos; R. M. Ewers, Vol. 121: 161–170.

and the effects may occur at different temporal scales (i.e., time lags between fragmentation and effect). The choice of scale assessment can affect the analysis of fragmentation and there is no single correct scale for analysis; the relevant scale being different for each species. Many forest fragmentation studies have been conducted to specifically investigate the effect of scale on the analysis of fragmentation.[29–33] It is widely accepted that any analysis of fragmentation and isolation should occur at multiple scales so the effect of scale on the response (e.g., species abundance, biodiversity, community indices) can be assessed at scales ranging from small (fine grain resolution) to large (landscape level extents). The temporal scale should also be considered by using long-term studies or meta-analyses that incorporate population-level responses over time to account for time lags between the fragmentation of habitat and the manifestation of species responses.

Response Measures to Fragmentation and Isolation

There are a multitude of response measures for assessing the effects of fragmentation and isolation ranging from molecular-level genetics to species' population-level effects to community dynamics to measures of biodiversity and landscape patterns and processes. Habitat fragmentation in both aquatic and terrestrial ecosystems has led to the genetic isolation of populations in numerous species, and can be assessed using measures of population genetic differentiation, dispersal and gene flow, genetic diversity, and effective population size. At higher levels of biological organization, population level measures, such as population density, dispersal, recolonization, abundance, fitness and productivity, extinction proneness, PVA, metapopulation dynamics, and patch model occupancy are used to predict a species' probability of extinction relative to the spatial arrangement across a network of habitat patches.[34–37] One conclusion from metapopulation modeling is that the probability of population extinction depends not only on habitat area or quality, but on the degree of isolation and spatial location within the metapopulation network.[38] The next response level of increasing complexity is the measure of species assemblages or community-level responses, including species richness and interactions. These community-level measures attempt to identify the general species distribution patterns from focal indicator and/or umbrella species,[39,36,40] seeking to strike a balance between detecting generalities that apply across landscapes and species with accurately predicting the effects of fragmentation for a single species. The highest level of response measure for evaluating the effects of fragmentation is the landscape level, where the altered ecological processes (e.g., flow of materials, magnitude, and timing of disturbance) are examined alongside altered patterns of habitat (e.g., forests, vegetative cover, and stream connectivity) to better understand how ecological processes and patterns emerge so that extrapolations from one landscape may be applied to another landscape. Landscape indices are used to characterize spatial patterns of habitat created by landscape alteration and typically include habitat patch metrics that describe the composition (patch habitat type), configuration (spatial arrangement of patches), and connectivity of habitat (from either a human or species-specific perspective).[36–41]

Habitat Loss Effects

Fragmentation involves two interrelated processes: habitat loss and isolation. As fragmentation occurs the total amount of habitat in the landscape is reduced and the remaining habitat is chopped up into fragments of varying sizes and degrees of isolation.[8] Empirical studies have demonstrated that it is difficult to separate these two interrelated processes as they generally co-vary.[42] Fahrig (2003)[42] concludes that habitat loss has a much greater negative effect on biodiversity than the breaking apart of habitat or fragmentation per se. This finding is only elucidated when empirical studies have been carefully designed to separate the effects of habitat loss and fragmentation. Because it has been challenging to separate the confounding effects of habitat loss and isolation from fragmentation, researchers have tried to experimentally control for habitat amount while varying fragmentation. Another research approach has been to compare many different landscapes extracting indices of fragmentation that are independent of the amount of habitat in each landscape.[43,44] A caveat to consider when measuring habitat loss is that the definition of habitat is species specific and therefore habitat loss is a species-specific phenomenon. Additionally, habitat loss often cooccurs with habitat degradation, isolation, and changes in species interactions so it becomes quite challenging to measure the effects of habitat loss alone. Another important consideration is that habitat loss is a highly non-random process,[45–47] and the habitat remnants are often a biased subset of the original landscape.[8] Given all of these caveats and confounding complications it appears that the most pragmatic approach to studying these effects is to use a conceptual framework that incorporates the interdependence in both effects and responses to fragmentation.[23]

Edge Effects

Human modified habitat often results in an increase in perimeter-to-area ratios of the remaining patches creating boundaries or marginal zones of altered microclimatic and ecological conditions (Figure 4.4). Edges can also occur naturally at the interface of two ecological communities or ecotones. Edge effects can be classified by their impacts in two broad categories: abiotic and biotic. Abiotic edge effects can modify microclimatic conditions such as light, air temperature and humidity, soil moisture, wind speed, and altered fire regimes. Other abiotic edge alterations could increase habitat fragment exposure to chemicals such as fertilizers, herbicides, road salt, and acid rain.[48] Biotic edge effects can be manifested in a number of ways, for example, increases in predators, invasive species, diseases, altered animal and plant dispersal, species interactions, reduced reproductive success, and habitat quality for individual species.[49–51] Also, not all species respond negatively to edges and some species' demographic rates are enhanced by edge habitat. Edge effects are species dependent and are directly influenced by the surrounding matrix.[33]

FIGURE 4.4 Example of anthropogenically induced edge effect.

Isolation and Measures of Connectivity

An obvious spatial consequence of habitat fragmentation is that remnant patches may become isolated in space and time from other patches of suitable habitat (Figure 4.5). There is a range of species-specific responses to isolation with some species having a naturally patchy distribution across the landscape (e.g., mountain top species, freshwater ponds), whereas other species have a more continuous distribution across the landscape and may be more affected by habitat isolation and loss of connectivity. "Effective isolation" is also scale and matrix dependent with a range of species tolerances for traversing the surrounding matrix that may allow dispersal to other remnant habitat patches.[5] Landscape connectivity can be defined as the degree to which a landscape facilitates or impedes movement among resources patches.[52] The two types of connectivity that are generally considered and measured by ecologists are structural and functional.[53,54] Structural connectivity assesses the spatial arrangement of habitat types and ecological systems in a landscape without considering the movement of organisms

FIGURE 4.5 **(See color insert.)** Example of intact naturally connected landscape and human-induced highly fragmented and isolated landscape.

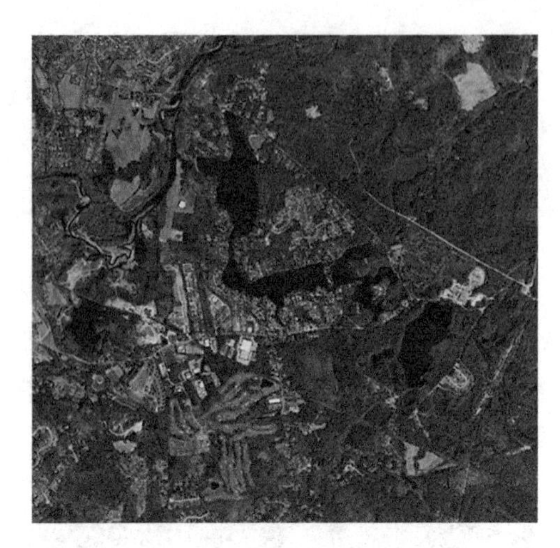

FIGURE 4.6 Highly fragmented landscape with marginal connectivity provided by remaining wetland fragments, surrounded by dense urban and recreational land uses.

through the landscape.[55] Alternatively, functional connectivity incorporates the likelihood of species movement and behavioral responses as well as ecological processes within the landscape. For example, corridors or narrow strips of lands that differ from the surrounding matrix can be used to connect patches of suitable habitat, facilitating movement of organisms and ecological flows, increasing patch connectivity. There have been many recent landscape-level studies and technical advancements for measuring and exploring the importance of maintaining and restoring functional connectivity within landscapes, including aquatic and terrestrial ecosystems.[56–63]

Matrix Effects

Forman (1995)[7] described the matrix as the largest background patch in a landscape, which exerts a dominant role on ecological processes. An increasing number of studies highlight the importance of matrix quality in modified landscapes as a major influence on fragment connectivity and as an influential force for mitigating biotic and abiotic edge effects.[5,54,64–66] Laurance (2004)[67] and others have found that a species tolerance or ability to exploit the surrounding matrix is positively associated with its tolerance of edge habitats, highlighting again a species-dependent response phenomenon. Ricketts (2001)[5] suggests that while many efforts have focused on reconnecting fragmented habitats using wildlife corridors and stepping stones (relatively small patches of native habitat facilitating connectivity across a fragmented landscape) (Figure 4.6), it may be more feasible to reduce effective isolation of fragments by altering management practices in the surrounding matrix. In the longest-running field experimental study of habitat fragmentation in Central Amazonia, Laurance et al. (2011)[33] found that the matrix in the study area had changed markedly and in turn has strongly influenced fragment dynamics and faunal persistence.

Conclusions

Habitat fragmentation and the resulting isolation of habitat remnants affects far more than biodiversity and interactions among species. As human-induced fragmentation increases many ecological processes are being altered within fragmented landscapes, including gene flow, hydrological flows, species dispersal patterns, forest biomass, and carbon and nitrogen storage. It is clear that to enhance the understanding of the effects of habitat fragmentation and isolation, researchers need to deploy an "interdependence" perspective for the multiple interacting drivers of habitat fragmentation, within a community and landscape context over larger spatial and temporal scales. Land managers will need to educate and elicit support from the general public to reduce the effects of fragmentation on privately owned land, as much of the remaining protected areas of natural habitat are surrounded by privately held land. Conservation conducted in a broader landscape context that includes incentives for private land owners has the potential to enhance the management objectives of protected areas and reduce the effects of fragmentation.[68] To mitigate the effects of anthropogenic fragmentation, ecologists and conservationists will need to expand beyond traditional reserve design principles (conserving large natural areas connected to other protected areas by corridors) and incorporate restorative practices, such as enhancing and managing the matrix surrounding habitat fragments.

Acknowledgments

The author wishes to acknowledge the far too many to list landscape ecologists, conservationists, field biologists, land practitioners, resource managers, bioreserve specialists who have contributed to the science and understanding of the effects of habitat fragmentation and isolation.

References

1. Wilcove, D. S.; McLellan, C.H.; Dobson, A.P. Habitat fragmentation in the temperate zone. In *Conservation Biology. The Science of Scarcity and Diversity*; Soulé, M. E. Ed.; Sinauer: Sunderland, 1986; 237–256.
2. Ranta, P.; Blom, T.; Niemela, J.; Joensuu, E.; Siitonen, M. The fragmented Atlantic rain forest of Brazil: size, shape and distribution of forest fragments. Biodivers. Conserv. **1998**, *7*, 385–403.
3. McGarigal, K.; McComb, W.C. Forest fragmentation effects on breeding birds in the Oregon Coast Range. In *Forest Fragmentation: Wildlife and Management Implications.* J.A. Rochelle, L.A. Lehman and J. Wisniewski, Eds.; Koninklijke Brill NV: Leiden, the Netherlands, 1999; 223–246.
4. Fahrig, L. Effects of habitat fragmentation on biodiversity. Annu. Rev. Ecol., Evol. Syst. **2003**, *34*, 487–515.
5. Ricketts, T. H. The matrix matters: effective isolation infragmented landscapes. Am. Nat. **2001**, *158*, 87–99.
6. Ewers, R. M.; R. K. Didham. Confounding factors in the detection of species responses to habitat fragmentation. Biol. Rev. **2006**, *81*, 117–142 pp.
7. Forman, R. T. T. *Land Mosaics: The Ecology of Landscapes and Regions*; Cambridge University Press: New York, USA, 1995; 632 p.
8. Laurance, W. F. Theory meets reality: how habitat fragmentation research has transcended island biogeographic theory. Biol. Conserv. **2008**, *141*, 1731–1744.
9. Haila, Y. A conceptual genealogy of fragmentation research: from island biogeography to landscape ecology. Ecol. Appl. **2002**, *12*(2), 321–334.
10. MacArthur, R.H.; Wilson, E. O. *The Theory of Island Biogeography*; Princeton University Press: Princeton, USA, 1967.
11. Preston, F. W. Time and space and the variation of species. Ecology **1960**, *41*, 611–627.
12. Powledge, F. Island biogeography's lasting impact. BioScience **2003**, *53*, 1032–1038.
13. Simberloff, D.S.; Abele, L. G. Island biogeography theory and conservation practice. Science, **1976**, *191*, 285–286.
14. Diamond, J.M. The island dilemma: lessons of modern biogeographic studies for the design of natural reserves. Biol. Conserv. **1975**, *7*, 129–146.
15. Diamond, J.M. Island biogeography and conservation: strategy and limitations. Science, **1976**, *193*, 1027–1029.

16. Saunders, D.A.; Hobbs, R.J.; Margules, C.R. Biological consequences of ecosystem fragmentation: a review. Conserv. Biol. **1991**, *5*, 18–32.
17. Ovaskainen, O. Long-term persistence of species and the SLOSS problem. J. Theor. Biol. **2002**, *218*, 419–433.
18. Wiens, J. Habitat fragmentation: island vs. landscape perspectives on bird conservation. Ibis **1994**, *137*, S97-S104.
19. Haila, Y. Implications of landscape heterogeneity for bird conservation. Int. Ornithol. Congr. **1991**, *20*, 2286–2291.
20. Lefkovitch, L.P.; Fahrig, L. Spatial characteristics of habitat patches and population survival. Ecol. Model. **1985**, *30*, 297–308.
21. Hanski, I. A.; M. E. Gilpin, Eds. *Metapopulation Biology. Ecology, Genetics, and Evolution*; Academic Press: New York, USA, 1997.
22. Didham, R. K. *Ecological Consequences of Habitat Fragmentation. Encyclopedia of Life Sciences*; Wiley: New York, USA, 2010.
23. Didham, R.K.; Kapos, V. and Ewers, R. M. Rethinking the conceptual foundations of habitat fragmentation research. Oikos **2012**, *121*, 161–170.
24. Lortie, C. J.; Brooker, R.W.; Choler, P.; Kikvidze, Z.; Michalet, R.; Pugnaire, F.I.; Callaway, R.M. Rethinking plant community theory. Oikos **2004**, *107*, 433–438.
25. Turner, M.G. Landscape ecology: the effect of pattern on process. Annu. Rev. Ecol. Syst. **1989**, *20*, 171–197.
26. Wiens, J.A. Spatial scaling in ecology. Funct. Ecol. **1989**, *3*, 385–397.
27. Levin, S.A. The problem of pattern and scale in ecology. Ecology **1992**, *73*(6), 1943–1967.
28. O'Neill, R.V.; Hunsaker, C.T.; Timmins, S.P.; Jackson, B.L.; Riitters, K.L.; Wickham, J.D. Scale problems in reporting landscape pattern at the regional scale. Landscape Ecol. **1996**, *11*(3), 169–180.
29. Campbell, D. J. Scale and patterns of community structure in Amazonian forests. In *Large-Scale Ecology and Conservation Biology*, P. J. Edwards, R. M. May, and N. R. Webb, Eds.; Blackwell: Oxford, 1994; 1–17.
30. Keitt, T. H.; D. L. Urban; B. T. Milne. Detecting critical scales in fragmented landscapes. Conserv. Ecol. **1997**, *1*, 4.
31. Riitters, K.; Wickham, J.; O'Neill, R.; Jones, B.; Smith, E. Global-scale patterns of forest fragmentation. Conserv. Ecol. **2000**, *4*, 3.
32. Radford, J.Q.; Bennett, A.F.; Cheers, G.J. Landscape-level thresholds of habitat cover for woodland-dependent birds. Biol. Conserv. **2005**, *124*, 317–337.
33. Laurance, W.; Camargo, J.L.C; Luizão, R.C.C. et al. The fate of Amazonian forest fragments: A 32-year investigation. Biol. Conserv. **2011**, *144*, 56–67.
34. Laurance, W.F.; Camargo, J.L.C.; Luizaoa, R.C.C.; Laurence, S.G.; Pimm, S.L.; Bruna, E.M.; Stouffer, P.C.; Williamson, G.B.; Benítez-Malvido, J.; Vasconcelos, H.L.; Houtan, K.S.V.; Zartman, C.E.; Boyle, S.A.; Didham, R.K.; Andrade, A.; Lovejoy, T.E. Ecological correlates of extinction processes in Australian tropical rainforest mammals. Conserv. Biol. **1991**, *5*, 79–89.
35. Simberloff, D.; Farr, J. A.; Cox, J.; Mehlman, D.W. Movement corridors: conservation bargains or poor investments? Conserv. Biol. **1992**, *6*, 493–504.
36. Lindenmayer, D.; Fischer, J. *Habitat Fragmentation and Landscape Change: An Ecological and Conservation Synthesis*; Island Press: Washington, D.C., USA, 2006.
37. Burns, C.E.; Grear, J. Effects of habitat loss on populations of white-footed mice: testing matrix model predictions with landscape-scale perturbation experiments. Landscape Ecol. **2008**, *23*, 817–831.
38. Ovaskainen, O.; Hanski, I. Metapopulation dynamics in highly fragmented landscapes. In *Ecology, Genetics, and Evolution of Metapopulations*; I. Hanski and O. E. Gaggiotti, Eds.; Elsevier Academic Press: San Diego, USA, 2004; 73–103.

39. Lambeck, R. Focal species: a multi-species umbrella for nature conservation. Conserv. Biol. **1997**, *11* (4), 849–856.
40. Watts, K.; Eycott, A.E.; Handley, P.; Ray, D.; Humphrey, J.W.; Quine, C.P. Targeting and evaluating biodiversity conservation action within fragmented landscapes: an approach based on generic focal species and least-cost networks. Landscape Ecol. **2010**, *25*, 1305–1318.
41. McGarigal, K.; Cushman, S.A.; Neel, M.C.; Ene, E. FRAGSTATS 3.0: Spatial Pattern Analysis Program for Categorical Maps; University of Massachusetts: Amherst, MA, 2002.
42. Fahrig, L. Effects of habitat fragmentation on biodiversity. Annu. Rev. Ecol. Syst. **2003**, *34*, 487–515.
43. McGarigal, K.; McComb, W.C. Relationships between landscape structure and breeding birds in the Oregon Coast Range. Ecol. Monogr. **1995**, *65*, 235–260.
44. Villard, M.A.; Trzcinski, M.; Merriam, G. Fragmentation effects on forest birds: relative influence of woodland cover and configuration on landscape occupancy. Conserv. Biol. **1999**, *13*, 774–783.
45. Laurance, W.F.; Cochrane, M.A. Synergistic effects in fragmented landscapes. Conserv. Biol., **2001**, *15*, 1488–1489.
46. Seabloom, E.W.; Dobson, A.P.; Stoms, D.M. Extinction rates under nonrandom patterns of habitat loss. Proc. Natl. Acad. Sci. USA **2002**, *99*, 1129–11234.
47. August, P.; Iverson, L.; Nugranad, J. Human conversion of terrestrial habitats. In *Applying Landscape Ecology in Biological Conservation*; Gutzwiller, K. J. Ed.; SpringerVerlag: New York, USA, 2002; 198–224.
48. Soulé, M.E.; Simberloff, D.S. What do genetics and ecology tell us about the design of nature reserves? Biol. Conserv. **1986**, *35*, 19–40.
49. Paton, P.W.C. The effect of edge on avian nest success: how strong is the evidence? Conserv. Biol. **1994**, *8*, 17–26.
50. Van Horn, M. A.; Gentry, R. M.; Faaborg, J. Patterns of Ovenbird (*Seiurus aurocapillus*) pairing success in Missouri forest tracts. Auk **1995**, *112*, 98–106.
51. Meentemeyer, R. K.; Haas, S.E.; Vaclavık, T. Landscape Epidemiology of emerging infectious diseases in natural and human-altered ecosystems. Annu. Rev. Phytopathol. **2012**, *50*, 19.1–19.24
52. Taylor, P. D.; Fahrig, L.; Henein, K.; Merriam, G. Connectivity is a vital element of landscape structure. Oikos **1993**, *68*, 571–573.
53. Crooks, K. R.; M. A. Sanjayan, Eds.; *Connectivity Conservation*. Cambridge University Press; New York, USA, 2006.
54. Watts K.; Handley P. Developing a functional connectivity indicator to detect changes in fragmented landscapes. Ecol. Indicators **2010**, *10*, 552–557.
55. Theobald, D. M.; Crooks, K. R.; Norman, J.B. Assessing effects of land use on landscape connectivity: loss and fragmentation of western U.S. forests. Ecol. Appl. **2011**, *21*(7), 2445–2458.
56. Urban, D. L.; Keitt, T. H. Landscape connectedness: A graph theoretic perspective. Ecology **2001**, *82*, 1205–1218.
57. Theobald, D. M. Exploring the functional connectivity oflandscapes using landscape networks. In *Connectivity Conservation: Maintaining Connections for Nature*; Crooks, K. R. and Sanjayan, M. A., Eds.; Cambridge University Press: Cambridge, UK, 2006; 416–443.
58. McRae, B. H.; Beier, P. Circuit theory predicts gene flow in plant and animal populations. PNAS, **2007**, *104*(50), 19885–19890.
59. Sterling, K.A.; Reed, D. H.; Noonan, B.P.; Warren Jr., M.L. Genetic effects of habitat fragmentation and population isolation on *Etheostoma raneyi* (Percidae). Conserv. Genet. **2012**, *13*, 859–872.
60. Hitt, N. P.; Angermeier, P. L. River-stream connectivity affects fish bioassessment performance. Environ. Manag., *42*, 132–150.
61. Cote, D.; Kehler, D. G.; Bourne, C.; Wiersma, Y.F. A new measure of longitudinal connectivity for stream networks. Landscape Ecol. **2009**, *24*, 101–113.
62. Pinto, N.; Keitt, T. H. Beyond the least-cost path: evaluating corridor redundancy using a graph-theoreticapproach. Landscape Ecol. **2009**, *24*, 253–266.

63. Pinto, N.; Keitt, T.H.; Wainright, M. LORACS: JAVA software for modeling landscape connectivity and matrix permeability. Ecography **2012**, *35*, 001–005.
64. Didham, R.K.; Lawton, J.H. Edge structure determines the magnitude of changes in microclimate and vegetation-structure in tropical forest fragments. Biotropica **1999**, *31*, 17–30.
65. Cook, W. M.; Lane, K. T.; Foster, B. L.; Holt, R. D. Island theory, matrix effects and species richness patterns inhabitat fragments. Ecol. Lett. **2002**, *5*, 619–623.
66. Perfecto, I.; Vandermeer, J. Quality of agroecological matrix in a tropical montane landscape: ants in coffee plantations in southern Mexico. Conserv. Biol. **2002**, *16*, 174–182.
67. Laurance, W.F. Forest-climate interactions in fragmented tropical landscapes. Phil. Trans. Roy. Soc. B, **2004**, *359*, 345–352.
68. Goetz, S.J.; Jantz, P.; Jantz, C.A. Connectivity of core habitat in the Northeastern United States: parks and protected areas in a landscape context. Remote Sensing Environ. **2009**, *113*, 1421–1429.

Bibliography

Gutzwiller, K. *Appl ying Landscape Ecology in Biological Conservation*; Springer-Verlag: New York, USA, 2002.

Hanski, I. *Metapopulation Ecology*; Oxford University Press: New York, USA, 1999.

Lindenmayer, D.B.; Fischer, J. *Habitat Fragmentation and Landscape Change: An Ecological and Conservation Synthesis;* Island Press: Washington, D.C., USA, 2006.

Losos, J.B.; Ricklefs, R.E. *The Theory of Island Biogeography Revisited*; Princeton University Press: Princeton, New Jersey, USA, 2010.

MacArthur, R.H.; Wilson, E.O. *The Theory of Island Biogeography*; Princeton University Press: Princeton, New Jersey, USA, 1967.

Quammen, D. *Song of the Dodo: Island Biogeography in an Age of Extinctions;* Scibner: New York, USA, 1996.

Tilman, D.; Kareiva, P. *Spatial Ecology: The Role of Space in Population Dynamics and Interspecific Interactions*; Princeton University Press: Princeton, New Jersey, USA, 1997.

Wiens, J.A.; Moss, M.R.; Turner, M.G.; Mladenoff, D.J. *Foundation Papers in Landscape Ecology*; Columbia University Press: New York, USA, 2007.

5

Land Surface Temperature: Remote Sensing

Yunyue Yu and
Jeffrey P. Privette
National Oceanic and Atmospheric Administration (NOAA)

Yuling Liu
University of Maryland

Donglian Sun
George Mason University

Introduction

Most of the solar energy incident on the Earth is absorbed at the surface; this absorbed energy is used up in the evaporation of surface water or heating of the air near the surface and finally drives our weather and climate. Land Surface Temperature (LST) is a key indicator of the fraction of the absorbed energy. It is fundamental for estimating the net radiation budget at the Earth's surface in climate research as well as for monitoring the state of crops and vegetation in environmental studies. Long-term trends of LST are indicative of the rate of greenhouse warming. Therefore, LST data are widely utilized in a range of hydrological, meteorological, environmental management and climatological applications. Some of the most important applications of remote sensed LST data are to be assimilated into weather and climate models for the weather and climate predictions, and to be input into mesoscale atmospheric and land surface models to estimate sensible heat flux and latent heat flux. Commercial applications include the use of LST to evaluate water requirements for crops in summer and estimate where and when a damaging frost may occur in winter. High LST is a warning sign for possible forest and grass fires, as well as an indicator of possible drought. In 2011, World Meteorological Organization discussed to include LST as one of essential climate variables in Global Climate Observation System.[1]

Traditionally, LST is measured (or can be estimated from the measurements) from weather stations, which are usually located in relatively densely populated regions and are sparse and unevenly distributed; only by remote sensing of LST from satellites are feasible on a regional or global scale. Satellite LSTs have been routinely produced from geostationary and polar-orbiting satellites. As of 2012, more than 30 years of global satellite LST data had been accumulated. Geostationary satellites provide high temporal resolution (up to 15 minutes) LST data that are ideal for diurnal variation study, while polar-orbiting satellites produce high spatial resolution (up to 100 m) LST data that are valued for daily variation monitoring and climatology study.

Physics of LST Remote Sensing

Major techniques of LST remote sensing were inherited from sea surface temperature (SST) remote sensing. Compared to the SST remote sensing, however, LST remote sensing is more difficult in terms of derivation and validation, because of heterogeneity and anisotropy of land surface.

LST remote sensing is based on physics of radiative transfer process from Earth surface to satellite sensor. Infrared channels or microwave channels are used for the LST sensing. Advantage of using microwave channels is for its all-weather feature since the microwave band is atmospheric transparency. However, spatial resolution of the microwave LST remote sensing is very poor (about several tens of kilometers), and quality of the sensed LST is not so good because of its relatively low signal-to-noise ratio. Routinely, satellite LSTs are sensed from infrared channels within 3–12 μm spectrum where emission energy from Earth surface reaches the maximum. In particular, the best satellite LST measurements are performed in thermal infrared channels around 10–12 μm spectrum where atmospheric transmittance reaches the maximum as well (see Figure 5.1). A disadvantage of the infrared LST remote sensing is that it must be under cloud-free sky condition as cloud is not transparent to infrared signal.

In infrared spectrum, satellite received radiance is mostly emission energy from Earth surface that is described by Planck's law:

$$B(\lambda,T)=\frac{2hc^2}{\lambda^5}\frac{1}{\dfrac{hc}{e\lambda kT-1}}$$

where B is the radiance at wavelength λ, T is the absolute temperature, k is the Boltzmann constant, h is the Planck constant, and c is the speed of light. Satellite received radiance can be explained in the following radiative transfer equation:

$$B(\lambda,T)=\varepsilon(\lambda)\tau(\lambda)B(\lambda,T_s)+I_{\text{atm}}(\lambda)^{\uparrow}+I_{\text{atm}}(\lambda)^{\downarrow}$$

where $B(\lambda,T_b)$ is the satellite sensor received radiance at wavelength λ, T_b, which is the so-called "brightness temperature" the satellite sensed, $\varepsilon(\lambda)$ and $\tau(\lambda)$ are the surface emissivity and the atmospheric transmittance at the wavelength, T_s is the surface temperature to retrieve; $I_{\text{atm}}(\lambda)^{\uparrow}$, and $I_{\text{atm}}(\lambda)^{\downarrow}$

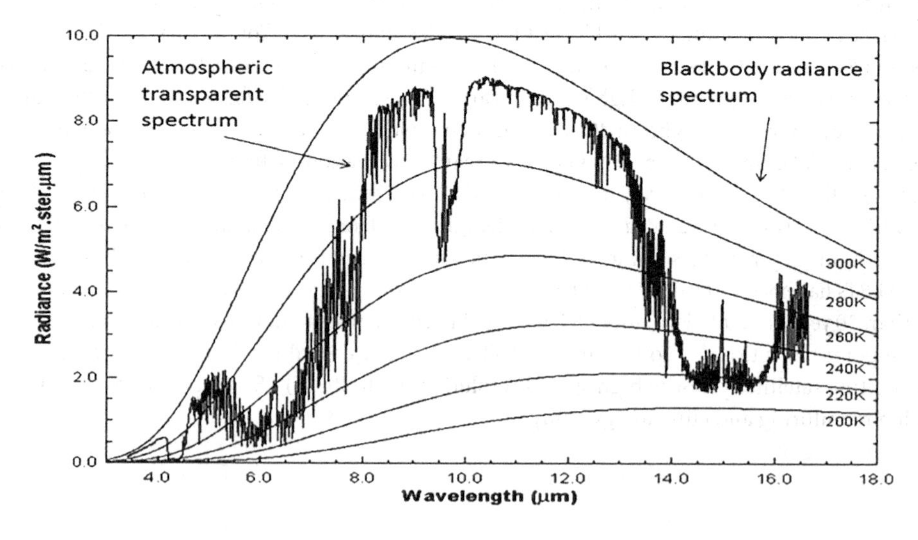

FIGURE 5.1 Infrared spectral radiance distribution of black body at different temperatures (smooth curves) and spectral transmittance from Earth surface to top of atmosphere.

are radiances of atmospheric emission and surface-reflected solar emission, respectively. Figure 5.2 illustrates the major components considered in the LST remote sensing.

All the LST remote sensing process is about how to derive surface temperature from the sensed brightness temperature. Note that satellite LST retrieval from infrared sensing is so-called "skin temperature" representing a few millimeter depth of land surface, as the infrared emission is blocked beyond the skin. Figure 5.3 shows a sample LST map derived from U.S. Earth Observation System (EOS) Terra satellite, processed at NASA Goddard Space Flight Center.

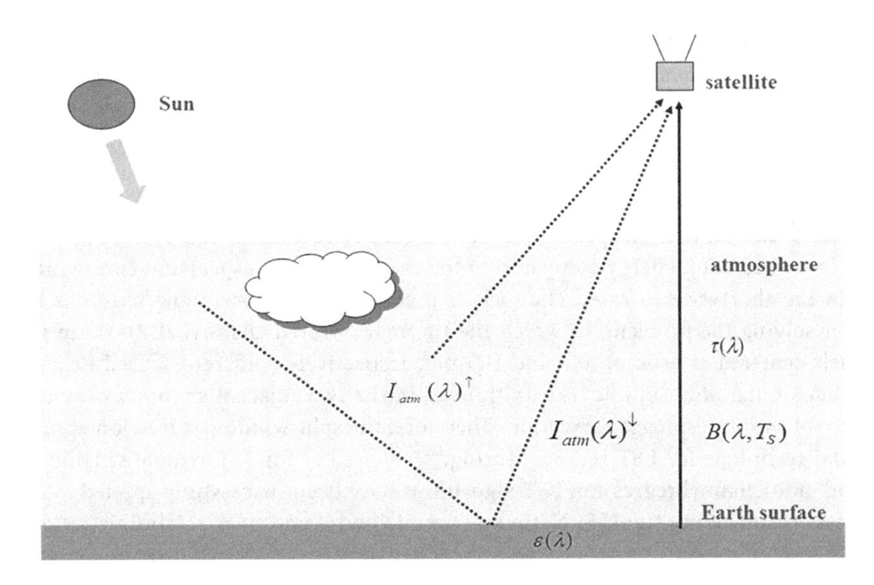

FIGURE 5.2 A sketch of infrared radiances satellite sensor received: surface emitted radiance $B(\lambda,T_s)$, reflected atmospheric radiance $I_{atm}(\lambda)$, direct atmospheric radiance $I_{atm}(\lambda)$, surface emissivity $\varepsilon(\lambda)$, and atmospheric transmittance $\tau(\lambda)$.

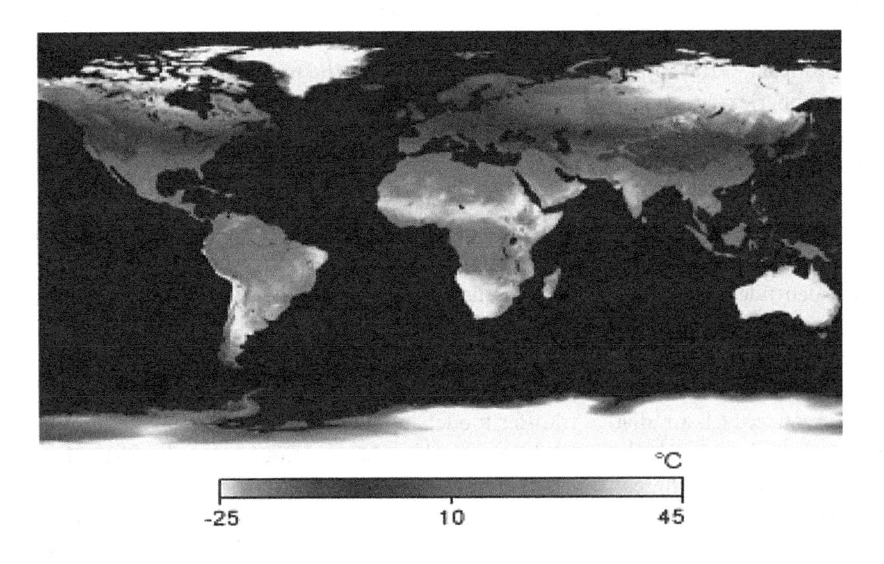

FIGURE 5.3 **(See color insert.)** Average global LST map obtained from the NASA Terra satellite MODIS (Moderate Resolution Imaging Spectroradiometer) sensor.

Source: From http://modis.gsfc.nasa.gov/gallery/individual.php?db_date=2009-04-30.

Technical Approaches

Difficulties of the LST retrieval are mostly due to nonlinearity of the Planck function, coupling of the emissivity and temperature in the radiative transfer equation, and correction of the atmospheric radiation and absorption. Linear relationship can be established between the remote sensed brightness temperature (T_b) and surface temperature (T_s) using Taylor expansion approach with the Planck function. This is a traditional regression technique of the LST remote sensing, in which coefficients of the linear relationship are determined from a training dataset, which can be generated from a radiative transfer model or from real measurements. Regression LST algorithms have been widely adopted in most satellite missions for LST production because of their simplicity and robustness.

The single-channel regression LST algorithm was developed in early 1980s,[2,3] which is a simple linear relationship between the satellite sensed brightness temperature at the thermal infrared channel and LST. Later, multichannel LST regression algorithms were developed, which in addition apply the shortwave infrared channels, in spectrum around 4 for better atmospheric correction.[4,5] However, because of lower radiation energy compared to the thermal infrared spectrum (see Figure 5.1), sensor noise in the shortwave infrared channels is a concern. A split-window sensor is specifically designed for solving the problem, in which the thermal infrared channel at 10–12 μm is split into two channels centered at around 10.5 and 11.5 pm, respectively. Different sensed brightness temperatures, but similar atmospheric transmittances, in the two adjacent channels provide an excellent measure of the atmospheric correction. Therefore, the split-window regression approach is the most popular technique for LST remote sensing,[6–8] as well as for SST remote sensing. The single channel and multichannel regression LST algorithms have been successfully applied to a variety of satellite missions including the U.S. National Oceanic and Atmospheric Administration (NOAA) Polar-orbiting Operational Environmental Satellite (POES) series, Geostationary Operational Environmental Satellite (GOES) series, Feng-Yun meteorological satellite series of China, and European Organization for the Exploitation of Meteorological Satellite (EUMETSAT) series for operational LST productions.

Another approach of the atmospheric correction is to measure the LST from two different viewing angles. The difference of the satellite sensed brightness temperatures of the two angles, representing the atmospheric radiation/ absorption difference along the two paths, provides excellent correction of the atmospheric impact.[9] The approach may be combined with the multichannel regression approach for optimal atmospheric correction. The multi-angle LST algorithm is applied to European Space Agency satellites, which are onboard a specifically designed sensor called Along Track Scanner Radiometer (ATSR) and later Advanced ATSR.[10,11]

Compared to the relative homogeneous ocean surface, emissivity of land surface may be significantly different from one surface species type to another. Coefficients of the regression LST algorithms can be emissivity dependent (such as the LST map shown in Figure 5.3, derived from the MODIS sensor), or surface type dependent (such as the LST map shown in Figure 5.4, derived from the Suomi-NPP VIIRS sensor), for correcting the emissivity variation impact.

Alternatively, atmospheric correction can be numerically computed using radiative transfer model. It requires a high performance radiative transfer model and prior accurate atmospheric profile information. In practice, such a radiative transfer model approach is applied in an emissivity-temperature separation technique,[12] in which the satellite-sensed brightness temperatures in a set of infrared channels are collected at two different times so that the number of spectral radiative transfer equations is doubled. Considering that LSTs vary significantly while land surface emissivity remains the same at those two different times, LSTs and land surface emissivities can be derived simultaneously by solving this well-posed problem. The approach is adopted in the U.S. EOS mission for the operational LST production.

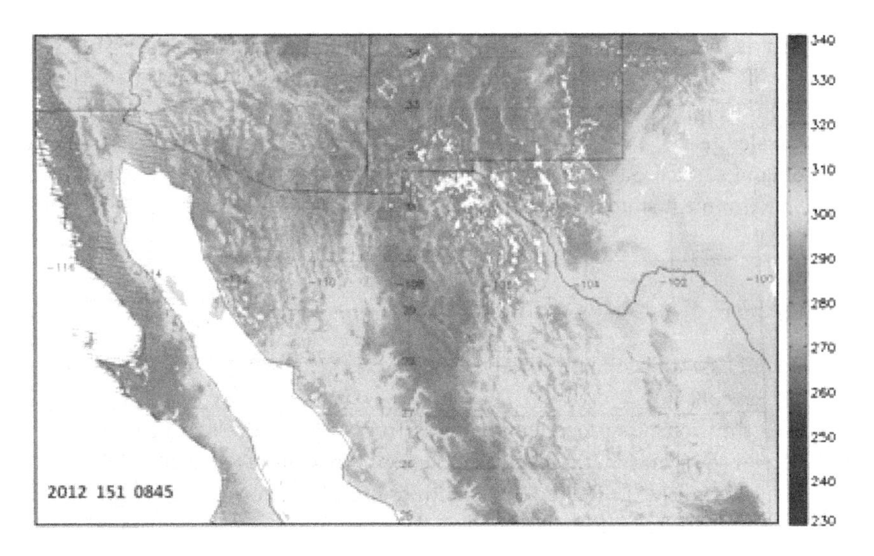

FIGURE 5.4 Land surface temperature map of the west coast of United States.
Source: Obtained from Suomi-NPP satellite, VIIRS (visible and infrared imager radiometer suite) sensor at 8:45 am, May 30, 2012.

Validation and Evaluation of Remote Sensed LST

Remote sensed LST needs to be validated before its applications. A four-stage validation process is defined by a land product validation subgroup of the Committee on Earth Observation Satellites working group on calibration and validation.[13] Traditionally, LST validation is performed by comparing the in situ LST measurements from ground stations or specific field campaigns. There are three major difficulties in the LST validation. First, within a satellite-sensed pixel, which is around a kilometer size in the infrared measurement, LST can vary significantly because of surface heterogeneity. In situ LST measurement must be evaluated or calibrated for its representativeness before it can be used for the satellite LST validation. Second, LST is coupled with surface emissivity in the satellite sensed signal. Direct or indirect prior knowledge of the surface emissivity is required, which is not easy to obtain. Third, LST may have rapid temporal variation so that in situ LST measurements must be well matched to the satellite measurements. A comprehensive in situ validation of LST is impossible if those problems cannot be solved.

The radiative transfer model can be used for satellite LST validation.[14] Assuming that high-quality atmospheric profile and surface emissivity property are available, the satellite-sensed brightness temperature can be calculated by setting a certain LST as the model input. Validating the satellite LST is to compare it to the set LST. It is believed that an error of the radiative transfer model in such process is ignorable.

Instead, satellite LST is mostly evaluated rather than validated using limited high-quality in situ data.[15] Also, LST data from a new satellite are usually evaluated using existing satellite LST products;[16] it is considered as well evaluated if the new LST data are close enough to a LST product that is well validated or evaluated. A successful application is also a good evaluation of the satellite LST.

Conclusion

LST is listed as a baseline climate data record. For example, the U.S. National Climate Data Center uses global and hemispheric LST data as important indicators in its monthly report of global climate highlight.[17] Satellite remote sensed LST data are the only resource for this purpose. Infrared imaging

LST measurement is the most popular manner for its accuracy and high spatial resolution. The accuracy requirement of the remote sensed LST data for many applications is about 1°C. Such a requirement is feasible through techniques (i.e., sensors and algorithms) developed in the past decades. However, validation of the satellite LST products is rather hard, mainly due to the variety, heterogeneity and emission isotropy of land surface. Satellite remote sensed LST products are widely applied in climatology, terrestrial and environment studies. The success of these applications lies in the best validation of the remote sensed LSTs.

References

1. Designation of Essential Climate Variables, GCOS SC-XIX, Doc. 10.1, GCOS Steering Committee Nineteenth Session, 2011.
2. Price, J.C. Estimation of surface temperatures from satellite thermal infrared data – a simple formulation for the atmospheric effect. Remote Sens. Environ. **1983**, *13*, 353–361.
3. Susskind, J.; Rosenfield, J.; Reuter, D.; Chahine, M.T. Remote sensing of weather and climate parameters from HIRS2/Msu on TIROS-N. J. Geophys. Res. **1984**, *89*, 4677–4697.
4. McClain, E.P.; Pichel, W.G.; Walton, C.C. Comparative performance of AVHRR-based multichannel sea surface temperatures. J. Geophys. Res. **1985**, *90* (11), 587–11, 601.
5. Ulivieri, C.; Cannizzaro, G. Land surface temperature retrievals from satellite measurements. Acta Astronaut. **1985**, *12* (12), pp. 985–997.
6. McMillin, L.M. Estimation of sea surface temperature from two infrared window measurements with different absorption. J. Geophys. Res. **1975**, *80*, 5113–5117.
7. Price, J.C. Land surface temperature measurements from the split window channels of the NOAA-7/AVHRR. J. Geophys. Res. **1984**, *89*, 7231–7237.
8. Wan, Z.; Dozier, J. A Generalized split-window algorithm for retrieving land-surface temperature from space, IEEE Trans. Geosci. Remote Sens. **1996**, *34*, 892–905.
9. Chedin, A.; Scott, N.; Berroir, A. A single-channel double viewing method for SST determination from coincident Meteosat and TIROS-N measurements. J. Appl. Meteorol. **1982**, *21*, 613–618.
10. Prata, A.J. Land surface temperatures derived from the AVHRR and the ATSR. 1, Theory J. Geophys. Res. **1993**, *98* (D9), 16,689–16,702.
11. Sobrino, J.A.; Li, Z.-L.; Stoll, M.P.; Becker, F. Multichannel and multi-angle algorithms for estimating sea and land surface temperature with ATSR data. Int. J. Remote Sens. **1996**, *17*, 2089–2114.
12. Wan, Z.; Li, Z.-L. A physics-based algorithm for land-surface emissivity and temperature from EOS/MODIS data. IEEE Trans. Geosci. Remote Sens. **1997**, *35*, 980–996.
13. CEOS Working Group on Calibration and Validation, Land Product Validation Subgroup Website: http://lpvs. gsfc.nasa.gov/.
14. Coll, C.; Wan, Z.; Galve, J.M. Temperature-based and radiance-based validations of the V5 MODIS land surface temperature product. J. Geophys. Res. **2009**, D20102, doi: 10.1029/2009JD012038.
15. Yu, Y.; Tarpley, D.; Privette, J.L.; Flynn, L.E.; Xu, H.; Chen, M.; Vinnikov, K.Y.; Sun, D.L.; Tian, Y.H. Validation of goes-r satellite land surface temperature algorithm using surfrad ground measurements and statistical estimates of error properties. IEEE Trans. Geosci. Remote Sens. **2012**, *50* (3), 704–713.
16. Yu, Y.; Privette, J.L.; Pinheiro, A.C. Analysis of the NPOESS VIIRS land surface temperature algorithm using MODIS data. IEEE Trans. Geosci. Remote Sens. **2005**, *43* (10), 2340–2350.
17. NOAA National Climate Data Center, State of Climate Global Analysis, http://www.ncdc.noaa.gov/sotc/global.

6

Landscape Connectivity and Ecological Effects

Marinés de la Peña-Domene and Emily S. Minor
University of Illinois at Chicago

Introduction

Landscape connectivity is broadly defined as the "degree to which the landscape facilitates or impedes movement among resource patches."[1] Movement across the landscape is crucial for a number of different ecological processes including migration,[2] dispersal,[3,4] and colonization of locally extinct habitat patches.[5,6] These movements have serious implications for gene flow and long-term evolutionary processes[7–9] as well as biological conservation.[10,11] For that reason, landscape connectivity has been of great interest to ecologists for at least two decades.[1,12,13] However, it has also been the source of confusion because, as Calabrese and Fagan[14] wrote, "connectivity comes in multiple flavors." Here, we give a brief overview of the different ways landscape connectivity has been defined and measured and highlight some important findings about the impact of connectivity on ecological processes, conservation, and natural resource management.

Definitions

Structural versus Functional Connectivity

Structural connectivity is based on the amount and physical configuration of landscape elements, including habitat patches, corridors, and stepping stones (Table 6.1, Figure 6.1). These elements are typically thought of as being embedded in a matrix of unsuitable habitat, through which movement may be limited or restricted.[15] Some studies have suggested that corridors or stepping stones can be established to increase connectivity between habitat patches.[3,16–20] However, these landscape elements may facilitate movement of some organisms more than others or more movement in some landscapes than others.[17,21,22] ***Functional connectivity*** incorporates an organism's behavior or response to elements such as corridors or stepping stones. Functional connectivity may also depend on a species' response to the matrix between landscape elements.[23,24] For example, corridors may be particularly important for species with a low capacity to move through the matrix.[25] Therefore, functional connectivity is

TABLE 6.1 Landscape Elements That Affect Landscape Connectivity

Element	Definition
Habitat patch	A contiguous area of relatively homogeneous land cover that is suitable for the species of interest.
Corridor	Continuous strips of habitat that structurally connect two or more patches.
Stepping stone	Ecologically suitable patch (typically small in size) that may act to connect larger patches of habitat. Stepping stones may function at two different time scales. First, at the shortest time scale, they may provide a temporary stopping location for an organism moving through the landscape. Second, at the longer time scale, they may provide a breeding location so that an organism's offspring can move to more distant patches.
Matrix	The intervening area between habitat patches in a landscape mosaic, usually characterized by being the most extensive cover, having high spatial continuity, or having a major influence on the landscape dynamic.

Source: Modified from Forman.[15]

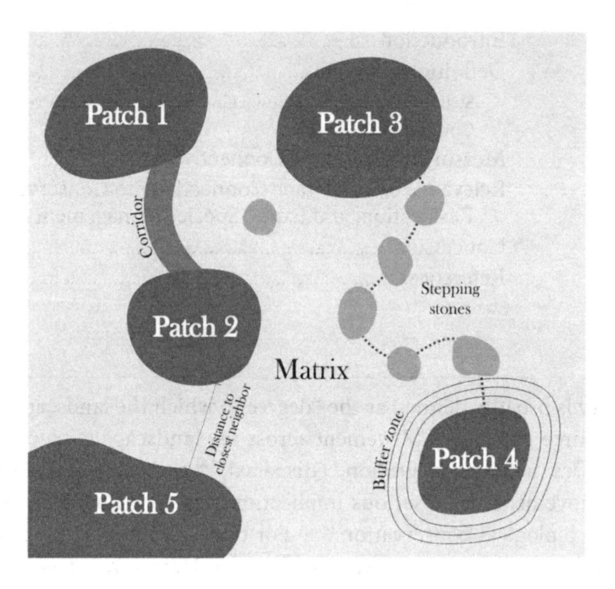

FIGURE 6.1 Landscape structure and elements.

species specific, and a single landscape can have different levels of connectivity depending on the species or process in question. Habitat that is functionally connected may not be structurally connected, and vice versa.[26]

Patch- versus Landscape-Level Connectivity

Connectivity can be measured at many different scales. On one hand, researchers or managers may want specific information about the connectivity of a single patch to the rest of the landscape. For example, studies of American pika ***(Ochotona princeps)*** have used incidence function models (IFMs)[27] to predict patch occupancy and population turnover in a large network of habitat patches.[28] On the other hand, researchers or managers may have questions about the connectivity of the entire landscape, perhaps to compare one landscape to another, or perhaps to identify whether connectivity is sufficient for a species of concern. For instance, a study of protected area networks in the United States found that certain ecoregions had greater landscape-level connectivity than others and that protected area networks were consistently more connected for large mammals than for small mammals.[29] Different metrics are used for patch-and landscape-level connectivity, and we describe some of these in the following section.

Measuring Landscape Connectivity

Dozens of different metrics exist for measuring landscape connectivity, and a number of reviews have been published on the topic.[26,30–35] Metrics may describe structural or functional connectivity, at the patch or the landscape level, or a combination of both (summarized in Table 6.2). One common approach that combines both patch-and landscape-level analyses involves patch-removal "experiments," in which researchers calculate a landscape-level connectivity metric, simulate the removal of a particular patch, and then recalculate the landscape-level metric once that patch has been removed. The difference between the first connectivity estimation and the second can be used to rank individual patches in terms of their importance for connectivity across the entire landscape.[36,37]

In general, structural connectivity measures may be calculated simply by examining a map. One of the simplest measures of patch-level structural connectivity is nearest-neighbor distance, which is the Euclidean distance to the nearest patch.[2] This measure was shown to be a useful predictor of the incidence of a frugivorous beetle in forest fragments.[38] This simple measure can be infused with more ecological information by measuring distance to the nearest occupied patch rather than distance to any patch,[2] but the more comprehensive measure is not always an improvement on the simpler one.[38] Another common way to measure patch-level structural connectivity is to examine the amount of habitat in a buffer around the focal patch. This measure was found to be much better than nearest-neighbor distance for predicting colonization events in two butterfly species but was very sensitive to buffer size.[32] At the landscape level, structural connectivity can be measured, for example, based on percolation or lacunarity. In these approaches, the landscape is considered as a two-dimensional grid where each cell in the grid is classified as either habitat or non-habitat. Landscape connectivity is then measured as the physical connection of habitat, in the case of percolation,[39] or as the variability and size of interpatch distances, in the case of lacunarity.[40] The lacunarity index was shown to be a good predictor of dispersal success for simulated organisms in fractal landscapes.[41]

Because functional connectivity is specific to the species or process of interest, it is necessary to know something about species' movement behavior to measure functional connectivity. Species may perceive and respond to landscape pattern differently according to their dispersal characteristics, their preferred habitat type, or other life history traits.[26] For example, seed-dispersing birds differ in their presence in remnant trees within the matrix based on their frugivory levels. Fruiting trees could represent stepping stones across the matrix for birds with a completely fruit-based diet, but may not for birds with mixed diets.[42] Functional connectivity measures are usually considered to be superior to structural connectivity measures.

TABLE 6.2 A Subset of Metrics Used to Measure Landscape Connectivity

Metric	Scale	Connectivity Type	References
Nearest-neighbor distance	Patch	Structural	2, 38
Buffer area	Patch	Structural	32
Incidence function models	Patch	Structural	27, 28
Lacunarity	Landscape	Structural	39, 40, 41
Network centrality metrics	Patch	Potential	59
Cohesion index	Landscape	Potential	44
Least-cost paths	Dispersal route	Potential	36
Multiple shortest paths	Dispersal route	Potential	67
Isolation by resistance	Dispersal route	Potential	65
Probability of connectivity (PC)	Landscape	Potential	63, 64
Cell or patch immigration	Patch or landscape	Actual	48
Animal homing time	Landscape	Actual	53

Depending on how it is measured, functional connectivity can be divided into ***potential connectivity***, where information about the movement ability of the organisms is limited, and ***actual connectivity***, where there is detailed movement data for the organism of interest.[14] Potential connectivity measures may be based on attributes such as body size of the animal or dispersal mechanism of the plant. At the patch level, measuring potential connectivity may be as simple as using an ecologically meaningful distance when counting nearest neighbors.[43] At the landscape level, potential connectivity can be measured using the cohesion index, for example. This index integrates habitat quality, amount and configuration of habitat, and permeability of the landscape matrix to indicate species persistence.[44] Potential connectivity of both patches and landscapes can be calculated based on resistance or cost surfaces, which represent the willingness of a focal organism to cross the environment between habitat patches.[45] This approach emphasizes the importance of the matrix to connectivity and is based on potential "costs" of movement (usually in terms of energetic expenditures or mortality risks) through different regions of the landscape. Cost surfaces have been used to identify potential conservation corridors for jaguars[46] and were shown to help predict patch occupancy for prairie dogs in Colorado, United States of America.[47] Dispersal routes connecting habitat patches can be identified using least-cost path tools. This estimates the route with the least resistance between two points; however, by identifying a single route, alternative paths with comparable resistance costs may be ignored. This approach has been extended to include multiple routes, with methods such as circuit theory or multiple shortest paths (MSPs), which are described in greater detail at the end of this section.

Actual connectivity measures are based on empirical, often spatially explicit, information about the movement of a particular organism. At the patch level, actual connectivity can be measured as the number of immigrants into a patch.[48] At the landscape level, actual connectivity may be measured as the number of patches visited by an organism or movement rates across the landscape[49,50] However, these approaches can present a paradox in that animals may move more frequently between patches in lower-quality habitat, counterintuitively resulting in higher connectivity indices in these less-desirable environments.[51] Another approach to measuring actual connectivity draws heavily from behavioral ecology. For instance, Bélisle[51] suggests several different experimental methods for determining the motivation underlying movement of individuals through the landscape, including translocations, playback experiments, and measuring giving-up densities. Playback and homing experiments have been used to study the effect of roads[52] and other barriers[53] on animal movement. More recently, playback techniques have been used to parameterize graph theory models (described below) to explain the occurrence pattern of an Atlantic rainforest bird.[54]

Functional connectivity can also be inferred from genetic information. Dispersal influences gene flow between subpopulations,[55,56] which results in genetic differences among organisms occupying different parts of the landscape. Landscape genetics is a rapidly growing field and will likely continue to make large contributions to measuring and understanding the consequences of landscape connectivity.[57]

Graph theory, also called network analysis, is a flexible method for measuring landscape connectivity that has gained traction over the last decade. A graph is a set of nodes connected by links, where a link between nodes indicates a connection between them. In landscape ecology, the nodes typically represent habitat patches, and links indicate dispersal between patches.[58] Commonly used metrics from graph theory include various centrality measures, which determine the importance of individual nodes in the graph (i.e., patch-level connectivity). Some examples include degree centrality, which measures the number of links of a given node (akin to the number of neighboring patches), and betweenness centrality, which measures the number of shortest paths that pass through a given patch.[59] Graph theory can measure both potential and actual connectivity, at either the patch or the landscape level or a combination of both. To date, most applications of graph theory define links based on either an ecologically relevant measure of Euclidean distance[60,61] or a distance that incorporates cost or resistance to movement.[36,62]

Graph-based metrics have also been developed around the concept of measuring habitat availability or reachability at a landscape level. In this approach, movement between and within patches is combined in a single metric that describes the ability of species to reach resources across the landscape whether those resources come from the same patch (intrapatch connectivity), from connected neighboring patches (interpatch connectivity), or from a combination of both. Habitat availability metrics combine topological features with ecological characteristics of landscape elements, which has helped to place connectivity considerations in a broader and more informative context for conservation management alternatives.[63,64]

Some graph theory-based methods explicitly model multiple paths between two points of interest, extending the least-cost path approach described earlier. Circuit theory, which comes from the field of electrical engineering,[65] can be used to model the movement of individuals across a landscape based on the idea of isolation by resistance[65] and incorporates the effect of the matrix on movement across the landscape. In circuit theory, landscape "circuits" are a kind of graph with links defined in terms of resistances between nodes. Circuit theory techniques can be combined with genetic information, for example, to examine the influence of landscape composition and configuration on gene flow.[66] Additional graph theory methods that account for multiple paths across the landscape include calculating conditional minimum transit costs (CMTCs) and MSPs. Both methods have been used to enable visualization of multiple dispersal routes that, together, are assumed to form a corridor.[67]

Relevance of Landscape Connectivity to Conservation, Restoration, and Exotic Species Management

Habitat fragmentation and loss put many species at risk of local or regional extinction[68] and persistence of many plant and animal populations depends on their ability to recolonize distant habitat patches.[6] One consequence of habitat fragmentation is that isolated populations tend to lose fitness though inbreeding depression and a loss of genetic diversity.[66] This reduces the ability of populations to adapt to environmental changes and could result in an increased risk of local extinction.[69-72] Habitat fragmentation may also prevent species from shifting their range in the case of climate change.[71] Recent studies have focused on how climate change will affect dispersal of individuals through the landscape and how populations will shift their distributions. For example, based on their movement capacities, it has been estimated that populations of many mammal species will be vulnerable following climate change.[73]

Many authors have suggested that increasing landscape connectivity is one of the best options for conservation in the face of habitat loss and climate change (reviewed in[74]). Corridors and stepping stones have been suggested as one way to increase connectivity[20,75] and have been shown to direct the movement of a number of different species.[17,76,77] Designing strategic networks of patches and corridors that allow for dispersal between environmentally similar habitats[78] or between different climatic areas based on expected changes in climates[79] may help to counterbalance the effects of climate change on natural populations. Where habitat fragmentation is prevalent, restoring functional connectivity in the landscape can broaden species distributions, rescue genetically isolated populations, and assist in the conservation of animal and plant species.[17,19,20,80-82] Connectivity measures based on graph theory in particular have been used to assist with conservation planning for many species, including the European bison[83] and the gray wolf,[84] and have also been applied to freshwater[85] and marine[86] environments.

One concern about increasing landscape connectivity is the potential for also increasing the risk of invasion from exotic species and pathogens.[87,88] However, the degree to which landscape configuration constrains the spread of an exotic species may depend on dispersal characteristics of the focal species. For example, species considered to be "invasive" and species with frequent long-distance dispersal events are likely to spread across a landscape regardless of the configuration of landscape elements[89,90] Fortunately, the methods described earlier may be useful for predicting and managing the spread of exotic species. For example, Minor and Gardner[89] used graph theory to identify critical points on the landscape where management could help contain the spread of invasive plants. Similarly, Wang et al. [91]

identified particular types and spatial arrangements of land cover that were conducive to the spread of the invasive rice water weevil ***(Lissorhoptrus oryzophilus)*** in eastern China.

Conclusion

There is a large and expanding body of literature on the topic of landscape connectivity. Connectivity is known to be important for a number of ecological processes and thus for long-term biological conservation. There are currently dozens of methods for measuring landscape connectivity and new methods are proposed on a regular basis. However, because movement is difficult to observe, and large-scale experiments are expensive and logistically challenging, the field has lagged behind on empirically testing the effect of landscape configuration on the movement of plants and animals. Therefore, much remains to be learned about how organisms move around the landscape, how these movements influence population processes and gene flow, and how we can improve connectivity for species of conservation concern while minimizing the movement of exotic species. Future research can help us to find the balance between "desirable" movement, like gene flow or seed and pollen dispersal of native species, with "undesirable" movement of invasive species and pathogens. In both cases, understanding how landscape connectivity influences population dynamics will allow us to identify better conservation strategies and management plans.

References

1. Taylor, P.D.; Fahrig, L.; Henein, K.; Merriam, G. Connectivity is a vital element of landscape structure. Oikos, **1993**, *68* (3), 571–573.
2. Matter, S.F.; Roslin, T.; Roland, J. Predicting immigration of two species in contrasting landscapes: effects of scale, patch size and isolation. Oikos, **2005**, *111* (2), 359–367.
3. Levey, D.J.; Bolker, B.M.; Tewksbury, J.J.; Sargent, S; Haddad, N.M. Effects of landscape corridors on seed dispersal by birds. Science **2005**, *309* (5731), 146–148.
4. Matlack, G.R.; Leu, N.A. Persistence of dispersal-limited species in structured dynamic landscapes. Ecosystems, **2007**, *10* (8), 1287–1298.
5. Alexander, H.M.; Foster, B.L.; Ballantyne, F.; Collins, C.D.; Antonovics, Jet al., Metapopulations and metacommunities: combining spatial and temporal perspectives in plant ecology. J. Ecol. **2012**, *100* (1), 88–103.
6. Gustafson, E.J.; Gardner, R.H. The effect of landscape heterogeneity on the probability of patch colonization. Ecology, **1996**, *77* (1), 94–107.
7. Amos, J.N.; Bennett, A.F.; Nally, R.M.; Newell, G. Pavlova, A. Radford, J.Q.; Thomson, J.R.; White, M; Sunnucks, P. Predicting landscape-genetic consequences of habitat loss, fragmentation and mobility for multiple species of woodland birds. Plos One **2012**, *7* (2), e30888.
8. Björklund, M.; Bergek, S. Ranta, E. Kaitala, V. The effect of local population dynamics on patterns of isolation by distance. Ecol. Info. **2010**, *5* (3), 167–172.
9. Hardesty, B.D.; Hubbell, S.P.; Bermingham, E. Genetic evidence of frequent long-distance recruitment in a vertebrate-dispersed tree. Ecol. Lett. **2006**, *9* (5), 516–525.
10. Berger, J. The last mile: How to sustain long-distance migration in mammals. Conserv. Biol. **2004**, *18* (2), 320–331.
11. Fahrig, L.; Merriam, G. Conservation of fragmented populations. Conserv. Biol. **1994**, *8* (1), 50–59.
12. Henein, K.; Merriam, G. The elements of connectivity where corridor quality is variable. Landscape Ecol. **1990**, *4* (2), 157–170.
13. Fahrig, L.; Merriam, G. Habitat patch connectivity and population survival. Ecology **1985**, *66* (6), 1762–1768.
14. Calabrese, J.M.; Fagan, W.F. A comparison-shopper's guide to connectivity metrics. Front. Ecol. Environ. **2004**, *2* (10), 529–536.

15. Forman, R.T.T. *Land Mosaics. The Ecology of Landscapes and Regions;* Cambridge University Press: Cambridge; 1995.
16. Loehle, C. Effect of ephemeral stepping stones on metapopulations on fragmented landscapes. Ecol. Complexity **2007**, *4* (1–2), 42–47.
17. Baum, K.A.; Haynes, K.J.; Dillemuth, F.P.; Cronin, J.T. The matrix enhances the effectiveness of corridors and stepping stones. Ecology, **2004**, *85* (10), 2671–2676.
18. Pardini, R.; de Souza, S.M.; Braga-Neto, R.; Metzger, J.P. The role of forest structure, fragment size and corridors in maintaining small mammal abundance and diversity in an Atlantic forest landscape. Biol. Conserv. **2005**, *124* (2), 253–266.
19. Tewksbury, J.J.; Levey, D.J.; Haddad, N.M.; Sargent, S.; Orrock, J.L.; Weldon, A. Danielson, B.J.; Brinkerhoff, J.; Damschen, E.I.; Townsend, P. Corridors affect plants, animals, and their interactions in fragmented landscapes. Proc. Nat. Acad. Sci. **2002**, *99* (20), 12923–12926.
20. Lander, T.A.; Boshier, D.H.; Harris, S.A. Fragmented but not isolated: Contribution of single trees, small patches and long-distance pollen flow to genetic connectivity for Gomortega keule, an endangered Chilean tree. Biol. Conserv. **2010**, *143* (11), 2583–2590.
21. Uezu, A.; Metzger, J.P.; Vielliard, J.M.E. Effects of structural and functional connectivity and patch size on the abundance of seven Atlantic forest bird species. Biol. Conserv. **2005**, *123* (4), 507–519.
22. Haddad, N.M.; Bowne, D.R.; Cunningham, A.; Danielson, B.J.; Levey, D.J.; Sargent, S. Spira, T. Corridor use by diverse taxa. Ecology **2003**, *84* (3), 609–615.
23. Ricketts, T.H. The matrix matters: Effective isolation in fragmented landscapes. Am. Nat. **2001**, *158* (1), 87–99.
24. Prevedello, J.A.; Vieira, M.V. Does the type of matrix matter? A quantitative review of the evidence. Biodivers. Conserv. **2010**, *19* (5), 1205–1223.
25. Martensen, A.C.; Pimentel, R.G.; Metzger, J.P. Relative effects of fragment size and connectivity on bird community in the Atlantic Rain Forest: Implications for conservation. Biol. Conserv. **2008**, *141* (9), 2184–2192.
26. Tischendorf, L.; Fahrig, L. On the usage and measurement of landscape connectivity. Oikos, **2000**, *90* (1), 7–19.
27. Hanski, I. A practical model of metapopulation dynamics. J. Anim. Ecol. **1994**, *63* (1), 151–162.
28. Moilanen, A.; Smith, A.T.; Hanski, I. Long-term dynamics in a metapopulation of the American pika. Am. Nat. **1998**, *152* (4), 530–542.
29. Minor, E.S.; Lookingbill, T.R. A multiscale network analysis of protected-area connectivity for mammals in the United States. Conserv. Biol. **2010**, *24* (6), 1549–1558.
30. Kindlmann, P.; Burel, F. Connectivity measures: a review. Landscape Ecol. **2008**, *23* (8), 879–890.
31. Prugh, L.R. An evaluation of patch connectivity measures. Ecol. Appl. **2009**, *19* (5), 1300–1310.
32. Moilanen, A.; Nieminen, M. Simple connectivity measures in spatial ecology. Ecology **2002**, *83* (4), 1131–1145.
33. Kool, J.T.; Moilanen, A.; Treml, E.A. Population connectivity: recent advances and new perspectives. Landscape Ecol. **2013**, *28* (2), 165–185.
34. Galpern, P.; Manseau M.; Fall, A. Patch-based graphs of landscape connectivity: A guide to construction, analysis and application for conservation. Biol. Conserv. **2011**, *144*(1) 44–55.
35. Pascual-Hortal, L.; Saura, S. Comparison and development of new graph-based landscape connectivity indices: towards the priorization of habitat patches and corridors for conservation. Landscape Ecol. **2006**, *21* (7), 959–967.
36. Bunn, A.G.; Urban D.L.; Keitt, T.H. Landscape connectivity: A conservation application of graph theory. J. Environ. Manag. **2000**, *59* (4), 265–278.
37. Bodin, O.; Saura, S. Ranking individual habitat patches as connectivity providers: Integrating network analysis and patch removal experiments. Ecol. Model. **2010**, *221* (19), 2393–2405.

38. Kehler, D.; Bondrup-Nielsen, S. Effects of isolation on the occurrence of a fungivorous forest beetle, *Bolitotherus cornutus*, at different spatial scales in fragmented and continuous forests. Oikos **1999**, *84* (1), 35–43.
39. Gardner, R.H.; Milne, B.T.; Turner, M.G.; O'Neill, R.V. Neutral models for the analysis of broad-scale landscape pattern. Landscape Ecol. **1987**, *1* (1), 19–28.
40. Plotnick, R.E.; Gardner, R.H.; Oneill, R.V. Lacunarity indexes as measures of landscape texture. Landscape Ecol. **1993**, *8* (3), 201–211.
41. With, K.A.; King, A.W. Dispersal success on fractal landscapes: a consequence of lacunarity thresholds. Landscape Ecol. **1999**, *14* (1), 73–82.
42. Lasky, J.R.; Keitt, T.H. The effect of spatial structure of pasture tree cover on avian frugivores in Eastern Amazonia. Biotrop. **2012**, *44* (4), 489–497.
43. Hanski, I.; Alho, J.; Moilanen, A. Estimating the parameters of survival and migration of individuals in metapopulations. Ecol. **2000**, *81* (1), 239–251.
44. Opdam, P.; Verboom, J.; Pouwels, R. Landscape cohesion: an index for the conservation potential of landscapes for biodiversity. Landscape Ecol. **2003**, *18* (2), 113–126.
45. Zeller, K.; McGarigal, K.; Whiteley, A. Estimating landscape resistance to movement: a review. Landscape Ecol. **2012**, *27* (6), 777–797.
46. Rabinowitz, A.; Zeller, K.A. A range-wide model of landscape connectivity and conservation for the jaguar, Panthera onca. Biol. Conserv. **2010**, *143* (4), 939–945.
47. Magle, S.; Theobald, D.; Crooks, K. A comparison of metrics predicting landscape connectivity for a highly interactive species along an urban gradient in Colorado, USA. Landscape Ecol. **2009**, *24* (2), 267–280.
48. Tischendorf, L.; Fahrig, L. How should we measure landscape connectivity? Landscape Ecol. **2000**, *15* (7), 633–641.
49. Jonsen, I.D.; Taylor, P.D. Fine-scale movement behaviors of calopterygid damselflies are influenced by landscape structure: an experimental manipulation. Oikos **2000**, *88* (3), 553–562.
50. Eycott, A.; Stewart, G.B.; Buyung-Ali, L.M.; Bowler, D.E.; Watts, K.; Pullin, A.S. A meta-analysis on the impact of different matrix structures on species movement rates. Landscape Ecol. **2012**, *27* (9), 1–16.
51. Bélisle, M. Measuring landscape connectivity: The challenge of behavioral landscape ecology. Ecology **2005**, *86* (8), 1988–1995.
52. Develey, P.F.; Stouffer, P.C. Effects of roads on movements by understory birds in mixed-species flocks in central Amazonian Brazil. Conserv. Biol. **2001**, *15* (5), 1416–1422.
53. Bélisle, M.; St. Clair, C.C. Cumulative effects of barriers on the movements of forest birds. Conserv. Ecol. **2002**, *5* (2), 9.
54. Awade, M.; Boscolo, D.; Metzger, J.P. Using binary and probabilistic habitat availability indices derived from graph theory to model bird occurrence in fragmented forests. Landscape Ecol. **2012**, *27* (2), 185–198.
55. Ibrahim, K.M.; Nichols, R.A.; Hewitt, G.M. Spatial patterns of genetic variation generated by different forms of dispersal during range expansion. Heredity **1996**, *77*, 282–291.
56. Sork, V.L.; Smouse, P.E. Genetic analysis of landscape connectivity in tree populations. Landscape Ecol. **2006**, *21* (6), 821–836.
57. Storfer, A.; Murphy, M.A.; Spear, S.F.; Holderegger, R.; Waits, L.P. Landscape genetics: where are we now? Molecular Ecol. **2010**, *19* (17), 3496–3514.
58. Urban, D.L.; Minor, E.S.; Treml, E.A.; Schick, R.S. Graph models of habitat mosaics. Ecol. Lett. **2009**, *12* (3), 260–273.
59. Estrada, E.; Bodin, O. Using network centrality measures to manage landscape connectivity. Ecol. Appl. **2008**, *18* (7), 1810–1825.
60. Minor, E.S.; Urban, D.L. Graph theory as a proxy for spatially explicit population models in conservation planning. Ecol. Appl. **2007**, *17* (6), 1771–1782.

61. Pascual-Hortal, L.; Saura, S. Integrating landscape connectivity in broad-scale forest planning through a new graph-based habitat availability methodology: application to capercaillie (*Tetrao urogallus*) in Catalonia (NE Spain). Eur. J. Forest Res. **2008**, *127* (1), 23–31.
62. Decout, S.; Manel, S.; Miaud, C.; Luque, S. Integrative approach for landscape-based graph connectivity analysis: a case study with the common frog (*Rana temporaria*) in human-dominated landscapes. Landscape Ecol. **2012**, *27(2)*, 267–279.
63. Saura, S.; Rubio, L. A common currency for the different ways in which patches and links can contribute to habitat availability and connectivity in the landscape. Ecogr. **2010**, *33* (3), 523–537.
64. Luque, S.; Saura, S.; Fortin, M.-J. Landscape connectivity analysis for conservation: insights from combining new methods with ecological and genetic data PREFACE. Landscape Ecol. **2012**, *27* (2), 153–157.
65. McRae, B.H.; Dickson, B.G.; Keitt, T.H.; Shah, V.B. Using circuit theory to model connectivity in ecology, evolution, and conservation. Ecology **2008**, *89* (10), 2712–2724.
66. Manel, S.; Schwartz, M.K.; Luikart, G.; Taberlet, P. Landscape genetics: combining landscape ecology and population genetics. Trends Ecol. Amp. Evo. **2003**, *18* (4), 189–197.
67. Pinto, N.; Keitt, T.H. Beyond the least-cost path: evaluating corridor redundancy using a graph-theoretic approach. Landscape Ecol. **2009**, *24* (2), 253–266.
68. Fischer, J.; Lindenmayer, D.B. Landscape modification and habitat fragmentation: a synthesis. Global Ecol. Biogeogr. **2007**, *16* (3), 265–280.
69. Frankham, R.; Ralls, D.K. Conservation biology: inbreeding leads to extinction. Nature **1998**, *392* (6675), 441–442.
70. Frankham, R.; Genetics and extinction. Biol. Conserv. **2005**, *126* (2), 131–140.
71. Davis, M.B.; Shaw, R.G. Range shifts and adaptive responses to quaternary climate change. Science **2001**, *292* (5517), 673–679.
72. Opdam, P.; Wascher, D. Climate change meets habitat fragmentation: linking landscape and biogeographical scale levels in research and conservation. Biol. Conserv. **2004**, *117*(2), 285–297.
73. Schloss, C.A.; Nunez, T.A.; Lawler, J.J. Dispersal will limit ability of mammals to track climate change in the Western Hemisphere. Proc. Nat. Acad. Sci. Am. **2012**, *109* (22), 8606–8611.
74. Heller, N.E.; Zavaleta, E.S. Biodiversity management in the face of climate change: A review of 22 years of recommendations. Biol. Conserv. **2009**, *142* (1), 14–32.
75. Beier, P.; Noss, R.F. Do habitat corridors provide connectivity? Conserv. Biol. **1998**, *12* (6), 1241–1252.
76. Brudvig, L.A.; Damschen, E.I.; Tweksbury, J.J.; Haddad, N.M.; Levey, D.J. Landscape connectivity promotes plant biodiversity spillover into non-target habitats. Proc. Nat. Acad. Sci. Am. **2009**, *106* (23), 9328–9332.
77. Herrera, J.M.; Garcia, D. The role of remnant trees in seed dispersal through the matrix: Being alone is not always so sad. Biol. Conserv. **2009**, *142* (1), 149–158.
78. Alagador, D.; Trivino, M.; Cerdeira, J.O.; Brás, R.; Cabeza, M.; Araújo, M.B. Linking like with like: optimising connectivity between environmentally-similar habitats. Landscape Ecol. **2012**, *27* (2), 291–301.
79. Nunez, T.A.; Lawler, J.J.; McRae, B.H.; Pierce, D.J.; Krosby, M.B.; Kavanagh, D.M.; Singleton, P.H.; Tewksbury, J.J. Connectivity planning to address climate change. Conserv. Biol. **2013**, *27* (2), 407–416.
80. Powell, G.V.N.; Bjork, R. Implications of intratropical migration on reserve design: a case study using *Pharomachrus mocinno*. Conserv. Biol. **1995**, *9* (2), 354–362.
81. Vogt, P.; Ferrri, J.P.; Lookingbill, T.R.; Gardner, R.H.; Riitters, K.H.; Ostapowicz, K. Mapping functional connectivity. Ecol. Indicators **2009**, *9* (1), 64–71.
82. Martínez-Garza, C.; Howe, H.F. Restoring tropical diversity: beating the time tax on species loss. J. Appl. Ecol. **2003**, *40*, 423–429.

83. Ziółkowska, E.; Ostapowicz, K.; Kuemmerle, T.; Perza-nowski, K.; Radeloff, V.C.; Kozak, J. Potential habitat connectivity of European bison (*Bison bonasus*) in the Carpathians. Biol. Conserv. **2012**, *146* (1), 188–196.
84. Carroll, C.; McRae, B.H.; Brookes, A. Use of linkage mapping and centrality analysis across habitat gradients to conserve connectivity of gray wolf populations in Western North America. Conserv. Biol. **2012**, *26* (1), 78–87.
85. Erös, T.; Olden, J.D.; Schick, R.S.; Schmera, D.; Fortin, M.-J. Characterizing connectivity relationships in freshwaters using patch-based graphs. Landscape Ecol. **2012**, *27*(2), 303–317.
86. Treml, E.A.; Halpin, P.N.; Urban, D.L.; Pratson, L.F. Modeling population connectivity by ocean currents, a graph-theoretic approach for marine conservation. Landscape Ecol. **2008**, *23*, 19–36.
87. Jules, E.S.; Kauffman, M.J.; Ritts, W.D.; Carroll, A.L. Spread of an invasive pathogen over a variable landscape: A nonnative root rot on Port Orford cedar. Ecology **2002**, *83*(11), 3167–3181.
88. Hess, G.R. Conservation corridors and contagious-disease-a cautionary note. Conserv. Biol. **1994**, *8* (1), 256–262.
89. Minor, E.S.; Gardner, R.H. Landscape connectivity and seed dispersal characteristics inform the best management strategy for exotic plants. Ecol. Appl. **2011**, *21* (3), 739–749.
90. Minor, E.S.; Tessel, S.M.; Engelhardt, K.A.M.; Lookingbill, T.R. The role of landscape connectivity in assembling exotic plant communities: a network analysis. Ecology **2009**, *90* (7), 1802–1809.
91. Wang, Z.; Wu, J.; Shang, H.; Cheng, J. Landscape connectivity shapes the spread pattern of the rice water weevil: a case study from Zhejiang, China. Environ. Manage. **2011**, *47* (2), 254–262.

7

Landscape Dynamics, Disturbance, and Succession

Hong S. He
University of Missouri

Introduction

Traditionally, a landscape is considered to be an expanse of scenery extending as far as the eye can see.[1] It usually implies an area so large that a person cannot readily traverse it on foot. With the advancement of technologies such as remote sensing and geographical information systems, the capacity to acquire, process, and analyze spatial information has been greatly expanded. Consequently, landscapes analyzed in contemporary scientific literature can cover millions of hectares, expanding the geographic extent of what is considered a landscape.[2]

Within a landscape, there may be numerous components such as forest, grassland, water bodies, houses, and roads. Or there may be one primary component divided into subclasses, such as forest divided into subclasses of young, mature, or old-growth forest. How to classify a landscape depends on the purpose of the study. For example, for an urban sprawl study, a landscape may be classified into different housing density classes;[3] for a wildlife habitat study, a landscape may be classified into habitat, hospitable matrix, and inhospitable matrix;[4] for a study of land cover change, a landscape may be classified as forest, pasture, cropland, and other.[5]

A landscape is a place where natural and human forces interact. Both forces lead to landscape changes (succession) over large spatial extents and long temporal spans. These forces are usually referred to as landscape disturbances and investigated as landscape processes by scientific communities.[6] In this entry, I discuss landscape succession, disturbances, landscape processes, and landscape modeling.

Landscape Succession

Succession is a series change from one stage to the next. As landscape components change over time landscape conditions change, which is often referred to as *landscape succession*. Landscape succession usually occurs slowly over decades or centuries and is driven by natural or anthropogenic forces or both. For example, in the Changbai Mountains of Northeast China, early successional aspen and birch dominate the open lands resulted from volcanic activities, after about 40–50 years mid-successional tree species such as maple, oak, and basswood replace aspen and birch to become dominant. In about another 40–50 years the maple, oak, and basswood will be outcompeted by late successional Korean pine, which will reach mixed Korean pine hardwood climax status and will maintain the status for the next 300 years.[7] In this example, young forests grow to become mature forests, mature forests grow to become old growth forests, and old-growth forests gradually regenerate to start a new forest succession cycle. The whole cycle takes hundreds of years and the main drivers are natural forces.

Increase in human population and the associated need for natural resources drive expansion, creation, and re-creation of human communities worldwide, and infrastructure developments also lead to more impervious surfaces. Through mapping historical impervious surface, researchers can reveal urban development for a region or state. For example, an urban sprawl study in the state of Missouri found that from 1980 to 2000, 129,853.2 ha of land were converted to impervious surface.[8] Although sprawl was very prominent on the urban fringe during 1980s in major metropolitan areas, the trend shifted to the rural landscapes in the 1990s and 2000s. Sprawl in rural areas (also called rural sprawl) may have greater impact on ecosystems due to the low density of development and larger affected areas. Increase in population and the expansion of human society can eventually cause a natural landscape to be transformed into an urban landscape.

Landscape changes are measured from two perspectives: landscape composition and landscape configuration. The former quantifies the proportions of various components (classes), the evenness, dominance, and diversity of the study landscape. The latter characterizes spatial arrangement of various landscape components such as degree of fragmentation, aggregation, association, and connectivity. The common indices for measuring landscape composition include fraction or proportion, relative richness, Shannon's Diversity index, and Shannon's Evenness index. The common indices for measuring landscape configuration include contagion, aggregation index, connectivity, and proximity index. Most landscape indices can be found in FRAGSTATS, a software package developed by McGarigal et al.[9]

Landscape Disturbance

Disturbances include environmental fluctuations and destructive events in an otherwise relatively stable system.[10] It is often characterized by a set of parameters, including distribution (spatial, geographic, topographic, environmental, and community gradients), frequency (mean number of events per time period), return interval (mean time needed to disturb an area equivalent to the study area), mean size (average area per event), intensity (physical force of the event per time such as rate of spread), and severity (impact on the organism, community, ecosystem, or landscape). Disturbance can have endogenous and exogenous causes, whereas in the former, change is driven by the biological properties of the system (e.g., insect susceptibility due to aging), and in the latter an outside driving force (e.g., extreme weather) is present. Disturbance can be either anthropogenic or natural, whereas the former is always exogenous and the latter is mostly endogenous.

Anthropogenic disturbances include forest harvesting, grazing, agriculture, or residential development. Their effects are gradual and cumulative, and are often profound on natural landscapes. Anthropogenic disturbances are found to be the main causes of landscape fragmentation, loss of species diversity, altered hydrological functions, or other types of natural resource degradation.

Natural disturbances include extreme weather, landslides, wildfires, insects, and diseases, which are forces that may cause abrupt changes in natural landscapes. They can change the path of landscape succession in many ways. For example, fire disturbance is an integral part of many forest ecosystems. In boreal forests (e.g., larch, spruce, fir forests), high-intensity crown fire can kill most of the trees where they occur and reset the successional stage to that of a newly regenerating forest.[11] In central hardwood forests (e.g., oak-pine forests), low-intensity ground fires can reduce understory density, increase herbaceous species diversity beneath a mature forest overstory, kill pathogens, and ultimately maintain a healthy condition for forest to grow and regenerate.[12] Many studies report that natural disturbances are crucial in maintaining landscape heterogeneity, which in turn determines biodiversity.

Landscape Processes

Biological organizations occur in a hierarchical manner across a wide spectrum of spatiotemporal scales. At a given level of resolution, a biological system is composed of interacting components (i.e., lower-level entities) and is itself a component of a larger system (i.e., higher level entity).[13] An equally wide range of ecological processes is associated with ecological components at each hierarchical level. For example, stomata conductance is an ecophysiological process at individual leaf level; growth and aging are at an individual organism level; competition is at the population level; inter-species competition is at the community level; water and nutrient cycling is at the ecosystem level; dispersal and fragmentation are at the landscape level; and adaptation, speciation, and extinction are at the biome level.[13] Processes at one scale often interact with processes at lower and higher scales.[14] An increasing level of organization leads to the increase in magnitude of both spatial and temporal scales. For example, at a single leaf scale, photosynthesis occurs on the order of minutes in a few square centimeters space, at a biome scale, species range distribution changes in response to climatic change, and species migration and extinction occur at a hundreds to thousands of years timeframe over a large region.[13]

In fact, any spatially continuous processes, including the above-mentioned natural and anthropogenic disturbances that are directly related to landscape position, spatial heterogeneity, and patch geometrics and adjacencies are landscape processes.[15] These processes operate across a range of spatial extents (10^3–10^6ha) and temporal spans (10^1–10^3year) and are often the main forces shaping the landscapes we have today.

Landscape Modeling

The challenges in predicting landscape change come from two fundamental aspects: the relevance of both long temporal and broad spatial dimensions of landscape processes. Temporally, a landscape may take hundreds of years to undergo significant successional change. Processes that operate on such long time spans may go undetected by many field experiments, which are often based on relatively short observation periods that may not capture the full range of the events. Spatially, landscape change can be strongly affected for centuries by environment heterogeneity; the current spatial pattern resulted from past human land use and natural disturbances. The interaction of spatial and temporal factors across the landscape can be so complicated that it is beyond human comprehension. Thus, computer simulation modeling becomes a useful tool for understanding these large (10^4–10^7ha), long-term, complex systems. With modeling techniques, it is possible to describe the modeled components and relationships mathematically and logically and deduce results that cannot otherwise be investigated.[16,17]

A specific definition of a landscape model is one that simulates spatiotemporal characteristics of at least one recurrent landscape process in a spatially interactive manner.[15] The term spatially interactive means that a simulated entity (e.g., a pixel or polygon) is a function of neighboring, or spatially related, entities. A landscape model under the specific definition has the following characteristics: 1) it is a simulation model, 2) it simulates one or more landscape processes repeatedly, and (c) it operates at a large spatial and temporal extent that is adequate to simulate the landscape process.

Landscape models applications generally fall in the three categories: (1) spatiotemporal patterns of landscape processes, (2) sensitivities of model object to input parameters, and (3) comparisons of model simulation scenarios.[15]

The direct outputs of landscape models are the spatiotemporal patterns of the model objects because the modeled spatial processes are stochastic and complex, and understanding their manifestations over space and time is necessary. Thus, the most effective method is to simulate spatiotemporal patterns of the model objects using built-in model relationships and parameters related to the model objects.

The lack of management experience at the landscape scales and the limited feasibility of conducting landscape scale experiments have resulted in increasing use of scenario modeling to analyze the effects of different management actions on focal forests and wildlife species. Model scenarios are created by altering input parameters to reflect changes in climate, disturbance, fuel and harvest alternatives. The built-in model relationships remain unchanged. Comparing results from different model scenarios provides relative measurements regarding the direction and magnitude of changes within the simulated landscape.

Conclusions

Landscape will continue to be a unit in nature where natural and anthropogenic disturbances operate and interact. Such interactions may alter natural landscape succession, lead to landscape fragmentation, result in loss of species diversity, reduce hydrological functions, and cause other types of natural resource degradation. Because of the large spatiotemporal dimensions and numerous variables and processes involved in landscape succession, landscape modeling becomes a necessary tool in defining modeled components and relationships mathematically and logically and deducing results that cannot otherwise be derived. Landscape models will be increasingly used in predicting the future landscape change under various climate change and management scenarios.

Acknowledgment

I thank Stephen Shifley for the helpful comments, which improved the clarity of this entry. Funding support of this work is come Missouri GIS Mission Enhancement Program and Chinese Academy of Sciences.

References

1. Troll, C. *The Geographic Landscape and Its Investigation.* Translated by Conny Davidsen, Studium Generale, **1950**, *3*, no. 4/5, 163–181.
2. Riitters, K.H.; O'Neill, R.V.; Hunsaker, C.T.; Wickham, J.D.; Yankee, D.H.; Timmins, S.P.; Jones, K.B.; Jackson, B.L. A factor analysis of landscape pattern and structure metrics. Landscape Ecol. **1995**, *10* (1), 23–39.
3. Radeloff, V.C.; Hammer, R.B.; Stewart, S.I. Rural and suburban sprawl in the U.S. Midwest from 1940 to 2000 and its relation to forest fragmentation. Conserv. Biol. **2005**, *19* (3), 793–805.
4. Tischendorf, L. Can landscape indices predict ecological processes consistently? Landscape Ecol. **2001**, *16* (3), 235–254.
5. Lambin, E.F.; Turner, B.L.; Geist, H.J.; Agbola, S.B.; Angelsen, A.; Bruce, J.W.; Coomes, O.T.; Dirzo, R.; Fischer, G.; Folke, C.; George, P.S.; Homewood, K.; Imbernon, J.; Leemans, R.; Li, X.; Moran, E.F.; Mortimore, M.; Ramakrishnan, P.S.; Richards, J.F.; Skånes, H.; Steffen, W.; Stone, G.D.; Svedin, U.; Veldkamp, T.A.; Vogel, C.; Xu, J. The causes of land-use and land-cover change: Moving beyond the myths. Global Environ. Change **2001**, *11* (4), 261–269.
6. Turner, M.G. Landscape ecology: the effect of pattern on process. Annu. Rev. Ecol. syst. **1989**, *20*, 171–197.
7. Wang, Z.; Xu, Z.; Tan, Z.; Dai, H.; Li, X. The main forest types and their features of community structure in Northern slope of Changbai Mountain. Forest Ecol. Res. **1980**, *1*, 25–42.
8. Zhou, B.; He, H.S.; Nigh, T. A.; Schulz, J.H. Mapping and analyzing change of impervious surface for two decades using multi-temporal Landsat imagery in Missouri. Int. J. Appl. Earth Observ. Geoinf. **2012**, *18*, 195–206.

9. McGarigal, K.; Cushman, S.A.; Ene, E. FRAGSTATS v4: Spatial Pattern Analysis Program for Categorical and Continuous Maps. Computer software program produced by the authors at the University of Massachusetts, Amherst, 2012 Available at the following web site: http://www.umass.edu/landeco/research/fragstats/fragstats.html.
10. Pickett, S.T.A., White, P.S. Eds.; *The Ecology of Natural Disturbance and Patch Dynamics*; Academic Press: New York, USA 1985.
11. Hunter Jr., M.L. Natural fire regimes as spatial models for managing boreal forests. Biol. Conserv. **1993**, *65* (2), 115–120.
12. Lorimer, C.G. Historical and ecological roles of disturbance in Eastern North American forests: 9,000 years of change. Wildlife Soc. Bull. **2001**, *29* (2), 425–439.
13. Delcourt, H.R.; Delcourt, P.A. Quaternary landscape ecology: Relevant scales in space and time. Landscape Ecol. **1988**, *2* (1), 23–44.
14. Currie, W. S. Units of nature or processes across scales? The ecosystem concept at age 75. New Phytol. **2011**, *190*, 21–34.
15. He, H.S. Forest landscape models: Definitions, characterization, and classification. Forest Ecol. Manag. **2008**, *254* (3), 445–453.
16. Mladenoff, D.J.; Baker, W.L. *Advances in Spatial Modeling of Forest Landscape Change: Approaches and Applications;* Cambridge University Press: Cambridge, UK 1999; 163–185.
17. Perry, G.L.W.; Enright, N.J. Spatial modelling of vegetation change indynamic landscapes: A review of methods and applications. Progr Phys Geogr. **2006**, *30*, 47–72.

8

Land-Use and Land-Cover Change (LULCC)

Burak Güneralp
Texas A&M University

Introduction

Land-use and land-cover change (LULCC), or more succinctly land change, is an integral component of global environmental change.[1] Land cover refers to the physical characteristics of Earth's surface such as water, grass, trees, or concrete. Land use, on the other hand, refers to how the land is utilized by humans for various social and economic purposes. The same land cover may be used for different purposes. For example, a forest may be used for timber extraction or set aside as a protected area for conservation of wildlife. Land-use change refers to a change in the management of land by humans, which may or may not lead to land cover-change. LULCC is important for biogeochemical cycles such as nutrient cycling,[2,3] for biodiversity through its impacts on habitats,[4,5] and for climate by changing sources and sinks of greenhouse gases and land-surface properties.[6,7] LULCC also has important implications for the provision of ecosystem services on which both rural and urban societies depend.[8]

There is a large body of both theoretically and empirically grounded work on LULCC. These studies fall under the umbrella of the relatively new but growing discipline of land-change science. Several international science efforts such as the Global Land Project,[9] a joint research project of the International Geosphere-Biosphere Programme (http://www.igbp.net) and the International Human Dimensions Programme (http://www.ihdp.org), are devoted to issues revolving around LULCC in recognition of the pivotal role it plays in the functioning of the Earth system. These issues are treated at length in the many books and reports published specifically on the subject of LULCC.[10–20]

Causes and Consequences

The processes that lead to land change are numerous. These processes in many cases interact and collectively lead to larger impacts on the land cover than they would individually.[21,22] Although biophysical processes and predominant climatic conditions broadly determine the land cover in a particular location, human activities play a key role in creating the observed land-use and land-cover patterns (Figure 8.1) and may even create land uses that could not be supported by the climatic conditions at that locale (e.g., irrigated agriculture in extremely arid environments).

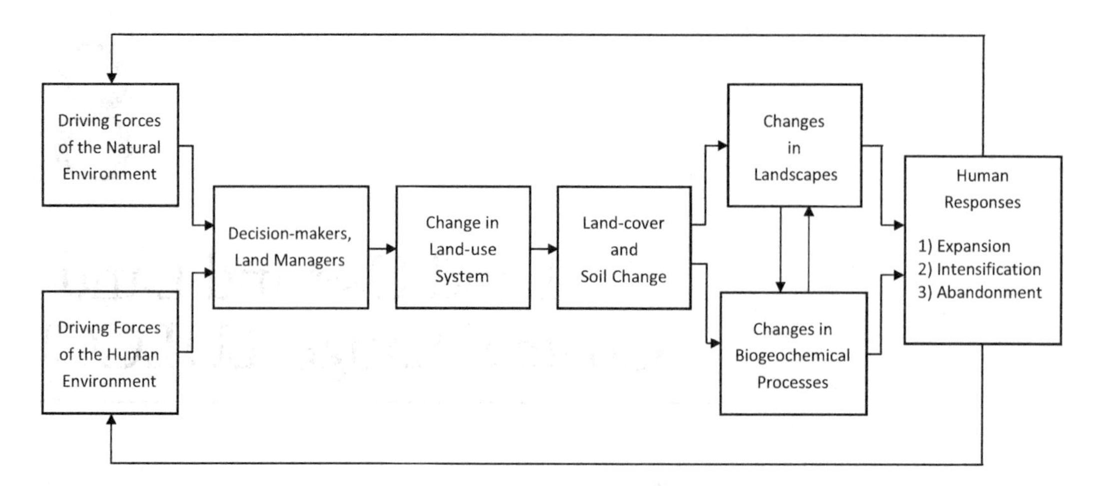

FIGURE 8.1 Conceptual framework showing the interaction of biophysical and human-related factors creating land-use and land-cover change. © The University of Arizona Press.
Source: Adapted from Redman.[23]

Over geological time scales, atmospheric, hydrologic, and tectonic processes continuously alter the land cover across the surface of Earth. Starting with the Paleolithic Age, and especially with the use of fire, humans have had an increasing role in land change on the face of Earth.[23] In particular, the changes in land use and land cover due to human activities have dramatically accelerated with the onset of agrarianization of human societies about 12,000 years ago. Another accelerated phase of LULCC with much larger geographical spread was initiated about 300 years ago with the onset of the Industrial Revolution; then yet again, another surge after the 1960s and the Green Revolution.

Whenever humans are involved, land change may become a highly political activity[24] involving—across various spatial and temporal scales—institutional as well as social and economic dynamics in interaction with biophysical dynamics. Within these webs of interactions, one way to explain land change is to refer to proximate and underlying causes. Proximate (or direct) causes are those human activities and immediate actions that originate from the intended manipulation of land cover, whereas underlying (or indirect) causes are fundamental processes that compel the more proximate processes.[25,26] Thus, proximate causes are generally limited to a small set of activities such as agricultural expansion or infrastructure development, whereas underlying causes can consist of, depending on the context of land change, many factors—biophysical, demographic, economic, technological, institutional, and cultural—acting across multiple spatial and temporal scales. These factors can play varying and often interacting roles in creating the observed land-use and land-cover patterns on the ground[27] that may be either in the form of changing one land cover to another or changing the biophysical conditions of the existing land cover (i.e., landuse modification).[28] Modification of existing land uses leads to subtle changes in the character of the land but not necessarily a change in the land-use classification (e.g., agricultural intensification).

There are relatively few global-scale studies on long-term historical changes in land cover that are due to human activities. These studies harbor significant uncertainties in their estimates; nevertheless, they offer a comprehensive perspective.[29] According to one estimate, 75% of the ice-free surface of Earth has undergone noticeable land change due to human activities.[28] Another suggests that between A.D. 1700 and 2000, 42–68% of the global land surface was impacted by various land-use activities.[30] Up to half of Earth's net primary productivity is estimated, albeit with large error bounds, to be appropriated every year by humans as a result of these land-use activities.[31] A study that reports historical changes in both natural vegetation and agricultural land cover estimates that nearly 5 million km^2 of natural vegetation were converted to agriculture (both cropland and pasture) between A.D. 800 and 1700 (Table 8.1).[32] In comparison, the agricultural lands are estimated to have increased more than 40 million km^2 between A.D. 1700 and 1992.[32,33]

Deforestation has historically been a prominent feature of human-induced land-cover change.[34] By the time of European colonization, in much of Eurasia, extensive forested areas had already been cut down either to meet timber and fuel wood demand or to open space for croplands and pastures. Evidence suggests that tropical forests in Central and South America had witnessed periodic and localized deforestation with the rise and fall of many civilizations.[23,35,36] However, across much of the world, deforestation rates increased with the onset of colonization and especially after the onset of the Industrial Revolution. Much of this deforestation had its ultimate cause in the demand originating from Europe and also from North America for forest products as well as for agro-industrial produce such as coffee, bananas, and rubber. More recently, palm plantations, ranching, and soybean production have appeared as the more dominant proximate causes of ongoing deforestation due to increased demand not only from Europe and North America but also from rapidly developing countries such as India and China.[37] From the late 19th to the early 20th century, there has been a trend of reforestation in many of the previously deforested lands of Europe and New England,[38] which is facilitated by the timber and food imports. There are also extensive afforestation efforts in previously deforested lands across the world with huge efforts expended to this end in China. These reforestation and afforestation dynamics are linked in complicated ways to the ongoing deforestation in much of the tropical belt as well as boreal forests in Asia.[39,401] However, these afforestation efforts do not always take place in the same geographical locations where deforestation had occurred. Also, the newly forested landscapes are not necessarily composed of similar species composition and/or richness as the pristine forests they replace.[41–43]

Most of Earth's original grassland and savanna cover have been converted to cropland or grazing land since humans took up agriculture about 10,000 years ago. Broadly following the accelerated increase in human populations, croplands increased dramatically in every continent except Antarctica. Especially during the past 300 years, large areas of grasslands have been converted to croplands in North America and Eurasia.[44] By one estimate, half of the total land converted to croplands since A.D. 1700 has been savanna-like vegetation.[45] Grasslands of South America are undergoing a similar transformation fueled by the growing demand for soybeans from China and the European Union.[37] Intensification of agriculture on croplands can partly meet the increased demand for food, but this needs to be done with utmost care to biodiversity and ecosystem services.[46–49] Although some of these croplands such as the rice paddies in China have generated high yields for millennia, intensive cultivation practices and heavy grazing may lead to severe land degradation; indeed, such practices may have contributed to the

TABLE 8.1 A Recent Estimate of Global Extent of Land-Cover Types over Time[a]

		Year				
Land Cover	Potential	800	1100	1400	1700	1992
Tropical forest	22.44	22.03	21.87	21.70	21.22	16.30
Temperate broadleaf forest	10.48	10.15	9.92	9.87	9.26	5.95
Temperate needle leaf forest	15.75	15.61	15.51	15.47	15.17	12.36
Total forest	*48.68*	*47.78*	*47.31*	*47.05*	*45.65*	*34.60*
Grass and shrubs	46.79	44.90	44.22	43.84	42.14	13.63
Tundra	4.08	4.07	4.06	4.06	4.05	2.94
Total natural vegetation	*99.55*	*96.75*	*95.60*	*94.95*	*91.84*	*51.16*
Crop	—	1.36	1.97	2.33	4.01	18.76
Pasture	—	1.44	1.98	2.27	3.70	29.63
Total agriculture	—	*2.80*	*3.95*	*4.60*	*7.71*	*48.39*

[a] Units are in $10^6 km^2$.
Source: Adapted from Pongratz et al.[31]

shift of large expanses to semiarid and even arid conditions across much of the Middle East, Central Asia, and Africa that were already predisposed due to the particular geophysical characteristics and larger-scale climatic changes in those regions.[50,51] About 20% of global drylands, which constitutes about 41% of global ice-free land area, is undergoing degradation.[52] However, as is typical in studies on land-use and land-cover dynamics, it is challenging to attribute causality to individual factors. For example, it took the innovative use of remotely sensed data to show that the dreaded expansion of the Sahara as widely believed until the 1980s is actually part of the multiyear cycle of expansion and contraction of the desert.[53,54]

Urban areas cover less than 3% of the ice-free land surface and yet are home to more than half the total human population. In addition to the direct impacts of urbanization such as expansion of urban land,[55,56] which is predominantly composed of impervious surfaces across the landscape, urbanization in a particular location may also influence land change elsewhere indirectly through demands for natural resources such as sand, gravel, and timber for both construction as well as demand for food.[37,57–59] Although urbanization leads to depopulation of extensive rural areas as more people move to urban areas that are relatively more densely populated[60] urbanization cannot by itself save land for nature because of the increased demands for natural resources that are needed to support urban activities[57,59] and also because subsequent suburbanization may again reverse the trend.[61]

Recent debates on LULCC are dominated by the expansion of croplands due to increased demand for food—driven primarily by dietary changes of more affluent and more urban populations but also by the sheer increase in the world populace[58] as well as the potential increase in demand for biofuels,[62] large-scale land leases in less developed countries by their more developed counterparts,[38] future changes in the extent and species composition of forests,[63,64] and impacts on coastal settlements of sea-level rise.[65,66] The critical aspects underlying much of these debates are the unprecedented urbanization, increasing occurrence of indirect land-use change symptomized by the further separation of the landscapes of consumption and those of production, and the ultimate scarcity of land prompting more efficient uses of this resource—all of which are inter-related in complicated, even complex ways.[38,59,67]

Theoretical Foundations

Although there is no overarching theory of land change, there have been many theories put forward to explain land-change dynamics.[68,69] While acknowledging the complicated nature of interactions among various processes that span multiple spatial and temporal scales to create the observed patterns of land use and land cover, each of these theories tends to highlight a particular subset of these processes and their interactions in bringing an explanation for the LULCC phenomenon in different places.[1] Thus, theories of land change collectively take a broad range of phenomena as their foci including, but not limited to, complexity and resilience concepts,[70] behavioral approaches based on economic theoretical frameworks,[71] globalization,[72] and institutions.[73]

Theorizing efforts can also be grouped under those focusing on particular types of land changes such as tropical deforestation, forest transition, desertification, agricultural change, and urban land-use change.[74–78] Yet, any empirical work, usually a case study, on land change is typically guided by more than one theory,[79] thus drawing upon the explanatory power of multiple perspectives.[80] The specific theories informing a particular case study are mainly determined by the disciplinary backgrounds of the researchers involved.[68] There are also differences in the theories of land change in the developed vs. developing world reflecting the differences between the two in terms of both the processes in question and the land-change outcomes.[24,81] None of these theories claim to have an explanatory power across the full spectrum of land-change dynamics. However, all bring something to the table that not only illuminates the particular conditions leading to observed land-change dynamics but also informs to an extent the land-change processes that are at play in other situations and places.

An overarching theory of land change will most likely bring together many aspects of these theories.[68] These include interactions within and between the biophysical and the human-related processes across spatial and temporal scales. Thus, it should reconcile the connections of local and regional with the broader world in which they are situated. Such a grand theory needs to be dynamic, meaning that it should explain land changes that occurred in the past as well as those that are being experienced in the present. The dynamic perspective will also have the desirable feature of allowing for anticipating potential land-change trajectories in the future in particular places under various scenarios pertaining to broader-scale changes.

Monitoring and Modeling

Satellite-based observations have transformed monitoring land cover as they provide a consistent way to observe the change in land cover both across space and over time. Among the numerous space-based land-change monitoring programs, Landsat by the National Aeronautics and Space Administration (NASA) of the United States, in operation since 1972, is the longest running. Among other satellites whose images are widely utilized to monitor land-change are the SPOT by the French corporation Spot Image and IRS by the Indian Space Research Organisation (ISRO).[82] In addition to the medium-resolution sensors on these satellites, there are sensors with lower spatial resolution—but with higher temporal resolution—such as MODIS that are used for large regional and global monitoring of land-cover change. There are also several commercial satellites that provide very high-resolution imagery such as IKONOS and QuickBird. These high-resolution images allow for very detailed land-cover classification; however, being very expensive, they are generally used for calibration and validation of land-cover maps derived from Landsat or other moderate-resolution sensors.[83,84]

However, satellite-based monitoring of land change faces a number of limitations; foremost among these are, depending on the application, challenges in terms of the spatial, temporal, or radiometric resolution, inability to differentiate different land uses, and its availability only for the past few decades.[82] Because of these limitations, ancillary information such as aerial photography, survey maps, household surveys, census data, and official statistics are still frequently used along with or instead of satellite-based data.[85,71] Novel methods such as those for fusion of information from sensors with different spatial, temporal, and radiometric resolutions are also explored to overcome some of these limitations.[86,87]

Land-change models help organize the present knowledge about observed land-change dynamics in particular places, and they are essential in evaluating various scenarios regarding demographic, economic, institutional, and biophysical changes in terms of land-change outcomes in the future. As such, these models both are informed by and potentially inform the body of theory on land change.[71,88] In line with the multitude of interacting processes that are involved in land change in different parts of the world, there is a wide range of modeling approaches to capture the different aspects of the dynamics of land use and land cover depending on the research questions. The particular methods employed reflect their theoretical underpinnings, which span from economic theoretical frameworks to the complexity theory.[70,71] Methodologically, these approaches span a wide spectrum from econometric to agent-based to cellular automata (CA) models[89,90] and can broadly be divided into three classes in terms of how they deal with interacting processes across spatial scales: i) ***Top-down*** methods emphasize the influence of broader-scale processes such as regional economic or demographic changes on land-use patterns. There are many CA models[91,92] as examples of this type of methods; ii) ***bottom-up*** methods emphasize the emergent properties of the land-change system from the interaction of many individual entities such as smallholder farmers. Agent-based modeling[93,94] and spatially explicit economic models of land change[71] are examples of such methods; and iii) ***hybrid*** methods tend to draw from existing top-down and bottom-up approaches using them in complementary ways to reflect the bidirectional influences of broader-scale processes and local-level interactions on each other.[95,96]

In addition to numerous land-change modeling studies focusing on local or regional dynamics of land change, there are global-scale land-cover change modeling efforts going back to the early 1990s.[97–101] A new generation of these models that incorporate both past changes going back to year 1500[33] and scenario projections out to year 2100 have recently been developed in preparation for the Fifth Assessment Report of the Intergovernmental Panel on Climate Change (IPCC).[102] Another recent effort focused specifically on future urban-land expansion[103] and its likely impacts on biodiversity.[104]

There are usually significant differences among the modeling applications that make comparisons of findings and sharing insights difficult, if not impossible. These differences stem from many reasons such as the availability of data, the inclination of modelers toward a particular modeling methodology, type of land change (e.g., urban growth, deforestation) under question, and case-study locations (e.g., tropical deforestation in Amazonia vs. Southeast Asia). One of the pressing challenges for land-change modelers is then to increase compatibility among the available land-change modeling approaches including validation protocols[105] but also improving integrated models that simulate the interactions among social, economic, institutional, and biophysical components of a land-change system.[1] Such improvements can also help increase the utility of these models as practical tools that inform land-use policy and planning.[88]

Conclusion

Improving our understanding of land change as an integral component of the Earth system's functioning is a critical initial step in addressing the interactions between people and their environments.[9] As we move well into the 21st century, human activities continue to change the surface of Earth directly and indirectly as they have been doing for at least the last 12,000 years.[33] The big difference is in the realization that human activities over the past 200 years or so dramatically intensified and reached such a scale to affect the functioning of the whole Earth system.[106] Perhaps because of the recency of this collective realization, humans are yet to come to grips with all the various kinds of linkages that enmesh our societies with the larger Earth system. Likewise, the diversity of factors involved in land dynamics and linkages among these across multiple spatial and temporal scales present immense challenges to the formulation of an overarching theory of land change. Formulation of such a grand theory hinges on the evolution of land-change science toward a more integrative endeavor that spans across the natural and the social sciences. Therefore, the challenge in the coming decades will be moving toward the formulation of such a theory that will surely evolve together with the constantly changing relation of humans with the land.

References

1. Turner, B.L., II; Lambin, E.F.; Reenberg, A. Land Change Science Special Feature: The emergence of land change science for global environmental change and sustainability. Proc. Natl. Acad. Sci. USA. **2007**, *104* (52), 20666–20671.
2. Fraterrigo, J.M.; Turner, M.G.; Pearson, S.M.; Dixon, P. Effects of past land use on spatial heterogeneity of soil nutrients in southern Appalachian forests. Ecol. Monogr. **2005**, *75* (2), 215–230.
3. Verchot, L.V.; Davidson, E.A.; Cattânio, J.H.; Ackerman, I.L.; Erickson, H.E.; Keller, M. Land use change and biogeochemical controls of nitrogen oxide emissions from soils in eastern Amazonia. Global Biogeochem. Cycles **1999**, *13* (1), 31–46.
4. Dupouey, J.L.; Dambrine, E.; Laffite, J.D.; Moares, C. Irreversible impact of past land use on forest soils and biodiversity. Ecology **2002**, *83* (11), 2978–2984.
5. Harding, J.S.; Benfield, E.F.; Bolstad, P.V.; Helfman, G.S.; Jones III, E.B.D. Stream biodiversity: The ghost of land use past. Proc. Natl. Acad. Sci. USA. **1998**, *95* (25), 14843–14847.
6. Pielke Sr., R.A.; Marland, G.; Betts, R.A.; Chase, T.N.; Eastman, J.L.; Niles, J.O.; Niyogi, D.D.S.; Running, S.W. The influence of land-use change and landscape dynamics on the climate system: Relevance to climate-change policy beyond the radiative effect of greenhouse gases. Philos. Trans. R. Soc. A **2002**, *360* (1797), 1705–1719.

7. Houghton, R.A. Revised estimates of the annual net flux of carbon to the atmosphere from changes in land use and land management 1850–2000. Tellus Ser. B **2003**, *55* (2), 378–390.
8. Naidoo, R.; Balmford, A.; Costanza, R.; Fisher, B.; Green, R.E.; Lehner, B.; Malcolm, T.R.; Ricketts, T.H. Global mapping of ecosystem services and conservation priorities. Proc. Natl. Acad. Sci. USA. **2008**, *105* (28), 9495–9500.
9. Global Land Project. *GLP Science Plan and Implementation Strategy;* IGBP Secretariat: Stockholm, 2005, 64.
10. Goudsblom, J.; de Vries, B. *Mappae Mundi: Humans and Their Habitats in a Long-Term Socio-Ecological Perspectives : Myths, Maps and Models;* Amsterdam University Press: Amsterdam, 2003.
11. Aspinall, R.J.; Hill, M.J., Eds.; *Land Use Change: Science, Policy, and Management*; CRC Press: Boca Raton, 2007; 185.
12. DeFries, R.S.; Asner, G.P.; Houghton, R.A., Eds.; *Ecosystems and land use change;* American Geophysical Union: Washington, D.C., 2004; 344.
13. Entwisle, B.; Stern, P.C., Eds.; *Population, Land Use, and Environment: Research Directions;* National Academies Press: Washington, D.C., 2005; 321.
14. Gutman, G.; Janetos, A.C.; Justice, C.O.; Moran, E.F.; Mustard, J.F.; Rindfuss, R.R.; Skole, D.; Turner, B.L.; Cochrane, M.A., Eds.; *Land Change Science: Observing, Monitoring and Understanding Trajectories of Change on the Earth's Surface;* Kluwer Academic Publishers: Dordrecht, 2004; 459.
15. Geist, H., Ed.; *Our Earth's Changing Land: An Encyclopedia of Land-Use and Land-Cover Change;* Greenwood Press: Westport, CT, 2006; 715.
16. Koomen, E.; Stillwell, J.; Bakema, A.; Scholten, H.J., Eds.; *Modelling Land-Use Change: Progress and Applications;* Springer: Dordrecht, 2007; 392.
17. Lambin, E.F.; Geist, H., Eds.; *Land-Use and Land-Cover Change: Local Processes and Global Impacts;* Springer: Berlin, 2006; 222.
18. Meyer, W.B.; Turner, B.L., Eds.; *Changes in Land Use and Land Cover: A Global Perspective*; Cambridge University Press: Cambridge, U.K., 1994; 537.
19. Walsh, S.J.; Crews-Meyer, K.A. *Linking People, Place, and Policy: A GIScience Approach*; Kluwer Academic Publishers: Dordrecht, 2002; 348.
20. Moran, E.F.; Ostrom, E., Eds.; *Seeing The Forest and The Trees: Human-Environment Interactions in Forest Ecosystems;* MIT Press: Cambridge, MA, 2005; 442.
21. Lambin, E.F.; Turner, B.L.; Geist, H.J.; Agbola, S.B.; Angelsen, A.; Bruce, J.W.; Coomes, O.T.; Dirzo, R.G; Fischer,; Folke, C.; George, P.S.; Homewood, K.; Imbernon, J.; Leemans, R.; Li, X.; Moran, E.F.; Mortimore, M.; Ramakrishnan, P.S.; Richards, J.F.; Skanes, H.; Steffen, W.; Stone, G.D.; Svedin, U.; Veldkamp, T.A.; Vogel, C.; Xu, J. The causes of land-use and land-cover change: Moving beyond the myths. Global Environ. Change 2001; *11* (4), 261–269.
22. Vitousek, P.M.; Mooney, H.A.; Lubchenco, J.; Melillo, J.M. Human domination of Earth's ecosystems. Science 1997; *277* (5325), 494–499.
23. Redman, C.L. *Human Impact on Ancient Environments;* University of Arizona Press: Tucson, 1999; 288.
24. Turner, B.L.; Robbins, P. Land-change science and political ecology: Similarities, differences, and implications for sustainability science. Ann. Rev. Environ. Resources **2008**, *33*, 295–316.
25. Geist, H.J.; Lambin, E.F. *What Drives Tropical Deforestation?: A Meta-Analysis of Proximate and Underlying Causes of Deforestation Based on Subnational Case Study Evidence, in LUCC Report Series;* LUCC International Project Office, University of Louvain: Louvain-la-Neuve, Belgium, 2001; 116.
26. Ojima, D.S.; Galvin, K.A.; Turner B.L., II. The global impact of land-use change. Bioscience 1994; *44* (5), 300–304.
27. Geist, H.; McConnell, W. Causes and trajectories of landuse/cover change. In *Land-Use and Land-Cover Change: Local Processes and Global Impacts;* Lambin, E.F., Geist, G, Eds.; Springer: Berlin, 2006; 41–70.

28. Ellis, E.C.; Ramankutty, N. Putting people in the map: anthropogenic biomes of the world. Front. Ecol. Environ. **2008**, *6* (8), 439–447.
29. Ellis, E.C.; Kaplan, J.O.; Fuller, D.Q.; Vavrus, S.; Goldewijk, K.K.; Verburg, P.H. Used planet: A global history. Proc. Natl. Acad. Sci. USA. **2013**, *110* (20), 7978–7985.
30. Hurtt, G.C.; Frolking, S.; Fearon, M.G.; Moore, B.; Shevliakova, E.; Malyshev, S.; Pacala, S.W.; Houghton, R.A. The underpinnings of land-use history: Three centuries of global gridded land-use transitions, wood-harvest activity, and resulting secondary lands. Global Change Biol. **2006**, *12* (7), 1208–1229.
31. Haberl, H.; Erb, K.H.; Krausmann, F.; Gaube, V.; Bondeau, A; Plutzar, C.; Gingrich, S.; Lucht, W.; Fischer-Kowalski, M. Quantifying and mapping the human appropriation of net primary production in earth's terrestrial ecosystems. Proc. Natl. Acad. Sci. USA. **2007**, *104* (31), 12942–12947.
32. Pongratz, J.; Reick, C.; Raddatz, T.; Claussen, M. A reconstruction of global agricultural areas and land cover for the last millennium. Global Biogeochemical Cycles **2008**, *22* (3): GB3018.
33. Goldewijk, K.K.; Beusen, A.; van Drecht, G.; de Vos, M. The HYDE 3.1 spatially explicit database of human-induced global land-use change over the past 12,000 years. Global Ecol. Biogeography **2011**, *20* (1), 73–86.
34. Williams, M. Forests. In *The Earth as Transformed by Human Action: Global and Regional Changes in the Biosphere over the Past 300 Years;* Turner, B.L., Clark, W.C., Kates, R.W., Richards, J.F., Mathews, J.T., Meyer, W.B., Eds.; Cambridge University Press with Clark University: New York, 1990, 179–201.
35. Heckenberger, M.J.; Russell, J.C.; Fausto, C.; Toney, J.R.; Schmidt, M.J.; Pereira, E.; Franchetto, B.; Kuikuro, A. PreColumbian urbanism, anthropogenic landscapes, and the future of the Amazon. Science **2008**, *321* (5893), 1214–1217.
36. McNeil, C.L.; Burney, D.A.; Burney, L.P. Evidence disputing deforestation as the cause for the collapse of the ancient Maya polity of Copan, Honduras. Proc. Natl. Acad. Sci. USA. **2010**, *107* (3), 1017–1022.
37. Grau, H.R.; Aide, M. Globalization and land-use transitions in Latin America. Ecol. Soc. **2008**, *13* (2), 16.
38. Lambin, E.F.; Meyfroidt, P. Global land use change, economic globalization, and the looming land scarcity. Proc. Natl. Acad. Sci. USA. **2011**, *108* (9), 3465–3472.
39. Meyfroidt, P.; Lambin, E.F. Forest transition in Vietnam and displacement of deforestation abroad. Proc. Natl. Acad. Sci. USA. **2009**, *106* (38), 16139–16144.
40. Mayer, A.L.; Kauppi, P.E.; Angelstam, P.K.; Zhang, Y.; Tikka, P.M. Importing timber, exporting ecological impact. Science **2005**, *308* (5720), 359–360.
41. Farley, K.A. Grasslands to tree plantations: Forest transition in the Andes of Ecuador. Ann. Assoc. Am. Geographers **2007**, *97* (4), 755–771.
42. Barlow, J.; Overal, W.L.; Araujo, I.S.; Gardner, T.A.; Peres, C. A. The value of primary, secondary and plantation forests for fruit-feeding butterflies in the Brazilian Amazon. J. Appl. Ecol. **2007**, *44* (5), 1001–1012.
43. Wilkie, D.S.; Bennett, E.L.; Peres, C.A.; Cunningham, A.A. The empty forest revisited. Ann. NY Acad. Sci. **2011**, *1223*, 120–128.
44. Ramankutty, N.; Foley, J.A. Estimating historical changes in global land cover: Croplands from 1700 to 1992. Global Biogeochemical Cycles **1999**, *13* (4), 997–1027.
45. Goldewijk, K.K. Estimating global land use change over the past 300 years: The HYDE Database. Global Biogeochemical Cycles **2001**, *15* (2), 417–433.
46. Godfray, H.C.J. Food and biodiversity. Science **2011**, *333* (6047), 1231–1232.
47. Tscharntke, T.; Klein, A.M.; Kruess, A.; Steffan-Dewenter, G.; Thies, C. Landscape perspectives on agricultural intensification and biodiversity - ecosystem service management. Ecol. Lett. **2005**, *8* (8), 857–874.

48. Zimmerer, K.S. The compatibility of agricultural intensification in a global hotspot of smallholder agrobiodiversity (Bolivia). Proc. Natl. Acad. Sci. USA. **2013**, *110* (8), 2769–2774.
49. Phelps, J.; Carrasco, L.R.; Webb, E.L.; Koh, L.P.; Pascual, U. Agricultural intensification escalates future conservation costs. Proc. Natl. Acad. Sci. USA. **2013**, *110* (19), 7601–7606.
50. Geist, H.J.; Lambin, E.F. Dynamic causal patterns of desertification. Bioscience **2004**, *54* (9), 817–829.
51. Issar, A.S. Climatic change and the history of the Middle East. Am. Sci. **1995**, *83* (4), 350–355.
52. Prince, S.D. Spatial and temporal scales for detection of desertification. In *Global Desertification: Do Humand Cause Deserts?;* Reynolds, J.F., Stafford Smith, D.M., Eds; Dahlem University Press: Berlin, 2002; 23–40.
53. Prince, S.D.; Wessels, K.J.; Tucker, C.J.; Nicholson, S.E. Desertification in the Sahel: A reinterpretation of a reinterpretation. Global Change Biol. **2007**, *13* (7), 1308–1313.
54. Tucker, C.J.; Dregne, H.E.; Newcomb, W.W. Expansion and contraction of the Sahara desert from 1980 to 1990. Science **1991**, *253* (5017), 299–301.
55. Seto, K.C.; Fragkias, M.; Guneralp, B.; Reilly, M.K. A meta-analysis of global urban land expansion. PLoS ONE **2011**, *6* (8), e23777.
56. Angel, S.; Parent, J.; Civco, D.L.; Blei, A.; Potere, D. The dimensions of global urban expansion: Estimates and projections for all countries, 2000–2050. Prog. Plann. **2011**, *75* (2), 53–107.
57. Fry, M. From crops to concrete: Urbanization, deagriculturalization, and construction material mining in central Mexico. Ann. Assoc. Am. Geographers **2011**, *101* (6), 1285–1306.
58. DeFries, R.S.; Rudel, T.; Uriarte, M.; Hansen, M. Deforestation driven by urban population growth and agricultural trade in the twenty-first century. Nat. Geosci. **2010**, *3* (3), 178–181.
59. Seto, K.C.; Reenberg, A.; Boone, C.G.; Fragkias, M.; Haase, D.; Langanke, T.; Marcotullio, P.; Munroe, D.K.; Olah, B.; Simon, D. Urban land teleconnections and sustainability. Proc. Natl. Acad. Sci. USA. **2012**, *109* (20), 7687–7692.
60. Díaz, G.I.; Nahuelhual, L.; Echeverría, C.; Marín, S. Drivers of land abandonment in Southern Chile and implications for landscape planning. Landscape Urban Plann. **2011**, *99* (3–4), 207–217.
61. Yackulic, C.B.; Fagan, M.; Jain, M.; Jina, A.; Lim, Y.; Marlier, M.; Muscarella, R.; Adame, P.; DeFries, R.; Uriarte, M. Biophysical and socioeconomic factors associated with forest transitions at multiple spatial and temporal scales. Ecol. Soc. **2011**, *16* (3), 23.
62. Wicke, B.; Verweij, P.; Van Meijl, H.; Van Vuuren, D.P.; Faaij, A.P.C. Indirect land use change: Review of existing models and strategies for mitigation. Biofuels **2012**, *3* (1), 87–100.
63. Wright, S.J. The future of tropical forests. Ann. N Y Acad. Sci. **2010**, *1195*, 1–27.
64. Meyfroidt, P.; Lambin, E.F. Global forest transition: Prospects for an end to deforestation. Ann. Rev. Environ. Resources **2011**, *36*, 343–371.
65. Syvitski, J.P.M.; Kettner, A.J.; Overeem, I.; Hutton, E.W.H.; Hannon, M.T.; Brakenridge, G.R.; Day, J.; Vorosmarty, C.; Saito, Y.; Giosan, L.; Nicholls, R.J. Sinking deltas due to human activities. Nat. Geosci. **2009**, *2* (10), 681–686.
66. Nicholls, R.J.; Cazenave, A. Sea-level rise and its impact on coastal zones. Science **2010**, *328* (5985), 1517–1520.
67. Foley, J.A.; DeFries, R.; Asner, G.P.; Barford, C.; Bonan, G.; Carpenter, S.R.; Chapin, F.S.; Coe, M.T.; Daily, G.C.; Gibbs, H.K.; Helkowski, J.H.; Holloway, T.; Howard, E.A.; Kucharik, C.J.; Monfreda, C.; Patz, J.A.; Prentice, I.C.; Ramankutty, N.; Snyder, P.K. Global consequences of land use. Science **2005**, *309* (5734), 570–574.
68. Lambin, E.F.; Geist, H.; Rindfuss, R.R. Introduction: Local processes with global impacts. In *Land-Use and Land-Cover Change: Local Processes and Global Impacts;* Lambin, E.F., Geist, H., Eds.; Springer: Berlin, 2006; 1–8.
69. Aspinall, R.J. Basic and applied land use science. In *Land Use Change: Science, Policy, and Management*; Aspinall, R.J., Hill, M.J., Eds.; CRC Press: Boca Raton, 2007; 3–16.

70. Manson, S.M. Challenges in evaluating models of geographic complexity. Environ. Plann. B **2007**, *34* (2), 245–260.
71. Irwin, E.G.; Geoghegan, J. Theory, data, methods: Developing spatially explicit economic models of land use change. Agriculture Ecosystems Environ. **2001**, *85* (1–3), 7–23.
72. Hecht, S. The new rurality: Globalization, peasants and the paradoxes of landscapes. Land Use Policy **2010**, *27* (2), 161–169.
73. Young, O.R.; Lambin, E.F.; Alcock, F.; Haberl, H.; Karlsson, S.I.; McConnell, W.J.; Myint, T.; Pahl-Wostl, C.; Polsky, C.; Ramakrishnan, P.S.; Schroeder, H.; Scouvart, M.; Verburg, P.H. A portfolio approach to analyzing complex human-environment interactions: Institutions and land change. Ecol. Soc. **2006**, *11* (2).
74. Barbier, E.B.; Burgess, J.C.; Grainger, A. The forest transition: Towards a more comprehensive theoretical framework. Land Use Policy **2010**, *27* (2), 98–107.
75. Geist, H. *The Causes and Progression of Desertification;* Ashgate Publishing: Aldershot, 2005, 258.
76. Webster, C.J.; Wu, F. Regulation, land-use mix, and urban performance. Part 1: Theory. Environ. Plann. A **1999**, *31* (7), 1433–1442.
77. Walker, R. Theorizing land-cover and land-use change: The case of tropical deforestation. Int. Reg. Sci. Rev. **2004**, *27* (3), 247–270.
78. Keys, E.; McConnell, W.J. Global change and the intensification of agriculture in the tropics. Global Environ. Change **2005**, *15* (4), 320–337.
79. VanWey, L.; Ostrom, E.; Meretsky, V. Theories underlying the study of human-environment interactions. In *Seeing The Forest and The Trees: Human-Environment Interactions in Forest Ecosystems;* Moran, E. and Ostrom, E., Eds.; MIT Press: Cambridge, MA, 2005; 23–56.
80. Chowdhury, R.R.; Turner II, B.L. Reconciling agency and structure in empirical analysis: Smallholder land use in the Southern Yucatán, Mexico. Ann. Assoc. Am. Geographers **2005** *96* (2), 302–322.
81. Lambin, E.F.; Meyfroidt, P. Land use transitions: Socioecological feedback versus socio-economic change. Land Use Policy **2010**, *27* (2), 108–118.
82. Ramankutty, N.; Graumlich, L.; Achard, F.; Alves, D.; Chhabra, A.; DeFries, R.S.; Foley, J.A.; Geist, H.; Houghton, R.A.; Goldewijk, K.K.; Lambin, E.F.; Millington, A.; Rasmussen, K.; Reid, R.S.; Turner, B.L. Global land-cover change: Recent progress, remaining challenges land-use and land-cover change. In *Land-Use and Land-Cover Change: Local Processes and Global Impacts*; Lambin, E.F., Geist, H., Eds.; Springer: Berlin, 2006; 9–39.
83. Goward, S.N.; Davis, P.E.; Fleming, D.; Miller, L.; Townshend, J.R. Empirical comparison of Landsat 7 and IKONOS multispectral measurements for selected Earth Observation System (EOS) validation sites. Remote Sensing Environ. **2003**, *88* (1–2), 80–99.
84. Steven, M.D.; Malthus, T.J.; Baret, F.; Xu, H.; Chopping, M.J. Intercalibration of vegetation indices from different sensor systems. Remote Sensing Environ. **2003**, *88* (4), 412–422.
85. Flint, E.P.; Richards, J.F. Historical-analysis of changes in land-use and carbon stock of vegetation in South and Southeast Asia. Can. J. Forest Res. **1991**, *21* (1), 91–110.
86. Gray, J.; Song, C. Mapping leaf area index using spatial, spectral, and temporal information from multiple sensors. Remote Sensing Environ. **2012**, *119*, 173–183.
87. Hilker, T.; Wulder, M.A.; Coops, N.C.; Linke, J.; McDermid, G.; Masek, J.G.; Gao, F.; White, J.C. A new data fusion model for high spatial-and temporal-resolution mapping of forest disturbance based on Landsat and MODIS. Remote Sensing Environ. **2009**, *113* (8), 1613–1627.
88. Verburg, P.H.; Schot, P.P.; Dijst, M.J.; Veldkamp, A. Land use change modelling: Current practice and research priorities. GeoJournal **2004**, *61*, 309–324.
89. Agarwal, C.; Green, G.; Grove, J.; Evans, T.; Schweik, C. *A Review and Assessment of Land-Use Change Models: Dynamics of Space, Time, and Human Choice, in Gen Tech Rep NE-2972002;* U.S. Dept Agric, Forest Service, Northeastern Research Station: Burlington, VT, 2002; 61.

90. Verburg, P.H.; Kok, K.; Pontius Jr, R.G.; Veldkamp, A. Modeling land-use and land-cover change. In *Land-Use and Land-Cover Change: Local Processes and Global Impacts*; Lambin, E.F., Geist, H., Eds.; Springer: Germany, **2006**, 117–135.
91. Veldkamp, A.; Fresco, L.O. CLUE: A conceptual model to study the conversion of land use and its effects. Ecol. Model. **1996**, *85* (2–3), 253–270.
92. Pontius, R.G.; Cornell, J.D.; Hall, C.A.S. Modeling the spatial pattern of land-use change with GEOMOD2: application and validation for Costa Rica. Agriculture Ecosystems Environ. **2001**, *85* (1–3), 191–203.
93. Parker, D.C.; Manson, S.M.; Janssen, M.A.; Hoffmann, M.J.; Deadman, P. Multi-agent systems for the simulation of land-use and land-cover change: A review. Ann. Assoc. Am. Geographers **2003**, *93* (2), 314–337.
94. Brown, D.G.; Robinson, D.T.; An, L.; Nassauer, J.I.; Zellner, M.; Rand, W.; Riolo, R.; Page, S.E.; Low, B.; Wang, Z. Exurbia from the bottom-up: Confronting empirical challenges to characterizing a complex system. Geoforum **2007,** *39* (2), 805–818.
95. Verburg, P.H.; Overmars, K.P. Combining top-down and bottom-up dynamics in land use modeling: Exploring the future of abandoned farmlands in Europe with the Dyna-CLUE model. Landscape Ecol. **2009**, *24* (9), 1167–1181.
96. Veldkamp, A.; Fresco, L.O. CLUE-CR: An integrated multiscale model to simulate land use change scenarios in Costa Rica. Ecol. Model. **1996**, *91* (1–3), 231–248.
97. Heistermann, M.; Muller, C.; Ronneberger, K. Land in sight? Achievements, deficits and potentials of continental to global scale land-use modeling. Agriculture Ecosystems Environ. **2006**, *114* (2–4), 141–158.
98. Alcamo, J.; Kreileman, G.J.J.; Bollen, J.C.; Van Den Born, G. J.; Gerlagh, R.; Krol, M.S.; Toet, A.M.C.; De Vries, H.J.M. Baseline scenarios of global environmental change. Global Environ. Change **1996**, *6* (4), 261–303.
99. Leemans, R.; Van Amstel, A.; Battjes, C.; Kreileman, E.; Toet, S. The land cover and carbon cycle consequences of large-scale utilizations of biomass as an energy source. Global Environ. Change **1996**, *6* (4), 335–357.
100. Woodward, F.I.; Smith, T.M.; Emanuel, W.R. A global land primary productivity and phytogeography model. Global Biogeochem. Cycles **1995**, *9* (4), 471–490.
101. Schaldach, R.; Alcamo, J.; Koch, J.; Kolking, C.; Lapola, D. M.; Schungel, J.; Priess, J.A. An integrated approach to modelling land-use change on continental and global scales. Environ. Model. Software **2011**, *26* (8), 1041–1051.
102. Hurtt, G.C.; Chini, L.P.; Frolking, S.; Betts, R.A.; Feddema, J.; Fischer, G.; Fisk, J.P.; Hibbard, K.; Houghton, R.A.; Janetos, A.; Jones, C.D.; Kindermann, G.; Kinoshita, T.; Goldewijk, K.K.; Riahi, K.; Shevliakova, E.; Smith, S.; Stehfest, E.; Thomson, A.; Thornton, P.; van Vuuren, D.P.; Wang, Y.P. Harmonization of land-use scenarios for the period 1500–2100: 600 years of global gridded annual land-use transitions, wood harvest, and resulting secondary lands. Climatic Change **2011**, *109* (1), 117–161.
103. Seto, K.C.; Guneralp, B.; Hutyra, L.R. Global forecasts of urban expansion to 2030 and direct impacts on biodiversity and carbon pools. Proc. Natl. Acad. Sci. USA. **2012**, *109* (40), 16083–16088.
104. Guneralp, B.; Seto, K.C. Futures of global urban expansion: uncertainties and implications for biodiversity conservation. Environ. Res. Lett. **2013**, *8*, 014025.
105. Pontius, R.; Boersma, W.; Castella, J.-C.; Clarke, K.; de Nijs, T.; Dietzel, C.; Duan, Z.; Fotsing, E.; Goldstein, N.; Kok, K.; Koomen, E.; Lippitt, C.; McConnell, W.; Mohd Sood, A.; Pijanowski, B.; Pithadia, S.; Sweeney, S.; Trung, T.; Veldkamp, A.; Verburg, P. Comparing the input, output, and validation maps for several models of land change. Ann. Regional Sci. **2008**, *42* (1), 11–37.
106. IPCC. *Climate change 2007. In Synthesis Report Contribution ofWorking Groups I, II and III to the Fourth Assessment Report of the Intergovernmental Panel on Climate Change;* Core Writing Team, Pachauri, R.K., Reisinger, A., Eds.; IPCC: Geneva, Switzerland, 2007; 104.

Bibliography

Baptista, S.R. Metropolitan land-change science: A framework for research on tropical and subtropical forest recovery in city-regions. Land Use Policy **2010**, *27* (2), 139–147.

Barnosky, A.D.; Koch, P.L.; Feranec, R.S.; Wing, S.L.; Shabel, A.B. Assessing the causes of late pleistocene extinctions on the continents. Science **2004**, *306* (5693), 70–75.

Blaikie, P. *The Political Economy of Soil Erosion in Developing Countries*; Longman Group Limited, Essex, 1985; 188.

Briassoulis, H. Land-use policy and planning, theorizing, and modeling: lost in translation, found in complexity? Environ. Plann. B **2008**, *35*, 16–33.

Brum, F.T.; Gonsalves, L.O; Cappelatti, L.; Carlucci, M.B.; Debastiani, V.J.; Salengue, E.V.; dos Santos Seger, G.D.; Both, C.; Bernardo-Silva, J.S.; Loyola, R.D.; da Silva Duarte, L. Land use explains the distribution of threatened new world amphibians better than climate. *PLoS ONE* **2013**, *8* (4), e60742.

Caldas, M.; Walker, R.; Arima, E.; Perz, S.; Aldrich, S.; Simmons, C. Theorizing land cover and land use change: The peasant economy of Amazonian deforestation. Ann. Assoc. Am. Geographers **2007**, *97* (1), 86–110.

Clarke, K.; Gazulis, N.; Dietzel, C.; Goldstein, N. A decade of SLEUTHing: Lessons learned from applications of a cellular automaton land use change model. In *Classics from IJGIS. Twenty Years of the International Journal of Geographical Information Systems and Science*; Fisher, P., Ed.; Taylor and Francis, CRC: Boca Raton, 2007; 413–425.

De Campos, C.P.; Muylaert, M.S; Rosa, L.P. Historical CO_2 emission and concentrations due to land use change of croplands and pastures by country. Sci. Total Environ. **2005**, *346* (1–3), 149–155.

Fischer, J.; Batary, P.; Bawa, K.S.; Brussaard, L.; Chappell, M.J.; Clough, Y.; Daily, G.C.; Dorrough, J.; Hartel, T.; Jackson, L.E.; Klein, A.M.; Kremen, C.; Kuemmerle, T.; Lindenmayer, D.B.; Mooney, H.A.; Perfecto, I.; Philpott, S.M.; Tscharntke, T.; Vandermeer, J.; Wanger, T.C.; Von Wehrden, H. Conservation: Limits of land sparing. Science **2011**, *334* (6056): 593.

Friis, C.; Reenberg, A. *Land Grab in Africa: Emerging Land System Drivers in a Teleconnected World*; Global Land Project (GLP)-IPO: Copenhagen, 2010; 42.

Geoghegan, J.; Pritchard, L.J.; Ogneva-Himmelberger, Y.; Chowdury, R.; Sanderson, S.; Turner, B.L.I. 'Socializing the pixel' and 'pixelizing the social' in land-use/cover change. In *People and Pixels*; Liverman, E.F.M.D., Rindfuss, R.R., Stern, P.C. National Research Council: Washington, DC, 1998; 51–69.

Goldewijk, K.K.; Beusen, A.; Janssen, P. Long-term dynamic modeling of global population and built-up area in a spatially explicit way: HYDE 3.1. Holocene **2010**, *20* (4), 565–573.

Goldewijk, K.K.; Ramankutty, N. Land cover change over the last three centuries due to human activities: The availability of new global data sets. *GeoJournal* **2004**, *61* (4), 335–344.

Gorissen, L.; Buytaert, V.; Cuypers, D.; Dauwe, T.; Pelkmans, L. Why the debate about land use change should not only focus on biofuels. Environ. Sci. Technol. **2010**, *44* (11), 4046–4049.

Guneralp, B.; Reilly, M.K.; Seto, K.C. Capturing multiscalar feedbacks in urban land change: a coupled system dynamics spatial logistic approach. Environ. Plann. B **2012**, *39* (5), 858–879.

Gutman, G.; Reissell, A., Eds.; *Eurasian Arctic Land Cover and Land Use in a Changing Climate;* Springer: Dordrecht, 2011; 306.

Heilig, G. Neglected dimensions of global land-use change: reflections and data. *Population and* Dev. Rev. **1994**, *20* (4), 831–859.

Houghton, R.A. The worldwide extent of land-use change. Bioscience **1994**, *44* (5), 305–313.

Irwin, E. G. New directions for urban economic models of land use change: Incorporating spatial dynamics and heterogeneity. J. Reg. Sci. **2010**, *50* (1), 65–91.

Kaimowitz, D.; Angelsen, A. *Economic Models of Tropical Deforestation: A Review*; Center International Forestry Research (CIFOR): Bogor, **1998**, 139.

Kalnay, E.; Cai, M. Impact of urbanization and land-use change on climate. Nature **2003**, *423* (6939), 528–531.

Klosterman, R.E. Simple and complex models. Environ. Plann. B **2012**, *39* (1), 1–6.

Krausmann, F.; Erb, K.-H.; Gingrich, S.; Haberl, H.; Bondeau, A.; Gaube, V.; Lauk, C.; Plutzar, C.; Searchinger, T.D. Global human appropriation of net primary production doubled in the 20th century. Proc. Natl. Acad. Sci. USA. **2013**, *110* (25), 10324–10329

Matthews, E. Global vegetation and land use: new high-resolution data bases for climate studies. J. Climate Appl. Meteorol. **1983**, *22* (3), 474–487.

Metzger, M.J.; Rounsevell, M.D.A.; Acosta-Michlik, L.; Leemans, R.; Schroter, D. The vulnerability of ecosystem services to land use change. Agriculture Ecosystems Environ. **2006**, *114* (1), 69–85.

Meyfroidt, P.; Rudel, T.K. Lambin, E.F. Forest transitions, trade, and the global displacement of land use. Proc. Natl. Acad. Sci. USA. **2010**, *107* (49), 20917–20922.

Nolte, C.; Agrawal, A.; Silvius, K.M.; Soares-Filho, B.S. Governance regime and location influence avoided deforestation success of protected areas in the Brazilian Amazon. Proc. Natl. Acad. Sci. USA. **2013**, *110* (13), 4956–4961.

Ostrom, E.; Nagendra, H. Insights on linking forests, trees, and people from the air, on the ground, and in the laboratory. Proc. Natl. Acad. Sci. USA. **2006**, *103* (51), 19224–19231.

Oswalt, P.; Rieniets, T. *Atlas of Shrinking Cities;* Hatje Cantz Publishers: Ostfildern, 2006; 160.

Petrosillo, I.; Semeraro, T.; Zaccarelli, N.; Aretano, R.; Zurlini, G. The possible combined effects of land-use changes and climate conditions on the spatial-temporal patterns of primary production in a natural protected area. Ecol. Indicators **2013**, *29*, 367–375.

Potere, D.; Schneider, A. A critical look at representations of urban areas in global maps. GeoJournal **2007**, *69*, 55–80.

Ramankutty, N.; Foley, J.A. Characterizing patterns of global land use: An analysis of global croplands data. Global Biogeochemical Cycles **1998**, *12* (4), 667–685.

Riebsame, W.E.; Parton, W.J.; Galvin, K.A.; Burke, I.C.; Bohren, L.; Young, R.; Knop, E. Integrated modeling of land use and cover change. Bioscience **1994**, *44* (5),350–356.

Rindfuss, R.R.; Walsh, S.J.; Turner, B.L.; Fox, J. Mishra, V. Developing a science of land change: Challenges and methodological issues. Proc. Natl. Acad. Sci. USA. **2004**, *101* (39), 13976–13981.

Robinson, D.T.; Brown, D.G.; Currie, W.S Modelling carbon storage in highly fragmented and human-dominated landscapes: Linking land-cover patterns and ecosystem models. Ecol. Model. **2009**, *220* (9–10), 1325–1338.

Rulli, M.C.; Saviori, A.; D'Odorico, P. Global land and water grabbing. Proc. Natl. Acad. Sci. USA. **2013**, *110* (3), 892–897.

Schneider, A.; Woodcock, C.E. Compact, dispersed, fragmented, extensive? A comparison of urban growth in twenty-five global cities using remotely sensed data, pattern metrics and census information. Urban Studies 2008, 45 (2), 659–692.

Seto, K.C.; de Groot, R.; Bringezu, S.; Erb, K.H.; Graedel, T.E; Ramankutty, N.; Reenberg, A.; Schmitz, O.; Skole, D. Stocks, flows, and prospects of land. In Linkages of Sustainability; Graedel, T., van der Voet, E., Eds.; MIT Press: Cambridge, MA, 2009, 71–96.

Seto, K.C.; Sanchez-Rodnguez, R.; Fragkias, M. The new geography of contemporary urbanization and the environment. Ann. Rev. Environ. Resources **2010**, *35* (1),167–194.

Stafford Smith, D.M.; McKeon, G.M.; Watson, I.W.; Henry, A. K.; Stone, G.S.; Hall, W.B.; Howden, S.M. Learning from episodes of degradation and recovery in variable Australian rangelands. Proc. Natl. Acad. Sci. USA. **2007**, *104* (52), 20690–20695.

Turner, B.L.; Clark, W.C.; Kates, R.W.; Richards, J.F.; Mathews, J.T.; Meyer, W.B., Eds.; *The Earth as Transformed by Human Action: Global and Regional Changes in the Biosphere over the Past 300 Years;* Cambridge University Press with Clark University: New York, 1990; 713.

9

Protected Area Management

Tony Prato
University of Missouri

Daniel B. Fagre
U.S. Geological Survey (USGS)

Introduction

The International Union for Conservation of Nature (IUCN) defines a protected area as a "clearly defined geographical space, recognized, dedicated and managed, through legal or other effective means, to achieve the long-term conservation of nature with associated ecosystem services and cultural values."[1] The Convention on Biological Diversity defines a protected area as a "geographically defined area which is designated or regulated and managed to achieve specific conservation objectives."[1] Protected areas encompass a broad range of ecosystems, including wetland, tropical or deciduous forest, alpine, savanna, and marine. Yellowstone National Park is the world's first public protected area. There has been increasing interest in transboundary protected areas that serve a dual purpose of protecting the landscape and generating cooperation between neighboring countries.

Designation, Extent, and Purposes of Protected Areas

The IUCN's World Commission on Protected Areas has established six protected area management categories: strict nature reserve, wilderness area, national park, natural monument or feature, habitat/species management area, protected landscape/seascape, and protected areas with sustainable use of natural resources.[2] Protected areas in the same category are managed based on similar objectives. For example, national parks are designated and managed to "(a) protect the ecological integrity of one or more ecosystems for present and future generations, (b) exclude exploitation or occupation inimical to the purposes of designation of the area, and (c) provide a foundation for spiritual, scientific, educational, recreational and visitor opportunities, all of which must be environmentally and culturally compatible."[3]

In addition to the IUCN categories, there are four other international designations of protected areas: Biosphere Reserve; World Heritage Site; Ramsar wetland; and marine. Biosphere Reserves serve three objectives: 1) conservation of landscapes, ecosystems, species, and genetic variation; 2) sustainable economic and human development; and 3) support for research, monitoring, education, and information exchange.[4] World Heritage Sites possess outstanding cultural, natural, or mixed (combination of cultural and natural) values that are protected for the benefit of all humanity through cooperative international arrangements.[5] The Convention on Wetlands[6] is an intergovernmental treaty whose member countries protect Wetlands of International Importance in their territories (Ramsar Convention on Wetlands undated). The United Nations Law of the Sea Treaty provides a legal framework for allocating rights to territorial seas and protecting marine-protected areas.[7]

In 2011, there were 157,897 protected areas in the world covering 24,236,479 km^2,[8] which amount to approximately 16% of the world's land surface area.[9] Locations of protected areas cover the gamut from densely populated urban areas to unpopulated wilderness areas. Protected areas range in size from a few square kilometers to thousands of square kilometers. Most protected areas are publicly owned and managed. The level of protection afforded protected areas depends on many factors, including adequate funding, diligent law enforcement, effective management practices, and the strength of citizen support.

Protected areas are established and managed for a wide range of social, economic, and ecological values, including preserving biodiversity, particularly for threatened and endangered species; contributing to regional conservation strategies; maintaining diversity of landscape or habitat and of associated species and ecosystems; ensuring the integrity and long-term maintenance of specified conservation targets; providing a range of ecosystem goods and services that support sustainable use of natural resources; providing research and educational opportunities; maintaining and protecting cultural sites and traditional values; providing opportunities for tourism and recreation; carbon storage; disaster mitigation, soil stabilization; water supply for human and agricultural uses; and wilderness preservation[1,2]

Nature and Types of Protected Areas

The establishment of public protected areas was motivated by setting aside novel or awe-inspiring resources for public enjoyment. Biodiversity or ecosystem integrity was not a relevant concern, and the size of many protected areas was the smallest necessary to protect unique features. With the passage of time, these areas became increasingly valuable for protecting species and accomplishing other societal goals that required relatively unaltered landscapes. Today protected areas are being considered in national strategic plans, such as carbon sequestration for mitigating global warming or the creation of continental conservation networks, purposes for which they were not created. Outside the United States, the concept of protected areas has expanded to include traditional uses deemed worthy of preservation. For that reason, historic agricultural practices are maintained in certain protected areas to continue the interaction between people and the landscape and increase public support for those areas. Some countries have established parks with the explicit aim of attracting tourists and increasing national income and employment (e.g., African wildlife parks). This demonstrates the important contributions of protected areas to local and regional economies and that the motivation for establishing protected areas has changed.

Publicly owned protected areas are managed by federal, state, and local governmental agencies. For example, at the federal level in the United States, the National Park Service manages the National Park System, the Forest Service manages the National Forest System, several federal agencies manage units in the National Wild and Scenic Rivers System and National Wilderness Preservation System, the National Oceanic and Atmospheric Administration manages Marine Protected Areas, and the U.S. Fish and Wildlife Service manages the National Wildlife Refuge System. Each federal agency operates under the authority of an organic act that dictates how certain types of protected areas should be managed. States also designate and manage protected areas. For instance, the Missouri Natural Areas System is an extensive statewide network of over 80 natural areas representing examples of the state's original natural landscape and outstanding biological and geological features.[10] Local park commissions manage urban parks, which are typically oriented towards outdoor sports, picnicking, hiking, swimming, and enjoyment of nature. Nongovernmental organizations purchase land and conservation easements to protect natural values and/or fish and wildlife habitat. For example, The Nature Conservancy works with landowners, communities, cooperatives, and businesses to establish local groups that protect land through the use of land trusts, conservation easements, private reserves, and incentives. Nature Conservancy lands, which are located primarily in the United States, constitute the world's largest system of private nature reserves covering 60,703 km^2.[11]

Threats to Protected Areas

As human populations, incomes, and leisure time increase, protected areas have come under greater pressure to serve more people and more diverse purposes often not envisioned when the protected areas were established. These human pressures vary greatly with country and the type of protected area but, in general, these dynamics have changed the role of protected area managers from caretakers to assertive advocates who manage their areas to provide benefits to local, regional, and national economies and minimize the impacts of threats to protected areas. Examples of such threats include rapid growth in visitation, changes in the mix of recreational and visitor uses (hunting, trapping, motorized/mechanized sports and transportation, hiking, camping, skiing, and air tours), development of roads and visitor facilities, spread of invasive species, residential and commercial development, legal resource extraction (logging, livestock grazing, farming, and energy development), illegal timber harvesting, wildlife poaching, air, water, and soil pollution, and climate change and variability.[12,13] Some threats, such as increases in visitation and invasive species, are internal to protected areas, whereas others such as residential and commercial development and climate change, are external. In terms of the nature of the threats, increases in visitation and invasive species exert pressure on natural resources of protected areas and make their preservation more difficult. Landscape fragmentation, caused by residential and commercial development adjacent to protected areas, and climate change reduce the ecological integrity of protected areas, particularly their ability to serve as refugia for threatened and endangered species. External threats to protected areas are typically more difficult to resolve than internal threats because protected area managers have limited legal authority and/or are unwilling to control events external to their protected areas.

Management of Protected Areas

A major challenge facing protected area managers is how to manage those areas so as to achieve multiple management objectives. This challenge occurs for several reasons. First, certain management objectives (e.g., resource protection and visitor satisfaction) are somewhat incompatible. Second, there is considerable uncertainty about the natural and/or socioeconomic processes influencing protected areas, how internal and external threats influence those processes, and how different management actions are likely to influence the achievement of management objectives. Third, some management objectives, such as minimizing adverse impacts on visitors and the local economy, are supported by one group of stakeholders, whereas other objectives, such as natural and cultural resource protection, are supported by another group of stakeholders. Fourth, many protected areas face declining financial and human resources, which makes it more difficult to develop and implement effective management plans.

Two methods can be used by protected area managers to evaluate and make management decisions under uncertainty, namely, scenario planning and adaptive management (AM). Scenario planning requires protected area managers to develop and evaluate decisions for all combinations of unknown future events (e.g., future climate scenarios) and management actions. The scenario planning approach used by the U.S. National Park Service to manage national parks for climate change requires park managers to: 1) choose the most probable climate scenario based on their subjective judgment; 2) determine the most preferred management action for that climate scenario; and 3) implement the most preferred management action.[14,15] A difficulty with scenario planning for climate change is that it requires protected area managers to choose the most probable climate scenario. This is problematic because scientists who develop climate scenarios have been unwilling or unable to assign probabilities to future climate scenarios. Thus, protected area managers need to develop management options for multiple scenarios.

AM postulates that "if human understanding of nature is imperfect, then human interactions with nature (e.g., management actions) should be experimental."[16] Kohm and Franklin state that "AM is the only logical approach under the circumstances of uncertainty and the continued accumulation of knowledge."[17] AM acknowledges that protected area managers cannot accurately predict the outcomes of management actions because of environmental, scientific, organizational, community, and political uncertainties.

AM can be either passive or active.[18] Application of passive AM to protected areas requires protected area managers to: 1) formulate a predictive model of how a protected area responds to alternative management actions; 2) select the most preferred management actions based on model predictions; 3) implement and monitor those management actions; 4) use monitoring results to revise the model; and 5) adjust management actions based on the revised model. Advantages of passive AM are that it is relatively simple to use and usually less expensive to implement than active AM. A disadvantage of passive AM is that it does not produce statistically reliable information about the impacts of management actions on protected areas because it does not utilize experimental methods.

Application of active AM to protected areas requires protected area managers to: 1) formulate hypotheses about how various management actions influence a particular feature of protected areas (e.g., biodiversity) when there is uncertainty about how future events (e.g., climate change) interact with management actions to influence that feature; 2) establish experiments to test hypotheses about the efficacy of management actions over time; 3) use test results to determine the best management action to implement in an upcoming management period; and 4) continue experiments and use experimental results to periodically retest the hypotheses to determine whether the best management action changes over time. Active AM typically involves major investments in research, monitoring, and modeling; has prerequisites that are not always satisfied; and is more complicated to implement than passive AM.[19–21]

Conclusion

Protected areas play increasingly important roles in our global society, including biodiversity conservation, providing ecosystem services, and supporting regional economies through recreational opportunities and tourism. These values of protected areas are reflected in the extensive distribution and diversity of protected areas and their increasing numbers. With a burgeoning global population and its demands on natural resources, protected areas will become even more critical to informed conservation planning and as scientific resources. Protected area managers must respond with vision and flexibility using new approaches such as scenario planning and AM.

Acknowledgment

Any use of trade, product, or firm names is for descriptive purposes only and does not imply endorsement by the U.S. Government.

References

1. Dudley, N., Ed. *Guidelines for Applying Protected Area Management Categories;* IUCN: Gland, Switzerland, and Cambridge, UK, 2008.
2. IUCN. *Guidelines for Protected Area Management Categories*; IUCN Commission on National Parks and Protected Areas with assistance of the World Conservation Monitoring Centre: Gland, Switzerland, and Cambridge, UK, 1994.

3. Davey, A.G. *Protected Areas Categories And Management Objectives (Appendix 2);* National System Planning for Protected Areas. IUCN: Gland, Switzerland, and Cambridge, UK, 1998.
4. UNESCO. Man and the Biosphere Program, http://www.unesco.org/new/en/natural-sciences/environment/ecological-sciences/man-and-biosphere-programme/(accessed October 2012).
5. UNESCO. Our World Heritage, http://whc.unesco.org/uploads/ activities/documents/activity-568-2.pdf (accessed October 2012).
6. Ramsar Convention on Wetlands. About the Ramsar Convention, http://www.ramsar.org/cda/en/ramsar-about-about-ramsar/main/ramsar/1-36%5E7687_4000_0__ (accessed October 10, 2012).
7. United Nations Law of the Sea Treaty. Law of the Sea Treaty (LOST) - Background, http://www.unlawoftheseatreaty.org/ (accessed October 2012).
8. IUCN and UNEP-WCMC. The World Database on Protected Areas (WDPA). Cambridge, UK, http://www.wdpa.org/ (accessed October 2012).
9. Coble, C.R.; Murray, E.G.; Rice, D.R. Earth Science 1987, 102.
10. Missouri Department of Natural Resources. About the Missouri State Park System, http://mostateparks.com/page/ 55047/about-missouri-state-park-system (accessed October 2012).
11. The Nature Conservancy. Public Lands Conservation, http://www.nature.org/about-us/private-lands-conservation/index.htm (accessed October 2012).
12. Prato, T.; Fagre, D. *National Parks and Protected Areas: Approaches for Balancing Social, Economic and Ecological Values;* Blackwell Publishers: Ames, Iowa. 2005.
13. Nolte, C.; Leverington, F.; Kettner, A.; Marr, M.; Nielsen, G.; Bastian, B.; Stoll-Kleemann, S.; Hockings, M. Protected Area Management Effectiveness Assessments in Europe: A Review of Application, Methods and Results; BfN-Skripten 271a. Bundesamt fur Naturschutz (BfN), Bonn, Germany, 2010.
14. Weeks, K.; Moore, P.; Welling, L. Climate change scenario planning: A tool for managing parks into uncertain futures. Park Sci. **2011**, *28*, 26–33.
15. Welling, L. National parks and protected areas management in a changing climate: Challenges and opportunities. Park Sci. **2011**, *28*, 6–9.
16. Lee, K.N. *Compass and Gyroscope: Integrating Science and Politics for the Environment;* Island Press: Washington, DC, USA. 1993.
17. Kohm, K.A.; Franklin, J.F. Introduction. In *Creating Forestry for the 21st Century: The Science of Ecosystem Management;* Kohm K.A.; Franklin J.F. Eds.; Island Press: Washington, DC, USA. 1997; 1–5.
18. Baron, J.S.; Gunderson, L.; Allen, C.D.; Fleishman, E.; McKenzie, D.; Meyerson, L.; Oropeza, J.; Stephenson, N. Options for national parks and reserves for adapting to climate change. Environ. Manag. **2009**, *44*, 1033–1042.
19. Prato, T. Bayesian adaptive management of ecosystems. Ecol. Model. **2005**, *183*, 147–156.
20. Prato, T. Accounting for risk and uncertainty in determining preferred strategies for adapting to future climate change. Mit. Adapt. Strateg. Global Change **2008**, *13*, 47–60.
21. Prato, T.; Fagre, D.B. Sustainable management of the Crown of the Continent Ecosystem. George Wright Forum **2010**, *27*, 77–93.

10

Protected Areas: Remote Sensing

Yeqiao Wang
University of Rhode Island

Introduction

The World Commission on Protected Areas adopted a definition that describes a protected area as clearly defined geographical space, recognized, dedicated, and managed, through legal or other effective means, to achieve the long-term conservation of nature with associated ecosystem services and cultural values.[1] In general, protected areas include national parks, national forests, national seashores, all levels of natural reserves, wildlife refuges and sanctuaries, and designated areas for conservation of native biological diversity and natural and cultural heritage and significance. Protected areas also include some of the last frontiers that have unique landscape characteristics and ecosystem functions in wilderness conditions.[2] Along the shoreline and over the ocean and sea, the International Union for the Conservation of Nature (IUCN) has defined marine protected areas (MPAs) as any area of intertidal or subtidal terrain, together with its overlying water and associated flora, fauna, historical, and cultural features, which has been reserved by law or other effective means to protect part or all of the enclosed environment.[3] As reported by the World Database on Protected Areas, under 15% of the world's terrestrial and inland waters, excluding Antarctica, is under protection. About 4.12% of the global ocean and 10.2% of coastal and marine areas under national jurisdiction are set as MPAs. Only 19.2% of key biodiversity areas are completely covered by protected areas.[4] Protected lands and waters serve as the fundamental building blocks of virtually all national and international conservation strategies, supported by governments and international institutions. These provide the essential and best of efforts to protect the world's threatened species and are increasingly recognized as essential providers of ecosystem services and biological resources; are key components in climate change mitigation strategies; and, in some cases, are also vehicles for protecting threatened human communities or sites of great cultural and spiritual value.[1]

Humans have created protected areas over past millennia for a multitude of reasons.[5] The establishment of Yellowstone National Park in 1872 by the U.S. Congress ushered in the modern era of governmental protection of natural areas, which catalyzed a global movement.[6,7] However, even with the implementation of a tremendous variety of monitoring programs and conservation planning efforts and achievements, species' population decline, biodiversity loss, extinction, system degradation, pathogen

FIGURE 10.1 **(See color insert.)** Climate change and human-induced disturbances impose uncertainty on the habitat conditions, so as the future of biodiversity. Field photos illustrate critically endangered Siberian cranes (*Leucogeranus leucogeranus*) wintering in Poyang Lake area (2017), China (**a**); endangered Hawaiian monk seals (*Neomonachus schauinslandi*) in Kauai (2018), endemic to the Hawaiian Islands (**b**); and polar bears in the Hudson Bay area (2016), Canada (**c**). (Photos: Yeqiao Wang.)

spread, and state change events are occurring at unprecedented rates.[8,9] The effects are augmented by continued changes in land use, invasive spread, alongside the direct, indirect, and interactive effects of climate change and disruption.[5] Protected areas become more important in serving as indicators of ecosystem conditions and functions either by their status and/or by comparison with unprotected adjacent areas. Protected lands are prized highly by the society with diversified representative characteristics. Earth's remaining wilderness areas are becoming increasingly important buffers against changing conditions in the Anthropocene. Yet they are not an explicit target in international policy frameworks.[10] The most recent report of United Nations concluded that up to 1 million animal and plant species are facing extinction, for which humans are to blame.[11] The most impacting drivers on global biodiversity scenarios toward the year 2100 include changes in land use, climate, nitrogen deposition, biotic exchange, and atmospheric CO_2.[12] Inventory, monitoring, understanding, and management of protected areas present challenges (Figure 10.1).

Remote sensing refers to art, science, and technology for Earth system data acquisition through nonphysical contact sensors or sensor systems mounted on space-borne, airborne, and other types of platforms; data processing and interpretation from automated and visual analysis; information generation under computerized and conventional mapping facilities; and applications of generated data and information for societal benefits and needs. Remote sensing can provide comprehensive geospatial information to map and study protected areas in different spatial scales (e.g., high spatial resolution and large area coverage), different temporal frequencies (daily, weekly, monthly, annual observations, etc.), different spectral properties (visible, near infrared, microwave, etc.), and with spatial contexts (e.g., immediate adjacent areas of protected areas vs. a broader background of land and water bases). Remote sensing, in combination with field-based studies, has created new and exciting opportunities to meet monitoring needs to study protected areas.[13]

Remote Sensing of Changing Landscape of Protected Areas

Park units of the National Park Service (NPS) of the United States are important components in a system of reserves that protect biodiversity and other natural and cultural resources. To meet the NPS mission to manage resources so they are left unimpaired for the enjoyment of future generations, it is essential to know what resources occur in parks and to monitor the status and trends in the condition of key resource indicators. The NPS Inventory and Monitoring Program (I&M) was designed to provide the infrastructure and staff to identify critical environmental indicators, so-called "vital signs," and to implement long-term monitoring of natural resources in more than 270 parks that contain significant natural resources.[10] The 270+ parks are organized into 32 ecoregional networks. The connectivity of natural landscapes illustrates magnitude of cumulative movement, assuming that animals avoid human-modified areas in the continental United States, with spatial distributions of the national park units (Figure 10.2).[14]

The Park AnaLysis and Monitoring Support (PALMS) project, for example, was to enhance the quality of natural resource management in parks by better integrating the routine acquisition and analysis of NASA Earth System Science products and other data sources into NPS I&M.[14] The study focused on four sets of national parks to develop and demonstrate the approach: the Sequoia-Kings Canyon and Yosemite National Parks (Sierra Nevada I&M Network), Yellowstone and Grand Teton National Parks (Greater Yellowstone I&M Network), Rocky Mountain National Parks (Rocky Mountain I&M Network), and a combination of Delaware Water Gap National Recreation Area and Upper Delaware Scenic and Recreational River (Eastern Rivers and Mountains I&M Network). The PALMS project developed a suite of indicators including measurements of weather and climate, stream health (water), land cover and land use, disturbances, primary production, and monitoring area.

Monitoring landscape dynamics of National Parks in the western United States by Landsat-based approaches was one of the examples.[15] The study was established by the approaches that translate ecologically based view of change into the spectral domain when the entire archive of Thematic Mapper (TM) spectral imagery is considered. A spectral index is used as a proxy for ecological attributes and that index can be tracked as a time-series trajectory. The developed algorithms, collectively known as "LandTrendr," use simple statistical fitting rules to identify periods of consistent progression in the spectral trajectory (segments) and turning points (vertices) that separate those periods. The change-monitoring methods capture a wide range of processes affecting vegetation inside and outside of western national parks, such as decline/mortality processes, growth/recovery processes, and combination of different processes, among others. The study concluded that even though national parks are protected from many forms of direct human intervention, their landscapes are anything but static. Vegetation is changing continuously in response to both endogenous and exogenous pressures, and monitoring these dynamic landscapes requires tools that capture a wide range of processes over large areas.

A study evaluated forest dynamics within and around the Olympic National Park using time-series Landsat observations.[16] The Olympic National Park (ONP) was established in the 1930s to protect the diversity of its ecosystems. It has some of the best examples of intact temperate rainforest in the Pacific Northwest and is home to endemic species. In 1976, it was designated by the UNESCO as an International Biosphere Reserve and was declared a World Heritage Site in 1981. However, due to extensive logging in the early 20th century, most of the unprotected old-growth forests outside the ONP and the Olympic National Forests were lost, and some of the private lands in the peninsular are among the most heavily logged in the United States. Viable protection of the unique biodiversity of the Olympic Peninsula requires continuous monitoring of forest dynamics. The time-series approach using remote sensing data allows reconstruction of disturbance history over the last decades by taking advantage of the temporal depth of the Landsat archive and can be used to provide continuous monitoring as new satellite images are acquired.

Remote sensing has been used for monitoring wildlife habitat changes in Kejimkujik National Park and the National Historic Site in southern Nova Scotia of the Canadian Atlantic Coastal Uplands Natural Region.[17] The study addressed the major goals in protection and maintenance of ecological integrity through two key measures of "effective habitat amount" and "effective habitat connectivity." The study employed the Automated Multitemporal Updating by Signature Extension protocol, known otherwise as AMUSE, to generate multitemporal land cover maps by Landsat remote sensing data. The resultant gradient maps showed areas of high and low effective habitat amount and connectivity throughout the park and greater park ecosystem for each species. Thresholds were then applied to identify species landscapes that were estimated to contain sufficient habitat. Statistics on the total amount and connectivity of effective habitat for Kejimkujik and the greater park ecosystem were then extracted for each 5-year time step and compared. The visual output from analyses provides powerful communication

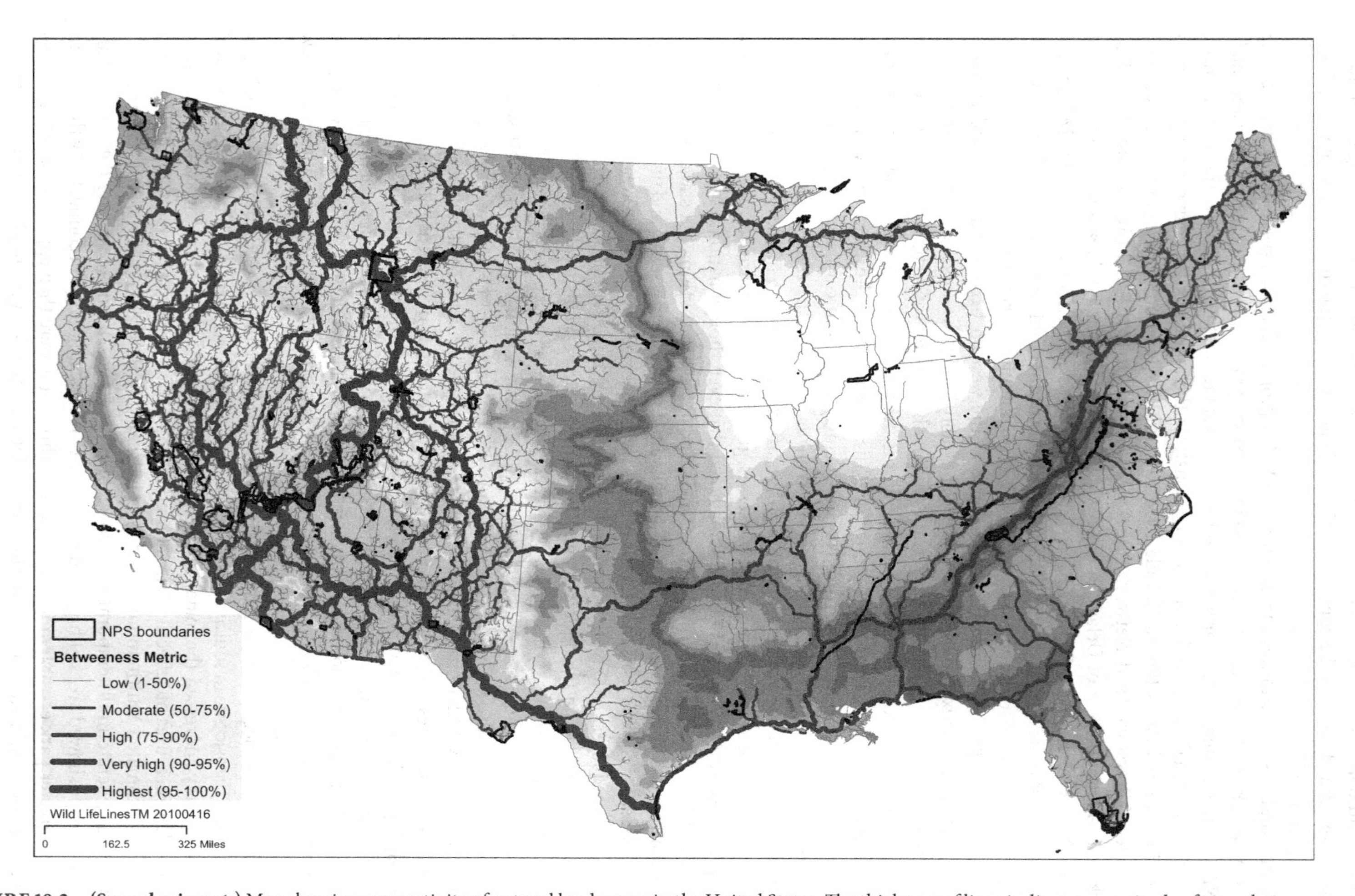

FIGURE 10.2 **(See color insert.)** Map showing connectivity of natural landscapes in the United States. The thickness of lines indicates magnitude of cumulative movement, assuming that animals avoid human-modified areas. The surface underneath the pathways depicts the averaged cost-distance surfaces, or the overall landscape connectivity surface. NPS units are outlined in black. (Gross et al.[14])

products on how land use can affect habitat patterns around the park. These communication products can support the park in engaging partner agencies and stakeholders in the development of collaborative conservation strategies that are vital in allowing Kejimkujik to manage toward achieving its long-term ecological integrity goals. Information is vital for sound decision-making at all levels. Remote sensing and the related information technology have been playing and will continue to play critical roles in natural resource monitoring and management.

Remote Sensing for Inventory, Mapping, and Conservation Planning of Protected Areas

One particular advantage that remote sensing can provide for inventory and monitoring of protected areas is the information for understanding the past and current status, the changes occurred under different impacting factors and management practices, the trends of changes in comparison with those in the adjacent areas and implications of changes on ecosystem functions.[18,19] Field survey and *in situ* observations are essential to identify protected habitats through remote sensing. Almost every remote sensing exercise requires field survey to define habitats, to calibrate remote sensing imagery, and to evaluate the accuracy out of remote sensing outputs.[20] With GPS-guided positioning and field survey becoming a routine operation, challenges remain for incorporation of *in situ* measurements with remote sensing observations for quantitative analyses of protected habitats.

A study employed Landsat-7 ETM+ data and GPS collars to provide detailed information for conservation of Matschie's tree kangaroos in Papua New Guinea.[21] Matschie's tree kangaroos (*Dendrolagus matschiei*) are arboreal marsupials endemic to the Huon Peninsula in Papua New Guinea. Due primarily to increased hunting pressure and loss of habitat from agricultural expansion, *D. matschiei* are currently listed as endangered by the IUCN. The study concluded that *Dacrydium nidulum*-dominant forests are the most widespread forest throughout the study area and are also where tree kangaroos located. However, additional research has shown that these are not the only forest type that is used by the species. Clustered and independent movement locations indicate that animals do not utilize their habitat uniformly. These data provide vital information toward a better understanding of the habitat, the requirements of the animals, and the long-term conservation of the Matschie's tree kangaroo habitat. Aerial remote sensing from Wildlife Conservation Society's Flight Program has been used to support biodiversity conservation in Madagascar and Eastern and Southern Africa with a focus on the Albertine Rift.[22] Many sites in the Albertine Rift are protected as national parks, wildlife reserves, or forest reserves. The aerial imagery has been used to map threats to biodiversity, to develop land-use plans for protected area management, and to measure vegetation cover and dynamics.

MPAs are among critical components of protected waters. Important factors that affect the way plants and animals respond to MPAs include distribution of habitat types, level of connectivity to nearby fish habitats, wave exposure, depth distribution, prior level of resource extraction, regulations, and level of with regulations.[23] Conservation benefits are evident through increased habitat heterogeneity at the seascape level, increased abundance of threatened species and habitats, and maintenance of a full range of genotypes.[24] Remote sensing data that quantify spatial patterns in habitat type, oceanographic conditions, and benthic complexity can be integrated with in situ ecological data for design, evaluation, and monitoring of MPA networks to design, assess, and monitor MPAs.[25,26] Combining remote sensing products with in situ ecological and physical data can support the development of a statistically robust monitoring program of living marine resources within and adjacent to marine protected areas.[27] Individual MPAs need to be networked in order to provide large-scale ecosystem benefits and to have the greatest chance of protecting all species, life stages, and ecological linkages if they encompass representative portions of all ecologically relevant habitat types in a replicated manner. High-resolution remote sensing data are capable of mapping physical and biological features of benthic habitat, such as monitoring of coral reef in the Hawaii Archipelago and near-shore protected areas in California and New England.[23]

Remote Sensing of Frontier Lands

Remote sensing has unique advantages in monitoring frontier lands, which are always in remote and difficult-to-reach locations and huge in area coverage. Different types of remote sensing data have been applied in the study of frontier lands, for example, using hyperspectral and radar data for monitoring of forests in the Amazon region[28–33] and Siberia,[34–36] and for hydrologic change detection in the lake-rich Arctic region.[37,38]

Case studies include satellite-observed endorheic lake dynamics across the Tibetan Plateau between 1976 and 2000; multisensor remote sensing of forest dynamics in Central Siberia; remote sensing and modeling for assessment of complex Amur tiger and Far Eastern leopard habitats in the Russian Far East; incorporating remotely sensed land surface properties to regional climate modeling; and InSAR for characterizing biophysical properties in protected tropical forests in Southeast Asia.

With a pronounced temperature rise, the Tibetan plateau is one of the world's most vulnerable areas to climate change. Tibetan warming has been accelerating since the 1950s, and the accelerated warming is expected to drive an array of complex physical and ecological changes in the region. Tibetan lakes in the endorheic basins serve as a sensitive indicator of regional climate and water cycle variability. The lakes are dynamic in their inundation area in response to climate and hydrological conditions. However, these lake dynamics at regional scales are not well understood due to inaccessibility and the inhospitable environment of this remote plateau, making satellite remote sensing the only feasible tool to detect lake dynamics across the plateau. A study monitors Tibetan lake changes using nearly a hundred Landsat scenes acquired around 1976 and 2000 and examined lake changes during ~25 years in endorheic basins across this broad plateau and analyzes the spatial patterns of the observed lake dynamics.[39]

The forested regions of Russian Siberia are vast and contain about a quarter of the world's forests that have not experienced harvesting. Monitoring the dynamics and mapping the structural parameters of the forest are important for understanding the causes and consequences of changes observed in these areas. Because of the inaccessibility and large extent of this forest, remote sensing data can play an important role in observing forest state and change. Ranson et al.[40] introduced a case study that used multisensor remote sensing data to monitor forest disturbances and to map aboveground biomass in Central Siberia. Radar images from the Shuttle Imaging Radar-C (SIR-C)/XSAR mission were used for forest biomass estimation in the Sayan Mountains. Radar images from the Japanese Earth Resources Satellite-1 (JERS-1), European Remote Sensing satellite-1 (ERS-1), and Canada's RADARSAT-1, and data from ETM+ on-board Landsat-7 were used to characterize forest disturbances from logging, fire, and insect damages.

Listed as endangered and critically endangered by the IUCN, respectively, fewer than 400 adult and subadult Amur (Siberian) tigers (*Panthera tigris altaica*) and only between 14 and 20 adult Amur (Far Eastern) leopards (*P. pardus orientalis*) exist in the wild in a topographically complex and biologically diverse landscape in the Russian Far East and in the Changbai Mountain area in northeastern China near the Russia–China border. These endangered species face threats mostly from loss of habitat due to forest fire and logging to illegal poaching and competing prey with man. Remote sensing is a valuable tool for characterizing the vast habitat of wide-ranging endangered species, such as the Amur tiger and the critically endangered Amur leopard. A study employed remote sensing for mapping, monitoring, and modeling tiger and leopard habitats related to vegetation and terrain.[41] The study helped target resources toward locations that are most important for ensuring a future in the wild for these species, as well as identify new areas that would be appropriate to connect good habitat, including that in established reserves.

In tropical forests in Southeast Asia, national parks have been established as a means of reducing deforestation in ecologically and economically favorable ways. Accurate estimates of forest biophysical properties in those national parks are important in assessing their current status and in supporting sustainable management of these protected lands. A study employed Landsat ETM+ and JERS-1 SAR images to estimate green and woody structures in a protected forest in northern Thailand.[29] Forest fractional cover and leaf area index are estimated from the ETM+ image using a linear unmixing model. A microwave/optical synergistic model has been modified to improve woody biomass estimation by removing leaf contribution from radar backscatter in tropical forests. Besides seasonal variations in different forest types, forest degradation by human disturbances is observed, which results in reduced green and woody biomass in forests closer to human settlements. The study demonstrates that a synergistic use of optical and synthetic aperture radar (SAR) images could extract quantitative biophysical information of protected forests in tropical mountains.

Conclusions

Protected areas and the networks provide critical habitats for biodiversity conservation, and yet their performances are challenged under climate change and shifting resource patterns.[42] The 1916 National Park Service Organic Act establishes the purpose of the U.S. national parks, i.e., to "conserve the scenery and the natural and historic objects and the wild life therein and to provide for the enjoyment of the same in such manner and by such means as will leave them unimpaired for the enjoyment of future generations." NPS Management Policies set the goal as "to ensure that park resources and values are passed on to future generations in a condition that is as good as, or better than, the conditions that exist today."[43] Effective management of protected area is fundamental. Evaluation of management effectiveness has been recognized as a vital component of responsive, pro-active protected area management.[44,45] Ecosystem indicators, whether process-based (e.g., productivity), pattern-based (e.g., land-use activities), or component-based (species populations), vary in space and time. A major limiting factor in comprehensive ecological models is a lack of explanatory geospatial data.[4] The issues conspire against the ready, standardized integration of remote sensing into ecological research for protected area management.

Remote sensing science is a universal tool for managers and researchers across many domains.[14] Remote sensing data and data products coupled with user-friendly data exploration, data management, analyses, and modeling tools in an accessible common platform allow both scientists and practitioners a better understanding of how environmental impacts affect species populations and the ecosystem goods and services that sustain them. The lessons learned and recommendations put forward for remote sensing of protected areas include the following: "allocate sufficient time to develop a genuine science - management partnership," "communicate results in a management-relevant context," "confirm or embellish existing frameworks and processes," "plan for persistence and change," and "build on existing, widely used data analysis tools and software frameworks, even if they seem inefficient."[13]

Remote sensing is among the most fascinating frontiers of science and technology that are constantly improving. Protected areas are by no means uniform entities and have a wide range of management aims and are governed by many stakeholders. Advances in remote sensing have helped gather and share information about the protected areas at unprecedented rates and scales. There are many new and exciting applications of remotely sensed data that contribute to better informing management of protected areas.[46,47] The achievements through the applications of science and technologies, the challenges, the lessons learned, and the recommendations in remote sensing of protected areas deserve an attention.

References

1. Dudley, N., Ed. *Guidelines for Applying Protected Area Management Categories*. IUCN: Gland, **2008**, p. 86.
2. Wang, Y. *Remote Sensing of Protected Lands*. CRC Press: Boca Raton, FL, **2012**, p. 604.
3. Kelleher, G. *Guidelines for Marine Protected Areas*. IUCN: Gland and Cambridge, **1999**.

4. UNEP-WCMC and IUCN. Protected planet report 2016. UNEP-WCMC and IUCN: Cambridge and Gland, **2016**.
5. Crabtree, R.; Sheldon, J. Monitoring and modeling environmental change in protected areas: Integration of focal species populations and remote sensing (Chapter 21). In: *Remote Sensing of Protected Lands*, Wang, Y., Ed. CRC Press: Boca Raton, FL, **2012**, pp. 495–524.
6. IUCN. Shaping a sustainable future. In: *The IUCN Programme 2009–2012*, IUCN: Gland, **2008**.
7. Heinen, J.; Hite, K. Protected natural areas. In: *Encyclopedia of Earth*, Cleveland, C.J., Ed. Environmental Information Coalition, National Council for Science and the Environment: Washington, DC, **2007**. www.eoearth.org/article/Protected_natural_areas (Last revised November 29, 2007; Retrieved February 5, 2011).
8. Hoffmann, M.; Hilton-Taylor, C.; Angulo, A. et al. The impact of conservation on the status of the World's vertebrates. *Science* **2010**, *330* (6010), 1503–1509.
9. Pereira, H.M.; Leadley, P.W.; Proença, V. et al. Scenarios for global biodiversity in the 21st century. *Science* **2010**, *330* (6010), 1496–1501.
10. Watson, J.E.M.; Allan, J.R. Protected the last of the wild. *Nature* **2018**, *563*, 27.
11. IPBES, Intergovernmental Science-Policy Platform on Biodiversity and Ecosystem Services (IPBES). Nature's dangerous decline 'unprecedented' species extinction rates 'Accelerating', Media Release, 6 May **2019**.
12. Sala, O.E.; Chapin, III, F.S.; Armesto, J.J.; Berlow, E.; Bloomfield, J.; Dirzo, R.; Huber-Sanwald, E.; Huenneke, L.F.; Jackson, R.B.; Kinzig, A.; Leemans, R.; Lodge, D.M.; Mooney, H.A.; Oesterheld, M.; LeRoy Poff, N.; Sykes, M.T.; Walker, B.H.; Walker, M.; Wall, D.H. Global biodiversity scenarios for the year 2100. *Science* **2010**, *287* (5459), 1770–1774.
13. Fancy, S.G.; Gross, J.E.; Carter, S.L. Monitoring the condition of natural resources in U.S. National Parks. *Environ. Monit. Asses.* **2009**, *151* (1–4), 161–174.
14. Gross, J.E.; Hansen, A.J.; Goetz, S.J.; Theobald, D.M.; Melton, F.M.; Piekielek, N.B.; Nemani, R.R. Remote sensing for inventory and monitoring of U.S. National Parks (Chapter 2). In: *Remote Sensing of Protected Lands*, Wang, Y., Ed. CRC Press: Boca Raton, FL, **2012**, pp. 29–56.
15. Kennedy, R.E.; Yang, Z.; Cohen, W.B. Detecting trends in forest disturbance and recovery using yearly Landsat time series: 1. LandTrendr—Temporal segmentation algorithms. *Remote Sens. Environ.* **2010**, *114* (12), 2897–2910.
16. Huang, C.; Schleerweis, K.; Thomas, N.; Goward, S.N. Forest dynamics within and around Olympic National Park assessed using time series landsat observations (Chapter 4). In: *Remote Sensing of Protected Lands*, Wang, Y., Ed. CRC Press: Boca Raton, FL, **2012**, pp. 75–94.
17. Zorn, P.; Ure, D.; Sharma, R.; O'Grady, S. Using earth observation to monitor species-specific habitat change in the Greater Kejimkujik National Park Region of Canada (Chapter 5). In: *Remote Sensing of Protected Lands*, Wang, Y., Ed. CRC Press: Boca Raton, FL, **2012**, pp. 95–110.
18. Hansen, A.J.; DeFries, R. Land use change around nature reserves: Implications for sustaining biodiversity. *Ecol. Appl.* **2007**, *17* (4), 972–973.
19. Hansen, A.J.; DeFries, R. Ecological mechanisms linking protected areas to surrounding lands. *Ecol. Appl.* **2007**, *17* (4), 974–988.
20. Wang, Y.; Traber, M.; Milestead, B.; Stevens, S. Terrestrial and submerged aquatic vegetation mapping in Fire Island National seashore using high spatial resolution remote sensing data. *Mar. Geod.* **2007**, *30* (1), 77–95.
21. Stabach, J.A.; Dabek, L.; Jensen, R.; Wang, Y.Q. Discrimination of dominant forest types for Matschie's tree kangaroo conservation in Papua New Guinea using high-resolution remote sensing data. *Int. J. Remote Sens.* **2009**, *30* (2), 405–422.
22. Ayebare, S.; Moyer, D.; Plumptre, A.J.; Wang, Y. Remote sensing for biodiversity conservation of the Albertine Rift in Eastern Africa (Chapter 10). In: *Remote Sensing of Protected Lands*, Wang, Y., Ed. CRC Press: Boca Raton, FL, **2012**, pp. 183–201.

23. Friedlander, A.M.; Wedding, L.M.; Caselle, J.E.; Costa, B.M. Integration of remote sensing and in situ ecology for the design and evaluation of marine protected areas: Examples from tropical and temperate ecosystems (Chapter 13). In: *Remote Sensing of Protected Lands*, Wang, Y., Ed. CRC Press: Boca Raton, FL, **2012**, pp. 245–280.
24. Edgar, G.J.; Russ, G.R.; Babcock, R.C. Marine protected areas. In: *Marine Ecology*, Connell, S.D.; Gillanders, B.M. Eds. Oxford University Press: South Melbourne, **2007**, pp. 533–555.
25. Wedding, L.; Friedlander, A.M. Determining the influence of seascape structure on coral reef fishes in Hawaii using a geospatial approach. *Mar. Geod.* **2008**, *31*, 246–266.
26. Wedding, L.; Friedlander, A.; McGranaghan, M.; Yost, R.; Monaco, M.E. Using bathymetric Lidar to define nearshore benthic habitat complexity: Implications for management of reef fish assemblages in Hawaii. *Remote Sens. Environ.* **2008**, *112*, 4159–4165.
27. Friedlander, A.M.; Brown, E.K.; Monaco, M.E. Coupling ecology and GIS to evaluate efficacy of màine protected areas in Hawaii. *Ecol. Appl.* **2007**, *17*, 715–730.
28. Sheng, Y.; Alsdorf, D. Automated ortho-rectification of Amazon basin-wide SAR mosaics using SRTM DEM data. *IEEE Trans. Geosci. Remote Sens.* **2005**, *43* (8), 1929–1940.
29. Arima, E.Y.; Walker, R.T.; Sales, M.; Souza, Jr., C., Perz, S.G. The fragmentation of space in the Amazon basin: Emergent road networks. *Photogram. Eng. Remote Sens.* **2008**, *74* (6), 699–709.
30. Walsh, S.J.; Messina, J.P.; Brown, D.G. Mapping and modeling land use/land cover dynamics in frontier settings. *Photogram. Eng. Remote Sens.* **2008**, *74* (6), 677–679.
31. Mena, C.F. Trajectories of land-use and land-cover in the northern Ecuadorian Amazon: Temporal composition, spatial configuration, and probability of change. *Photogram. Eng. Remote Sens.* **2008**, *74* (6), 737–751.
32. Wang, C.; Qi, J.; Cochrane, M. Assessment of tropical forest degradation with canopy fractional cover from Landsat ETM+ and IKONOS imagery. *Earth Interact.* **2005**, *9* (22), 1–18.
33. Wang, C.; Qi J. Biophysical estimation in tropical forests using JERS-1 SAR and VNIR Imagery: II-aboveground woody biomass. *Int. J. Remote Sens.* **2008**, *29* (23), 6827–6849.
34. Sun, G.; Ranson, K.J.; Kharuk, V.I. Radiometric slope correction for forest biomass estimation from SAR data in Western Sayani mountains, Siberia. *Remote Sens. Environ.* **2002**, *79*, 279–287.
35. Bergen, K.M.; Zhao, T.; Kharuk, V.; Blam, Y.; Brown, D.G.; Peterson, L.K.; Miller, N. Changing regimes: Forested land cover dynamics in central Siberia 1974–2001. *Photogram. Eng. Remote Sens.* **2008**, *74* (6), 787–798.
36. Kharuk, V.I.; Ranson, K.J.; Im, S.T. Siberian silkmoth outbreak pattern analysis based on SPOT VEGETATION data. *Int. J. Remote Sens.* **2009**, *30* (9), 2377–2388; *Sens. Environ. 114*, 2897–2910.
37. Stow, D.A.; Hope, A.; McGuire, D.; Verbyla, D.; Gamon, J.; Huemmrich, F.; Houston, S.; Racine, C.; Sturm, M.; Tape, K.; Hinzman, L.; Yoshikawa, K.; Tweedie, C.; Noyle, B.; Silapaswan, C.; Douglas, D.; Griffith, B.; Jia, G.; Epstein, H.; Walker, D.; Daeschner, S.; Petersen, A.; Zhou, L.; Myneni, R. Remote sensing of vegetation and land-cover change in Arctic tundra ecosystems. *Remote Sens. Environ.* **2004**, *89* (3), 281–308.
38. Sheng, Y.; Shah, C.A.; Smith, L.C. Automated image registration for hydrologic change detection in the lake-rich arctic. *IEEE Geosci. Remote Sens. Lett.* **2008**, *5* (3), 414–418.
39. Sheng, Y.; Li, J. Satellite-observed endorheic lake dynamics across the Tibetan plateau between circa 1976 and 2000 (Chapter 15). In: *Remote Sensing of Protected Lands*, Wang, Y., Ed. CRC Press: Boca Raton, FL, **2012**, pp. 305–319.
40. Ranson, K.J.; Sun, G.; Kharuk, V.I.; Howl, J. Multisensor remote sensing of forest Dynamics in Central Siberia (Chapter 16). In: *Remote Sensing of Protected Lands*, Wang, Y., Ed. CRC Press: Boca Raton, FL, **2012**, pp. 321–377.

41. Sherman, N.J.; Loboda, T.V.; Sun, G.; Shugart, H.H. Remote sensing and modeling for assessment of complex Amur (Siberian) Tiger and Amur (Far Eastern) Leopard Habitats in the Russian Far East (Chapter 17). In: *Remote Sensing of Protected Lands*, Wang, Y., Ed. CRC Press: Boca Raton, FL, **2012**, pp. 379–407.
42. Thomas, C.; Gillingham, P.K. The performance of protected areas for biodiversity under climate change. *Biol. J. Linn. Soc.* **2015**, *115*, 718–730.
43. National Park Service. Management policies, **2006**, www.nps.gov/policy.
44. Hockings, M.; Stolton, S.; Leverington, F.; Dudley, N.; Courrau, J. *Evaluating Effectiveness: A Framework for Assessing Management Effectiveness of Protected Areas*, 2nd edition. IUCN: Gland and Cambridge, **2006**, pp. xiv+105.
45. Gaston, K.J.; Charman, K.; Jackson, S.F.; Armsworth, P.R.; Bonn, A.; Briers, R.A.; Callaghan, C.; Catchpole, R.; Hopkins, J.; Kunin, W.E.; Latham, J.; Opdam, P.; Stoneman, R.; Stroud, D.A.; Tratt, R. The ecological effectiveness of protecte areas: The United Kingdom. *Biol. Conserv.* **2006**, *132*, 76–87.
46. Fan, L.; Zhao, J.; Wang, Y.; Ren, Z.; Zhang, H.; Guo, X. Assessment of nighttime lighting for global terrestrial protected and wilderness areas. *Remote Sens.* **2019**, *11*, 2699.
47. Campbell, A.; Wang, Y. Salt marsh monitoring along the Mid-Atlantic coast by Google Earth Engine enabled time series. *PLOS One* **2020**, *15*(2), e0229605.

II

Genetic Resource and Land Capability

11

Genetic Diversity in Natural Resources Management

Thomas Joseph McGreevy
Boston University

Jeffrey A. Markert
Providence College

Introduction

What Is Genetic Diversity?

Genetic diversity is the amount of variation in the sequence of four nucleotides (adenine, thymine, cytosine, and guanine) that comprise deoxyribonucleic acid (DNA) within individuals, populations, or species. The particular combination of nucleotides within an individual is called a genotype.[1] From an information theoretic approach, population genetic diversity represents the amount of information stored in DNA within a population above and beyond the information common to all individuals.[2,3] Thus, genetic diversity represents the maximum amount of information available to code for distinct physical characteristics (phenotypes) within a group. Less information means fewer distinct phenotypes, decreasing the likelihood that some individuals in a population have characteristics suited to the next environmental challenge. These challenges may include things like new diseases, introduced predators, or altered physical environments. In diploid organisms, inbreeding depression-like effects also may reduce individual fitness when both copies of a segment of DNA within an individual are identical. The probability of this occurring is higher in populations with lower genetic diversity.

From a wildlife management perspective, we would expect that groups with lower levels of genetic diversity are at greater risk for extinction. Experimental studies support this expectation. For example, when genetic diversity is experimentally manipulated in isolated plant populations, those with lower genetic diversity are dramatically less fit.[4] Similarly, work using short-lived crustaceans as a model shows that even minor losses of genetic diversity greatly increase extinction risk when low-diversity populations are challenged with suboptimal environments.[5]

The importance of conserving genetic diversity is recognized by the International Union for Conservation of Nature (IUCN) as a major component of biodiversity.[1] In the United States, genetic diversity has been recognized as an important resource that should be conserved since the drafting of the Endangered Species Act (ESA) of 1973. Further rationale for why genetic diversity matters was explained by the House of Representatives in House Resolution 37, the precursor to the ESA of 1973:[6]

> "From the most narrow possible point of view, it is in the best interests of mankind to minimize the losses of genetic variations. The reason is simple: they are potential resources. They are keys to puzzles which we cannot yet solve, and may provide answers to questions which we have not yet learned to ask."

There are still important puzzles to solve and unasked questions in conservation genetics. In the past 40 years, researchers have developed and refined a strong set of genetic tools that address key aspects of conservation biology. These techniques have been used to inform management decisions for wild and captive populations since the 1980s, but the full utility of these tools is yet to be realized.[7] One major limitation has been the communication disconnect between research scientists and resource managers in some jurisdictions. The first goal of this entry is to help bridge this gap by providing an overview of the type of information that can be gained with molecular genetic techniques to inform a wide range of management decisions. The second goal is to discuss future directions that will reduce the cost and expand the utility of molecular genetic techniques in the management of natural resources.

How Can Genetic Techniques Inform Management Decisions?

The development of molecular genetic tools to assist management decisions has increased with the emergence of the field of conservation genetics in the early 1990s. Conservation geneticists are concerned with measuring genetic diversity within groups and determining how it is partitioned among groups. These techniques are used to address a number of important wildlife management issues, including the following:

Demographics and Extinction Risk

Molecular markers can be used to detect the presence of a species, estimate census and effective population sizes, determine sex ratios, and measure other key demographic parameters. Detecting the presence of a species is important to determine their geographic distribution and potential range decline. The concept of an effective population size was introduced by Wright[8] and in general terms is the number of individuals in an ideal population that would lose genetic diversity through sampling losses at the same rate as the observed population. It is generally smaller than the census size because not all individuals in a population contribute the same number of offspring to the next generation. It is critical to understand how quickly genetic diversity will be lost in a population, which in part is dictated by their effective population size.[1] Two important demographic parameters for managing a population are census size and density. Molecular genetic techniques can be used to identify individuals by determining their unique genotype and quantify a population size using traditional mark recapture methods, which also can be used to estimate the density of individuals per area. This technique can be used with non-invasively collected samples and replace the physical marking of individuals using unique markers, such as ear tags. Sex chromosome-specific markers can be used to determine sex ratios and model population dynamics.[7]

The demographic history of a population can impact its population dynamics and adaptive potential. When the size of a population is severely reduced, it is said to have undergone a bottleneck.[1] Similarly, founding events involve the establishment of a new population from a small number of individuals. Both bottlenecks and founding events result in populations with low genetic diversity and can impact their evolutionary trajectory because genetic information is lost randomly.[9] Lost genetic diversity in a population also can lead to inbreeding depression, which occurs when related or genetically similar animals mate and produce offspring. Inbreeding depression can manifest as reduced fitness of quantitative traits such as number of offspring produced or offspring survival rate.[10] All these forces can result in reduced adaptive potential and increased risk of extinction.[1]

Two well-known examples of the negative effects of inbreeding depression that resulted from a bottleneck occurred in the African cheetah (*Acinonyx jubatus*)[11] and the Florida panther (*Puma concolor coryi*).[12] The genetic diversity of African cheetahs was reduced during the Pleistocene because their population size was dramatically reduced over several generations. Their extremely low levels of genetic diversity are manifested in extant populations by their high infant mortality rates and sperm deformities. Due to the genetic similarity among individuals, cheetahs are unable to reject skin grafts from nominally unrelated individuals.[11]

The Florida panther endured two bottlenecks, one during the same time as the African cheetah and a second bottleneck during the 19th and early 20th century as a result of human activities.[11] By the early 1990s, the Florida panther population had declined to an estimated 20 to 25 individuals and had very low levels of genetic diversity. The animals showed many signs of inbreeding depression, including undescended testicles and low sperm quality, all of which suggested that extinction was very likely. Because genetic diversity takes a long time to increase by mutation, the only source of additional alleles (variable forms of a gene) was from a sister subspecies from Texas (P *c. stanleyana*). Females from this sister group were transported to Florida where they bred with Florida panthers with positive initial results. Genetic diversity and apparent fitness both increased; however, the long-term success of this genetic rescue remains to be demonstrated. Further, because the rescue was effected by introducing alien genes, and the new genomes may swamp those of the original subspecies, philosophical questions remain about precisely what has been conserved.[13]

Resolving Taxonomic Status and Determining Relevant Conservation Units

One of the top priorities of a conservation program should be to resolve any existing taxonomic uncertainties. Funding resources and personnel time that can be devoted to a given project are limited, and it is vital to ensure that the species status of a focus organism is well established. If a focal organism is really the same species as another organism that is not of conservation concern, then precious funds and resources will be wasted. This is fundamentally a question of understanding how genetic diversity is distributed among groups. The expectation is that distinct species will have higher levels of genetic diversity between the taxa than within each species because of the homogenizing effect of gene flow.

One example is the dusky seaside sparrow (*Ammodramus maritimus nigrescens*), which was initially one of nine recognized subspecies.[14] Genetic analyses, however, only supported the designation of two subspecies.[15,16] Conversely, if a species is not recognized as a separate entity, then a species could be lost or identified at a point where the number of organisms remaining precludes conserving the species. This was the case with the New Zealand tuatara, a lizard that was once widely distributed throughout New Zealand but became extinct on the mainland before the arrival of European settlers. Initially, it was managed as a single species. More recently, genetic and morphological analyses identified two species, *Sphenodon punctatus* and *S. guntheri*,[16,17] that are now managed as separate taxa because they do not currently exchange genetic material.

The identification of conservation units is used to determine the fundamental units that will be the focus of management efforts. Relevant conservation units include consideration of species boundaries as described in the preceding text, but it goes further by identifying subgroups that may be on distinct evolutionary trajectories from other populations within the species. One example is the mountain yellow-legged frog (*Rana muscosa*) that has undergone a rapid decline in recent decades due to predation by an introduced species, the virulent chytrid fungus (*Batrachochytrium dendrobatidis*), and habitat destruction related to wildfires. Currently, these frogs exist as isolated mountain populations. When once-widespread species are reduced to a collection of isolated populations, it is useful to understand whether these populations have always been isolated (and are therefore potentially adapted to local conditions) or whether they are merely relics of a previously widespread but uniform population. Genetic analyses determined that existing populations have long been isolated from each other, suggesting that they should be managed as separate units,[18] and this will need to be considered when frogs from captive breeding programs are reintroduced to the wild.

Determining Population Structure

An important part of identifying conservation units is testing for population structure, which involves the level of dispersal between geographically distinct areas. When widely distributed populations consist of several distinct genetic subunits, population structure is said to exist. It is also important to understand metapopulation dynamics (periodic local extinction and recolonization events) when they are present. The traditional approach is to group individuals into researcher-determined geographic units and measure the distribution of genetic diversity within and between these groupings.[19] Assignment test and clustering programs like STRUCTURE[20] are being used to determine whether cryptic population structure exists without any *a priori* groupings of individuals.

The presence of distinct population units has important consequences for wildlife management plans because a widely distributed species with distinct and isolated population segments might have little potential to recolonize habitat after a local population crash. Sometimes the presence of structured populations is indicated by obvious biogeographic barriers or the presence of distinct phenotypes. However, population structure often exists without these telltale signs. Cryptic population structure is rather common in nature, even in large, mobile animals such as the Eurasian lynx (*Lynx lynx*) in Scandinavia where genetic markers indicate that the species is divided into three distinct genetic units.[21]

Captive Breeding Programs

Molecular techniques can be used to inform management decisions during all phases of a captive breeding program. Captive-bred populations ideally should be representative of the genetic diversity found in wild populations. Genetic techniques can be used to define population structure in wild populations to help identify appropriate source populations, confirm the correct species identification for animals included in the captive breeding program, assist initial pairing decisions to ensure that closely related individuals are not mated, help determine appropriate release sites, and monitor the success of reintroduced animals by monitoring their population size and reproductive contributions. Genetic techniques also can be used several years after a captive breeding program is established to reconstruct pedigrees for organisms that live in groups[22] or determine the amount of genetic diversity that a captive population has retained from wild population.[23,24]

A combination of genetic and morphological analyses were used to discover that the Association of Zoos and Aquarium's Asiatic lion *(Panthera leo persica)* captive breeding program included African lion (P. leo leo) *founders*.[25] Asiatic lions have low levels of genetic diversity, so the interbreeding of a closely related subspecies increased their genetic diversity. However, the crossing of two subspecies could reduce adaptations that may have developed in each taxon. The previous example of interbreeding Texas puma with Florida panther may have reduced some of their adaptations to their environment—but if nothing had been done, extinction was the likely outcome.

Managing Invasive Species

Invasive species can have a tremendous negative impact on biodiversity.[1] Molecular genetic techniques can be used to identify the source population of an invasive species, monitor potential hybridization with closely related taxa, determine the geographic distribution of an invasive species, and monitor their population dynamics. Genetic techniques also can be used to predict whether an invasive species will have a good chance at becoming established and resisting control agents. Sometimes, invasive species have more genetic diversity than the suspected source populaiton, suggesting that there may have been multiple source populations. The higher amount of genetic diversity created by interbreeding multiple source populations in turn could increase the adaptive potential of the invasive species to resist strong selective pressures, such as herbicides and pesticides.[26]

Wildlife Forensics

The field of wildlife forensics has used molecular genetic techniques to identify the species of origin of unknown samples, identify the gender of an individual, and determine the geographic location from which an organism was obtained.[26] These techniques have been applied to a wide range of animals that include cetaceans,[27] sharks,[28] elephants,[29] and sturgeon.[30] DNA sequence information can be used like grocery store barcodes to match an unknown sample to a database of known samples to determine the species identity. If the population structure of an organism is known, then the most likely geographic source of the organism can be determined by identifying the source population with the most similar genotype to the organism's genotype of interest, and this information could be used to determine if an animal was taken from a protected population.

Crop and Livestock Management

Genetic diversity is a key aspect of food security. The industrial-scale food production necessary to feed the increasing human population requires a high level of standardization and automation, leading to a focus on a relatively few varieties of crops and animals.[31] While standardization and mass production are important in feeding the world's population, this strategy is not without risk. When a single variety dominates the food production system in a region, there is a risk of total crop failure when something goes wrong.[32] In low-diversity varieties, all individuals are likely equally vulnerable to diseases or environmental challenges.

A classic manifestation of this risk is the southern U.S. corn blight that occurred in 1970. Commercial corn seeds are produced by crossing different inbred lines of corn. One of these lines was used in seed corn that was widely distributed in the 1960s, which initially contributed to robust and healthy corn crops. However, parasites with short generation times can often out-evolve their relatively long-lived hosts. One such parasite is the fungus *Helmithosporium maydis,* which was a minor agricultural pest until 1970. It is believed that natural mutation created a more virulent form of the fungus which, when combined with the warm, wet summer weather, reduced U.S. corn yields by perhaps 15% that year.[33]

Strategies to mitigate these effects in the long term include seed banks and the preservation of heirloom varieties and landraces.[34] Genetic and genomic technologies are useful in mitigating these risks. For example, molecular markers have been used to determine that Iberian landraces of pigs (*Sus scrofa domesticus*) represent important reservoirs of genetic diversity.[35]

Genetic Ecotoxicology

The field of genetic ecotoxicology studies the impact of contaminants on an organism's DNA.[36] Pollutants can increase the mutation rate of DNA and increase the amount of genetic diversity in a population. However, most mutations that affect fitness will be harmful.[26] The genetic diversity of populations from control and contaminated sites can be compared to quantify the impact of a particular contaminant.[37,38] Ellegren et al.[39] compared the occurrence of a mutation that causes partial albinism in barn swallows (*Hirundo rustica*) from a site near the nuclear accident at Chernobyl to barn swallows from uncontaminated sites. They found a higher incidence of partial albinism in barn swallows that bred near Chernobyl compared to their control sites. Quantifying the genetic diversity of a population that was exposed to a pollutant can be used to predict the potential negative impacts of the pollutant on the viability of the population.

How Is Genetic Diversity Measured and Interpreted?

DeYoung and Honeycutt[38] provided an overview of the different types of molecular genetic tools or molecular markers that can be used to generate genetic data. These molecular markers include 1) expressed proteins; 2) Restriction Fragment Length Polymorphisms (RFLPs); 3) Random Amplified Polymorphic DNA (RAPDs); 4) Amplified Fragment Length Polymorphisms (AFLPs[40]); 5) nucleotide sequence data from nuclear, mitochondrial, or chloroplast DNA; 6) Simple Sequence Repeats (SSRs); and 7) Single-Nucleotide Polymorphisms (SNPs).

The first three methods are now obsolete.[41] While they yielded important data in early studies, they are expensive per individual, are difficult to automate, and require considerable skill to generate reproducible results. AFLPs, SSRs, and sequencing are all still widely used in conservation genetics. The AFLP technique uses restriction enzymes to break the genome into small fragments and a modified version of the polymerase chain reaction (PCR) to visualize a subset of these fragments. As with RFLPs, pairs of homologous restriction sites will generate identically sized fragments. The PCR primers used contain fluorescent labels that allow fragment sizes to be measured using high-resolution DNA capillary electrophoresis. A high degree of automation increases the objectivity of the analysis.

Similarly, genotypes determined using microsatellite loci (also known as SSR loci, Single Tandem Repeat loci, Variable Number of Tandem Repeat loci, and other names) are quite reproducible and easy to automate. SSR loci are repeated nucleotide motifs, typically one to six nucleotides that are repeated dozens of times. In the nuclear genome, they are usually about 75 to 300 base pairs long. These short motifs are repeated, and mutations occur frequently, manifesting themselves as different numbers of motif repeats that are called alleles.[26] Microsatellite markers quickly gained in popularity in the mid-1990s, and they are one of the most widely used markers in conservation genetics.[41] A key advantage of this technique is that when combined with high-resolution electrophoresis, the data are highly reproducible and the scoring of genotypes is quite objective. Traditionally, the cost of developing a set of microsatellite loci and testing their reliability for a new species has been quite high; however, next-generation DNA sequencing methods (also known as massively parallel or deep sequencing methods) have made locus discovery more affordable.[42]

Direct sequencing is perhaps the most objective and reproducible method for measuring genetic diversity. However, the level of diversity in a short region is generally low; so long stretches of DNA must be sequenced to provide useful information. The cost of determining sequences has historically been high relative to the other markers. One popular strategy is to sequence a small number of individuals for a large number of base pairs to identify positions where more than one nucleotide is variable within a population. These SNPs are single base pair differences either between homologous chromosomes within an individual or between individuals. Once these positions are identified, PCR-based techniques may be used to assay these SNPs without having to resequence long stretches of non-variable DNA. The identification of SNPs using PCR or other assay techniques is labor intensive. However, the reduced cost of next-generation DNA sequencing methods suggests in the future that broad and representative sections of the genomes of many individuals may be sequenced at a reasonable cost, thus collapsing the processes of locus discovery and population assays.[43] In the near future, SNP markers will likely replace microsatellite markers as the most common type of markers used in conservation genetic studies, and these will be derived from next-generation DNA sequences.

Next-generation DNA sequencing is only the latest innovation to reduce the cost of generating sequencing data, which has been steadily falling since the development of Sanger sequencing in the 1970s.[44] New sequencing technologies in the field of human genomics are the main drivers for the reduction in cost for generating enormous amounts of nucleotide sequence data.[45] Sequence information from partial or entire genomes can be generated for multiple individuals. Miller et al.[46] developed a restricted site-associated DNA (RAD) marker method that uses restriction enzymes to digest genomic DNA and determine the presence or absence of a restriction enzyme's recognition site across an entire genome. This method was substantially improved by Baird et al.[47] by adding adapters to the fragmented pieces of DNA and using next-generation DNA sequencing to determine the nucleotides in each fragmented piece of DNA. This RAD- sequenced (RAD-seq) method can generate tens of thousands of SNP markers. An initial limitation of next-generation methods has been the requirement of substantial genomic resources, such as a genome map, to fully use these methods. Peterson et al.[48] have greatly improved the RAD-seq method by developing a cost-effective double-digest RAD-seq method for identifying tens to hundreds of thousands of SNP markers in non-model organisms that have no previous genomic resources available. We expect that going forward, these partial genome sequences or RAD-seq markers will be widely used to directly assay both SNP and microsatellite variation in population studies.

Conservation Genomics

Genomics is a field concerned with the study of an organism's whole genome (or at least substantial portions of it). The major components of genomics are data acquisition, bioinformatics, and archiving data in databases.[49] The field of genetics in general is rapidly expanding with the advances in technology. As a consequence, conservation genetics is transitioning into the field of conservation genomics.[41,50–52] Advances in human genomics are greatly reducing the cost of generating genomic data. A challenge grant was initiated in 2003 by the J. Craig Venter Science Foundation to generate an entire human genome for $1,000.[53] Once this goal is achieved, the whole genome analysis of wild populations (i.e., population genomics) will become routine. This transition represents a shift from analyzing a limited number of genes to a more comprehensive survey of the entire genome of an organism. The conceptual difference between a conservation genetic and genomic approach is depicted by Ouborg et al.[41] One main difference is that a genomic approach can quantify the impact of selection and include gene expression. One of the major contributors to the founding of the field of conservation genetics and its transformation into conservation genomics is Dr. Stephen J. O'Brien, former Chief of the Genomic Diversity Laboratory at the National Cancer Institute. O'Brien was directly involved with applying the latest developments in human genomics to endangered species[25] and began an internationally recognized annual workshop focused on conservation genetics in 1996. In 2010, the American Genetics Association held a conference on integrating genomics into the field of conservation genetics.[54]

Conservation genetics is mainly interested in characterizing levels of quantitative variation or how individuals differ in traits that are important for reproductive fitness. A major assumption of conservation genetics is that neutral genetic variation correlates with levels of quantitative genetic variation, and if neutral genetic variation is preserved, then an organism's adaptive potential is retained. However, the correlation between neutral and quantitative genetic variation is low.[55] The development of genomic tools has exponentially increased the number of genetic markers that can be produced from typically dozens to more than tens of thousands and has catapulted the field of conservation genetics beyond just neutral markers to direct measures of quantitative genetic variation. The enormous increase in the number of markers will greatly improve the conclusions made from genetic analyses by providing more comprehensive measures of genetic diversity across the entire genome of a given species.[52] This also has allowed for the direct measurement of detrimental and adaptive genetic variation[51] and the development of adaptive potential metrics.[56]

The transition of conservation genetics to conservation genomics represents a major shift in the types of questions that can be addressed and how the questions can be addressed. Conservation genetic studies have mainly been limited to correlative studies. The development of genomic tools has allowed for a shift to a causative approach that can directly investigate evolutionary processes.[41] Genomic tools can be used to investigate gene expression using transcriptomics,[57] rather than just measuring nucleotide variation. Transcriptomics is a field of study that investigates genes that are actually expressed in an organism. Ouborg et al.[41] identified four main questions that can be addressed using genomic tools that involve: 1) measuring the impact of small population size on levels of adaptive variation; 2) identifying the underlying mechanisms that connect adaptive genetic variation and fitness; 3) determining how genes and the environment interact; and 4) estimating the impact of inbreeding and genetic drift on gene expression. Genomic tools will greatly improve the quantity and quality of information gained from genetic analyses; however, genomic tools will not be a panacea. Identifying the genetic mechanisms of adaptation is a difficult endeavor and will require a multidisciplinary approach.[50]

Genomic techniques also have so far been restricted to the analysis of relatively high-quality DNA, which will limit their application in cases where DNA is obtained from non-invasively collected hair and scat samples or specimens in museum collections. However, recently developed whole-genome amplification techniques are being used to analyze low quantities of extracted DNA in the field of human forensics. Giardina et al.[58] tested the efficacy of using whole-genome amplification on low quantities of extracted DNA to produce accurate SNP marker genotypes.

The use of genomic tools to inform management decisions is beginning to be used in fishery management.[59–61] Neutral genetic markers have been the main tool used to identify conservation units; however, with the advent of genomic tools, adaptive genetic variation can now be included in these analyses.[62] For example, Tymchuk et al.[60] used microsatellite markers and gene expression assays (i.e., transcriptomics) to identify potential adaptive differences in Atlantic salmon *(Salmo salar)* from different rivers in the Bay of Fundy. Hansen[60] advocates using Tymchuk et al.'s[59] approach to identify conservation units. Rusello et al.[61] used a genome scan of over 10,000 nuclear markers to demonstrate the efficacy of using adaptive loci (i.e., outlier loci) to identify ecologically divergent populations of Okanagan Lake kokanee, landlocked sockeye salmon *(Oncorhynchus nerka)*. These researchers have identified eight potentially adaptive loci that identify the two kokanee ecotypes more reliably than 44 neutral markers. Miller et al.[46] used 4590 SNP markers to quantify potentially adaptive differences in rainbow trout *(O. mykiss)* and recommend the use of adaptive genetic variation to identify conservation units.

Spatial Analyses of Genetic Diversity

The management of wild populations is inherently a spatial endeavor that is directly connected to an organism's environment. The spatial analysis of genetic diversity has increased with the development of the field of landscape genetics. Landscape genetics is a nascent field that combines two other fields, population genetics and landscape ecology, to study the processes that influence the spatial patterns of an organism's genetic diversity in the environment.[63] Landscape genetic studies can be used to detect discontinuities in an organism's spatial pattern of genetic diversity and correlate these gene flow discontinuities to environmental variables.[64] The spatial analyses of genetic diversity patterns also can be used to identify areas that should be designated as reserves. Vandergast et al.[65] spatially analyzed the pattern of genetic diversity of multiple taxa to identify areas that harbor a large amount of genetic discontinuities relative to other areas. These "hotspots" of evolutionary potential can then be the focus of management efforts to conserve biodiversity.

Understanding and mitigating the effects of habitat fragmentation are among the chief concerns of applied conservation genetics. Humans continue to impact the environment by increasing the amount of fragmentation. A landscape genetics approach can be used to quantify the impact of fragmentation on the movement of an organism's genes. One example is the Jerusalem cricket species *Stenopelmatus* n. sp.

"*santa monica*" described by Vandergast et al.[66] from Los Angeles and Ventura Counties in southern California. These large, wingless insects are thought to be poor dispersers, and current taxonomy suggests that the family Stenopelmatidae is particularly species rich. Most human development in southern California has occurred relatively recently, and the origin dates of major settlements and highways are well documented. Using both mitochondrial DNA sequences and nuclear sequences between microsatellite loci as genetic markers, the authors identified highway presence and age as a critical contributor to levels of population isolation above and beyond that predicted based on geographic distance alone. It is intuitive to expect that small, earthbound creatures will find it difficult to cross a multilane freeway, but why should the age of the freeways be a factor? The answer is genetic drift. When populations are small, random factors cause allele and haplotype frequencies to vary. When a patch of habitat is cut in half by a barrier, each new subpopulation is more susceptible to the effects of genetic drift than the previous, larger population. Because the new parts of a bisected population are likely to drift in different directions, the genetic distance between them will increase over time. Newly isolated subpopulations, however, will be similar until the effects of genetic drift have manifested themselves, and so populations on either side of a new highway will be similar at first but then diverge over time. The same is true for lost genetic diversity, as new population fragments may be highly diverse but eventually lose allelic diversity over time through drift and selection. Another key aspect of genetic diversity and divergence is that they may reflect divergent selection between distinct types of habitat. When local adaptation occurs, it may be important to preserve it.

The field of landscape genetics is at the cusp of rapidly transforming into the field of landscape genomics, similar to the transformation that is occurring in the field of conservation genetics to conservation genomics. The advantages of this transformation are similar to the ones discussed for the transformation from conservation genetics to conservation genomics. Landscape genetic studies have been focused on the patterns of a limited number of neutral molecular markers. The expansion to landscape genomics will allow researchers to analyze thousands of neutral markers and include adaptive loci.[67] Joost et al.[68] developed a method for identifying adaptive loci in a wild population by using a landscape genomics approach. Landscape genomic studies have the potential to identify historic processes that have influenced the currently observed patterns of adaptive genetic variation in wild populations and predict how future processes, such as climate change, will influence a population's viability.

Examples of applied landscape genomic studies are limited. Several studies have described their approach as a landscape genomics approach, but they have only used a limited number of AFLPs,[69] microsatellites,[70] or SNPs[71] and should still be described as landscape genetic studies. The best example of a landscape genomics study is by Poncet et al.[72] This group identified potentially adaptive loci in the alpine plant *Arabis alpina* by correlating their AFLP genotypes to environmental variables. This allows management efforts to focus on preserving *A. alpina* adaptive genetic variation. This study included 825 AFLP markers, a significant improvement upon the typical number or AFLP markers; however, using this number of markers in such a study will soon become outdated with the near eventuality that the analysis of tens to hundreds of thousands of markers in a study will be affordable.

Conclusion

Conservation genetics is a young field;[52] however, genetic techniques already have been used to inform a wide range of management decisions. For the field to reach its full potential, conservation geneticists need to continue to educate resource managers about the important contributions that genetic analyses can make to improve the efficacy of management decisions. Collaborations should be established at the initial stage in a project between research scientists and resource managers to bridge the communication barrier and make full use of the genetic toolbox.

The quality and quantity of genetic information are expanding exponentially with the transition of conservation genetics into conservation genomics. The increased amount of information will significantly increase the accuracy and precision of conclusions made using genomic tools. The transition to a genomics approach also will expand the questions that can be addressed and fundamentally change the approach that is used to answer questions. Genomic tools will allow the direct quantification of a population's adaptive potential and transcend the field beyond a correlative field to one that uses a causative approach to discover ultimate rather than proximate causes of natural processes.

Acknowledgments

We thank Matthew DeBlois, Kelsey Garlick, and the reviewers for their comments. McGreevy is supported by the National Science Foundation under Award Number 1003226. Any opinions, findings, and conclusions or recommendations expressed in this material are those of the authors and do not necessarily reflect the views of the National Science Foundation.

References

1. Frankham, R.; Ballou, J.D.; Briscoe, D.A. *Introduction to Conservation Genetics*, 2nd Ed.; Cambridge University Press: New York, 2010.
2. Lewontin, R.C. The apportionment of human diversity. Evol. Biol. **1972,** *6* (381), 391–398.
3. Sherwin, W.B.; Jabot, F.; Rush, R.; Rossetto, M. Measurement of biological information with applications from genes to landscapes. Mol. Ecol. **2006,** *15* (10), 2857–2869.
4. Newman, D.; Tallmon, D.A. Experimental evidence for beneficial fitness effects of gene flow in recently isolated populations. Conserv. Biol. **2001,** *15,* 1054–1063.
5. Markert, J.A.; Champlin, D.M.; Gutjahr-Gobell, R.; Grear, J.S.; Kuhn, A.M., McGreevy, T.J.; Roth, A.; Bagley, M.J.; Nacci, D.E. Population genetic diversity and fitness in multiple environments. BMC Evol. Biol. **2010,** *10* (205), http://www.biomedcentral.com/1471-2148/1410/1205
6. Waples, R.S. *Definition of "species" under the endangered species act: Application to pacific salmon. NOAA Technical Memorandum NMFS F/NWC-194.* 1991.
7. Schwartz, M.K.; Luikart, G.; Waples, R.S. Genetic monitoring as a promising tool for conservation and management. Trends Ecol. Evol. **2007,** *22* (1), 25–33.
8. Wright, S. Evolution in Mendelian populations. Genetics. **1931,** *16* (2), 97–159.
9. Kolbe, J.J.; Leal, M.; Schoener, T.W.; Spiller, D.A.; Losos, J.B. Founder effects persist despite adaptive differentiation: a field experiment with lizards. Science **2012,** *335* (6072), 1086–1089.
10. Ralls, K.; Ballou, J.D.; Templeton, A. Estimates of lethal equivalents and the cost of inbreeding in mammals. Con- serv. Biol. 1988, *2* (2), 185–193.
11. O'Brien, S.J.; Johnson, W.E. Big cat genomics. In *Annual Review of Genomics and Human Genetics;* Chakravarti, A., Green, E., Eds.; Annual Review, Inc.: Palo Alto, California, 2005; pp. 407–429.
12. Johnson, W.E.; Onorato, D.P.; Roelke, M.E.; Land, E.D.; Cunningham, M.; Belden, R.C.; McBride, R.; Jansen, D.; Lotz, M.; Shindle, D.; Howard, J.; Wildt, D.E.; Penfold, L.M.; Hostetler, J.A.; Oli, M.K.; O'Brien, S.J. Genetic restoration of the Florida panther. Science **2010,** *329* (5999), 1641–1645.
13. Hedrick, P. Genetic future for Florida panthers. Science **2010,** *330* (6012), 1744; author reply 1744.
14. Avise, J.C. Toward a regional conservation genetics perspective: phylogeography of faunas in the southeastern United States, In *Conservation Genetics Case Histories from Nature;* Avise, J., Hamrick, J., Eds.; Chapman and Hall: New York, New York, 1996; pp. 431–470.
15. Avise, J.C.; Nelson, W.S. Molecular genetic relationships of the extinct dusky seaside sparrow. Science **1989,** *243* (4891), 646–648.
16. Mills, L.S. *Conservation of Wildlife Populations*; Blackwell Publishing: Malden, Massachusetts, 2007.
17. Daugherty, C.H.; Cree, A.; Hay, J.M.; Thompson, M.B. Neglected taxonomy and the continuing extinctions of tua- tara *(Sphenodon).* Nature **1990,** *347* (6289), 177–179.

18. Schoville, S.D.; Tustall, T.S.; Vredenburg, V.T.; Backin, A.R.; Gallegos, E.; Wood, D.A.; Fiser, R.N. Conservation genetics of evolutionary lineages of the endangered mountain yellow-legged frog, *Rana muscosa* (Amphibia: Ranidae) in southern California. Biol. Conserv. **2011,** *144* (7), 2031–2040.
19. Wright, S. *Evolution and the Genetics of Populations, Volume 4: Variability within and among Natural Populations;* University of Chicago Press: Chicago, 1978.
20. Pritchard, J.K.; Stephens, M.; Donnelly, P. Inference of population structure using multilocus genotype data. Genetics **2000,** *155* (2), 945–959.
21. Rueness, E.K.; Jorde, P.E.; Hellborg, L.; Stenseth, N.C.; Ellegren, H.; Jakobsen, K.S. Cryptic population structure in a large, mobile mammalian predator: The Scandinavian lynx. Mol. Ecol. **2003,** *12* (10): 2623–2633.
22. Ivy, J.A.; Lacy, R.C. Using molecular methods to improve the genetic management of captive breeding programs for threatened species, In *Molecular Approaches in Natural Resource Conservation and Management*; DeWoody, J.A., Bickham, J.W., Michler, C.H., Eds.; Cambridge University Press: New York, 2010; pp. 267–295.
23. McGreevy, T.J.; Dabek, L.; Gomez-Chiarri, M.; Husband, T.P. Genetic diversity in captive and wild Matschie's tree kangaroo (*Dendrolagus matschiei*) from Huon Peninsula, Papua New Guinea, based on mtDNA control region sequences. Zoo Biol. **2009,** *28* (3), 183–196.
24. McGreevy, T.J.; Dabek, L.; Husband, T.P. Genetic evaluation of the association of zoos and aquariums Matschie's tree kangaroo (*Dendrolagus matschiei*) captive breeding program. Zoo Biol. **2011,** *30* (6), 636–646.
25. O'Brien, S.J.; Wienberg, J.; Lyons, L.A. Comparative genomics: Lessons from cats. Trends Genet. **1997,** *13* (10), 393–399.
26. Allendorf, F.W.; Luikart, G. *Conservation and the Genetics of Populations*; Blackwell Publishing: Malden, Massachusetts, 2007.
27. Baker, C.S.; Cipriano, F.; Palumbi, S.R. Molecular genetic identification of whale and dolphin products from commercial markets in Korea and Japan. Mol. Ecol. **1996,** *5* (5), 671–685.
28. Magnussen, J.E.; Pikitch, E.K.; Clarke, S.C.; Nicholson, C.; Hoelzel, A.R.; Shivji, M.S. Genetic tracking of basking shark products in international trade. Anim. Conserv. **2007,** *10* (2), 199–207.
29. Wasser, S.K.; Joseph Clark, W.; Drori, O.; Stephen Kisamo, E.; Mailand, C.; Mutayoba, B.; Stephens, M. Combating the illegal trade in African elephant ivory with DNA forensics. Conserv. Biol. **2008,** *22* (4), 1065–1071.
30. Birstein, V.J.; Desalle, R.; Doukakis, P.; Hanner, R.; Ruban, G.I.; Wong, E. Testing taxonomic boundaries and the limit of DNA barcoding in the Siberian sturgeon, *Acipenser baerii.* Mitochondrial DNA. **2009,** *20* (5–6), 110–118.
31. Groeneveld, L.F.; Lensstra, J.A.; Eding, H.; Toro, M.A.; Scherf, B.; Pilling, D.; Negrini, R.; Finlay, E.K.; Jianlin, H.; Groeneveld, E.; Weigend, S.; Consortium, G. Genetic diversity in farm animals - A review. Anim. Genet. **2010,** *41* (Suppl. 1), 6–31.
32. Esquinas-Alcazar, J. Science and society: Protecting crop genetic diversity for food security: Political, ethical and technical challenges. Nat. Rev. Genet. **2005,** *6* (12), 946–953.
33. NRC. *Genetic Vulnerability of Major Crops,* N.R.C.N.A.O. Sciences; Ed.; Washington, D.C, 1972.
34. Food and Agriculture Organization of the United Nations. Report of the commission on genetic resources for food and agriculture. CGRFA-12/09/Report. 2009, ftp://ftp.fao.org/docrep/fao/meeting/017/k6536e.pdf.
35. Vicente, A.A.; Carolino, M.I.; Sousa, M.C.; Ginja, C.; Silva, F.S.; Martinez, A.M.; Vega-Pla, J.L.; Carolino, N.; Gama, L.T. Genetic diversity in native and commercial breeds of pigs in Portugal assessed by microsatellites. J. Anim. Sci. **2008,** *86* (10), 2496–2507.
36. Shugart, L.R.; Theodorakis, C.W.; Bickham, J.W. Evolutionary toxicology, In *Molecular Approaches in Natural Resource Conservation and Management,* DeWoody, J.A., Bickham, J.W., Michler, C.H., Eds.; Cambridge University Press: New York, 2010; pp. 320–362.

37. Bickham, J.W.; Sandhu, S.; Hebert, P.D.; Chikhi, L.; Athwal, R. Effects of chemical contaminants on genetic diversity in natural populations: implications for biomonitoring and ecotoxicology. Mutat. Res. **2000,** *463* (1), 33–51.
38. DeYoung, R.W.; Honeycutt, R.L. The molecular toolbox: genetic techniques in wildlife ecology and management. J. Wildlife Manag. **2005,** *69* (4), 1362–1384.
39. Ellegren, H.; Lindgren, G.; Primmer, C.R.; Moller, A.P. Fitness loss and germline mutations in barn swallows breeding in Chernobyl. Nature **1997,** *389* (6651), 593–596.
40. Vos, P.; Hogers, R.; Bleeker, M.; Reijans, M.; van de Lee, T.; Hornes, M.; Frijters, A.; Pot, J.; Peleman, J.; Kuiper, M.; Zabeau, M. AFLP: A new technique for DNA fingerprinting. Nucleic Acids Res. **1995,** *23* (21), 4407–4414.
41. Ouborg, N.J.; Pertoldi, C.; Loeschcke, V.; Bijlsma, R.K.; Hedrick, P.W. Conservation genetics in transition to conservation genomics. Trends Genet. **2010,** *26* (4), 177–187.
42. Davey, J.W.; Hohenlohe, P.A.; Etter, P.D.; Boone, J.Q.; Catchen, J.M.; Blaxter, M.L. Genome-wide genetic marker discovery and genotyping using next-generation sequencing. Nat. Rev. Genet. **2011,** *12* (7), 499–510.
43. Glenn, T.C. Field guide to next-generation DNA sequencers. Mol. Ecol. Resour. **2011,** *11* (5), 759–769.
44. Sanger, F.; Coulson, A.R. A rapid method for determining sequences in DNA by primed synthesis with DNA polymerase. J. Mol. Biol. **1975,** *94* (3), 441–446.
45. Pareek, C.S.; Smoczynski, R.; Tretyn, A. Sequencing technologies and genome sequencing. J. Appl. Genet. **2011,** *52* (4), 413–35.
46. Miller, M.R.; Brunelli, J.P.; Wheeler, P.A.; Liu, S.; Rexroad, C.E., 3rd; Palti, Y.; Doe, C.Q.; Thorgaard, G.H. A conserved haplotype controls parallel adaptation in geographically distant salmonid populations. Mol. Ecol. **2012,** *21* (2), 237–249.
47. Baird, N.A.; Etter, P.D.; Atwood, T.S.; Currey, M.C.; Shiver, A.L.; Lewis, Z.A.; Selker, E.U.; Cresko, W.A.; Johnson, E.A. Rapid SNP discovery and genetic mapping using sequenced RAD markers. PLoS One **2008,** *3* (10), e3376.
48. Peterson, B.K.; Weber, J.N.; Kay, E.H.; Fisher, H.S.; Hoekstra, H.E., Double digest RADseq: an inexpensive method for de novo SNP discovery and genotyping in model and non-model species. PLoS One **2012,** *7* (5), e37135.
49. Amato, G.; DeSalle, R. *Conservomics? The Role of Genomics in Conservation Biology. Conservation Genetics in the Age of Genomics, New Directions in Biodiversity Conservation;* Columbia University Press: New York, 2009; 169–178.
50. Allendorf, F.W.; Hohenlohe, P.A.; Luikart, G. Genomics and the future of conservation genetics. Nat. Rev. Genet. **2010,** *11* (10), 697–709.
51. Kohn, M.H.; Murphy, W.J.; Ostrander, E.A.; Wayne, R.K. Genomics and conservation genetics. Trends Ecol. Evol. **2006,** *21* (11), 629–637.
52. Primmer, C.R. From conservation genetics to conservation genomics. In *The Year in Ecology and Conservation Biology, Annals of the New York Academies of Sciences*, Ostfeld, R.S., Schlesinger, W.H. Eds.; Wiley-Blackwell: Hoboken, New Jersey, 2009; pp. 357–368.
53. Service, R.F., Gene sequencing. The race for the $1000 genome. Science **2006,** *311* (5767), 1544–1546.
54. AGA, American Genetics Association Annual Symposium, 2010.
55. Reed, D.H.; Frankham, R. How closely correlated are molecular and quantitative measures of genetic variation? A meta-analysis. Evolution **2001,** *55* (6), 1095–1103.
56. Bonin, A.; Nicole, F.; Pompanon, F.; Miaud, C.; Taberlet, P. Population adaptive index: a new method to help measure intraspecific genetic diversity and prioritize populations for conservation. Conserv. Biol. **2007,** *21* (3), 697–708.
57. Anonymous. Proteomics, transcriptomics: What's in a name? Nature **1999,** *402* (6763), 715.

58. Giardina, E.; Pietrangeli, I.; Martone, C.; Zampatti, S.; Marsala, P.; Gabriele, L.; Ricci, O.; Solla, G.; Asili, P.; Arcudi, G.; Spinella, A.; Novelli, G. Whole genome amplification and real-time PCR in forensic casework. BMC Genomics **2009,** *10,* 159.
59. Tymchuk, W.; O'Reilly, P.; Bittman, J.; Macdonald, D.; Schulte, P. Conservation genomics of Atlantic salmon: variation in gene expression between and within regions of the Bay of Fundy. Mol. Ecol. **2010,** *19* (9), 1842–1859.
60. Hansen, M.M. Expression of interest: transcriptomics and the designation of conservation units. Mol. Ecol. **2010,** *19* (9), 1757–1759.
61. Rusello, M.A.; Kirk, S.L.; Frazer, K.K.; Askey, PJ. Detection of outlier loci and their utility for fisheries management. Evol. Appl. **2011,** *5* (1), 39–52.
62. Nosil, P.; Funk, D.J.; Ortiz-Barrientos, D. Divergent selection and heterogeneous genomic divergence. Mol. Ecol. **2009,** *18* (3), 375–402.
63. Manel, S.; Schwartz, M.K.; Luikart, G.; Taberlet, P. Landscape genetics: combining landscape ecology and populai- ton genetics. Trends Ecol. Evol. **2003,** *18* (4), 189–197.
64. McKelvey, K.S.; Cushman, S.A.; Schwartz, M.K. Landscape genetics. In *Spatial Complexity, Informatics, and Wildlife Conservation,* Cushman, S.A., Huettmann, F., Eds.; Springer: New York, 2010; pp. 313–328.
65. Vandergast, A.G.; Bohonak, A.J.; Hathaway, S.A.; Boys, J.; Fisher, R.N. Are hotspots of evolutionary potential adequately protected in southern California? Biol. Conserv. **2008,** *141* (6), 1648–1664.
66. Vandergast, A.G.; Lewallen, E.A.; Deas, J.; Bohanak, A.J.; Weissman, D.B.; Fisher, R.N. Loss of genetic connectivity and diversity in urban microreserves in a southern California endemic Jerusalem cricket (Orthoptera: Senopelmati- dae: *Stenopelmatus* n. sp. "santa monica"). J. Insect Conserv.*13* (3), 329–345.
67. Schwartz, M.K.; Luikart, G.; McKelvey, K.S.; Cushman, S.A. Landscape genomics: A brief perspective. In *Spatial Complexity, Informatics, and Wildlife Conservation*, Cushman, S.A., Huettmann, F., Eds.; Springer: New York, 2010; pp. 165–174.
68. Joost, S.; Kalbermatten, M.; Bonin, A. Spatial analysis method (sam): a software tool combining molecular and environmental data to identify candidate loci for selection. Mol. Ecol. Resour. **2008,** *8* (5), 957–960.
69. Parisod, C.P.; Christin, P-A. Genome-wide association to fine-scale ecological heterogeneity within a continuous population of *Biscutella laevigata* (Brassicaceae). New Phytol. **2008,** *178* (2), 436–447.
70. Meier, K.; Hansen, M.M.; Bekkevold, D.; Skaala, O.; Mens- berg, K-LD. An assessment of the spatial scale of local adaptation in brown trout (*Salmo trutta* L.): Footprints of selection at microsatellite DNA loci. Heredity **2011,** *106* (3), 488–499.
71. Pariset, L.; Joost, S.; Marsan, P.A.; Valentini, A.; Consortium, E. Landscape genomics and biased Fst approaches reveal single nucleotide polymorphisms under selection in goat breeds of North-East Mediterranean. BMC Genet. **2009,** *10* (7), 1–8.
72. Poncet, B.N.; Herrmann, D.; Gugerli, F.; Taberlet, P.; Holderegger, R.; Gielly, L.; Rioux, D.; Thuiller, W.; Aubert, S.; Manel, S. Tracking genes of ecological relevance using a genome scan in two independent regional population samples of *Arabis alpina*. Mol. Ecol. **2010,** *19* (14), 2896–2907.

Bibliography

Bertorelle, G.; Bruford, M.W.; Hauffe, H.C.; Rizzoli, A.; Vernesi, C. *Population Genetics for Animal Conservation, Conservation Biology 17;* Cambridge University Press: New York, 2009.

Cassidy, B.G.; Gonzales, R.A. DNA testing in animal forensics. J. Wildl. Manage. **2005,** *69* (4), 1454–1462.

Leberg, P. Genetic approaches for estimating the effective size of populations. J. Widl. Manage. **2005,** *69* (4), 1385–1399.

Scribner, K.T.; Blanchong, J.A.; Bruggeman, D.J.; Epperson, B.K.; Lee, C-Y.; Pan, Y-W.; Shorey, R.I.; Prince, H.H.; Winterstein, S.R.; Luukkonen, D.R.; Geographical genetics: conceptual foundations and empirical application of spatial genetic data in wildlife management. J. Wildl. Manage. **2005,** *69* (4), 1434–1453.

Waits, L.P.; Paetkau, D. Noninvasive genetic sampling tools for wildlife biologists: a review of applications and recommendations for accurate data collection. J. Wildl. Manage. **2005,** *69* (4), 1419–1433.

Wilson, E.O.; Bossert, W.H. *A Primer of Population Biology;* Sinauer Associates, Inc.: Sunderland, Massachusetts, 1971.

12

Genetic Resources Conservation: Ex Situ

Mary T. Burke
University of California

Introduction

Botanical gardens and arboreta are collections of living plants used for research, study, and education. Arboreta traditionally focus on collections of large trees and shrubs (woody plants) while botanical gardens include research and display collections of both herbaceous (bulbs, biennials, perennials) and woody (trees and shrubs) plants. In actual practice, the boundaries between these two collection types have become less clear during the past century; for instance, many arboreta have expanded their collections to include herbaceous plants but retain their historical names. Both botanical gardens and arboreta are distinguished from public parks and other recreational landscapes by collection policies that emphasize wild-collected plants, by collections organized for scientific and educational purposes, and by extensive documentation; that is, a plant record system that includes information on provenance, nomenclature, and other taxonomic and cultural information of interest to researchers. A herbarium (herbaria, pl.) is a collection of dried, pressed, or preserved plant specimens with associated relevant collection information. Many herbaria are associated with botanical gardens and arboreta; others are associated with universities and other plant science research facilities. Some botanical gardens also maintain or are closely associated with seed banks, facilities where seeds are stored under cold and dry conditions in order to preserve the seed viability for future use.[1] As extinction rates for plant species have increased, these institutions have begun to play important roles in ex situ conservation, or conservation outside the native habitat.

Centers for Education

Botanical gardens and arboreta are "living museums"; in addition to their scientific mission, these plant collections are important and popular centers of informal science education. In the last 20 years, botanical gardens and arboreta have begun to switch their focus away from simply displaying and identifying plants, and more toward conveying critical scientific ideas and significant issues. Exhibits and labels are being redesigned to teach not only about plants as beautiful or interesting organisms, but about the role plants play in the real world: their ecosystem functions, the dependence of all life upon them, information about invasive species and their threats, the need to survey and understand unknown areas, and how people can manage native plant ecosystems. Effective conservation exhibits give visitors more of the fascinating and complex story of which plants are a critical part, rather than just a little information about a particular plant.

Globally, botanical gardens and arboreta are visited by more than 150 million people a year.[2] Recognizing the opportunity this represents, botanical gardens have assumed a new responsibility to educate people in a way that will help local citizens go on to influence policy makers and create a locally active group of people that have global concerns.

Examples of gardens and displays organized around conservation themes include the two-acre New England Garden of Rare and Endangered Plants at the New England Wild Flower Society's Garden in the Woods; the large display of rare and endangered California native plants, as well as 210 rare taxa from around the world, at the University of California Botanical Garden at UC Berkeley; and the exhibits highlighting the impact of invasive plants in Australia at the Botanic Gardens of Adelaide. Even the work of tissue culture as a method for preserving genetic resources has been showcased as a public exhibit at the Center for Research of Endangered Wildlife, the research program of the Cincinnati Zoo and Botanical Gardens. Here, tour groups can walk through the scientific facility and look through glass-lined walls at tissue culture incubators, a working greenhouse, and into a laboratory with liquid nitrogen storage tanks—The Frozen Garden. These exhibits introduce the general public to some aspects of science and technology currently focused on the preservation of rare and endangered species.[3]

Centers for Research and Discovery

As scientific institutions, botanical gardens and arboreta serve another important conservation role by conducting research or plant surveys on critical habitats, endangered plants, and invasive species. Collaborating with universities and other centers for plant studies, botanical gardens and arboreta are a place where people come together to study species of special concern. In many countries, public gardens are among the leading, and sometimes the only, institutions capable of undertaking the extensive work needed in plant research and conservation.[2]

There are about 2200 botanical gardens and arboreta in the world; it is estimated that about one-fourth of the world's flowering plants and ferns are included within their collections.[4] The extensive ex situ taxonomic and geographic collections that botanical gardens have amassed also play a critical role in botanical research, as they gather together in one place a wide variety of plants that would be difficult to study in the wild. These documented collections provide scientists with accurately identified plants of known provenance. Information about the locality in which the plant was originally collected is maintained by the scientific staff to enhance the value of the collection for researchers. The genetic resource collections in botanical gardens are often supplemented at major plant research centers by ancillary collections, including extensive herbarium collections of preserved plant specimens and seed and tissue banks. Herbaria deserve special mention as repositories of plant genetic resources: these museum specimens serve as a permanent global reference on the plant diversity of the world.

Efforts are currently underway to consolidate the inventories of ex situ plant genetic resource collections around the world, held in seed banks, field collections, herbaria, and other collections. These efforts have been hampered to some extent by not simply the scale of the undertaking, but the wide variety of ways in which collection managers have independently tracked information about their scientific holdings. Protocols for information management in natural history collections are gradually evolving in the scientific community (Darwin protocol,[5] Dublin metadata core[6]). Currently, some botanical gardens and arboreta that use a common database standard (BG-Base) list their holdings on the web in a common format that can be searched in all the consolidated inventories through a single search query.

Traditionally, botanical gardens, arboreta, and herbaria have provided open access to their collections and holdings for research or conservation purposes. In the spirit of willing compliance with the Convention on Biologic Diversity (1993) on the issue of ownership of genetic resources,[7] many gardens have begun to implement policies and procedures to address obligations regarding access to their collections and sharing the benefits of research based on them with the nations of origin. An estimated 90% of all living plant collections in botanical gardens predate the Convention on Biological Diversity (CBC). Although, technically, such collections are not covered by its provisions,[2] most scientific collections attempt to meet the CBC's goals and comply with the spirit of the convention.

Centers for Plant Protection

In addition to education and research, botanical gardens have another critical role to play in plant conservation: they often serve as the sole source of horticultural expertise for specialty plant groups, particularly for wild plants that are rarely used for display or ornamental purposes.[2] Highly accomplished nursery staff at botanical gardens often are experienced in the propagation of unusual plants. The value of this highly specialized skill to global plant conservation cannot be overemphasized, for knowing how to grow a plant may be a key to its survival or its return from the brink of extinction. In Hawaii, tissue culture and micropropagation have also been critical tools in the attempt to reestablish plants from in situ populations with less than 10 or 20 individuals.

Working within national or local consortiums or with state or federal agencies, botanical gardens may grow large collections of endangered plants, sometimes receiving them in the wake of new development or urbanization of former wildlands, holding them safely in cultivation or in seed banks, and occasionally reintroducing them back into the wild as part of species-recovery programs. Other endangered plants are featured in display gardens as part of the educational program of the garden. These displays vividly illustrate the disappearing flora to visitors and educate them about their own disappearing local ecosystems. For example, the Coco De Mer Gardens Reserve on Praslin Island and the National Botanic Garden on Mahe in the Seychelles both feature the unusual rare and endemic plants native to these islands.

Many botanic gardens are also involved in in situ conservation efforts, as they manage natural reserves or work with associated scientists and activists to study, monitor, and conserve plants in the wild. Restoration ecology and its practical applications is an important area of study and application for many botanical gardens in the United States.

The Royal Botanical Garden in Hamilton, Ontario, Canada, has been working for 10 years on Project Paradise, the largest habitat restoration project in North America, to restore the wetlands and fisheries included within the 2200 acres of natural land the garden manages on the shores of Lake Ontario.[8] Staff of botanical gardens often work as champions of natural reserve systems that protect the flora of their region. Examples include the work of Dr. Charles Lamoureux, Director of the University of Hawaii Lyon Arboretum, in establishing the 19 reserves, encompassing more than 109,000 acres, of the State of Hawaii Natural Reserves System, and the work of Dr Mildred Mathias, Director of the UC Irvine Botanical Garden (now named in her honor), in establishing the University of California Natural Reserve System that protects 130,000 acres.

A New Era of International and National Collaboration

During the last 20 years, botanical gardens, arboreta, and public gardens worldwide have begun to link together in a highly focused effort to safeguard the world's genetic resources. Clearly, in an era of declining habitats and disappearing plant species, intensive and deliberate collaboration on an international scale is needed to identify plants most in need of protection in ex situ collections. In 1987, Botanical

Gardens Conservation International (BGCI) was founded as a network of cooperating botanical gardens dedicated to effective plant conservation.

At present, with over 450 member gardens in 100 countries, BGCI works to establish clear and effective plant conservation goals for a worldwide community of scientists and professions and to assist gardens in working collaboratively to meet these goals. In its first 15 years, BCGI has supported a wide range of activities for the international community: it provides technical guidance and support for botanical gardens engaged in plant conservation efforts; assists in the development and implementation of the worldwide Botanical Gardens Conservation Strategy for plant conservation (the "International Agenda for Botanic Gardens in Conservation," June 2000); helps to create and strengthen national and regional networks of gardens in many parts of the world; and organizes major meetings, workshops, and training courses. Recognizing the primary role public gardens play in conservation education, BCGI also publishes a newsletter, *Roots*, to share experience and information about plant conservation specifically written for educators.

Just as importantly, BGCI has a support group, the Plant Charter Group, to reach out to the business community to educate them about the importance of plants and help develop partnerships for plant conservation projects. These efforts have led to establishment of a unique $50 million "eco-partnership" between HSBC, an investment bank, and BCGI, Earthwatch, and the World Wildlife Federation to assist botanical gardens internationally in conserving and managing plant genetic resources.[9]

In addition to working tirelessly to increase communication and collaboration between the botanical gardens and arboreta internationally, BGCI also conducts inventories and surveys on issues in plant conservation, supplying much needed data for good decision making. A critical first step was the 2001 survey of botanical gardens of the world and their ex situ plant collections. This survey documented the status of 2178 botanical gardens in 153 countries. These research and teaching gardens include approximately 6.13 million accessions in their living collections and another 42 million herbarium specimens in botanical garden herbaria. The majority of these botanical gardens are in the developed countries in Europe and North America (850), while another 200 are found in East and Southeast Asia. There are relatively few botanical gardens or centers of plant conservation in North and Southern Africa, the Caribbean islands, South West Asia, and the Middle East. In some tropical countries, new gardens have been created in conjunction with national parks and are designed to play roles in conservation, sustainable development, and public education.[4]

Similar collaborative efforts are underway in the United States. The American Association of Botanical Gardens and Arboreta has an active Plant Conservation Committee and regularly publishes articles on the plant conservation activities of its member gardens in its journal, *The Public Garden*. The North American Plant Collections Consortium is a voluntary association of botanical gardens and arboreta in North America that focuses each participating garden's collection development in way that will ensure effective conservation of genetic resources.[10] National collections of important genera are protected in a coordinated and complementary network of public gardens across North America, where each garden is encouraged to enrich its collection with a targeted list of plants that have been identified as top priority for conservation and are appropriate to that garden's climate and site. The Center for Plant Conservation conducts similar collaborative efforts, enjoining its participant gardens to step forward and assume coordinated ex situ protection of the rare, endangered, and threatened plants in its local flora.

Conclusions

Botanical gardens, arboreta, and herbaria hold nearly 50 million documented ex situ plant specimens in scientific collections. As educators, scientists, and conservationists, the staff of these institutions are in a special position to reach the public with ideas, sound information, and inspiration. In addition to their considerable scientific merit, these collections help educate a community of people worldwide and engage with them to preserve and restore the damaged ecosystems of the world.

References

1. Schoen, D.J.; Brown, A.H.D. The conservation of wild plant species in seed banks. BioScience **2001**, *51*, 960–966.
2. *Introduction: Botanic Gardens & Conservation.* BGCI Online. Botanic Gardens Conservation International website, 2003; http://raq43.nildram.co.uk/bgci/action/viewDocument?id¼43&location¼searchresults#Introduction (accessed August 2003).
3. Marinelli, J. Bringing plant conservation to life. Public Garden **2001**, *16* (1), 8–11.
4. Wyse Jackson, P.S. *An International Review of the Ex Situ Plant Collections of the Botanic Gardens of the World*, Botanic Gardens Conservation International, 2001; http://www.biodiv.org/programmes/socio-eco/benefit/bot-gards.asp (accessed August 2003).
5. Darwin Core (minimum set of standards for search and retrieval of natural history collections and observation databases), 2001; http://tsadev.speciesanalyst.net/DarwinCore/darwin_core.asp (accessed August 2003).
6. Dublin Core Metadata Initiative, 2003; http://www.dublincore.org/index.shtml (accessed August 2003).
7. Convention on Biological Diversity, United Nations Environment Programme; *Global Strategy for Plant Conservation*, 1993; http://www.biodiv.org/convention/articles.asp (accessed August 2003).
8. Stewart, S. Project paradise: Restoring biodiversity to LakeOntario. Public Garden **2001**, *16*(1), 14–17.
9. Bond, J. Investing in Nature, website home page, 2003; http://www.investinginnature.com/ (accessed August 2003).
10. North American Plant Collections Consortium, website home page, 2003; http://www.aabga.org/napcc/ (accessed August 2003).

References

1. [illegible] the conservation of wild plants [illegible]
2. [illegible] Conservation [illegible]
3. [illegible]
4. [illegible]
5. [illegible]
6. [illegible]
7. [illegible]

13

Genetic Resources Conservation: In Situ

V. Arivudai Nambi
and L. R. Gopinath
M.S. Swaminathan Research Foundation

Introduction

The concept of "reserve" with regard to natural resources has been to set aside areas for exclusive current use as well as unperceived future uses. Precolonial states had several reserves where royalty protected valuable species like elephants, deer, teak, or sandalwood. Systematic and organized conservation efforts have largely been undertaken only in the last two centuries or so. Such conservation efforts closely followed conquest, expanding trade, commerce, and capital accumulation by colonial powers. Colonial expansion clubbed with the Industrial Revolution opened up the possibility of large-scale movement of raw material and goods from one region to another. What was considered an inexhaustible resource at one point in time turned out to be exhaustible, leading to threats of scarcity. Such a threat led to the emergence of systematic and organized efforts at conservation of valuable resources.[1,2]

Conservation efforts can broadly be classified into the following: managed in situ conservation was pursued through National Parks, Protected Areas, Biosphere Reserves, and World Heritage Sites while ex situ conservation was through botanical gardens and gene banks. The above are widely recognized efforts; conservation in the public domain, community conservation, or in situ on-farm conservation by rural and tribal women and men remain largely unrecognized (Figure 13.1).[3,4] These communities continue to possess multifaceted traditional knowledge related to biodiversity, such as medicinal properties or food value of plants and animals.[3]

Conservation Strategies in Agriculture and Forestry Sectors

The last century saw worldwide conservation efforts that focused on a package of physical and biological entities such as soil, water, and individual species within ecosystems and excluded humans. The present century poses human societies with a formidable challenge: sustainable management of physical and biological resources where agriculture and forestry sectors move in opposite directions with regard to conservation. The agriculture sector has moved from the level of species to varieties and gene, while the

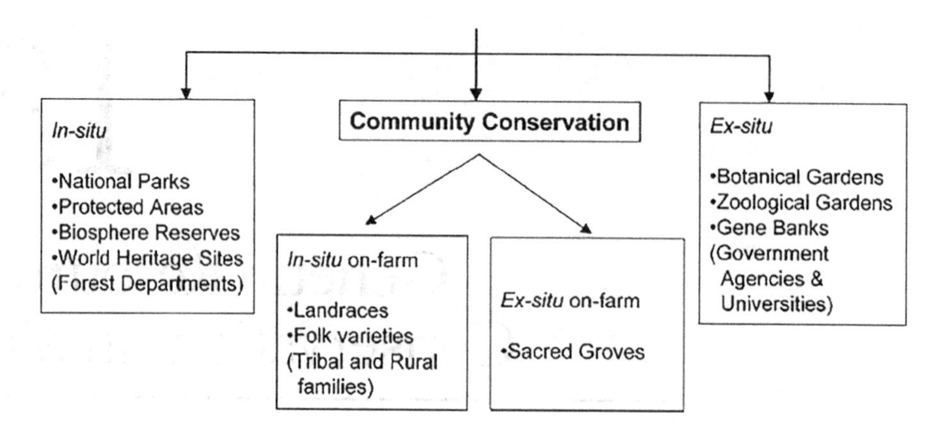

FIGURE 13.1 The integrated gene management system.
Source: Adapted from Swaminathan.[3]

forestry sector has moved from the level of species to ecosystem and landscape. A balance of competing interests of individuals and society in the multiple uses of natural resources is required.

Community conservation encompasses not only genetic resources such as landraces, folk varieties, cultivars, and breeds, but also ecosystems and landscapes in which sacred species, groves, and landscapes are embodied. There is a growing realization about the biodiversity–cultural diversity link and its importance. Changes in the social values with regard to traditional lifestyles, emerging new economies, livelihoods, and lifestyles have weakened community conservation to a significant extent. There is a pressing need to revitalize on-farm in situ conservation, as it is the key link between traditional knowledge, livelihoods, and lifestyles. Human beings always remained as managers of natural resources, through which they gained knowledge on animals, plants, climate, soil, and water and evolved various techniques and technologies for survival. This process continues to the present time. Communities living in hilly and inaccessible forested regions continue to hold traditional knowledge as well as generate new knowledge for the management of natural resources. It was through centuries of conservation and use by communities across continents that facilitated transfer of plants, such as rubber, tea, coffee, cotton, and groundnut to new locations for human use. The noted Russian scientist N.I. Vavilov undertook theoretical and applied work on economically important plants, to identify center of origin climatic analogies and wild relatives.

Since the Second World War, community-conserved genetic material was the feedstock for the crop varieties of the Green Revolution. The Green Revolution and the global environmental movement since 1960 together shaped conservation strategies for the domesticated and the wild species. With the passage of time, a number of global agreements were signed on realization that community-conserved genetic wealth (domesticated and wild) and its associated traditional knowledge would leave more options for the morrow.

Ecology in Community Conservation

Species coexist under different relationships like competition, predation, parasitism, symbiosis, commensalism, and mutualism. Although humans are a part of this relationship, they have been able to control nature because of their intelligence. In the light of the above, community- conserved biological diversity has three important functions (Figure 13.2): 1) ecological functions that maintain soil fertility and conserve water, leading to synchronized utilization of natural resources; 2) economic functions include food, fuel, fodder, timber, fiber, and medicine that contribute to local incomes both on-farm and off-farm; and 3) socio-functions that lead to diversity in resource use, tenures (private or public property), and social customs. Such relationships have a major impact on the demography of the region,

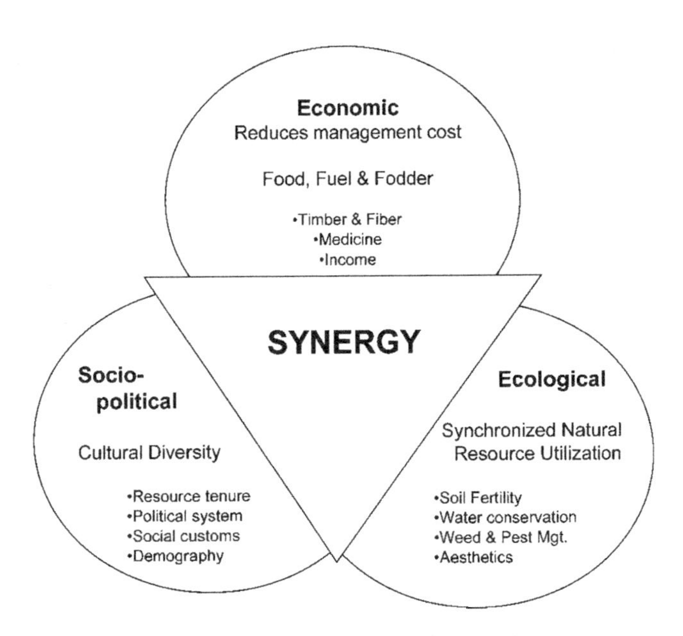

FIGURE 13.2 Functional role of biodiversity of a system.

political system, and social customs that are reflected in cultural diversity. Long-term sustainability of a species or diversity of species in a system depends upon the level of synergy among these three functions.[5]

However, the economic function of a system assumes a vital role, since transactions among various social groups are determined by economics. The synergy of the above system is lost when there is a reduction in the economic function. A reduction in the local economy leads to direct negative environmental impacts that further reduce local biodiversity. It is within this context that institutions can contribute to biodiversity conservation.

Natural Resource Management: The Role of Communities

Although initial conservation at the species level largely excluded human activity, the establishment of national parks (Figure 13.3) entailed some amount of inclusion of the human element. The evolution of the concept of the biosphere at the level of the landscape considers humans and their activities as part of nature. Such conservation efforts at the landscape level undertaken in the recent past have proved to be far more effective and sustainable.[6]

In general, biosphere reserves act as havens for endangered animals and plants with a core zone in the center surrounded by buffer and transition zones. Traditional communities live in the core zone, and the buffer zone depends on forest resources such as wild food, minor forest products, and traditional agriculture. The buffer zone and the core zones are restricted in access to nonlocal people. The buffer zone withstands changes from inner core and the outer transition zones. The transition zone is sandwiched between the human modified modern industrial world and the buffer zone. It is this zone that assumes importance with regard to community conservation of crop genetic resources.

An Integrated Approach to Conservation

The biosphere concept considers humans as a natural component of the landscape, but there is no legal basis for its implementation. As a result, biosphere reserves share the rules and regulations of national parks, reserves, and sanctuaries as their management strategy. However, such overlap of different

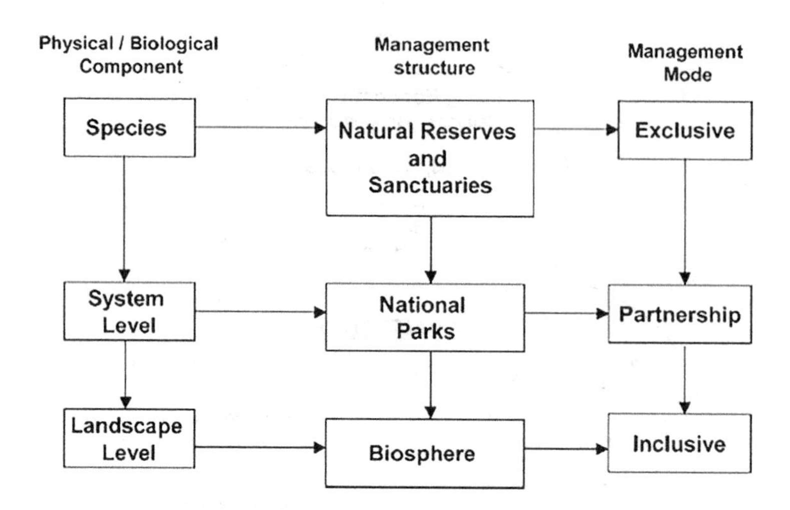

FIGURE 13.3 Management strategies in natural resources: historical perspective.

management strategies in the same region may lead to conflicts. Therefore, the concept of biosphere reserve needs a more synchronized pathway for operationalization.

A multi stakeholder approach trusteeship model and participatory conservation system (PCS) evolved by the M.S. Swaminathan Research Foundation (MSSRF) is an attempt to balance several issues addressed above through a process of share and care. This model is being implemented in the Gulf of Mannar Biosphere Reserve, a coastal aquatic system in southern India. For example, in the Gulf of Mannar Biosphere Trust in the Indian state of Tamil Nadu, MSSRF partners with the local communities; fisheries, forestry, and agricultural departments; and research institutes such as the Central Marine Fisheries Research Institute to enhance sustainable harvest of natural resources. The activities in the area concentrate on providing land-and water-based income generation activities to local communities to reduce overexploitation of the bio resources of the Gulf of Mannar. Charcoal production, dairy farming, and renovation of freshwater tanks are the land-based activities. Community-based agar production and pickled fish unit, creation of artificial reefs, and pearl culture are some of the marine-based livelihoods.

An integrated conservation strategy that combines in situ and ex situ conservation by blending traditional knowledge with modern conservation science should be developed in centers of high natural biological and agro biodiversity. In situ conservation cannot be pursued without the component of traditional knowledge of communities associated with genetic resources. Moreover, given the present genetic erosion due to industrial agriculture, it makes practical sense to create agro biodiversity sanctuaries in areas that are rich in agro biodiversity and have a living traditional knowledge. In addition, eco-agriculture strategies[7] may be required to integrate at landscape-level agricultural and forestry concerns.

Conclusions

Natural systems are complex and multifunctional, providing differential services to different sections of the populations. While financial capital is crucial to the continued management of plant and animal genetic resources, social capital involving multiple stakeholders to manage natural capital at the landscape level is essential for sustainable natural resource management. This will be possible through a three-pronged strategy for the conservation and use of genetic resources: 1) regulations on conservation, enhancement, sustainable use, and equitable sharing of benefits; 2) social mobilization and community participation in conservation, enhancement, use, and sharing of benefits; and 3) public education on the importance of conservation, enhancement, sustainable use, and equitable sharing of benefits.

Acknowledgments

The authors wish to express their profound gratitude to Prof. M.S. Swaminathan, Chairman, who gave us an opportunity to write this entry and offered his valuable comments and suggestions on the it and to Prof. P.C. Kesavan, Executive Director, M.S. Swaminathan Research Foundation, Chennai, for his constant encouragement and support and critical comments on an earlier version.

References

1. Rangarajan, M. *Fencing the Forest: Conservation and Ecological Change in India's Central Province 1866–1914;* Oxford University Press: New Delhi, 1996; 245.
2. Poffenberger, M. *Keepers of the Forest: Land Management Alternatives in Southeast Asia;* Kumarian Press: Connecticut, 1990; 292.
3. Swaminathan, M.S. Government-Industry-Civil Society: partnerships in integrated gene management. Ambio **2000**, *29*(2), 115–121.
4. Qualsat, C.O.; Damania, A.B.; Zanatta, A.C.A.; Brush, S.B. Locally-based crop plant conservation. In *Plant Conservation: The In Situ approach;* Maxted, N., Ford-Lloyd, B.V., Hawkes, J.G., Eds.; Chapman and Hall: London, 1997; 160–175.
5. Gopinath, L.R. *SHGs: Effective Pathway to Biodiversity Conservation*; M.S. Swaminathan Research Foundation: Chennai, 2002 (Unpublished report).
6. Ramakrishnan, P.S.; Rai, R.K.; Katwal, R.P.S.; Mehmdiratta, S. *Traditional Ecological Knowledge for Managing Biosphere Reserves in South and Central Asia;* Oxford & IBH Publishers: New Delhi, 2002.
7. McNeely, J.; Scherr, J.S. *Ecoagriculture: Strategies to Feed the World and Save Wild Biodiversity*; Island press: Washington, 2002; Vol. 323.

14

Genetic Resources: Farmer Conservation and Crop Management

Daniela Soleri and
David A. Cleveland
University of California

Introduction

Food production is essential to support human society, yet agriculture is one of the largest contributors to global environmental destruction through loss of habitat and diversity, and greenhouse gas emissions that are driving the anthropogenic climate crisis.[1] Identifying options for more sustainable production requires consideration of diverse strategies, including understanding farmer management and conservation of crop varieties that have been the basis of our food system for nearly all of settled human history. About 2 billion people live on 500 million small-scale farms (under 2 ha) globally,[2] most of these in Traditional-Based agricultural systems (TBAS), and the number will grow dramatically with population growth in the coming decades. Many farmers in TBAS save their own seed to grow at least a portion, or most, of the food they eat, conserving valuable genetic resources in the process. Plant breeders working with TBAS farmers consider diversity at many spatial levels of the agrifood system a key to alternatives such as organic and low-input agriculture.[3]

The beginning of agriculture with the Neolithic revolution initiated a dramatic reduction in the diversity of species humans used for food. After domestication, crop species were often transported widely, and many genetically distinct farmers' varieties (FVs, crop varieties traditionally maintained and grown by farmers) developed in specific locations, greatly increasing intraspecific diversity.[4] As Simmonds stated, "Probably, the total genetic change achieved by farmers over the millennia was far greater than that achieved by the last hundred or two years of more systematic science-based effort,"[5] an insight verified by a genome-wide review of maize wild relatives, FVs, and modern varieties (MVs) created by professional plant breeders.[6] FVs continue to be grown today by many small-scale farmers in TBAS, providing for both local consumption and the conservation of genetic diversity for global society.[7]

Crop genetic variation (V_G) is a measure of the number of alleles and degree of difference between them, and their arrangement in plants and populations. For our purposes, a cumulative change in crop population V_G over generations is called microevolution (E_V). Farmers and the biophysical environment select plants within populations based on their phenotypic variation (V_P). Farmers also choose between populations or varieties. This phenotypic selection and choice together determine the degree to which varieties change between generations, evolve over generations, or stay the same. With in situ conservation in farmers' fields, specific alleles and genetic structures contributing to V_G may evolve in response to changing local selection pressures, while still maintaining a high level of V_G.[8] In contrast, ex situ conservation in gene banks is more narrowly defined as conserving the specific alleles and structures of V_G present at a given location and moment in time. Thus, different forms of conservation include different amounts and forms of change.

Sometimes, farmers carry out selection or choice intentionally to change or conserve V_G. However, much of farmer practice is intended to further production and consumption goals, affecting crop evolution unintentionally if at all. Thus, in order to understand farmer selection and conservation, it is important to understand the relationship between production, consumption, selection, and conservation in TBAS.[4] This in turn involves understanding the relationship between farmer knowledge and practice in terms of the basic genetics of crop populations and their interactions with growing environments (genetic variation, environmental variation, variation due to genotype-by-environment interaction [$V_{G \times E}$], and response to selection[R])[7,9] (Table 14.1).

Farmers and Farmer Varieties in Traditional-Based Agriculture Systems

TBAS are characterized by integration within the household or community of conservation, improvement, seed multiplication, production, distribution, and consumption, whereas in industrial agriculture, these functions are spatially and structurally separated. Farm households in TBAS typically rely on their own food production for a significant proportion of their consumption; this production is essential for feeding the population in TBAS now and in the future, even with production increases in industrial agriculture.[10]

TBAS are also characterized by marginal growing environments (relatively high stress, high temporal and spatial variability, and low external inputs) and by the continued use of FVs, even when MVs are available.[11] FVs include landraces, traditional varieties selected by farmers, MVs adapted to farmers' environments by farmer and natural selection, and progeny from crosses between landraces and MVs (sometimes referred to as "creolized" or "degenerated" MVs).

TBAS farmers value FVs for agronomic traits, such as drought resistance, pest resistance, and photoperiod sensitivity. Because farmers grow some or most of the food they eat, storage and culinary criteria are also frequently important; for example, families who make the traditional maize beverage *tejate* maintain more varieties of maize than families who do not, using them in preparation of that drink.[12]

The V_G of farmer-managed FVs is often much higher than that of MVs and is presumed to support broad resistance to multiple biotic and abiotic stresses.[13,14] This makes FVs valuable not only for farmers, because they decrease the production risks in marginal environments especially with climate change,[15] but also for plant breeders and conservationists as the basis for future production in industrial agriculture.[8,16]

TABLE 14.1 Farmer Selection and Choice and the Change and Conservation of Crop Varieties

Farmer Knowledge (Including Values) on Which Practice May Be Based	Farmer Practice	Potential Effect of Farmer Practice on Selection and Conservation of Populations/Varieties	Example
Indirect selection/conservation by farmer-managed growing and storage environment			
Understanding of $G \times E$	Allocation of varieties to spatial, temporal, and management environments	Selection pressures in environments result in maintenance of existing, or development of new populations/varieties, including evolution of wide or narrow adaptation	*Spatial*: varieties specified for different soil or moisture types; rice, Nepal; pearl millet, India *Temporal*: varieties with different cycle lengths, maize Mexico
	Management of growing environments	Changing selection pressures	Changes in fertilizer application, maize, Mexico
Risk, values, $G \times E$	Choice of environments for testing new populations/ varieties	↑ or ↓ V_G	High stress, rice, Nepal; optimal conditions, barley, Syria
Escape from economic or political pressure; desire for different ways of life	Abandonment of fields or farms, reduced field size	↑ V_G within species due to pooled seeds and reduced area for planting, or ↓ effective population size, genetic drift	Pooling of subvarieties, maize, Hopi and Zuni Native Americans Reduction in area, potatoes, Peru; maize, Mexico
Direct selection/conservation, intentional re. population change			
Low discount rate (value the future), altruism (value community)	Conservation of varieties for the future, for other farmers	↑ intraspecific V_G	Rice, Thailand; maize, Hopi
Interest and expertise in experimentation	Deliberate crossing	↑ V_G and heterozygosity	Maize-teosinte, Mexico; MV-FV pearl millet, India; MV-FV and FV-FV, maize, Mexico
Understanding of h^2	Selection of individuals (plants, propagules) from within parent population	↓ or maintenance of V_G via R	Among seedlings, cassava, Guyana; among panicles, pearl millet, India
Direct, selection/conservation, unintentional re. population change, but intentional re. other goals, as result of production/ consumption practices			
Attitudes toward risk re. yield stability	Adoption and abandonment of FVs, MVs	↑ or ↓ intraspecific diversity	Maize, Hopi; rice, Nepal
	Adoption and abandonment of lines in multiline varieties of self-pollinated crops; seed lots in cross-pollinating crops	↑ or ↓ intravarietal diversity	Common bean, East Africa; maize, Mexico
Agronomic, storage, culinary, aesthetic and ritual criteria, implicit and explicit	Selection or choice based on production, consumption, aesthetic, historic criteria	↑ or ↓ intra- and intervarietal diversity	Storage and culinary criteria: maize, Mexico; and ritual criteria, rice, Nepal
Choice criteria	Acquisition of seed, seed lots	Gene flow via seed, then pollen flow, hybridization, recombination within varieties	Cycle length, maize, Mexico; cuttings and seedlings, cassava, Guyana

Abbreviations: FV, farmer developed crop variety; $G \times E$, genotype-by-environment interaction; h^2, heritability in the narrow sense; MV, modern crop variety, product of formal breeding system; R, response selection; V_G, genetic variation; ↑ increase; ↓ decrease.

Farmer Choice: Phenotypic and Genetic Variation, Classification, Genotype-by-Environment Interaction, and Risk

Farmers classify and value traits in their crops, and this can vary between women and men,[7] and between households in a community.[17–19] This variation affects their definition of varieties and populations, and thus the degree of intraspecific V_P (and V_G) they are willing to accept, as a result, for example, of intravarietal gene flow. These definitions in turn affect farmers' choice, such as which crops, varieties, and populations to adopt or abandon, and thus the total V_G they manage, and the number of populations from which they select plants. Experimental evidence indicates that farmers can choose among large numbers of genotypes. In Syria, farmers were able to effectively identify high-yielding barley populations from among 208 entries, including 100 segregating populations.[20]

Farmers' choice of varieties and populations without discriminating between individual propagules, when adopting or abandoning them from their repertoires, saving seed for planting, and in seed procurement, does not change the genetic makeup of those units directly, and there is no evidence that farmers expect to change them. However, genetic structure may be altered due to sampling error, if the number of seeds required to plant an area is small, and many of these may be half sibs in a crop like maize, with <143 ears ha^{-1} in the case of Oaxaca, Mexico, and many farmers planting much smaller areas.[21]

FV crop mating systems in combination with farmers' propagation methods are important determinants of inter- and intraspecific V_G. These also affect differences in phenotypic consistency over generations, and therefore farmers' perception and management.[22] Apart from low-frequency somatic mutations, V_G in vegetatively propagated outcrossing crops, such as cassava, is unchanged between generations, and discrete, types (clones) or groups of types are maintained as distinct varieties[23,24] that may be either homo- or heterogeneous. Intrapopulation V_G increases, and genetic structures become more variable and dynamic with the intentional inclusion by farmers of sexually propagated individuals into clonal populations based on morphological similarity or heterosis.[24]

The same increase in dynamism occurs with increasing rates of outcrossing in sexually propagated crops, because variation can be continuous within a population. Moreover, segregation, crossing-over, recombination, and other events during meiosis and fertilization result in much change in V_G between generations. In highly allogamous crops, such as maize, heterozygosity can be high, making it difficult to discern discrete segregation classes, particularly in the presence of environmental variation, and retaining distinguishing varietal characteristics requires maintenance selection[25] (see below). Highly autogamous crops such as rice are predominantly homozygous, making exploitation of V_G and retention of varietal distinctions easier, even if varieties are composed of multiple, distinct lines. Farmers' choices depend in part on the range of spatial, temporal, and management environments present, the V_G available to them, and the extent to which genotypes are widely versus narrowly adapted. In turn, environmental variation (V_E) in these growing environments interacts with V_G to produce variation in yield of grain, straw, roots, tubers, leaves, and other characteristics over space and time. As a result, farmers may have different choice criteria for different environments, as in Rajasthan, India, where pearl millet farmers realize there is a trade-off between panicle size and tillering ability. So, farmers in a less stressful environment prefer varieties producing larger panicles, whereas those in a more stressful environment prefer varieties with high tillering.[26]

Patterns of variation in yield affect farmers' choice of crop variety via their attitude toward risk. In response to scenarios depicting varietal $V_{G \times E}$ and temporal variation, farmers from more marginal growing environments were more risk-averse compared to those from more favorable environments. The former preferred a crop variety with low but stable yields across environments, while the latter chose a variety highly responsive to favorable conditions but with poor performance under less favorable conditions.[27] Sorghum farmers in Mali tend to choose varieties to optimize outputs in the face of variation in rainfall, level of Striga infestation, and availability of labor and other production resources, especially cultivator and seeder plows.[28] As a result, they choose combinations of long- and short-cycle sorghum varieties to optimize yield, yield stability, and post-harvest traits like taste. For example, when rains are better, farmers choose a greater number of long-cycle varieties.

Farmer Selection: Phenotypic Variability, Heritability, Phenotypic Selection Differential, and Response

Phenotypic selection, operating on V_P, is identification of the individual plants within a population that will contribute genetic material to the next generation. Phenotypic selection of FVs in TBAS can be classified according to the agent of selection (natural environment, farmer-managed environment, or farmer) and according to farmers' goals for selection (Figure 14.1). Farmer selection can also be classified according to the outcome (Figure 14.2). Geneticists and plant breeders tend to think of phenotypic selection as seeking to produce genetic change, but farmers often do not.[29] Whether or not farmer

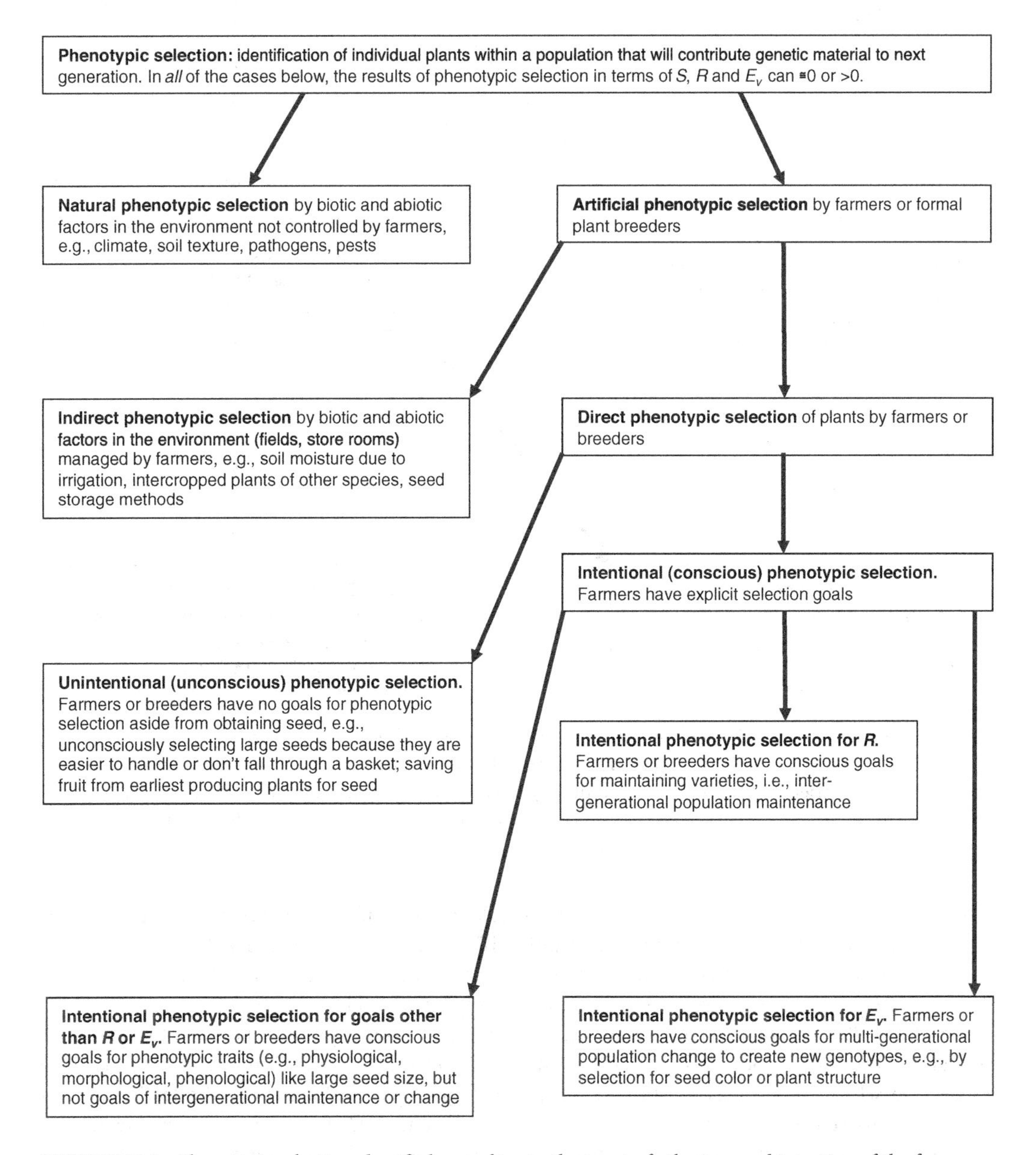

FIGURE 14.1 Phenotypic selection classified according to the agent of selection, and intention of the farmer or plant breeder as agent. See text for the definition of abbreviations. © D. Soleri & D.A. Cleveland, 2013.

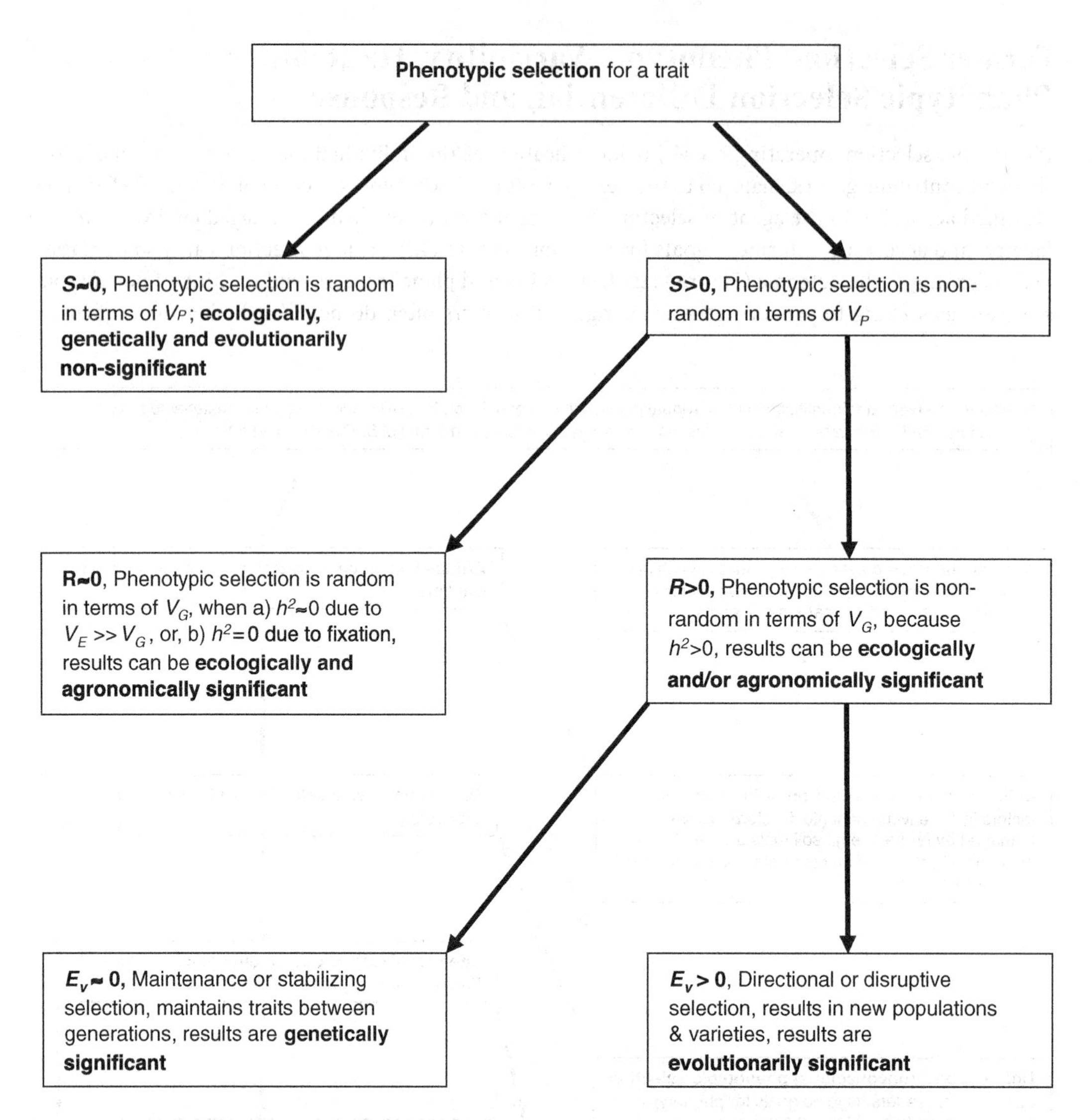

FIGURE 14.2 Phenotypic selection classified according to outcome of selection. See key to symbols below. © D.A. Cleveland & D. Soleri, 2013. Abbreviations: FV, farmer developed crop variety; h^2, heritability in the narrow sense; MV, modern crop variety, product of formal breeding system; R, response to selection; V_E, environmental variation; V_G, genetic variation; $V_{G\times E}$, genotype-by-environment interaction variation; V_P, phenotypic variation; ↑, increase; ↓, decrease.

selection does change the genetic makeup of the population (i.e., effects response between generations [R] or cumulative multigenerational microevolution [E_V]) depends on heritability (h^2), or the proportion of phenotypic variation that is genetic and can be inherited; and the selection differential (S), or the difference in mean between parental population and sample selected from it: $R = h^2S$.

$S > 0$, $R \approx 0$. Heritability is often understood by farmers who distinguish between high and low heritability traits, consciously selecting the former, while often considering it not worthwhile or even possible to select the latter, especially in cross-pollinating crops.[27] When farmers' selection criteria center on relatively low heritability traits such as large ear and seed size in maize, they may achieve high S, and little or no R. However, they persist with that selection because their goal is high-quality seed for planting.[25,30] A study across four sites, each with different crops, found that often a majority of farmers at a site did not see their seed selection as a process of cumulative, directional change.[27] However, intentional phenotypic selection for goals other than genetic response was practiced by nearly all farmers in that study. Selection exercises with maize in Oaxaca, Mexico,[30] found farmers' selections to be significantly different from that of the original population in their selection criteria, resulting in high S values. However, R values were zero for these as well as other morphophenological traits. The reasons documented to date are for seed quality (germination and early vigor) and purity, and because of "custom," that is, not to change or improve a variety. To understand this from the farmers' perspective, it is necessary to take into account the multiple functions of crop populations in TBAS (production of food and seed, consumption, conservation, improvement).

$R > 0, E_V \approx 0$. Farmers also select seed to maintain defining, desirable, heritable varietal traits that change as a result of gene flow and indirect selection by environmental factors in fields and storage containers, especially in allogamous species. When successful, this results in R and, over time, prevents unwanted E_V. Selection exercises with maize in Jalisco, Mexico, found that farmers' selection served to diminish the impact of gene flow and maintain varieties' morphological characteristics, but not to change the population being selected on. Indian farmers were able to maintain the distinct ideotypes of introduced FVs of their allogamous crop pearl millet via intentional selection of panicles for their unique phenotypes.[31]

$E_V > 0$. In seeking cumulative genetic response, i.e., microevolution or E_V, farmers may practice intentional selection either to create new varieties, best documented in vegetatively propagated and self-pollinating crops,[32] or for varietal improvement, although much evidence for this is anecdotal. Most often, this is selection for heritable, qualitative traits; for example, farmers in central Mexico have selected for and maintained a new landrace, based on seed and ear morphology, among segregating populations resulting from the hybridization of two existing landraces.[33]

Conclusions

Crop selection and conservation in TBAS contrast substantially with industrial agricultural systems. Therefore, understanding farmers' practices, and the knowledge and goals underlying them, is critical for supporting food production, food consumption, crop improvement, and crop genetic resource conservation for farm communities in TBAS and for long-term global food security. The urgency of understanding farmer selection and conservation will increase in the future with the ongoing loss of genetic resources, the rapid spread of transgenic crop varieties with limited genetic diversity, the development of a global system of intellectual property rights in crop genetic resources, and the movement to make formal plant breeding more relevant to farmers in TBAS through plant breeding and conservation based on direct farmer and scientist collaboration.

At present, attention and investment in transgenic genetic engineering dominate crop improvement globally. Yet, schemes that are to some extent modeled on and make use of farmer management and the V_G of their FVs show good potential for increasing yields, conserving genetic resources and supporting adaptation to growing environments that are changing at an accelerating pace.[34] Farmer management and conservation of crop varieties developed in situ is a form of precision agriculture that, when combined with formal scientific methods and research support, may be the strategy most likely to address multiple criteria of environmental, economic, and social sustainability in the global food system.[35] It will be important for scientists collaborating with farmers to be aware of their assumptions, and open to learning from the local experts, the farmers who have been managing and conserving their crop populations long before the advent of modern breeding.[36]

Acknowledgments

This entry was originally published in 2014 in this *Handbook*,[37] as an update of an entry in the *Encyclopedia of Plant and Crop Science*.[38] The research on which some of this entry is based was supported in part by NSF Grant Nos. SES-9977996 and DEB-0409984, and a grant from the Wallace Genetic Foundation.

References

1. Power, A.G. Ecosystem services and agriculture: Tradeoffs and synergies. *Philos. Trans. R. Soc. B-Biol. Sci.* **2010**, *365* (1554), 2959–2971.
2. IFAD. Conference on new directions in *smallholder agriculture*, proceedings of the conference, 24–25 January 2011; Rome, IFAD HQ. In IFAD: Rome, 2011.
3. SOLIBAM Newsletter 1. www.solibam.eu/modules/addresses/viewcat.php?cid=1 (2011 November 11).
4. Harlan, J.R. *Crops and Man*. Second ed.; American Society of Agronomy, Inc. and Crop Science Society of America, Inc.: Madison, WI, 1992; p. 284.
5. Simmonds, N.W.; Smartt, J. *Principles of Crop Improvement*. Second ed.; Blackwell Science Ltd.: Oxford, UK, 1999; pp. xii, 412.
6. Hufford, M.B.; Xu, X.; van Heerwaarden, J.; Pyhajarvi, T.; Chia, J.-M.; Cartwright, R.A.; Elshire, R.J.; Glaubitz, J.C.; Guill, K.E.; Kaeppler, S.M.; Lai, J.; Morrell, P.L.; Shannon, L.M.; Song, C.; Springer, N.M.; Swanson-Wagner, R.A.; Tiffin, P.; Wang, J.; Zhang, G.; Doebley, J.; McMullen, M.D.; Ware, D.; Buckler, E.S.; Yang, S.; Ross-Ibarra, J. Comparative population genomics of maize domestication and improvement. *Nat. Genet.* **2012**, *44* (7), 808–811.
7. Smale, M. Economics perspectives on collaborative plant breeding for conservation of genetic diversity on farm. In *Farmers, Scientists and Plant Breeding: Integrating Knowledge and Practice*, Cleveland, D.A., Soleri, D., Eds. CAB International: Oxon, UK, 2002; pp. 83–105.
8. Brown, A.H.D. The genetic structure of crop landraces and the challenge to conserve them in situ on farms. In *Genes in the Field: On-Farm Conservation of Crop Diversity*, Brush, S.B., Ed. Lewis Publishers; IPGRI; IDRC: Boca Raton, FL; Rome; Ottawa, 1999; pp. 29–48.
9. Soleri, D.; Cleveland, D.A. Farmers' genetic perceptions regarding their crop populations: An example with maize in the Central Valleys of Oaxaca, Mexico. *Econ. Bot.* **2001**, 55, 106–128.
10. Heisey, P.W.; Edmeades, G.O. Part 1. Maize production in drought-stressed environments: Technical options and research resource allocation. In World Maize Facts and Trends 1997/98, CIMMYT, Ed. CIMMYT: Mexico, D.F., 1999; pp. 1–36.
11. Brush, S.B.; Taylor, J.E.; Bellon, M.R. Technology adoption and biological diversity in Andean potato agriculture. *J. Dev. Econ.* **1992**, *39*, 365–387.
12. Soleri, D.; Cleveland, D.A.; Aragón Cuevas, F. Food globalization and local diversity: The case of tejate, a traditional maize and cacao beverage from Oaxaca, Mexico. *Curr. Anthropol.* **2008**, 49, 281–290, www.journals.uchicago.edu.proxy.library.ucsb.edu:2048/doi/full/10.1086/527562?cookieSet=1#apb.
13. Ceccarelli, S. Landraces: Importance and use in breeding and environmentally friendly agronomic systems. Agrobiodiversity Conservation: Securing the Diversity of Crop Wild Relatives and Landraces 2012; 103.
14. Cooper, H.D.; Spillane, C.; Hodgkin, T. Broadening the genetic base of crops: An overview. In *Broadening the Genetic Base of Crop Production*, Cooper, H.D., Spillane, C., Hodgkin, T., Eds. CABI: Wallingford, Oxon, UK, 2001; pp. 1–23.
15. Ceccarelli, S.; Grando, S.; Maatougui, M.; Michael, M.; Slash, M.; Haghparast, R.; Rahmanian, M.; Taheri, A.; Al-Yassin, A.; Benbelkacem, A.; Labdi, M.; Mimoun, H.; Nachit, M. Plant breeding and climate changes. *J. Agri. Sci.* 2010, 148, 627–637.
16. Newton, A.C.; Akar, T.; Baresel, J.P.; Bebeli, PJ.; Bettencourt, E.; Bladenopoulos, K.V.; Czembor, J.H.; Fasoula, D.A.; Katsiotis, A.; Koutis, K.; Koutsika-Sotiriou, M.; Kovacs, G.; Larsson, H.; de Carvalho, M.; Rubiales, D.; Russell, J.; Dos Santos, T. M.M.; Patto, M.C.V. Cereal landraces for sustainable agriculture. A review. *Agron. Sustain. Dev.* **2010**, 30 (2), 237–269.
17. Barry, M.B.; Pham, J.L.; Courtois, B.; Billot, C.; Ahmadi, N. Rice genetic diversity at farm and village levels and genetic structure of local varieties reveal need for in situ conservation. *Genet. Resour. Crop. Evol.* **2007**, 54, 1675–1690.

18. Delêtre, M.; McKey, D.B.; Hodkinson, T.R. Marriage exchanges, seed exchanges, and the dynamics of manioc diversity. *Proc. Nat. Acad. Sci.* **2011**, 108 (45), 18249–18254.
19. Worthington, M.; Soleri, D.; Aragón-Cuevas, F.; Gepts, P. Genetic composition and spatial distribution of farmer-managed Phaseolus bean plantings: An example from a village in Oaxaca. *Mexico. Crop. Sci.* **2012**, 52, 1721–1735.
20. Ceccarelli, S.; Grando, S.; Tutwiler, R.; Bahar, J.; Martini, A.M.; Salahieh, H.; Goodchild, A.; Michael, M. A methodological study on participatory barley breeding I. *Selection Phase. Euphytica* **2000**, 111, 91–104.
21. Cleveland, D.A.; Soleri, D.; Aragón Cuevas, F.; Crossa, J.; Gepts, P. Detecting (trans)gene flow to landraces in centers of crop origin: Lessons from the case of maize in Mexico. *Env. Biosafety Res.* 2005, 4, 197–208.
22. Cleveland, D.A.; Soleri, D.; Smith, S.E. A biological framework for understanding farmers' plant breeding. *Econ. Bot.* 2000, 54, 377–394.
23. Gibson, R.W. A review of perceptual distinctiveness in landraces including an analysis of how its roles have been overlooked in plant breeding for low-input farming systems. *Econ. Bot.* 2009, 63 (3), 242–255.
24. McKey, D.; Elias, M.; Pujol, B.; Duputie, A. The evolutionary ecology of clonally propagated domesticated plants. *New Phytol.* 2010, 186 (2), 318–332.
25. Louette, D.; Smale, M. Farmers' seed selection practices and maize variety characteristics in a traditional Mexican community. *Euphytica* 2000, 113, 25–41.
26. Weltzien R.E.; Whitaker, M.L.; Rattunde, H.F.W.; Dhamotharan, M.; Anders, M.M. Participatory approaches in pearl millet breeding. In *Seeds of Choice*, Witcombe, J., Virk, D., Farrington, J., Eds.; Intermediate Technology Publications: London, 1998; pp. 143–170.
27. Soleri, D.; Cleveland, D.A.; Smith, S.E.; Ceccarelli, S.; Grando, S.; Rana, R.B.; Rijal, D.; Ríos Labrada, H. Understanding farmers' knowledge as the basis for collaboration with plant breeders: Methodological development and examples from ongoing research in Mexico, Syria, Cuba, and Nepal. In *Farmers, Scientists and Plant Breeding: Integrating Knowledge and Practice*, Cleveland, D.A., Soleri, D., Eds.; CAB International: Wallingford, Oxon, UK, 2002; pp. 19–60.
28. Lacy, S.; Cleveland, D.A.; Soleri, D. Farmer choice of sorghum varieties in southern Mali. *Human Ecol.* 2006, 34, 331–353.
29. Cleveland, D.A.; Soleri, D. Extending Darwin's analogy: Bridging differences in concepts of selection between farmers, biologists, and plant breeders. *Econ. Bot.* 2007, 61, 121–136.
30. Soleri, D.; Smith, S.E.; Cleveland, D.A. Evaluating the potential for farmer and plant breeder collaboration: A case study of farmer maize selection in Oaxaca, Mexico. *Euphytica* 2000, 116, 41–57.
31. vom Brocke, K.; Weltzien, E.; Christinck, A.; Presterl, T.; Geiger, H.H. Farmers' seed systems and management practices determine pearl millet genetic diversity in semiarid regions of India. *Crop Sci.* 2003, 43, 1680–1689.
32. Boster, J.S. Selection for perceptual distinctiveness: Evidence from Aguaruna cultivars of Manihot esculenta. *Econ. Bot.* 1985, 39, 310–325.
33. Perales, H.; Brush, S.B.; Qualset, C.O. Dynamic management of maize landraces in Central Mexico. *Econ. Bot.* 2003, 57 (1), 21–34.
34. IAASTD synthesis report with executive summary: A synthesis of the global and sub-global IAASTD reports; Washington, DC, 2009; p. 97.
35. De Schutter, O. Agroecology and the right to food. Report Submitted by the Special Rapporteur on the Right to Food; Human Rights Council, United Nations General Assembly; 2010.
36. Soleri, D.; Cleveland, D.A. Investigating farmers' knowledge and practice regarding crop seeds: Beware your assumptions! In *Indigenous knowledge: Enhancing Its Contribution to Natural Resources Management*, Sillitoe, P., Ed.; CABI Publishing: Wallingford, Oxfordshire, UK, 2017; pp. 158–173.

37. Soleri, D.; Cleveland, D.A. Farmer selection and conservation of crop varieties. In *Encyclopedia of Plant & Crop Science*, Goodman, R.M., Ed.; Marcel Dekker: New York, 2004; pp. 433–438.
38. Soleri, D.; Cleveland, D.A. Genetic resources: Farmer conservation and crop management. In *Encyclopedia of Natural Resources: Land*, Wang, Y.Q. Ed.; Taylor and Francis: New York, 2014; pp. 256–262.

15

Genetic Resources: Seeds Conservation

Florent Engelmann
Institute of Research for Development (IRD)

Introduction

In the 1950s and 1960s, major advances in plant breeding brought about the "green revolution," which resulted in wide-scale adoption of high-yielding varieties and genetically uniform cultivars of staple crops, particularly wheat and rice. Consequently, global concern about the loss of genetic diversity in these crops increased, as farmers abandoned their locally adapted landraces and traditional varieties, replacing them with improved, yet genetically uniform modern ones. The International Agricultural Research Centers (IARCs) of the Consultative Group on International Agricultural Research (CGIAR) started to assemble germplasm collections of the major crop species within their respective mandates. The International Board for Plant Genetic Resources (IBPGRs) was established in 1974 in this context to coordinate the global effort to systematically collect and conserve the world's threatened plant genetic diversity. Today, as a result of this effort, over 1750 genebanks and germplasm collections exist around the world, maintaining ~7,400,000 accessions, largely of major food crops including cereals and some legumes, i.e., species that can be conserved easily as seed.[1]

This entry reviews the current storage technologies and management procedures developed for seeds, describes the problems and achievements with seed storage, and identifies priorities for improving the efficiency of seed conservation.

Conservation of Orthodox Seeds

Many of the world's major food plants produce so-called orthodox seeds, which tolerate extensive desiccation and can be stored dry at low temperature. Storage of orthodox seeds is the most widely practiced method of ex situ conservation of plant genetic resources,[1] as 90% of the accessions stored in genebanks are maintained as seed. Following drying to low moisture content (3–7% fresh weight basis, depending on the species), such seeds can be conserved in hermetically sealed containers at low temperature, preferably at −18°C or cooler, for several decades.[1] All relevant techniques are well established, and practical documents covering the main aspects of seed conservation, are available, including design of seed storage facilities for genetic conservation, principles of seed testing for monitoring viability of seed accessions maintained in genebanks, methods for removing dormancy and germinating seeds, and suitable methods for processing and handling seeds in genebanks.[2]

In addition to being the most convenient material for genetic resource conservation, seeds are also a convenient form for distributing germplasm to farmers, breeders, scientists, and other users. Moreover, since seeds are less likely to carry diseases than other plant material, their use for exchange of plant germplasm can facilitate quarantine procedures.

Conservation of Nonorthodox Seeds

In contrast to orthodox seeds, a considerable number of species, predominantly from tropical or subtropical origin, such as coconut, cacao, and many forest and fruit tree species, produce so-called nonorthodox seeds, which are unable to withstand much desiccation and are often sensitive to chilling. Nonorthodox seeds have been further subdivided in recalcitrant and intermediate seeds based on their desiccation sensitivity, which is high for the former and lower for the latter group. Nonorthodox seeds cannot be maintained under the storage conditions described above, i.e., low moisture content and temperature, and have to be kept in moist, relatively warm conditions to maintain viability. Even when stored in optimal conditions, their lifespan is limited to weeks, occasionally months. Of over 7000 species for which published information on seed storage behavior exists,[3] ~10% are recorded as nonorthodox or possibly nonorthodox.

Genetic resources of nonorthodox species are traditionally conserved as whole plants in field collections. This mode of conservation is faced with various problems and limitations, and cryopreservation (liquid nitrogen (LN), –196°C) currently offers the only safe and cost-effective option for the long-term conservation of genetic resources of problem species. However, cryopreservation research is still at a preliminary stage for nonorthodox species.[4]

Management of Seed Collections

Only a limited number of genebanks operate at very high standards, whereas many others face difficulties owing to inadequate infrastructures, lack of adequate seed processing and storage equipment, unreliable electricity supply, funding and staffing constraints, and inadequate management practices. Therefore, seeds are often stored under suboptimal conditions and require more frequent regeneration, thus bearing additional costs on often already insufficient operating budgets of genebanks. Great difficulties are faced, in particular, by many countries with regeneration of seed collections.[1] Seed storage technologies are relatively easy to apply. The problems relate more to resource constraints that impact the performance of essential operations. Efficient and cost-effective genebank management procedures have now become key elements for long-term ex situ conservation of plant genetic resources.

Other important aspects of genebank operations concern germplasm characterization and documentation. The extent to which germplasm collections are characterized varies widely between genebanks and species[1] but is far from complete in many instances. Concerning documentation, the situation is highly contrasted.[1] Some, mainly developed, countries have fully computerized documentation systems and relatively complete accession data, while many others lack information on the accessions in their collections, including the so-called passport data.

Main Challenges and Priorities for Improving Seed Conservation

Various priority areas have been identified for orthodox and nonorthodox seed conservation research and for improving seed genebank management procedures.

Research Priorities for Orthodox Seed Species

Longevity of Seeds under Standard Genebank Storage Conditions

One priority research topic is the longevity of seeds under standard genebank storage conditions. Indeed, over the past 30 years, relatively widespread evidence has emerged of less than expected longevity at conventional seed bank temperatures.[5] Across ~200 species, species from drier (total rainfall) and warmer temperature (mean annual) locations tended to have greater seed P_{50} (time taken in storage for viability to decline to 50%) under accelerated aging conditions than species from cool, wet conditions.[6] Moreover, species P_{50} values were correlated with the proportion of accessions (not necessarily the same species) in that family that lost a significant amount of viability after 20 years under conditions for long-term seed storage, that is, seeds pre-equilibrated with 15% relative humidity air and then stored at −20°C.[7] Such relative underperformance at −20°C was apparent in 26% of the accessions.[6] Similarly, it has been estimated that half-lives for the seeds of 276 species held for an average of 38 years under cool (−5°C) and cold (25 years at −18°C) temperature was >100 years only for 61 (22%) of the species.[8] Nonetheless, cryogenic storage did prolong the shelf life of lettuce (*Lactuca*) seeds with projected half-lives in the vapor and liquid phases of LN of 500 and 3,400 years, respectively,[9] up to 20 times greater than that predicted for that species in a conventional seed bank at −20°C.[10,11] Although 25 species (from 19 genera) of Cruciferae had high germination (often >90%) after ~40 years storage at −5 to −10°C,[12] it is hard not to conclude that, as an extra insurance policy for conservation, cryopreservation should be considered appropriate for all orthodox seeds, and one subsample of any accession systematically stored in LN, in addition to the samples stored under classical genebank conditions.[13] As highlighted by,[14] such loss of viability during seed storage under standard conditions reflects molecular mobility within the system, i.e., relaxation of glassy matrixes. Stability of biological glasses is currently not fully understood and research in the thermodynamic principles, which contribute to temperature dependency of glassy relaxation as the context for understanding potential changes in viability of cryogenically stored germplasm is urgently needed.

Determining Critical Seed Moisture Content

The preferred conditions recommended for long-term seed storage are 3–7% moisture content, depending on the species, at −18°C or lower.[7] However, it has been shown that drying seeds beyond a critical moisture content provides no additional benefit to longevity and may even accelerate seed aging rates, and that interactions exist between the critical relative humidity and storage temperature. Research should be pursued on this topic of high importance.

Developing Low-Input Storage Techniques

Various research projects have focused on the development of the ultra-dry seed technology, which allows storing seeds desiccated to very low moisture contents at room temperature, thereby suppressing the need for refrigeration equipment. Although drying seed to very low moisture prior to storage seems to have fewer advantages than was initially expected, ultra-dry storage is still considered to be a useful, practical, low-cost technique in circumstances where no adequate refrigeration can be provided.[15] Research on various aspects of the ultra-dry seed storage technology and on its applicability to a broader number of species should therefore be continued.

Improving and Monitoring Viability

The viability of conserved accessions depends on their initial quality and how they have been processed for storage, as well as on the actual storage conditions. There is evidence that a very small decrease in initial seed viability can result in substantial reduction in storage life. This needs further investigation, and research should be performed notably on the effect of germplasm handling in the field during regeneration and during subsequent processing stages prior to its arrival at the genebank, as well as on growing conditions, disease status, and time of harvest of the plants.

Research Priorities for Nonorthodox Seed Species

Understanding Seed Recalcitrance

A number of physical and metabolic processes or mechanisms have been suggested to confer, or contribute to, desiccation tolerance.[16] Different processes may confer protection against the consequences of loss of water at different hydration levels, and the absence, or ineffective expression, of one or more of these could determine the relative degree of desiccation sensitivity of seeds of individual species. Additional research is needed to improve our understanding of the mechanisms involved in seed recalcitrance.

Developing Improved Conservation Techniques

Various technical options exist for improving storage of nonorthodox species. Especially with species for which no or only little information is available, it is advisable, before undertaking any "high-tech" research, to examine the development pattern of seeds and to run preliminary experiments to determine their desiccation sensitivity as well as to define germination and storage conditions. IPGRI, in collaboration with numerous institutions worldwide, has developed a protocol for screening tropical forest tree seeds for their desiccation sensitivity and storage behavior,[17] which might be applicable to seeds of other species after required modification and adaptation. For long-term storage of nonorthodox species, cryopreservation represents the only option. Numerous technical approaches exist, including freezing of seeds, embryonic axes, shoot apices sampled from embryos, adventitious buds, or somatic embryos, depending on the sensitivity of the species studied.[4]

Germplasm Management Procedures

The objective of any genebank management procedure is to maintain genetic integrity and viability of accessions during conservation and to ensure their accessibility for use in adequate quantity and quality at the lowest possible cost. As mentioned earlier, many genebanks face financial constraints that hamper their efficient operation. It is therefore very important to improve genebank management procedures to make them more efficient and cost-effective. In this aim, the use of molecular markers should be increased in order to improve characterization and evaluation, improved seed regeneration and accession management procedures should be established, the use of collections should be enhanced through the development of core collections, germplasm health aspects including new biotechnological tools for detection, indexing, and eradication of pathogens should be better integrated in routine genebank operations, and improved documentation tools should be developed. Special attention should be given to developing appropriate techniques for genebanks in developing countries, where specialized equipment is frequently lacking and resources are usually limited.

Conclusion

The improvements in the seed storage techniques and genebank management procedures resulting from the research performed on the priority areas identified earlier will further increase the key role of seeds in the ex situ conservation of genetic resources of many species. However, it is now well recognized that an appropriate conservation strategy for a particular plant genepool requires a holistic approach, combining the different ex situ and in situ conservation techniques available in a complementary manner.[18] Selection of the appropriate methods should be based on a range of criteria, including the biological nature of the species in question, practicality and feasibility of the particular methods chosen (which depends on the availability of the necessary infrastructures), their efficiency, and the cost-effectiveness and security afforded by their application. An important area in this is the linkage between in situ and ex situ components of the strategy, especially with respect to the dynamic nature of the former and the static, but potentially more secure, approach of the latter.

References

1. FAO, The Second Report on the State of the World's Plant Genetic Resources for Food and Agriculture; Food and Agriculture Organization of the United Nations: Rome, 2010.
2. FAO/IPGRI Genebank Standards; Food and Agriculture Organization of the United Nations: Rome and International Plant Genetic Resources Institute: Rome, 1994.
3. Hong, T.D.; Linington, S.; Ellis, R.H. Seed Storage Behaviour: A Compendium; Handbooks for Genebanks; International Plant Genetic Resources Institute: Rome, 1996; Vol. 4.
4. Engelmann, F.; Takagi, H. Cryopreservation of Tropical Plant Germplasm—Current Research Progress and Applications; Japan International Center for Agricultural Sciences: Tsukuba, Japan and International Plant Genetic Resources Institute: Rome, 2000.
5. Li, D.Z.; Pritchard, H.W. The science and economics of ex situ plant conservation. Trends Plant Sci. **2009**, *14*, 614–621.
6. Probert, R.J.; Daws, M.I.; Hay F. Ecological correlates of ex situ seed longevity: a comparative study on 195 species. Ann. Bot. **2009**, *104*, 57–69.
7. http://www.ipgri.cgiar.org/system/page.asp?theme=7 (accessed March 2003).
8. Walters, C.; Wheeler, L.J.; Grotenhuis, J.M. Longevity of seeds stored in a genebank: species characteristics. Seed Sci. Res. **2005**, *15*, 1–20.
9. Walters, C.; Wheeler, L.J.; Stanwood, P.C. Longevity of cryogenically stored seeds. Cryobiology **2004**, *48*, 229–244.
10. Roberts, E.H.; Ellis, R.H. Water and seed survival. Ann. Bot. **1989**, *63*, 39–52.
11. Dickie, J.B.; Ellis, R.H.; Kraak, H.L.; Ryder, K.; Tompsett, P.B. Temperature and seed storage longevity. Ann. Bot. **1990**, *65*, 197–204.
12. Perez-Garcia, F.; Gonzalez-Benito, M.E.; Gomez-Campo, C. High viability recorded in ultra-dry seeds of 37 species of Brassicaceae after almost 40 years of storage. Seed Sci. Technol. **2007**, *35*, 143–155.
13. Pritchard, H.W.; Ashmore, S.; Berjak, P.; Engelmann, F.; González-Benito, M.E.; Li D.Z.; Nadarajan, J.; Panis, B.; Pence, V.; Walters, C. Storage stability and the biophysics of preservation. In: Proc. Plant conservation for the next decade: a celebration of Kew's 250th anniversary; Royal Botanic Garden Kew: London, UK, 2009; 12–16.
14. Walters, C.; Volk, G.M.; Stanwood, P.C.; Towill, L.E.; Koster, K.L.; Forsline, P.L. Long-term survival of cryopreserved germplasm: contributing factors and assessments from thirty-year-old experiments. Acta Hort. **2011**, *908*, 113–120.
15. Walters, C. Ultra-dry seed storage. Seed Sci. Res. **1998**, *8*, Suppl. No 1.
16. Pammenter, N.W.; Berjak, P. A review of recalcitrant seed physiology in relation to desiccation-tolerance mechanisms. Seed Sci. Res. **1999**, *9*, 13–37.
17. IPGRI/DFSC Desiccation and Storage Protocol. The Project on Handling and Storage of Recalcitrant and Intermediate Tropical Forest Tree Seeds, Newsletter, 1999; 23–39.
18. Maxted, N.; Ford-Lloyd, B.V.; Hawkes, J.G. Complementary Conservation Strategies. In *Plant Genetic Resources Conservation;* Maxted, N.; Ford-Lloyd, B.V.; Hawkes, J. G., Eds.; Chapman & Hall: London, 1997; 15–39.

16

Herbicide-Resistant Crops: Impact

Micheal D. K. Owen
Iowa State University

Introduction

The most important constraint on efficient production of food, fiber, and biofuel is the ability to manage weeds effectively. Weeds represent the most widespread and consistent pest to economic crop production. As such, weeding crops has required more global expenditure of energy and time than any other agronomic task. Historically, mankind has managed weeds by hand. Mechanical weed management began a new era in crop production efficiency. In the mid-20th century, the control of weeds with herbicides dramatically improved weed management and thus crop production efficiency. The next technological step in weed management was the development of herbicide-resistant (HR) crops, which were adopted at an unprecedented rate compared to any previous agricultural technology. HR crops, specifically those with resistance(s) attributable to genetic engineering (GE) have become a dominant feature in global agriculture since the mid-1990s, and their impact on mankind has been important.

Transgenes and Crops

The ability to transform plants by the introduction of selected genetic characteristics has dramatically changed the ability of plant breeders to improve agronomic crops.[1] The first GE crops contained transgenes that conferred resistance to herbicides, diseases, or insects.[2] Commercial utilization of GE crops has increased to over 160 million hectares worldwide of which 126 million hectares are GE crops with HR traits.[3,4] Although most of the GE HR crops are utilized in the industrialized countries, a total of 29 countries worldwide have planted GE crops.[4] Proponents of biotechnology suggest that increases in yields and food quality attributable to GE crops are critically important to developing countries.[3,5–9]

Although GE crops have been positioned by some to be environmentally benign and more beneficial than conventional crops, there continues to be concerns and debate that there may be undesirable characteristics and resultant negative impacts with the commercial-scale production of GE crops that has dominated agriculture in the last 16 years.[2] Specifically, the concerns reflect increased pesticide use, movement of transgenes to compatible weedy species, evolved resistance(s) in pest complexes, transgene contamination of food, concerns about soil and plant health attributable to pesticide use, and other perceived risks.[10–15] The concerns about increased pesticide use must be considered from the perspective of what is the appropriate comparison. In the case of HR crops, specifically those that include a transgene that provides resistance to glyphosate compared to previously used herbicides were used at a fraction of the active ingredient that glyphosate is used; so it is intuitively obvious that the herbicide amount used per treated field increased. However, the number of hectares treated with herbicides did not change.[3,15–18] Consider that inclusion of transgenes that code of the *Bacillus thuringiensis* (Bt) protein(s) has resulted in reduced use of insecticides. However, the recent evolution of resistance to Bt in specific insects pests suggests that there are issues with the technology.[14]

Transgenes that accrue glyphosate resistance in crops allow growers to use glyphosate more effectively and efficiently for weed management. Glyphosate is suggested to be an environmentally safer herbicide. However, there are concerns that the inclusion of transgenes that code for proteins may cause a hypoallergenic reaction in some people and other issues reported to be of great import.[12,19,20] Interestingly, there has not been a societal consensus that reflects the attitudes toward GE crops despite the unprecedented adoption of the technology by industrialized agriculture.[3,11,18,21,22]

As previously stated, one of the most important and widely successful concepts resulting from transgene technology is the introduction of herbicide resistance to crops. However, HR crops have also been developed with traditional breeding techniques. HR maize (*Zea mays*), soybean (*Glycine max*), canola (*Brassica napus*), rice (*Oryza sativa*), cotton (*Gossypium hirsutum*), and sugar beets (*Beta vulgaris*) are widely planted in North America, whereas HR wheat (*Triticum aestivum*) and sunflowers (*Helianthus annuus*) are commercially available albeit not widely adopted.[3] A primary benefit attributed to some HR crops is the ability to use environmentally benign herbicides such as glyphosate.

Crops with HR cultivars resistant to glyphosate include maize, soybean, cotton, canola, sugar beet, and alfalfa. Given the public concerns about herbicide use, the ability to use a herbicide with a favorable environmental profile is an important consideration supporting the adoption of HR crops.[3,12,17,21,23,24] Other pecuniary and non-pecuniary benefits supporting the adoption of HR crops include GE-conferred herbicide selectivity where no selective herbicides were previously available, choices of cheaper herbicides, ease or simplicity of weed management or both, increased adoption of tillage practices that minimize soil erosion and improve soil quality, and the elimination of herbicide injury to crops.[3,24–26]

Herbicides are the primary, if not sole tactic used to control weeds in most crops including HR crops. Farmers in the United States applied herbicides to almost 9 million crop hectares at a cost of $6.6 billion in 2003.[27] Herbicides are used annually on more than 90% of the row crop acres, and it has been suggested that without herbicides, crop production would decline by more than 20% and accounts for the efforts of 70 million laborers.[16] The adoption of HR crops has increased the emphasis on herbicidal weed control and minimized the use of alternative weed management tactics.[15,17,18] The unprecedented adoption of GE HR crops has changed herbicide use practices and the number of herbicides used, and resulted in a limited number of herbicides (e.g., glyphosate) being applied recurrently to a majority of the row crop acres.[28,29] This entry will provide an overview of the implications of GE HR crops on agriculture and describe in a broad sense the impact of these technologies on current agroecosystems and the socioeconomic considerations of an agriculture that is based on GE HR crop technologies.

Weed Management with HR Crops

Fundamentally, weed management with HR crops is no different than using other selective herbicide technologies—a herbicide is applied to the crop, and the weeds are selectively controlled. The difference is how selectivity between crops and weeds is achieved.[30] For most herbicides, selectivity between crops and weeds is the result of differential abilities to metabolize the herbicide to nonactive products. Crops are able to more effectively and efficiently metabolize the herbicide than the weeds.

Differential metabolism of herbicides is influenced by a number of factors including application timing relative to crop and weed stage of development and the overall growth conditions of the plants. The insertion of transgenes, or the selection for critical plant enzymes through traditional breeding programs, greatly elevates the differential response of the HR crop and weeds compared to conventional crops. The mechanisms by which the HR crop is protected from the herbicide is either by an alteration of the target enzyme (i.e., glyphosate resistance in soybean) or by the addition of genetic code that allows the crop plant to metabolize the herbicide (i.e., glufosinate in maize).[3,17,31,32]

As the utilization of HR crops does nothing to change the emphasis on herbicides for weed control and in fact has lessened the use of alternative tactics, public concerns for the negative impacts of herbicides on health and the environment cannot be discounted.[18,33] Furthermore, whereas the use of herbicides and HR crops is described to be generally beneficial to the environment, provides lower cost of weed management, and allows the farmer to use simpler weed management tactics, there are consequences from using HR crops and herbicides on weed management and weed populations.[15]

These consequences include weed population shifts that favor weed species not effectively managed by the target herbicide, the evolution of weed biotypes resistant to the target herbicide, the hybridization of the HR crops with closely related weed species and the potential to confer resistance to the resultant progeny, the likelihood that volunteer crops will become HR weeds in rotational crops, and the potential for the HR crops to contaminate non-HR and organically produced crops because of the inability to segregate grain or via pollen movement.[34–42] Weeds with evolved resistance to glyphosate have increased significantly since the introduction of HR crops with GE resistance to glyphosate and now represent a major problem for HR crop production in specific crop production systems[18,43–47] In addition, given that multiple applications of the target herbicides (e.g., glyphosate) are typically required to provide acceptable weed control, there is a greater opportunity for off-target movement of the herbicide.[37]

Generally, the adoption of HR crops and subsequent weed control has been an unprecedented major commercial success.[31] The ability to use registered herbicides that previously did not provide sufficient selectivity between crops and weeds provides an excellent potential to improve the profitability of agriculture and increase weed management options for farmers.[3,17] Furthermore, many crops do not represent a large-enough market opportunity for the agrochemical industry either because of the limited hectares grown or an unacceptable risk of crop injury relative to the profit potential from selling an existing herbicide. Thus, weed control options are limited in these crops.

Developing HR varieties of these crops will improve weed management options and thus increase yields, grain quality, and profitability.[39] Developing HR cultivars in "minor" crops could represent an important opportunity to growers who farm smaller areas and grow minor crops to achieve more efficient and less expensive weed management.

Impact of HR Crops on Crop Production and Economics

The adoption of HR crops has largely been driven by the perceived production benefits ascribed to the HR traits. For example, in Western Canada, an estimated 90% of the hectares seeded to canola are HR varieties.[48] The overall adoption of GE crops has increased 94-fold since the commercial introduction of the first GE crop in 1996 and represented an accumulated hectarage of 1.25 billion hectares.[4]

However, most of the HR crop adoption reflects the unprecedented acceptance of HR soybean. An estimated 73.8 million hectares of HR soybeans were seeded in 2010 and represented 50% of the GE crops grown worldwide.[4] HR alfalfa, maize, canola, cotton, sugar beet, and soybean accounted for 83% or 122 million hectares of the total 148 million hectares planted to GE crops in 2010.[4] Other HR crops that are approved for production include carnation (*Dianthus caryophyllus*), chicory (*Cichorium intybus*), linseed (*Linum usitatissimum*), and tobacco (*Nicotiana tabacum*).[49] As many as 14 other crops have transgenic HR cultivars currently under development worldwide.[2]

In total, the inclusion of HR crops represents a major change in crop production systems and specifically in weed management tactics, and has required major effort by regulative groups in the federal government to insure that food safety has not changed because of the inclusion of transgenes.[40] From a regulatory perspective, HR and non-HR crops are identical. Importantly, the Environmental Protection Agency has determined that the major HR transgenes are safe, and food developed from HR crops was determined to be no different than food derived from non-HR crops. Canadian regulatory authorities determined that the glyphosate-resistant canola was the same as other cultivars after conducting an extensive series of compositional, processing, and feeding studies.[50] There can be, however, pleiotropic consequences to the addition of the transgene in HR crops.[51]

The economic impact of HR crops on the global agricultural economy is clearly positive.[4] Farmers and manufacturers of the HR crops report that there are clear and consistent savings in weed management costs compared with weed management not based on HR technology. The 2011 global market value of GE crops was US$13.3 billion, of which 77% was in industrial countries and 23% in developing countries.[4] However, movement of the HR pollen into non-HR crops can result in contamination of the grain.[34,35,52] In addition, the grain-handling industry is not currently able to segregate HR grain from non-HR grain, further increasing the risk of contamination and subsequent economic penalty to the producers.

Conclusions

There can be no question that the development of HR technologies has been one of the most important and yet contentious changes in global agriculture. Whereas the adoption of HR technology by farmers has been unprecedented worldwide, acceptance of the technology by consumers is less clear.[53–55] In fact, the issues about the benefits and risks of HR crops are similar to those voiced by consumers about the adoption of pesticide technologies.[56] On the other hand, the introduction of HR traits by transgene technology could not be readily or efficiently accomplished by traditional crop-breeding tactics, with the possible exception of resistance to imidazolinone herbicides in several crops.

Regardless, the resultant HR traits are portrayed as environmentally and economically beneficial;[24,28,57] a major component of the debate is that the HR trait is based on herbicidal weed management.

The historic baggage of the herbicide-use debate has not placed the HR crops in a positive position with many consumers, particularly in the European Union.[20,22] However, it is clear that farmers have decided that HR crops represent an important benefit to agriculture.[25]

It is interesting to note that the technology that allowed the development of HR crops is predicted to provide new weed management tools not based on herbicides.[30,58] Three novel approaches to utilizing transgene technology in weed management were reported by Gressel.[59] These were the alteration of biocontrol agents to improve their effectiveness, the use of transgenes to develop more competitive crop either by improving the growth habit or allowing the production of natural allelochemicals, or the alteration of cover crops to make them more applicable in weed management.

Thus, whereas it is unclear what the future impact of HR crops will be on world food production, it is clear that transgene technology will continue to play an important role.

References

1. Noteborn, H.P.J.M.; A.A.C.M. Peijnenburg, and R. Zeleny, Strategies for analysing unintended effects in transgenic food crops. In *Genetically Modified Crops: Assessing Safety;* Atherton, K.T., Ed.; Taylor and Francis: London, 2002; 74–93.
2. Snow, A.A.; Palma, P.M. Commercialization of transgenic plants: potential ecological risks. BioScience **1997**, *47* (2), 86–96.
3. Green, J.M. The benefits of herbicide-resistant crops. Pest Manag. Sci. **2012**, *68* (10), 1323–1331.

4. James, C. *Global Status of Commercialized Biotech/GM crops:* 2011; ISAAA Brief No. 43: Ithaca, NY, 2011.
5. Chrispeels, M.J. Biotechnology and the poor. Plant Physiol. **2000,** *124* (1), 3–6.
6. Evenson, R.E.; Gollin, D. Assessing the impact of the green revolution, 1960 to 2000. Science **2003,** *300,* 758–762.
7. Gressel, J. Biotech and gender issues in the developing world. Nat. Biotechnol. **2009,** *27,* 1085–1086.
8. Gressel, J. Global advances in weed management. J. Agrlcultural Sci. **2011,** *149,* 47–53.
9. Qaim, M.; Zilberman, D. Yield effects of genetically modified crops in developing countries. Science **2003,** *299,* 900–902.
10. Benbrook, C.M. Impacts of genetically engineered crops on pesticide use in the U.S.—the first sixteen years. Environ. Sci. Eur. **2012,** *24,* 24.
11. Bonny, S. Herbicide-tolerant transgenic soybean over 15 years of cultivation: pesticide use, weed resistance and some economic issues. The case of the USA. Sustainability **2011,** *3* (9), 1302–1322.
12. Duke, S.O.; Lydon, J.; Koskinen, W.C.; Moorman, T.B.; Chaney, R.L.; Hammerschmidt, R. Glyphosate effects on plant mineral nutrition, crop rhizosphere microbiota, and plant disease in glyphosate-resistant crops. J. Agricultural Food Chem. **2012,** *60,* 10375–10397.
13. Ellstrand, N.C. Evaluating the risks of transgene flow from crops to wild species. In *Gene Flow among Maize Landraces, Improved Maize Varieties, and Teosinte: Implications for Transgenic Maize*; CIMMYT: Mexico City, 1997.
14. Gassman, A.J.; Petzold-Maxwell, J.L.; Keweshan, R.S.; Dunbar, M.W. Field-evolved resistance to Bt maize by western corn rootworm. Plos ONE **2011,** *6* (7), e22629.
15. Owen, M.D.K.; Young, B.G.; Shaw, D.R.; Wilson, R.G.; Jordan, D.L.; Dixon, P.M.; Weller, S.C.; Benchmark study on glyphosate-resistant cropping systems in the USA. II. Perspective. Pest Manag. Sci. **2011,** *67* (7), 747–757.
16. Gianessi, L.P.; Reigner, N.P. The value of herbicides in U.S. crop production. Weed Technol. **2007,** *21,* 559–566.
17. Green, J.M.; Owen, M.D.K. Herbicide-resistant crops: Utilities and limitations for herbicide-resistant weed management. J. Agricultural Food Chem. **2011,** *59,* 5819–5829.
18. Owen, M.D.K. Weed resistance development and management in herbicide-tolerant crops: experiences from the USA. J. Consum. Prot. Food Safety **2011,** *6* (1), 85–89.
19. Johal, G.S.; Huber, D.M. Glyphosate effects on diseases of plants. Eur. J. Agronomy. **2009,** *31,* 144–152.
20. Madsen, K.H.; Sandoe, P. Ethical reflections on herbicide-resistant crops. Pest Manag. Sci. **2005,** *61,* 318–325.
21. Duke, S.O. Comparing conventional and biotechnology-based pest management. J. Agric. Food Chem. **2011,** *59,* 5793–5798.
22. Ehlers, U. Interplay between GMO regulations and pesticide regulation in the EU. J. Consum. Prot. Food Safety. **2011,** *4,* 4.
23. Kudsk, P.; Streibig, J.C. Herbicides - a two-edged sword. Weed Res. **2003,** *43* (2), 90–102.
24. Cerdeira, A.L.,Duke, S.O., The current status and environmental impacts of glyphosate-resistant crops; a review. J. Environ. Qual. **2006,** *35,* 1633–1658.
25. Ervin, D.E.; Carriere, Y.; Cox, W.J.; Fernandez-Cornejo, J.; Jussaume R.A., Jr.; Marra, M.C.; Owen, M.D.K.; Raven, P.H.; Wolfenbarger, L.L.; Zilberman, D. *The Impact of Genetically Engineered Crops on Farm Sustainability in the United States, in National Research Council Report;* Grossblatt, N., Ed.; National Research Council: Washington, D.C., 2010; 250.
26. Frisvold, G.B.; Hurley, T.M.; Mitchell, P.D. Overview: Herbicide resistant crop - diffusion, benefits, pricing, and resistance management. AgBioForum **2009,** *12* (3, 4), 244–248.
27. Gianessi, L.P.; Sankula, S. Executive Summary: The value of herbicides in U.S. crop production. [computer website] 2003. http://www.ncfap.org (cited 2003 10/6/03).
28. Gianessi, L.P. Economic impacts of glyphosate-resistant crops. Pest Manag. Sci. **2008,** *64* (4), 346–352.

29. Young, B.G. Changes in herbicide use patterns and production practices resulting from glyphosate-resistant crops. Weed Technol. **2006**, *20*, 301–307.
30. Duke, S.O.; Weeding with transgenes. Trends Biotechnol. **2003**, *21* (5), 192–195.
31. Duke, S.O.; Powles, S.B. Glyphosate: a once-in-a-century herbicide. Pest Manag. Sci. **2008**, *64* (4), 319–325.
32. Green, J.M.; Hale, T.; Pagano, M.A.; Andreassi II, J.L.; Gutteridge, S.A. Response of 98140 corn with gat4621 and hra transgenes to glyphosate and ALS-inhibiting herbicides. Weed Sci. **2009**, *57*, 142–148.
33. Phipps, R.H.; Park, J.R. Environmental benefits of genetically modified crops: Global and European perspectives on their ability to reduce pesticide use. J. Anim. Feed Sci. **2002**, *11*, 1–18.
34. Legere, A. Risks and consequences of gene flow from herbicide-resistant crops: canola (*Brassica napus* L.) as a case study. Pest Manag. Sci. **2005**, *61*, 292–300.
35. Mallory-Smith, C.; Zapiola, M. Gene flow from glyphosate-resistant crops. Pest Manag. Sci. **2008**, *64* (4), 428–440.
36. Mallory-Smith, C.A.; Olguin, E.S. Gene flow from herbicide-resistant crops: It's not just for transgenes. J. Agricultural Food Chem. **2010**, *59*, 5813–5818.
37. Owen, M.D.K. Current use of transgenic herbicide-resistant soybean and corn in the USA. Crop Prot. **2000**, *19*, 765–771.
38. Owen, M.D.K. Weed species shifts in glyphosate-resistant crops. Pest Manag. Sci. **2008**, *64*, 377–387.
39. Orson, J. *Gene Stacking in Herbicide Tolerant Oilseed Tape: Lessons from the North American Experience*; English Nature: Peterbrough, UK, 2002; 17.
40. Goldman, K.A. Bioengineered foods - safety and labeling. Science **2000**, *290* (5491), 457–459.
41. Hodgson, E. Genetically modified plants and human health risks: can additional research reduce uncertainties and increase public confidence? Toxicol. Sci. **2001**, *63*, 153–156.
42. Kimber, I.; Dearman, R.J. Approaches to assessment of the allergenic potential of novel proteins in food from genetically modified crops. Toxicol. Sci. **2002**, *68*, 4–8.
43. Culpepper, A.S. Glyphosate-induced weed shifts. Weed Technol. **2006**, *20*, 277–281.
44. Culpepper, A.S.; Webster, T.M.; Sosnoskie, L.M.; York, A.C. Glyphosate-resistant Palmer amaranth in the United States. In *Glyphosate Resistance in Crops and Weeds;* Nadula, V.K., Ed.; John Wiley & Sons: New York, NY, 2010; 195–212.
45. Heap, I. The international survey of herbicide resistant weeds. 2012. http://www.weedscience.com.
46. Powles, S.B. Evolution in action: glyphosate-resistant weeds threaten world crops. Outlooks Pest Manag. **2008**, *12* (December), 256–259.
47. Powles, S.B. Evolved glyphosate-resistant weeds around the world: lessons to be learnt. Pest Manag. Sci. **2008**, *64* (4), 360–365.
48. Zhu, B.; Ma, B.-l. Genetically-modified crop production in Canada: agronomic, ecological and environmental considerations. Am. J. Plant Sci. Biotechnol. **2011**, *5*(1), 90–97.
49. Nap, J.-P.; Metz, L.J.; Escaler, M.; Conner, A.J. The release of genetically modified crops into the environment. Part I. Overview of current status and regulations. Plant J. **2003**, *33*, 1–18.
50. Taylor, M.; Stanisiewski, E.; Riordan, S.; Nemeth, M.; George, B.; Hartnell, G. Comparison of broiler performance when fed diets containing roundup ready-event RT73-, nontransgenic control or commercial canola meal. Poult. Sci. **2004**, *83*, 456–461.
51. Pline, W.A.; Wu, J.; Hatzios, K.K. Effects of temperature and chemical additives on the response of transgenic herbicide-resistant soybeans to glufosinate and glyphosate applications. Pest Biochem. Physiol. **1999**, *65* (2), 119–131.
52. Snow, A.A.; Andow, D.A.; Gepts, P.; Hallerman, E.M.; Power, A.; Tiedje, J.M.; Wolfenbarger, L.L.; Genetically engineered organisms and the environment: current status and recommendations. Ecol. Appl. **2005**, *15* (2), 377–404.
53. Jayaraman, K.; Jia, H. GM phobia spreads to South Asia. Nat. Biotechnol. **2012**, *30*, 1017–1019.

54. Meldolesi, A. Media leaps on French study claiming GM maize carcinogenicity. Nat. Biotechnol. **2012**, *30,* 1018.
55. Nording, L. Opposition thaws for GM crops in Africa. Nat. Biotechnol. **2012**, *30,* 1019.
56. William, R.D.; Ogg, A.; Baab, C.; My view. Weed Sci. **2001**, *49,* 149.
57. Strauss, S.H. Genomics, genetic engineering, and domestication of crops. Science **2003**, *300* (5616), 61–62.
58. Gressel, J.; Levy, A.A. Stress, mutators, mutations and stress resistance. In *Abiotic Stress Adaptation in Plants: Physiological, Molecular and Genomic Foundation*; Pareek, A., Sopory, S.K., Bohnert, H.J., Govindjee, Eds.; Springer Science: Dordrecht, 2009; 471–483.
59. Gressel, J. *Molecular Biology of Weed Control. Frontiers in Life Science;* Taylor and Francis: London, 2002; Vol. 1, 504.

17

Herbicide-Resistant Weeds

Carol
Mallory-Smith
Oregon State University

Background Information

Herbicide resistance is the inherited ability of a biotype to survive and reproduce following exposure to a dose of a herbicide that is normally lethal to the wild type.[1] Herbicide resistance is an evolved response to selection pressure by a herbicide.

Herbicide-resistant biotypes may be present in a weed population in very small numbers. The repeated use of one herbicide or herbicides with the same site of action allows these resistant plants to survive and reproduce. The number of resistant plants increases until the herbicide is no longer effective. There is no evidence that the herbicide causes the mutations that lead to resistance.

Cross-resistance is the expression of one mechanism that provides plants with the ability to withstand herbicides from different chemical classes.[2] For example, a single point mutation in the enzyme acetolactate synthase (ALS) may provide resistance to five different chemical classes including the widely used sulfonylurea and imidazolinone herbicides.[3] However, cross-resistance at the whole-plant level is difficult to predict because a different point mutation in the ALS enzyme may provide resistance to one chemical class and not others. Cross-resistance also can result from increased metabolic activity that leads to detoxification of herbicides from different chemical classes.

Multiple-resistance is the expression of more than one mechanism that provides plants with the ability to withstand herbicides from different chemical classes.[2] Weed populations may have simultaneous resistance to many herbicides. For example, a common waterhemp (*Amaranthus rudis*) population in Illinois is resistant to triazine and ALS-inhibiting herbicide classes. However, resistance to these two different classes of herbicides is endowed by two different mechanisms within the same plant.[4] The weed species has two target site mutations, one for each herbicide class. An annual ryegrass (*Lolium rigidum* Gaud.) population in Australia is resistant to at least nine different herbicide classes.[5] In this case, herbicide options may become very limited. As with cross-resistance, multiple resistance is difficult to predict; therefore, management of weeds with these types of resistance is complicated.

History of Herbicide Resistance

It was not long after the commercial use of herbicides that Abel[6] and Harper[7] discussed the potential for weeds to evolve resistance. However, the first well-documented example for the selection of a herbicide-resistant weed was triazine-resistant common groundsel (*Senecio vulgaris* L.), which was identified in 1968 in Washington State.[8] The resistant biotype was found in a nursery that had been treated once or twice annually for 10 years with triazine herbicides. There were earlier reports of differential responses within weed species, but these variable responses to herbicides were not necessarily attributed to resistance.[8]

To date, 281 herbicide-resistant biotypes from 168 species have been identified.[9] The reason that the biotype number and the species number are different is that the same species has been identified with resistance to different herbicides in different locations. Of the 168 species, 100 are dicots and 68 monocots. Resistance has occurred to most herbicide chemical families.

Mechanisms Responsible for Resistance

Several mechanisms theoretically could be responsible for herbicide resistance. Those mechanisms include reduced herbicide uptake, reduced herbicide translocation, herbicide sequestration, herbicide target-site mutation, and herbicide detoxification. In the cases where the resistance mechanism has been determined, the mechanism responsible in most instances has been either target-site mutations or detoxification by metabolism.[10]

Target-site mutations have been identified in weeds resistant to herbicides that inhibit photosynthesis, microtubule assembly, or amino acid production. Most often there is a point mutation, a single nucleotide change, which results in an amino acid change and is responsible for the resistance.[10] The shape of the herbicide binding site is modified and the herbicide can no longer bind.

Metabolism-based resistance does not involve the binding site of the herbicide, but instead the herbicide is broken down by biochemical processes that make it less toxic to the plant. Several groups of enzymes are involved in the process. Enzymes that are thought to be most important in herbicide metabolism are glutathione *s*-transferases and cytochrome P450 monooxygenases.[11,12]

Factors That Influence the Selection of Resistant Biotypes

Resistance usually occurs when a herbicide has been used repeatedly, either year after year or multiple times during a year, and the herbicide is highly effective, killing more than 90% of the treated weeds. The more effective a herbicide, the higher the selection pressure for resistance. Therefore, all herbicides do not exert the same selection pressure.

If a herbicide has only one site of action, it is easier to select a resistant biotype because only one mutation is needed. A herbicide that has soil residual activity will provide more selection pressure because the herbicide remains in the environment and any new seedlings will be exposed to the herbicide. If a herbicide has a very short residual, repeated applications during one growing season can have the same effect.

Herbicide factors influence the selection of herbicide-resistant weeds, but agronomic factors can also be important. Many resistant weed species have been selected in monoculture production systems. These systems result in the repeated use of the same herbicide. The increased reliance on herbicides for weed control along with a concurrent decrease in other weed management tactics further increases selection of resistant biotypes. The introduction of herbicide-resistant crops also increased the use of a single herbicide with decreased alternative controls.

Some weed species seem to be more prone to herbicide resistance than others. Some species have increased mutation rates or increased genetic variability. Increased variability is usually found in cross-pollinating species. Selection of resistant biotypes varies depending upon how likely it is for the resistance mutation to be lethal.

Weeds that produce more than one generation per year may be exposed to herbicides with the same site of action more often than those that produce only one generation per year. Because the selection of a resistant biotype is dependent on the number of individuals that are exposed to the herbicide, those weed species that produce more seeds may also produce a resistant individual more quickly.

Breeding systems and inheritance of the trait will influence how fast resistance spreads once it occurs. If the trait is controlled by one recessive gene, the heterozygote and the homozygote dominant plants will be susceptible, so only ¼ of the population will survive herbicide treatment. If the trait is dominant, the heterozygote and the homozygote dominant plants will be resistant, and the population will build quickly because ¾ of the plants will survive herbicide treatment. The trait will be readily moved within and between populations if the weed species outcrosses. In a selfing population, the trait will have reduced movement with pollen. Maternal inheritance will prevent herbicide resistance from moving with the pollen. An example of maternal inheritance is triazine resistance.[13]

Fitness is the reproductive ability of an individual, and competitive ability is the capacity of a plant to acquire resources. Initially, researchers assumed that herbicide-resistant weed species would have reduced fitness and competitive ability. Indeed, triazine-resistant weed species do have reduced growth and competitive ability.[14–16] However, weeds resistant to ALS-inhibiting herbicides have not consistently had reduced fitness or competitive ability.[17,18]

Refuges are a common tactic for the management of insect and pathogen resistance. Susceptible populations are maintained in surrounding areas and can be used to swamp resistance alleles. Generally, pollen and seed are not sufficiently mobile to swamp resistant weed populations. Migration is probably more important in the movement of resistance to a susceptible population than vice versa. This is particularly true with the tumbleweeds. Long-distance movement of a herbicide resistance gene out of an area is most likely to occur through seed movement.

Prevention and Management of Herbicide-Resistant Weeds

Any management strategy that reduces the selection pressure from a herbicide will reduce the selection for a herbicide-resistant weed in the system. Recommendations for the prevention or management of herbicide-resistant weeds are often the same.[1] The recommendations from the herbicide industry and university personnel include many common factors. Common recommendations are to rotate herbicides with different sites of action to reduce selection pressure, to use short-residual herbicides so that selection pressure during a cropping season is reduced, to use multiple weed-control methods in conjunction with herbicides, and to plant certified crop seed so that herbicide-resistant weed seeds are not introduced into a field. Growers need to keep accurate records of herbicides that have been used on a field so that they can adopt a weed management plan that reduces the selection of herbicide-resistant weeds. The integration of these recommendations will reduce selection pressure for herbicide-resistant weeds.

Conclusion

Herbicide-resistant weeds will continue to be an issue for weed management as long as herbicides are used for weed control. Herbicide-resistant weeds can be managed, and no herbicides have been removed from the marketplace because of resistance. When growers use multiple weed-management techniques, the selection of herbicide-resistant weeds is reduced. An integrated approach using crop rotation, herbicides with different sites of action, and alternative weed control such as physical and mechanical weed control is useful in both the prevention of herbicide-resistant weeds and the management of herbicide-resistant weeds if they do occur.

References

1. Retzinger, E.J.; Mallory-Smith, C. Classification of herbicides by site of action for weed resistance management strategies. Weed Sci. **1997**, *11* (2), 384–393.
2. Hall, L.M.; Holtum, J.A.M.; Powles, S.B. Mechanisms responsible for cross resistance and multiple resistance. In *Herbicide Resistance in Plants*; CRC Press: Boca Raton, FL, 1994; 243–261.
3. Tranel, P.J.; Wright, T.R. Resistance of weeds to ALSinhibiting herbicides: What have we learned? Weed Sci. **2002**, *50* (6), 700–712.
4. Foes, M.J.; Liu, L. A biotype of common waterhemp (*Amaranthus rudis*) resistant to triazine and ALS herbicides. Weed Sci. **1998**, *46* (5), 514–520.
5. Burnet, M.W.M.; Hart, Q.; Holtum, J.A.M.; Powles, S.B. Resistance to nine herbicide classes in a population of rigid ryegrass (*Lolium rigidum*). Weed Sci. **1994**, *42* (3), 369–377.
6. Abel, A.L. The rotation of weedkillers. Proc. Brit. Weed Control Conf. **1954**, *2*, 249–255.
7. Harper, J.C. The evolution of weeds in relation to resistance to herbicides. Proc. Brit. Weed Control Conf. **1956**, *3*, 179–188.
8. Ryan, G.F. Resistance of common groundsel to simazine and atrazine. Weed Sci. **1970**, *18* (5), 614–616.
9. Heap, I. The International Survey of Herbicide Resistant Weeds, http://www.weedscience.com (accessed August 2003).
10. Preston, C.; Mallory-Smith, C.A. Biochemical mechanisms, inheritance, and molecular genetics of herbicide resistance in weeds. In *Herbicide Resistant Weed Management in World Grain Crops*; CRC Press: Boco Raton, FL, 2001; 23–60.
11. Barrett, M. The role of cytochrome P450 enzymes in herbicide metabolism. In *Herbicides and Their Mechanisms of Action*; CRC Press: Boca Raton, FL, 2000; 25–37.
12. Edwards, R.; Dixon, D.P. The role of glutathione in herbicide metabolism. In *Herbicides and Their Mechanisms of Action*; CRC Press: Boca Raton, FL, 2000; 38–71.
13. Souza-Machado, V.; Bandeen, J.D.; Stephenson, G.R.; Lavigne, P. Uniparental inheritance of chloroplast atrazine tolerance in *Brassica campestris*. Can. J. Plant Sci. **1978**, *58* (4), 977–981.
14. Conard, S.G.; Radosevich, S.R. Ecological fitness of *Senecio vulgaris* and *Amaranthus retroflexus* biotypes susceptible and resistant to atrazine. J. Appl. Ecol. **1979**, *16* (1), 171–177.
15. Marriage, P.B.; Warwick, S.I. Differential growth and response to atrazine between and within susceptible and resistant biotypes of *Chenopodium album*. L. Weed Res. **1980**, *20* (5), 9–15.
16. Williams, M.M.I.I.; Jordan, N. The fitness cost of triazine resistance in jimsonweed (*Datura stramonium* L.). Am. Midl. Nat. **1995**, *133* (1), 131–137.
17. Alcocer-Rutherling, M.; Thill, D.C.; Shafii, B. Differential competitiveness of sulfonylurea resistant and susceptible prickly lettuce (*Lactuca serriola*). Weed Tech. **1992**, *6* (2), 303–309.
18. Thompson, C.R.; Thill, D.C.; Shafii, B. Growth and competitiveness of sulfonylurea-resistant and -susceptible kochia (*Kochia scoparia*). Weed Sci. **1994**, *42* (2), 172–179.

18

Herbicides in the Environment

Kim A. Anderson
and Jennifer
L. Schaeffer
Oregon State University

Introduction

Herbicides are an integral part of our society; they are used by the general public, governments, institutions, foresters, and farmers. The benefits of herbicides need to be delivered without posing unacceptable risk to non-target sites. Therefore, understanding (and predicting) the environmental fate of herbicides is critically important. Environmental fate is determined by the individual fate processes of transformation and transport. Chemical and physical processes are primary determinants of transformation and transport of herbicides applied to soils and plants. Transformation determines what herbicides are degraded to in the environment and how quickly, while transport determines where herbicides move in the environment and how quickly. Jointly, these processes affect how much of a pesticide and its metabolites (degradation products) are present in the environment, where, and for how long. Herbicide fate also varies in response to changes in environmental conditions and application management practices (e.g., spray drift, volatilization), so it is important to understand these variables as well.

Major environmental compartments for herbicides can be considered as surface waters, the subsurface (soil and groundwater), and the atmosphere. Each medium has its own unique characteristics; however, there are many similarities when considering herbicide movement. Herbicides are rarely restricted to only one medium; therefore, chemical exchange among the compartments must be considered.

Wind erosion, volatilization, photo-degradation, runoff, plant uptake, sorption to soil, microbial or chemical degradation and leaching are potential pathways for loss from an application site. Chemical and microbial degradation are critically important factors in the fate of herbicides. Chemical conditions in soil are important secondary determinants of herbicide transport and fate. The importance of interactions between herbicides and solid phases of soils, soil water, and air within and above soil depends on a variety of chemical factors. Adsorption of herbicides from soil water to soil particle is one of the most important chemical determinants that limit mobility in soils. Environmental fate of herbicides depends on the chemical transformations, degradation, and transport in each environmental compartment.

Transformation

Transformations determine how long a herbicide will stay in the environment. Molecular interactions of herbicides are based in part on the herbicide's chemical nature and are predicated on the physical-chemical properties and reactivities of the herbicide. Several generalized exchange processes between compartments are shown in Figure 18.1. In order to understand (and therefore predict) environmental fate, the physical-chemical properties of herbicides must be known. Physical-chemical properties such as molecular formula, molecular weight, boiling point, melting point, decomposition point, water solubility, organic solubility, vapor pressure (V_p), Henry's law constant (K_H), octanol/ water partitioning (K_{ow}), acidity constant (pKa), and soil sorption (K_d, K_{oc}) are important in predicting transformations. How some of these chemical-physical properties affect herbicide fate is briefly discussed (Figure 18.2). Table 18.1,[1-4] demonstrates the wide variation of chemical-physical properties of a selected group of herbicides. Water solubility, the solubility of a herbicide in water at a specific temperature, determines the affinity for aqueous media and affects movement between air, soil, and water compartments. Vapor pressure, the pressure of the vapor of a herbicide at equilibrium with its pure condensed phase, measures a herbicide's tendency to transfer to and from gaseous environmental phases. Vapor pressure is critical for predicting either the equilibrium distribution or the rate of exchange to and from natural waters. Henry's law constant (K_H), the air–water distribution ratio for neutral compounds in dilute solutions in pure water, determines how a chemical will distribute between the gas and aqueous phase at equilibrium. Henry's law constant, therefore, only approximates the air–water partition in natural waters. Octanol–water partition coefficient (K_{ow}) the partition of organic compounds between water and octanol, is used to estimate the equilibrium partitioning of nonpolar organic compounds between water and organisms. The octanol–water partition coefficient is also proportional to partitioning of organic compounds (herbicides) into soil humus and other naturally occurring organic phases.

In the environment, many herbicides are present in a charged state (not neutral). Charged species have different properties and reactivities as compared to their neutral counterparts. Therefore, the extent to which a compound forms ions in environmental ecosystems is important. The pKa is a measure of the strength of an acid relative to water. Strong organic acids (pKa $\cong$ 0–3) in ambient natural waters (pH 4–9)

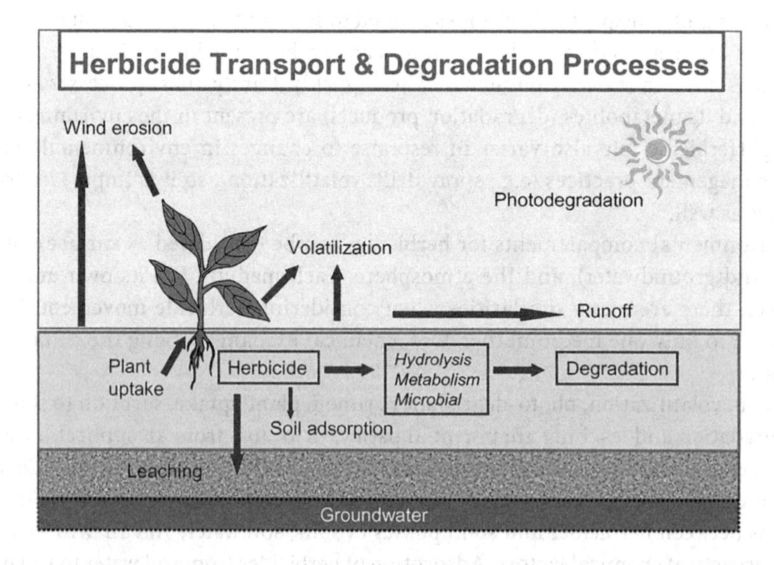

FIGURE 18.1 Illustrated are pathways for loss of herbicide from a herbicide application site. Each pathway (arrow) illustrates a particular loss process occurring at some rate constant (k). The importance of each pathway and magnitude of each rate constant will vary substantially between herbicides and environmental conditions.

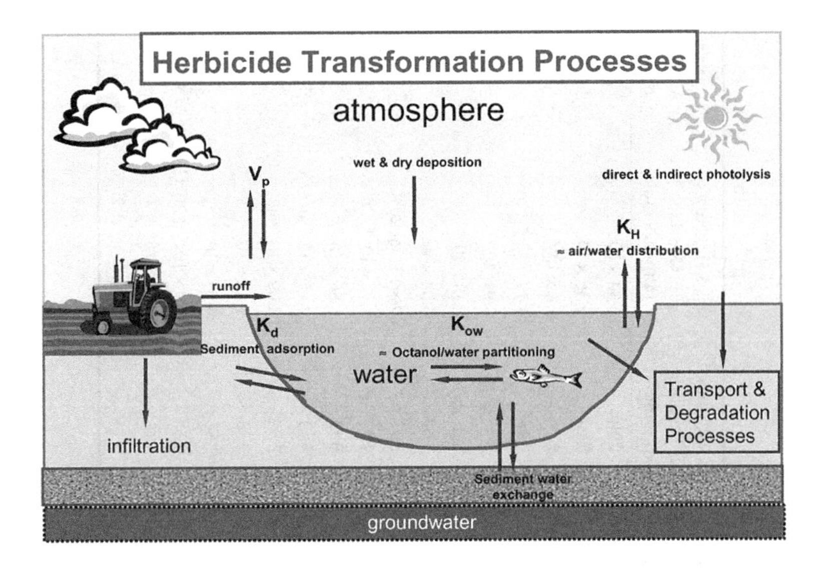

FIGURE 18.2 Illustrated are pathways for transformation processes for loss of herbicide from a herbicide application site. Each pathway (arrow) illustrates a particular loss process occurring at some rate constant (k). The importance of each pathway and magnitude of each rate constant will vary substantially between herbicides and environmental conditions.

will be present predominantly as anions. Conversely, very weak acids (pKa ≥ 12) in ambient natural waters will be present in their associated form (neutral). In an analogous fashion, strong bases (pKa ≥ 11) will be present as ions.[5] Examples of weak acids are 2,4-D and triclopyr, and examples of weak bases are atrazine, dicamba, and simazine (Table 18.1).

Sorption of a herbicide is when the herbicide binds to the soil or sediment particles (K_d). Soil sorption is an important process since it can dramatically affect herbicide fate. Figure 18.2 illustrates the relationship between herbicides sorbed to soil particles versus those dissolved in soil water. Soil sorption is dependent on soil type and influenced heavily by the organic matter content. Typically, soils high in clay and organic matter have a higher sorption capacity. To account for organic matter content, sorption is sometimes normalized for % organic matter, referred to as K_{oc}. K_d, K_H, and V_p describe the potential for exchange of herbicide compounds among soil, water, and air over short distances by diffusion. Transport over longer distances involves mass transfer (advection-dispersion).

Transport

Transport of herbicides from runoff from soil is an important route of entry into surface waters. Herbicide runoff can be classified into two categories: herbicides that are either dissolved or suspended in runoff waters. Dissolved herbicides are characterized by low adsorption and high water solubility, while suspended herbicides are characterized by high soil sorption. Leaching depends strongly on local environmental conditions, such as percolation rates through local soils. Adsorption decreases herbicide leaching mobility by reducing the amount of herbicide available to the percolating soil water. However, herbicide runoff and leaching also are controlled by the amount of herbicide degradation.

Degradation

Herbicide degradation ultimately ends with the formation of simple stable compounds, such as carbon dioxide; however, there are intermediates of varying stability on the way to complete mineralization (e.g., H_2O and CO_2). The rate of degradation of a particular herbicide can vary widely. The chemical

TABLE 18.1 Physiochemical Properties of Selected Herbicides[a]

Chemical Name	CAS Number	Chemical Class	Solubility (mg/L)	Vapor Pressure (mm Hg)	Kd Soil Sorption Koc Normalized for Organic Intent	Log Kow Octanol/ Water Partition Constant	KH Henry's Constant (atm-m^3/ mole)	pKa Acidity Constant
2,4-D	94-75-7	Chlorophenoxy acid	677	8.25×10^{-5} at 20°C	0.08–0.94	2.81	3.54×10^{-8}	2.73
Atrazine	1912-24-9	Triazine	34.7	2.89×10^{-7}	K_{oc} 100	2.61	2.36×10^{-9}	1.7
Bromoxynil	1689-84-5	Hydroxybenzonitrile	130	4.8×10^{-6}	K_{oc} 2.48	2	1.4×10^{-6}	4.06
Carbaryl	63-25-2	Carbamate	110 at 22°C	1.36×10^{-6}	2.45–1.69	2.36	3.27×10^{-9}	NA
Chlorsulfuron	64902-72-3	Sulfonylurea	2.8	2.3×10^{-11}	K_{oc} 1.02	2	3.9×10^{-15}	3.6
Dicamba	1918-00-9	Benzoic acid	8310	3.38×10^{-5}	0.07–0.53	2.21	2.18×10^{-9}	1.97
Diuron	330-54-1	Substituted urea	42	6.9×10^{-8}	2.9–14	2.68	5.04×10^{-10}	–1
Metolachlor	051218-45-2	Chloroacetanilide	530 at 20°C	3.14×10^{-5}	1.5–10	3.13	9×10^{-9} at 20°C	NA
Metsulfuron	74223-64-6	Sulfonylurea	9500	2.5×10^{-12}	0.36–1.40	2.2	1.32×10^{-16}	3.64
Pendimethenalin	40487-42-1	2,4-Dinitroanaline	0.3 at 20°C	3×10^{-5}	30–380	5.18	8.56×10^{-7}	NA
Pentachlorophenol	87-86-5	Chlorinated phenol	14	1.1×10^{-4}	K_{oc} 262–38905	5.12	2.45×10^{-8} at 22°C	4.7
Prometon	1610-18-0	Triazine	750	2.3×10^{-6} at 20°C	0.373–2.61	2.99	9.09×10^{-10} at 20°C	4.3
Simazine	122-34-9	Triazine	6.2 at 22°C	2.21×10^{-8}	0.48–1.31	2.18	9.42×10^{-10}	1.62
Thifensulfuron	79277-27-3	Sulfonylurea	230	1.28×10^{-10}	0.08–1.38	1.56	4.08×10^{-14}	4
Triasulfuron	82097-50-5	Sulfonylurea	815	1.5×10^{-8}	K_{oc} 73.4–190.6	0.58	9.9×10^{-7} at 20°C	4.64
Tribenuron	101200-48-0	Sulfonylurea	2040	3.97×10^{-10}	0.19–2.0	1.17	1.01×10^{-8}	5
Trifluralin	1582-09-8	Dinitro analine	0.184	4.58×10^{-5}	18.6–54.8	5.34	1.03×10^{-4} at 20°C	NA

[a] Data at 25°C unless otherwise noted.

nature of the herbicide is important, as discussed previously, but the degradation rate also depends on the availability of other reactants, as well as environmental factors. There are three major degradation pathways for herbicides: photo-degradation, chemical degradation, and microbial degradation.

Photo-degradation is the breakdown of a herbicide by sunlight at the plant, soil, or water surface. Direct photolysis occurs when the herbicide absorbs light (some portion of the solar spectrum), and this leads to dissociation of some kind (e.g., bonds break). Indirect photolysis occurs when a sensitizer molecule is radiatively excited and is sufficiently long-lived to transfer energy, such as an electron, a hydrogen atom, or a proton to another "receptor" molecule. The receptor molecule (herbicide), without directly absorbing radiation, can be activated via the "sensitizer" molecule to undergo dissociation or other kinds of chemical reactions leading to photo-degradation.

Chemical degradation is the breakdown of herbicides by processes not involving living organisms (abiotic). Hydrolysis may be one of the more important mechanisms of degradation for herbicides. Hydrolysis is the reaction of a herbicide where water interacts with the herbicide, replacing a portion of the molecule with OH. The following functional groups and chemical classes are known to be susceptible to hydrolysis: ethers, amides, phenylurea compounds, nitrile, carbamates, thiocarbamates, and triazines. Hydrolysis is influenced by environmental conditions including pH, water hardness, dissolved organic matter, dissolved metals, and temperature. Half life ($t_{1/2}$) is the amount of time it takes the parent compound (herbicide) to decay to half its original concentration. The half life for chemical degradation via the hydrolysis pathway is strongly dependent on environmental conditions. For example, the $t_{1/2}$ for 2,4-D at pH 6 is four years, while the $t_{1/2}$ at pH 9 is 37 hours.[6]

Microbial degradation is the breakdown of compounds (herbicides) by microorganisms. Bacteria and fungi are the primary microorganisms responsible for biotransformations (biotic reactions). Rates of microbial degradation are largely determined by environmental conditions such as temperature, pH, reduction-oxidation conditions, moisture, oxygen, organic matter, and food. Complete biodegradation, mineralization, yields carbon dioxide, water, and minerals.

Conclusion

The perfect herbicide would be one that controls weeds as necessary and then quickly breaks down and never moves off-site. The development of new herbicides over the last decade has focused in part on minimizing unacceptable risk to non-target sites. This has resulted in the development of more biologically active herbicides that greatly reduce application rates. Other research developments include minimizing the leaching of herbicides. When herbicides are applied in the environment, many transport and transformation processes are involved in their dissipation. Predicting herbicide environmental fate and behavior is complicated and is determined by the chemical processes for each environmental compartment.

In addition to understanding the fate of herbicides in the environment, emerging research also endeavors to look at the risk to non-target organisms. The risk of herbicides to organisms is relative to the bioavailability of the herbicide. Bioavailability may be approximately inversely proportional to K_d. Herbicides with larger K_ds tend to have lower bioavailability. Along with the determination of transport and transformation chemical processes, understanding the complete risk to non-target sites, therefore, should include the determination of the bioavailability of the herbicide.

Acknowledgments

The authors thank Dr. Jeffery Jenkins, Oregon State University, for helpful discussions.

References

1. University of Guelph Ontario Agricultural College. http://www.uoguelph.ca/OAC/env/bio/PROFILES.pdf, (accessed on March 2003).

2. The Extension Toxicology Network. http://ace.orst.edu/info/extoxnet/pips/ghindex.html, (accessed March 2003).
3. United States Department of Agriculture, Agricultural Research Services. http://wizard.arsusda.gov/acsl/ppdb3.html (accessed March 2003).
4. Environmental Science Syracuse Research Corporation. http://esc.syrres.com/interkow/physdemo.htm (accessed March 2003).
5. Montgomery, J.H. *Agrochemicals Desk Reference: Environmental Data;* Lewis Publishers: Ann Arbor, 1993.
6. Kamrin, M.A. *Pesticide Profiles: Toxicity, Environmental Impact, and Fate;* CRC Press LCC: New York, 1997.

19

Insects: Economic Impact

David Pimentel
Cornell University

Introduction

Insects, plant pathogens, and weeds are major pests of crops in the United States and throughout the world. Approximately 70,000 species of pests attack crops, with about 10,000 species being insect pests worldwide. This entry focuses on the economic consequences for agriculture, but it is important to establish that there are serious consequences for humans and other animals in the food chain from insects—topics that cannot be explored in detail in this entry.

Scope of Problem

There are ~2500 insect and mite species that attack U.S. crops. About 1500 of these species are native insect/mite species that moved from feeding on native vegetation to feeding on our introduced crops. Thus, an estimated 60% of the U.S. insect pests are native species and ~40% are introduced invader species.[1,2] Both groups of pests contribute almost equally to the current annual 13% crop losses to insect pests, despite all the insecticides that are applied plus other types of controls practiced in the United States. Most or ~99% of U.S. crops are introduced species; therefore, many U.S. native insect species find these introduced crops an attractive food source.[3] For example, the Colorado potato beetle did not feed originally on potatoes but fed on a native weed species. Although the center of origin of the potato is the Andean regions of Bolivia and Peru, the plant was introduced into the United States from Ireland, where it has been cultivated since the mid-17th century.[3] Today, the Colorado potato beetle is the most serious pest of potatoes in the United States and elsewhere in the world where both the potato and potato beetle have been introduced.

An estimated 3 billion kg of pesticides are applied annually in the world in an attempt to control world pests.[4] Approximately 40% of the pesticides are insecticides, 40% herbicides, and 20% fungicides. About 1 billion kg of pesticides are applied within the United States; however, only 20% are insecticides.[4] Most or 68% of the pesticides applied in the United States are herbicides.

Despite the application of 3 billion kg of pesticides in the world, >40% of all crop production is lost to the pest complex, with crop losses estimated at 15% for insect pests, 13% weeds, and 12% plant pathogens.[4]

In the United States, total crop losses are estimated at 37% despite the use of pesticides and all other controls, with crop losses estimated at 13% due to insect pests, 12% weeds, and 12% plant pathogens.

The 3 billion kg of pesticides applied annually in the world costs an estimated $32 billion per year or slightly >$10 per kg. With ~1.5 billion ha of cropland in the world, it follows that ~2 kg of pesticides is applied per ha, or $20 per ha is invested in pesticide control.

Insect Pests

Note that despite a tenfold increase in total quantity (weight) of insecticide used in the United States since about 1945 when synthetic insecticides were first used, crop losses to insect pests have nearly doubled from ~7% in 1945 to 13% today.[1] Actually, the situation is a great deal more serious because during the past 40 years, the toxicity of pesticides has increased 100–200 fold. For example, in 1945 many of the insecticides were applied at ~1 to 2 kg per ha; however, today many insecticides are applied at dosages of 10–20 grams per ha.[1]

The increase in crop losses to insect pests from 7% in 1945 to 13% today, despite a 100–200-fold increase in the toxicity of insecticides applied per ha, is due to changes in agricultural technologies over the past 50 years.[2] These changes in agricultural practices include the following: 1) the planting of some crop varieties that are more susceptible to insect pests than those planted previously; 2) the destruction of natural enemies of some pests by insecticides (e.g., destruction of cotton bollworm and bud-worm natural enemies), thereby creating the need for additional insecticide applications; 3) insecticide resistance developing in insect pests, thus requiring additional applications of more toxic insecticides; 4) reduction in crop rotations, which caused further increases in insect pest populations (e.g., corn rootworm complex); 5) lowering of the Food and Drug Administration (FDA) tolerances for insects and insect parts in foods, and enforcement of more stringent "cosmetic standards" for fruits and vegetables by processors and retailers; 6) increased use of aircraft application of insecticides for insect control, with significantly less insecticide reaching the target crop; 7) reduced field sanitation and more crop residues for the harboring of insect pests; 8) reduced tillage that leaves more insect-infested crop residues on the surface of the land; 9) culturing some crops in climatic regions where insect pests are more severe; and 10) the application of some herbicides that improve the nutritional makeup of the crop for insect invasions.[2]

The 13% of potential U.S. crop production represents a crop value lost to insect pests of ~$35 billion per year. In addition, nearly $2 billion in insecticides and miticides are applied each year for control. Ignoring other control costs, combined crop losses and pesticides thus total ~$37 billion per year.[3]

Total crop losses to insects and mites worldwide are estimated to be ~$400 billion per year. Given that >4.5 billion of the 7 billion people on earth are malnourished, this loss of food to insect and mite pests each year, despite the use of 3 billion kg of pesticides, is an enormous loss to society.

Insecticide and Miticide Pest Controls

In general, insecticide and miticide control of insects and mites return ~$4 for every $1 invested in chemical control.[1] This is an excellent return, but not as high as some of the nonchemical controls. For example, biological pest control has reported earnings of $100–$800 per $1 invested in pest control. It must be recognized that the development of biological controls, although highly desirable, are not easy to develop and implement.

Both in crops and nature, host plant resistance and natural enemies (parasites and predators) play an important role in pest control. In nature, seldom do insect pests and plant pathogens remove >10% of the resources from the host plant. Host plant resistance, consisting of toxic chemicals, hairiness, hardness, and combinations of these, prevents insects from feeding intensely on host plants.[5] Some of the chemicals involved in plants resisting insect attack include cyanide, alkaloids, tannins, and others. Predators and parasites that attack insects play an equally important role in controlling insect attackers on plants in nature and in agro ecosystems.

Insect Transmission of Plant Pathogens

Insects with sucking mouth parts, such as aphids and plant bugs, play a major role in the transmission of plant pathogens from plant to plant. It is estimated that ~25% of the plant pathogens are transmitted by insects. The most common pathogens transmitted are viruses. These pathogens include lettuce yellows and pea mosaic virus. Fungal pathogens are also transmitted by insects. For example, Dutch elm disease is transmitted by two bark beetle species that live under the bark of elm trees. In infected trees, the bark beetles become covered with fungal spores. When the beetles disperse and feed on uninfected elm trees, they leave behind fungal spores that in turn infect the healthy elm trees.

Environmental and Public Health Impacts of Insects

Some insect species have become environmental and public health pests. For example, the imported red fire ant kills poultry chicks, lizards, snakes, and ground-nesting birds. Investigations suggest that the fire ant has caused a 34% decline in swallow nesting success as well as a decline in the northern bob white quail populations in the United States. The ant has been reported to kill infirm people and people who are highly sensitive to ant sting. The estimated damages to wildlife, livestock, and public health in the United States is >$1 billion per year, with these losses occurring primarily in southern United States.[3]

In another example, the Formosan termite that was introduced into the United States has been reported to cause >$1 billion per year in property damage, repairs, and controls. As it spreads further in the nation, the damages will increase.[3]

Conclusion

Insect and mite pests in the United States and throughout the world are causing significant crop, public health, and environmental damages. Just for crop losses in the United States, it is estimated that insect and mite species are causing $37 billion per year, if control costs are included. Worldwide crop losses to insects and mites are estimated to be $400 billion per year. The public health and environmental damages in the United States and throughout the world are estimated to be valued at several hundred billion dollars per year.

References

1. Pimentel, D. *Handbook on Pest Management in Agriculture*; Three Volumes; CRC Press: Boca Raton, FL, 1991; 784 pp.
2. Pimentel, D.; Lehman, H. *The Pesticide Question: Environment, Economics and Ethics*; Chapman and Hall: New York, 1993; 441 pp.
3. Pimentel, D.; Lach, L.; Zuniga, R.; Morrison, D. Environmental and economic costs of non-indigenous species in the United States. BioScience **2000**, *50*(1) 53–65.
4. Pimentel, D. *Techniques for Reducing Pesticides: Environmental and Economic Benefits*; John Wiley & Sons: Chichester, UK, 1997; 456 pp.
5. Pimentel, D. Herbivore population feeding pressure on plant host: Feedback evolution and host conservation. Oikos **1988**, *53*, 289–302.

20

Insects: Flower and Fruit Feeding

Antônio R. Panizzi
Embrapa Trigo

Introduction

Flowers are frequently visited by insects searching for food, such as pollen and nectar.[1] The impact of such insects can be detrimental, beneficial, or mutualistic. Some insects will feed on flowers and damage them. Fruits are damaged by a great variety of insects, which look for seeds that are very rich in nitrogen.[2] Others will feed on the fruit itself.[3,4]

Every year, millions of dollars are spent to control the insect pests of flowers, and, in particular, those feeding on fruits and seeds. However, insects visiting flowers are also beneficial, by dispersing seeds and pollinating several major crops and vegetables.[5] Mutualistic interactions occur between flowers and honey bees, *Apis mellifera* L. Whereas bees facilitate pollination, flowers provide pollen and nectar for developing bees in hives. This mutualistic interaction led to domestication of the honey bee.

Detrimental Insects

Hundreds of different species of insects feed on flowers and fruits. These include larvae and adults of beetles (Coleoptera), butterflies and moths (Lepidoptera), flies (Diptera), sawflies, wasps, ants, and bees (Hymenoptera), and the "true bugs" (Heteroptera) (Figure 20.1). In addition, scale insects (Homoptera) feed on the surface of fruits, and the so-called thrips (Thysanoptera) damage flowers.[3,4,6-9]

Mouthparts

Insects feeding on flowers, fruits, and seeds have different types of mouthparts that allow them to obtain different types of food. The most common one is the chewing mouthparts, when insects use their mandibles to cut the plant tissue to penetrate the hard surface of fruits and seeds. Chewing mouthparts also allow some beetles to feed on flowers, by cutting the petals.

A second type is the sucking mouthparts where mouthparts are modified into a long sharp tube with two internal openings, one used to pump in saliva into the plant tissue, and the other is used to suck up the fluids. Sucking insects will feed in the following ways: stylet-sheath feeding, lacerate-and-flush

FIGURE 20.1 The southern green stink bug, *Nezara viridula* (L.) (Heteroptera). It feeds on seeds and on fruits of a wide variety of crops, vegetables, and fruit trees worldwide.

feeding, macerate-and-flush feeding, and osmotic pump feeding.[7] In the stylet-sheath feeding the bugs insert their stylets in the tissue, mostly in the phloem, destroying few cells, and a stylet sheath is produced, which remains in plant tissues. In the lacerate-and-flush feeding, the bugs move their stylets vigorously back and forth, and several cells are lacerated. In the macerate- and-flush feeding type the cells are macerated by the action of salivary enzymes. In the last two cases, the cell contents are injected with saliva, resulting in several cells damaged. Finally, the osmotic pump feeding occurs through the secretion of salivary enzymes injected in the plant tissue to increase osmotic concentration of intercellular fluids, which are then sucked, leaving empty cells around the stylets.

A third type of mouthparts is the rasping-sucking. In this case the insects scratch the plant tissue and suck up the emerging plant fluid. This type of mouthpart is less common and is typical of the small insects called "thrips."

Parts Damaged

Flowers, fruits, and seeds are damaged in many different ways. For instance, flowers may be damaged by chewing insects that either destroy the petals or may feed on the pollen. Many insects feed on the nectar, but in this case they do not damage the flowers.

Fruits are damaged in many ways, such as on the surface, by scale insects (Homoptera) or chewing insects, or to the inner parts, in particular by boring pests.[3,4]

Seeds on the plant and in storage are a preferred feeding site, because of their great nutritional value.[2] Seed damage is variable, depending on the type of feeding. For instance, sucking insects feed on the cotyledons, and if the feeding punctures reach the radicule or the hypocotyl it can prevent the seed germination.[2,10] Chewing insects will consume the seed/fruit tissues either by feeding on the outer surface or by boring into the seeds and fruits.

Types of Damage

There are several types of damage caused by insects feeding on flowers, fruits, and seeds. Larvae of chewing bruchids bore into the seeds and the adults emerge through round holes cut in the seed. The "sap beetles" cause damage by feeding and are also known to transmit fruit-degrading microorganisms, such as brown rot *Monolinia fruticola*, in stored fruits.[3,4]

Sucking insects feed by inserting their stylets into the food source to suck up the nutrients. By doing this, they cause injury to plant tissues, resulting in plant wilt and, in many cases, abortion of fruits and seeds. The damage to plant tissues, including seeds and fruits, results from the frequency of stylet penetration and feeding duration, associated with salivary secretions that can be toxic and cause tissue necrosis.[2,6,7,10]

During the feeding process, sucking insects also may transmit plant pathogens, which increase their damage potential. For instance the "cotton stainers" transmit a series of fungi and bacteria, the most important one called the internal boll disease is caused by *Nematospora gossypii*. The feces of adults are yellowish and stain the cotton lint, which is another type of damage they cause. The direct damage varies with the age of the boll, with young bolls having the seeds destroyed and drying up as a result of the feeding activity.[7]

Beneficial Insects

Beneficial insects involved with flowers and fruits include mostly species associated with pollination (see "Mutualistic Associations" section) and production of honey. These insects in general belong to the Order Hymenoptera.

This order includes all different species of bees, such as the honey bee (A. *mellifera* L.), the stingless bees (*Melipona* and *Trigona*), the bumble bees (*Bombus*), the carpenter bees (*Xylocopa*), the mining and cuckoo bees (Anthophoridae), the leaf cutting bees (Megachilidae), and others, which are frequent visitors of flowers.[1,5] Among the hymenopterans there are the so-called pollen wasps (masarine wasps) which are the only wasps that provision their nest cells with pollen and nectar, as bees do.[11]

Honey is perhaps the greatest benefit produced by insects as they visit the reproductive structure of plants. Honey is used as food by bees, and it is explored by man as one of the most popular and beneficial food.

Another benefit that insects associated with flowers and fruits may do to plants is by disperse their seeds. Small seeds might get attached to the insects' body and be taken elsewhere, helping the dispersion of the plant species.

Mutualistic Associations

Insects may have a mutualistic relationship with plants, where both insects and plants have advantages of the association.[1,5] Plants are the source of food for insects, and in return they pollinate the flowers. Pollination is the transfer of the male parts (anthers) of a flower to the female part (stigma) of the same or different flowers. More seeds develop when large numbers of pollen grains are transferred.

Insects, in general, play an important role in pollination (entomophily), as well as wind (anemophily). Beyond the hymenopterans (bees, wasps and ants), dipterans (flies), lepidopterans (butterflies and moths), coleopterans (beetles), thysanopterans (thrips), and other minor orders are also engaged in pollination.[5] Other pollination agents include birds and bats.

About 130 agricultural plants are pollinated by bees in the United States. The annual value of honey bee pollination to U.S. agriculture has been estimated at over $9 billion. The honey bee pollination in Canada is estimated at Can$443 million, and over 47,000 colony rentals are taken every year. In the United Kingdom at least 39 crops grown for fruit and seed are insect pollinated, by honey bees and bumble bees. The European Union has an estimated annual value of 5 billion euros, with 4.3 billion attributable to honey bees.[5]

As flowers are visited by insects and get pollinated, flowers provide proteinaceous pollen and sugary nectar, which are used by insects as food source to larvae and adults.

Conclusion

Insects feeding on flowers/fruits/seeds can cause severe damage to many plants of economic importance worldwide. Despite the many efforts to control these insects, they still remain a challenge to economic entomologists. Because of their feeding habits, i.e., feeding on the reproductive structures of plants, their economic importance is much greater than that of insect pests that feed on other plant parts, such as leaves, from whose damage some plants can compensate for at least partially. Moreover, for some pests such as the heteropterans that suck plant juices, the damage is difficult to detect early on compared to damage by leaf chewing insects. Therefore, sometimes, damage by sucking insects is noticed only at harvest, when control measures are too late to avoid economic losses.

In conclusion, while insects feeding on reproductive structures of plants cause enormous damage, they can also be of enormous benefit to humankind. Honey production is a source of a healthy and nutritional food for humans. Pollination caused by insects feeding on flowers is responsible for the development of fruit/seeds of hundreds of plant species, several of them of highly economic importance. Moreover, without pollination, many of the plant species we know today would have disappeared a long time ago.

References

1. Barth, F.G. *Insect and Flowers. The Biology of a Partnership*; Princeton University Press: Princeton, NJ, USA, 1996.
2. Slansky, F., Jr.; Panizzi, A.R. Nutritional Ecology of Seedsucking Insects. In *Nutritional Ecology of Insects, Mites, Spiders and Related Invertebrates*; Slansky, F., Jr, Rodriguez, J.G., Eds.; Wiley: New York, 1987; 283–320.
3. Borror, D.J.; Triplehorn, C.A.; Johnson, N.F. *An Introduction to the Study of Insects,* 6th Ed.; Saunders College Publishing: Philadelphia, USA, 1989.
4. Hill, D.S. *The Economic Importance of Insects*; Chapman & Hall: London, UK, 1997.
5. Delaplane, K.S.; Mayer, D.F. *Crop Pollination by Bees*; CABI Publishing: New York, USA, 2000.
6. McPherson, J.E.; McPherson, R.M. *Stink Bugs of Economic Importance in America North of Mexico*; CRC Press: Boca Raton, Florida, 2000.
7. *Heteroptera of Economic Importance*; Schaefer, C.W., Panizzi, A.R., Eds.; CRC Press: Boca Raton, Florida, USA, 2000.
8. Schuh, R.T.; Slater, J.A. *True Bugs of the World (Hemiptera: Heteroptera). Classification and Natural History*; Cornell University Press: Ithaca, New York, USA, 1995.
9. White, I.M.; Elson-Harris, M.M. *Fruit Flies of Economic Significance: Their Identification and Bionomics*; CAB International: Oxon, UK, 1992.
10. Panizzi, A.R. Wild hosts of pentatomids: Ecological significance and role in their pest status on crops. Annu. Rev. Entomol. **1997**, *42,* 99–122.
11. Gess, S.K. *The Pollen Wasps. Ecology and Natural History of the Masarinae*; Harvard University Press: Cambridge, Massachusetts, USA, 1996.

21

Integrated Pest Management

Marcos Kogan
Oregon State University

Introduction

In the context of integrated pest management (IPM), a pest is any living organism whose life system conflicts with human interests, economy, health, or comfort. Pests belong to three main groups: (i) invertebrate and vertebrate animals, (ii) disease-causing microbial pathogens, and (iii) weeds or undesirable plants growing in desirable sites.

The concept of pest is anthropocentric. There are no pests in nature, in the absence of humans. An organism becomes a pest only if it competes with crop plants or it causes direct injury to the plants in the field or to their products in storage or, outside an agricultural setting, if it affects structures built to serve human needs or are vectors of disease organisms. This chapter focuses on the impact of pests on agricultural plants.

Background

Since the beginnings of agriculture some 10,000 years ago, humans competed with other organisms for the same food resources. Primitive tools to fight pests were probably limited to handpicking and methods such as flooding or fire; although at times of despair, agriculturists resorted to magic incantations or religious proclamations.[1] But as early as 2500 BC, Sumerians used sulfur to control insects, and by 1000 BC, Egyptians and Chinese were using plant-derived insecticides to protect stored grain.[2] Real progress in the fight against agricultural pests occurred with advances in biological sciences, resulting in a better understanding of arthropod pests and the relationships between microbial pathogens and plant diseases. Until the middle of the 20th century, insect pest control ingeniously combined knowledge of pest biology and cultural and biological control methods. Trichlorodiphenyltrichloroethane (DDT) in the early 1940s ushered in an era of organosynthetic insecticides. Initial results were spectacular, and many entomologists foresaw the day when all insect pests would be eradicated. This optimistic view was premature, as insects quickly evolved mechanisms to resist the new chemicals, requiring more frequent applications, higher dosages, or the replacement of old insecticides with more powerful ones. This cycle has become known as the insecticide treadmill.[1] At the end of the 1950s, scientists around the world became aware of the risks associated with the application of powerful insecticides as resistance rendered one chemical after the other useless chemical and as beneficial organisms were indiscriminately killed along with target pest species. Such killing of insect parasitoids and predators caused the resurgence of pests at even higher levels. In addition, populations of other plant-eating species, usually balanced by the pests' natural enemies, exploded in outbreak proportions, causing upsurges of secondary pests and pest replacements. As a result of mounting concern about resistance, resurgence, and upsurges of secondary pests, the concept of integration emerged—an attempt to reconcile pesticides with the preservation of natural enemies.[3]

Integration

The next phase in the evolution of IPM was the expansion of the term "integrated" to include all available methods (biological, cultural, mechanical, physical, and legislative, as well as chemical) and to span all pest categories of importance in the cropping system (invertebrates, vertebrates, microbial pathogens, and weeds). Because of the multiple meanings, integration in IPM was conceived at three levels of increasing complexity. Level I integration refers to multiple control tactics against single pests or pest complexes; level II is integration of all pest categories and the methods for their control; level III integration applies the steps in level II to the entire cropping system. At level III, the principles of IPM and sustainable agriculture converge.[4]

The Ecological Bases of IPM

Ecology offered the tools to describe how populations expand and contract under natural conditions; how biotic communities are organized and how their component organisms interact; how abiotic (nonliving) factors (climate, soil, topography, hydrology) shape biotic communities; how biodiversity stabilizes ecological systems; and, perhaps most significantly, how to understand the effects of disturbances in the dynamics of biotic communities.[5,6] Thus, IPM should rest upon solid ecological foundation.[4,5,7,8]

The Tools of Pest Management

Management of pests in an IPM system is accomplished through selection of control tactics singly or carefully integrated into a management system. These tactics usually fall into the following main classes: (i) chemical, (ii) biological, (iii) cultural/mechanical, (iv) physical control methods (these four methods apply to all pests), (v) host plant resistance (applicable to invertebrate and microbial pests), (vi) behavioral, and (vii) sterile insect technique, a genetic control method (relevant only to insects).

Chemical Control

Chemical pesticides are used to kill or interfere with the development of pest organisms. Pesticides are called insecticides, acaricides, nematicides, fungicides, or herbicides if the target pests are insects, mites, nematodes, fungi, or weeds, respectively. Pesticides are useful tools in IPM if used selectively and if the pest population threatens to reach the economic injury level (EIL) (described later). Much care is necessary to handle pesticides and avoid their potentially negative effect on the environment.[9]

Biological Control

Biological control refers to the action of parasitoids, predators, and pathogens in maintaining another organism's population density at a lower average than would occur in their absence.[10] In IPM practice, biological control includes intentional introduction of natural enemies to reduce pest population levels (classical biological control), the mass release of biocontrol organisms (inundative biocontrol), or the timely inoculation of natural enemies (augmentative biocontrol).[10]

Cultural/Mechanical Control

Cultural practices become cultural controls when adopted for intentional effect on pests. Thus, tillage and other operations for soil preparation have a direct impact on weeds and on soilborne insect pests. The manipulation of the crop environment to favor natural enemies is called habitat management and is of growing interest as a form of conservation biological control.[11,12]

Physical Control

Fire, water, electricity, and radiation are some of the main physical forces used in IPM. Flooding in paddy rice cultivation is a major adjuvant method in many parts of the world.[13]

Host Plant Resistance

This means of control is based on breeding crop plants to incorporate genes whose products cause plants to be unsuitable for insects feeding on them (antibiosis) or impair insects' feeding behavior (antixenosis).[14] Resistance is the only effective method for control of plant viral diseases. Techniques of genetic engineering have opened new opportunities to expand plant resistance in IPM, but the approach is not without risks and should be adopted cautiously, within a strict IPM framework.[15]

Behavioral Control

Behavioral control uses chemicals (allelochemicals) to interfere with normal patterns of mainly sexual (mating) and feeding behaviors of arthropods. Sex pheromones are used for mating disruption.[16,17] Feeding excitants or deterrents are used to disrupt normal feeding behavior or to attract and kill insects.[18]

Sterile Insect Technique

This genetic control method disrupts normal progeny production of the target species by the mass release of sterile insects.[19]

The Scale of IPM Systems

Most IPM programs focus on single fields because of local variability in physical, crop, and pest conditions. Some pests, however, are highly mobile and require a regional approach for their control. Furthermore, certain control tactics are effective only if deployed over large areas. Mating disruption for control of the codling moth in apple and pear orchards in the western United States used an area-wide approach that required a minimum operational unit of about 160 hectares. Advancement of IPM to higher levels of integration will require planning and implementation at the landscape or even ecoregional levels. Advanced technologies of geographic information systems (GISs) and remote sensing are essential for the development of such programs. Such planning is still at its infancy in IPM.

Decision Support Systems for IPM Implementation

IPM uses objective criteria for making decisions about the need for a control action and selection of the most appropriate tactics. In arthropod pest control, the criteria are based on assessments of the extant pest population in the field, its damage potential (if not controlled), and the relative costs of treatment and crop value. The main parameters are the EIL and economic threshold (ET). ET is the population level that will trigger a control action. Figure 21.1 describes these parameters. To use ETs, it is necessary to monitor pest populations in the field using well-defined sampling methods, which vary with the pest species. Although useful, the concept of EIL is less effective as a decision-making support tool for weed management and is partially applicable to microbial disease management.[20]

IPM Education and Extension

Since the establishment of the Land Grant Universities (1890), teaching of entomology and plant pathology was gradually introduced in the curricula. Courses on insect science that included the control of agricultural pests were variously designated as applied, economic, or agricultural entomology. As faculties expanded in plant protection departments, specialized courses were introduced in chemical pesticides, biological control, host plant resistance, and others. With the introduction of IPM in the late 1960s, most entomology departments started offering basic IPM courses at the undergraduate level and advanced IPM for graduate students. One of the first entomology textbooks aimed at teaching IPM was by Metcalf and Luckmann.[21] Other texts followed. Pedigo's[22] *Entomology and Pest Management*, first published in 1989, became the standard for introductory courses in insect IPM in the United States, and Dent's[23] *Integrated Pest Management* probably had the same role in the United Kingdom. Although IPM continues to be taught mostly along disciplinary boundaries, a few universities offer multidisciplinary courses combining entomology, plant pathology, and weed science within integrated curricula. The book by Norris, Caswell-Chan, and Kogan[2] was the first to offer an interdisciplinary text for IPM teaching.

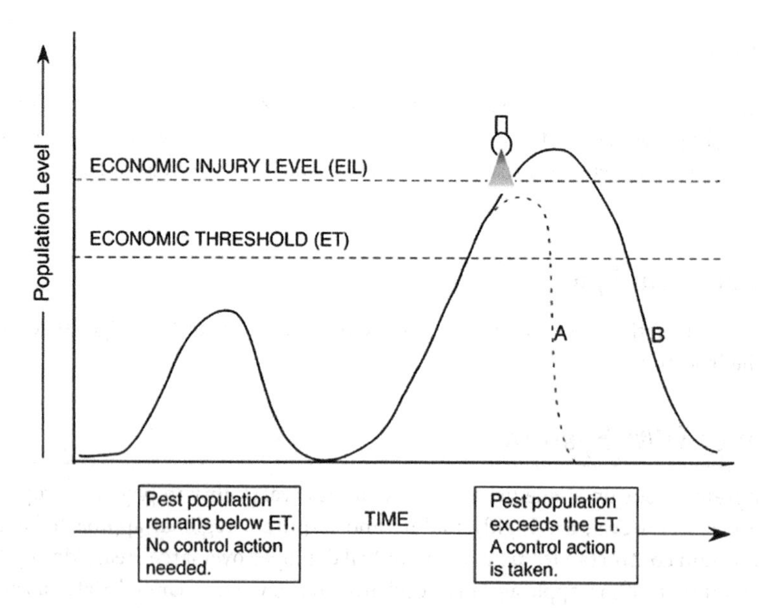

FIGURE 21.1 Fundamental decision-making concepts for insect pest management: ET and EIL. Illustrated is an insect pest population with two peaks in 1 year. The first peak remains below ET, and no treatment is needed; the second peak exceeds ET. If a treatment is applied, the population crashes (curve A); if a treatment is not applied, the population exceeds EIL and an economic crop loss occurs (curve B).

It is the role of agricultural extension specialists to bring to growers the most up-to-date information on IPM technological advances and promote adoption of effective IPM systems. Extension specialists use demonstration plots, field days, intensive short courses, special publications, and increasingly, the power of the Internet, to expose growers to the latest in IPM. Formerly printed IPM information is now available online, which allows for more timely updates and ease of access to the information. An example is the *Pacific Northwest IPM Handbooks* that had been traditionally issued once a year in three volumes for insect pests, plant diseases, and weeds. These are now entirely published online.[24] One critical area for IPM implementation involves decision support systems. Online degree day-driven pest development models are available to help growers in the optimization of sampling and timing of control actions.[25] In developing regions of the world where computers are still rare in rural areas, extension has been highly effective with a model developed by specialists of the Food and Agriculture Organization (FAO) in collaboration with international research centers. The model known as farmer field schools is a participatory approach in which extension specialists and farmers meet in the field to identify pest problems and discuss possible solutions. First tested in Southeast Asia,[26] the approach has been successful in Africa also.[27] Maintaining the flow of information from research laboratories and experimental fields into the classroom and the farm is essential to expand IPM adoption.

Conclusion

It is difficult to assess the global impact of IPM. It is safe, however, to state that the concept drastically changed the approach to pest control among progressive growers, the chemical industry, research establishments, and policymakers. One of the early economic analyses of the impact of IPM was done in Brazil on soybean production costs. The study suggested that technologies available in 1980 (date of the study), if adopted throughout the entire production area (8.5 million ha), would have resulted in a saving of U.S. $216 million annually mainly in the cost of insecticides and application costs. A study on the environmental benefits of changes in pesticide use conducted in Ontario, Canada, estimated benefits to amount to $711 million.[28] Farmers in most agricultural production regions of the world have benefited from the adoption of IPM. Although IPM is still in its infancy, major advances are to be expected as modern technologies are incorporated into the management of all pests.[29]

References

1. Van den Bosch, R. *The Pesticide Conspiracy*; Doubleday: Garden City, NY, 1978; Vol. 8, 226.
2. Norris, R.F.; Caswell-Chen, E.P.; Kogan, M. *Concepts in Integrated Pest Management*; Prentice Hall: Upper Saddle River, NJ, 2003; 586.
3. Kogan, M. Integrated pest management: Historical perspectives and contemporary developments. *Annu. Rev. Entomol.* 1998, *43*, 243–277.
4. Kogan, M. Integrated pest management theory and practice. *Entomol. Exp. Appl.* 1988, *49*, 59–70.
5. Levins, R.L.; Wilson, M. Ecological theory and pest management. *Annu. Rev. Entomol.* 1980, *25*, 287–308.
6. Kogan, M. Ecological theory and integrated pest management practice. In *Environmental Science and Technology*; Wiley: New York, 1986; Vol. 84, 362.
7. Huffaker, C.B.; Gutierrez, A.P. *Ecological Entomology*; Wiley: New York, 1999; Vol. 19, 756.
8. Kogan, M. Integration and integrity in IPM: The legacy of Leo Dale Newsom (1915–1987). *Amer. Entomol.* 2013 (Fall), 150–160.
9. Wheeler, W.B. *Pesticides in Agriculture and the Environment*; Marcel Dekker: New York, 2002; Vol. 10, 330.
10. Bellows, T.S.; Fisher, T.W. *Handbook of Biological Control: Principles and Applications of Biological Control*; Academic Press: San Diego, CA, 1999; Vol. 23, 1046.

11. Gurr, G.M.; Wratten, S.D. Integrated biological control: A proposal for enhancing success in biological control. *Int. J. Pest Manag.* 1999, *45*, 81–84.
12. Landis, D.A.; Wratten, S.D.; Gurr, G.M. Habitat management to conserve natural enemies of arthropod pests in agriculture. *Annu. Rev. Entomol.* 2000, *45*, 175–201.
13. Vincent, C., Weintraub, P.; Hallman, G. Physical control of insect pests. In *Encyclopedia of Insects*. V.H. Resh and R.T. Carde, Eds.; Elsevier: USA, 2009; 794–797.
14. Smith, C.M. *Plant Resistance to Insects: A Fundamental Approach*; Wiley: New York, 1989; Vol. 8, 286.
15. Atherton, K.T. *Genetically Modified Crops: Assessing Safety*; Taylor & Francis: London, UK, 2002, 256.
16. Carde, R.T.; Bell, W.J. *Chemical Ecology of Insects* 2; Chapman & Hall: New York, 1995; Vol. 8, 433.
17. Carde, R.T.; Minks, A.K. *Insect Pheromone Research: New Directions*; Chapman & Hall: New York, 1997; Vol. 22, 684.
18. Metcalf, R.L.; Deem-Dickson, L.; Lampman, R.L. Attracticides for the control of diabroticite rootworms. In *Pest Management: Biologically Based Technology*; J.L. Vaughn, Ed.; American Chemical Society: Beltsville, MD, 1993; 258–264.
19. Calkins, C.O.; Klassen, W.; Liedo, P. *Fruit Flies and the Sterile Insect Technique*; CRC Press: Boca Raton, FL, 1994; 258.
20. Higley, L.G.; Pedigo, L.P.; Eds. *Economic Thresholds for Integrated Pest Management*; The University of Nebraska Press: Lincoln, NE, 1996,; 327.
21. Metcalf, R.L.; Luckmann, W.H. *Introduction to Insect Pest Management*, 1st ed.; Wiley: New York, 1975; 587.
22. Pedigo, L.P. *Entomology and Pest Management*, 1st ed.; Macmillan: New York, 1989.
23. Dent, D.; Ed. *Integrated Pest Management*. Springer: London, 1995; 372.
24. Integrated Plant Protection Center. *On-line IPM Handbooks*; Corvallis, Oregon, 2012. http://ipmnet.org/IPM_Handbooks.htm (Accessed December 2013).
25. Coop, L. IPM pest and plant disease models and forecasting for agricultural pest management, and plant biosecurity decision support in the U.S. In *Pest and Crop Models*. pnwpest.org/wea (Accessed December 2013).
26. Pontius, J.C.; Dilts, R.; Bartlett, A. *From Farmer Field School to Community IPM: Ten Years of IPM Training in Asia*. RAP/2002/15; FAO Regional Office for Asia and the Pacific: Bangkok, 2002; 106.
27. SUSTAINET EA. Technical Manual for farmers and Field Extension Service Providers: Farmer Field School Approach. Sustainable Agriculture Information Initiative, Nairobi. 2010. www.fao.org/ag/ca/CA-Publications/Farmer_Field_School_Approach.pdf (Accessed December 2013).
28. Brethour, C.; Weersink, A. An economic evaluation of the environmental benefits from pesticide reduction. *Ag. Econ.* 2001, 25, 219–226.
29. Kogan, M.; Jepson, P. *Perspectives in Ecological Theory and Integrated Pest Management*; Cambridge University Press: Cambridge, UK, 2007; 570.

22

Land Capability Analysis

Michael J. Singer
University of California

Introduction

All soils are not the same and they are not of the same capability for every use. Because the Earth's human population exceeds seven billion on its way to nine or ten billion by the mid-21st century, the proper use of land becomes a critical component in human sustainability. Land capability implies that the choice of land for a particular use contributes to the success or failure of that use. It further implies that the choice of land for a particular use will determine the potential impact of that use on surrounding resources such as air and water. To make the best use of land and to minimize the potential for negative impacts on surrounding lands, land capability analysis is needed. The assessment of land performance for specific purposes is land evaluation.[1] A system that organizes soil and landscape properties into a form that helps differentiate among useful and less useful soils for a purpose is land capability classification. Land capability is a broader concept than soil quality, which has been defined as the degree of fitness of a soil for a specific use.[2] Bouma[3] points out that land capability or land potential needs to be evaluated via various scales and gives the example of precision agriculture, which requires land capability analysis in more detail than that required to determine if an investment should be made to initiate agriculture.

Land capability or suitability classification systems have been designed to rate land and soil characteristics for specific uses (Table 22.1). Huddleston[4] has reviewed many of these systems.

TABLE 22.1 Examples of Land Capability Systems

System	Purpose	Property	References
FAO framework	Development potential	Used for large-scale development of agriculture	[1,13]
USBR irrigation suitability	Potential for irrigation development	Used for determining potential to repay costs of developing irrigation	[11,12]
USDA land capability classification	Land capability for agriculture	Uses 13 soil, climate, and landscape properties to determine agricultural capability	[4,15]
SIR	Land capability for agriculture	Uses nine soil and management factors to determine agricultural capability	[20–22]
Soil potential	Soil suitability for specific uses	Uses a cost index to rate land for any potential use	[19]
Soil quality	Determining status of soil profile	Used for determining the status of selected soil properties	[6,19,24]

They may also rate land qualities. Land qualities have been defined by the U.N. Food and Agricultural Organization[1] as "attributes of land that act in a distinct manner in their influence on the function of land for a specific kind of use." An example of a land quality is the plant available water stored in soil, and an example of a soil characteristic is the clay content that contributes to the plant available water holding capacity. It is generally recognized that a single soil characteristic is of limited use in evaluating differences among soils,[5] and that use of more than one quantitative variable requires a system for combining the measurements into a useful index.[6] Gersmehl and Brown[7] advocate regionally targeted systems.

Kinds of Systems

Land rating systems for agriculture include those that are used to evaluate the potential for agricultural development of new areas, and others that evaluate the potential for agriculture in already developed areas, for example, land capability for grazing in Australia[8] and fertility capability classification for the tropics.[9] Many other land capability assessments exist to help planners rate suitability of agricultural lands for nonagricultural uses; for example, see Steiner et al.[10] Some examples of agricultural land capability systems are described in this entry. All systems have, in common, a set of assumptions on which the analysis is based and each system answers the question "capability for what use."

Development Potential

Examples of systems designed to determine the potential for agricultural development include the FAO framework for land evaluation and the U.S. Bureau of Reclamation (USBR) irrigation suitability classification. The FAO framework combines soil and land properties with a climatic resources inventory to develop an agroecological land suitability assessment.

The USBR capability classification was frequently used to evaluate land's potential for irrigation in the Western U.S. during the period of rapid expansion of water delivery systems.[11–14] It combines social and economic evaluations with soil and other ecological variables to determine whether the land has the productive capacity, once irrigated, to repay the investment necessary to bring water to an area. It recognizes the unique importance of irrigation to agriculture and the special qualities of soils that make them irrigable.

Land Capability

The USDA Land Capability Classification (LCC) is narrower in scope than either the FAO or USBR capability rating systems. The purpose of the LCC is to place arable soils into groups based on their ability to sustain common cultivated crops that do not require specialized site conditioning or treatment.[15] Nonarable soils, unsuitable for long-term, sustained cultivation, are grouped according to their ability to support permanent vegetation, and according to the risk of soil damage if mismanaged.

Several studies have shown that lands of higher LCC have higher productivity with lower production costs than lands of lower LCC.[16–18] In a study of 744 alfalfa-, corn-, cotton-, sugar beet- and wheat-growing fields in the San Joaquin Valley of California, those with LCC ratings between 1 and 3 had significantly lower input/output ratios than fields with ratings between 3.01 and 6.[18] The input/ output ratio is a measure of the cost of producing a unit of output and is a better measure of land capability than output (yield) alone. This suggests that the LCC system provides an economically meaningful assessment of agricultural soil capability.

Quantitative systems result in a numerical index, typically with the highest number being assigned to the land or soil with the highest capability for the selected use. The final index may be additive, multiplicative, or more complex functions of many land or soil attributes. Quantitative systems have two important advantages over nonquantitative systems: 1) they are easier to use with

GIS and other automated data retrieval and display systems and 2) they typically provide a continuous scale of assessment.[19] No single national system is presently in use, but several state or regional systems exist.

One example of a quantitative system is the Storie Index Rating (SIR). Storie[20] determined that land productivity is dependent on 32 soil, climate, and vegetative properties. He combined only nine of these properties into the SIR, to keep the system from becoming unwieldy. The nine factors are soil morphology (A), surface texture (B), slope (C), and management factors drainage class (X_1), sodicity (X_2), acidity (X_3), erosion (X_4), micro-relief (X_5), and fertility (X_6). Each factor is rated from 1 to 100%. These are converted to their decimal value and multiplied together to yield a single rating for a soil map unit:[20–22]

$$\mathrm{SIR} = \left(A \times B \times C \times \prod_{i=1}^{6} X_i \right) \times 100$$

An area-weighted SIR can be calculated by multiplying the SIR for each soil map unit within a parcel by the area of the soil unit within the parcel, followed by summing the weighted values and dividing by the total area:

$$\text{Area} - \text{weighted SIR} = \frac{1}{\text{totalarea}} \times \sum_{i=1}^{n} \text{SIRsoil}_i \times \text{area soil}_i$$

Values for each factor were derived from Storie's experience mapping and evaluating soils in California, and in soil productivity studies in cooperation with California Agricultural Experiment Station cost-efficiency projects relating to orchard crops, grapes, and cotton. Soils that were deep, which had no restricting subsoil horizons, and held water well had the greatest potential for the widest range of crops. More recently, much of what was nonquantitative scoring has been modernized using computer-based decision support systems.[23]

Reganold and Singer[18] found that area-weighted average SIR values between 60 and 100 for 744 fields in the San Joaquin Valley had lower but statistically insignificant input/output ratios than fields with indices <60. The lack of statistical significance is scientifically meaningful, but may not be interpreted the same by planners or farmers for whom small differences in profit can be significant.

Designers of these systems have selected thresholds or critical limits that separate one land capability class from another. Frequently, little justification has been given for the critical limit selected. The issue of critical limits is a difficult one in soils because of the range of potential uses and the interactions among variables.[24]

Soil Potential

Soil potential is an interpretive system that is used for both agricultural and nonagricultural capability assessments. It is an example of a land capability system that is both quantitative and highly specific. Soil potential is defined as the usefulness of a site for a specific purpose using available technology at an indexed cost. Experts in the location where the system will be used determine the soil properties that most influence the successful use of a soil for a particular purpose. The well-suited soils for the use are given a rating of 100. Soil limitations that can be removed or reduced are identified and continuing limitations that occur even after limitations have been addressed are also identified. The cost of available practices and technologies used to remove limitations and the cost of continuing limitations are indexed, and these index values are subtracted from 100 to yield a soil potential rating for each soil.

$$\mathrm{SPI} = \mathrm{PI} - (\mathrm{CI} + \mathrm{CL})$$

The best soils in an area are those that have the lowest indexed cost for operating or maintaining use, while ensuring the lowest possible environment.

Conclusions

All soils are not the same and are not equally capable of sustained use. Various systems have been created that rate soil properties for these uses. Examples include FAO land suitability, USBR irrigation suitability, USDA LCC, SIR, and soil potential. Each has strengths and weaknesses that the user must recognize before using the system.

References

1. *A Framework for Land Evaluation; Soils Bulletin;* Food and Agricultural Organization: FAO: Rome, Italy, 1976; 32.
2. Gregorich, E.G.; Carter, M.R.; Angers, D.A.; Monreal, C.M.; Ellert, B.H. Towards a minimum data set to assess soil organic matter quality in agricultural soils. Can. J. Soil Sci. **1994**, *74*, 367–386.
3. Bouma, J. Land evaluation for landscape units. In *Hand-Book of Soil Science;* Sumner, M.E., Ed.; CRC Press: Boca Raton, FL, 2000; E-393–E-412.
4. Huddleston, J.H. Development and use of soil productivity ratings in the United States. Geoderma **1984**, *32*, 297–317.
5. Reganold, J.P.; Palmer, A.S. Significance of gravimetric versus volumetric measurements of soil quality under biodynamic, conventional, and continuous grass management. J. Soil Water Conserv. **1995**, *50*, 298–305.
6. Halvorson, J.J.; Smith, J.L.; Papendick, R.I. Integration of multiple soil parameters to evaluate soil quality: a field example. Biol. Fert. Soils **1996**, *21*, 207–214.
7. Gersmehl, P.J.; Brown, D.A. Geographic differences in the validity of a linear scale of innate soil productivity. J. Soil Water Conserv. **1990**, *45*, 379–382.
8. Squire, V.R.; Bennett, F. Land capability assessment and estimation of pastoral potential in semiarid rangeland in South Australia. Arid Land Res. Manage. **2004**, *18*, 25–39.
9. Sanchez, P.A.; Palm, C.A.; Buol, S.W. Fertility capability soil classification: a tool to help assess soil quality in the tropics. Geoderma **2003**, *114*, 157–185.
10. Steiner, F.; McSherry, L.; Cohen J. Suitability analysis for the upper Gila River watershed. Landscape Urban Plan. **2000**, *50*, 199–214.
11. USBR. Land Classification Handbook; Bur. Recl. Pub. V, Part 2; USDI: Washington, DC, 1953.
12. Maletic, J.T.; Hutchings, T.B. Selection and classification of irrigable lands. In *Irrigation of Agricultural Lands;* Hagen, R.M., Ed.; Soil Science Society of America: Madison, Wisconsin, 1967; 125–173.
13. Food and agricultural organization. *Land Evaluation Criteria for Irrigation.* Report of an Expert Consultation; World Soil Resourc. Rep. 50; FAO: Rome, Italy, 1979.
14. McRae, S.G.; Burnham, C.P. *Land Evaluation;* Clarendon Press: Oxford, UK, 1981.
15. Klingebiel, A.A.; Montgomery, P.H. *Land-Capability Classification; Agriculture Handbook No. 210;* Soil Conservation Service USDA: Washington, DC, 1973.
16. Patterson, G.T.; MacIntosh, E.E. Relationship between soil capability class and economic returns from grain corn production in southwestern Ontario. Can. J. Soil Sci. **1976**, *56*, 167–174.
17. Van Vliet, L.J.; Mackintosh, E.E.; Hoffman, D.W. Effects of land capability on apple production in southern Ontario. Can. J. Soil Sci. **1979**, *59*, 163–175.
18. Reganold, J.P.; Singer, M.J. Comparison of farm production input/output ratios of two land classification-systems. J. Soil Water Conserv. **1984**, *39*, 47–53.
19. Ewing, S,A.; Singer, M.J. Soil quality. In *Handbook of Soil Science;* Huang, P.M.; Li, Y.; Sumner, M.E., Eds.; CRC Press: Boca Raton, Florida, 2012; 26-1–26-28.

20. Storie, R.E. *An Index for Rating the Agricultural Value of Soils;* California Agr. Exp. Sta. Bull. 556; Berkeley, CA, 1932.
21. Wier, W.W.; Storie, R.E. *A Rating of California Soils;* Bull. 599; California Agr. Exp. Sta.: Berkeley, CA, 1936.
22. Storie, R.E. *Handbook of Soil Evaluation;* Associated Students Store: U. Cal. Berkeley, CA, 1964.
23. O'Geen, A.T.; Southard, S.B.; Southard, R.J. *A Revised Storie Index for Use with Digital Soils Information;* ANR Publication: 8355; 2008.
24. Arshad, M.A.; Coen, G.M. Characterization of soil quality: physical and chemical criteria. Am. J. Altern. Agric. **1992**, *7*, 25–31.

23

Land Capability Classification

Thomas E. Fenton
Iowa State University

Introduction

A common definition of land is the surface of the Earth and all its natural resources. This is interpreted to include the atmosphere, the soil, and underlying geology, hydrology, and plants on the Earth's surface.[1] In the *1938 Yearbook of Agriculture,* land is defined as the total natural and cultural environment within which production must take place. Its attributes include climate, surface configuration, soil, water supply, subsurface conditions, etc., together with its location with respect to centers of commerce and population. It should not be used as synonymous with soil or in the sense of the Earth's surface only.[2] Other definitions included the results of the past and present human activities as well as the animals within this area when they exert a significant influence on the present and future uses of land by man. Thus, the concept of land is much broader than soil. However, because soils are a major component of terrestrial ecosystems,[3] they integrate and reflect internal and external environmental factors. They have been considered an important factor—if not the major factor—in many land classification systems.

Classifications

Land classification may be defined as a classification of specific bodies of land according to their characteristics or to their capabilities for use. A natural land classification system is one in which natural land types are placed in categories according to their inherent characteristics. A land classification according to its capabilities for use may be defined as one in which the bodies of land are classified on the basis of physical characteristics with or without economic considerations according to their capabilities for man's use.[2]

Olson[4] defined land classification as the assignment of classes, categories, or values to the areas of the Earth's surface (generally excluding water surfaces) for immediate or future practical use. The project, product, or proposal resulting from this activity may be also generally referred to as land classification. Any land classification involves two parts or phases: resource inventory and analysis and categorization. The inventory consists of gathering data and delineating land characteristics on maps.

The analysis and categorization put the basic data into a form that can be used generally for a specific use. Thus, there are potentially many land classification systems that could be developed depending on the objectives of the system.

Historical Perspective

Some sort of land classification system was used in the transfer of public lands to private owners and users as the United States settled. However, a scientific approach was not possible until soil surveys, topographic maps, economic analyses of production, and other activities became available in the early 1900s. These sources provided essential data for a scientific approach to land classification. This approach gained added momentum in the mid-1930s due to the need for land classification for many new government programs.

A report submitted to the President of the United States by the National Resources Board dated December 31, 1934,[5] included a section that dealt with the physical classification of the productivity of the land. The total land area of the United States was rated and divided into five grades based on the factors thought to affect productivity—soil type, topography, rainfall, and temperature. Norton[6,7] defined five classes of land in the regions of arable soils according to their use capability. Two other classes were defined for land that should not be cultivated. A National Conference on Land Classification (NCLC) was held in the campus of the University of Missouri in October 1940. The proceedings of this conference were published in the Bulletin 421 of the Missouri Agricultural Experiment Station.[8] At this conference, it was recognized that land classification in the United States had begun many years prior to 1940. Simonson[9] summarized the purpose of the NCLC meeting as follows: to encourage and provide opportunity for the discussion of land classification in all of its different aspects.

In the years following the conference, there were many entries published that examined questions related to land classification. General principles associated with technical grouping[10] or solution of soil management problems[11] using natural soil classification systems were discussed in detail. Hockensmith and Steele[12] presented a land capability classification developed for conservation and development of land and for land-use adjustments. The system used eight land capability classes with the first four suited for cultivation and the remaining four not suited for cultivation. Within a capability class, the subclasses were determined by the kind of limitation. Within each subclass, the land that required the same kind of management and the same kind of conservation treatment was called a land capability unit. The highest level of abstraction, the capability class, provided a quick, easy understanding of the general suitability of the land for cultivation. Hockensmith[13] discussed the use of soil survey information in farm planning. A land capability map was an essential part of the farm plan. This map was an interpretative map of the soil map together with other pertinent facts. It was used to help the farmers understand the limitations of their soils and to aid in selection of those practices that would preserve their soils and keep them productive. Klingebiel[14] defined capability classification as the grouping of individual kinds of soils, called soil mapping unit, into groups of similar soils, called capability units, within a framework of eight general capability classes that were divided into subclasses representing four kinds of conservation problems. This land classification system has been used as a basis for all the conservation farm plans developed by the Soil Conservation Service (now Natural Resources Conservation Service) in the United States since the 1950s and continues to be used today. Because of its widespread use, this system is discussed in more detail in the following section.

Land Capability Classification

One of the most commonly used land classification systems for land management is the land capability classification. Hockensmith[15] defined this classification as a systematic arrangement of different kinds of land according to those properties that determine the ability of the land to produce permanently. Classification was based on the detailed soil survey (scale of 1:15,840 or 1:20,000). The system was described in detail in the Agriculture Handbook No. 210[16] and continues to be in use, with no significant changes except that presently Arabic numerals are used to identify the capability classes rather than Roman numerals. There are a number of land capability classifications used throughout the world but all these methods are patterned after the U.S. system.[17]

The capability grouping of soils is designed to

1. Help landowners and others use and interpret the soil maps
2. Introduce users to the detail of the soil map itself
3. To make possible broad generalizations based on soil potentialities, limitations in use, and management problems.

There are three major groupings: 1) capability classes; 2) capability subclasses; and 3) capability units. All map unit components, including miscellaneous areas, are assigned a capability class and subclass.[18]

Capability Classes

The broadest unit is the capability class and there are eight classes. The probability of soil damage or limitations in use become progressively greater from Class I to VIII. There are no additional subdivisions of soils in Class I because by definition the soils in this category have no limitations in use. The proper grouping of the soils assumes that the recommended use will not deteriorate the soil over time.

- Class I soils have few limitations that restrict their use.
- Class II soils have some limitations that reduce the choice of plants or require moderate conservation practices.
- Class III soils have severe limitations that reduce the choice of plants or require special conservation practices or both.
- Class IV has very severe limitations that restrict the choice of plants, require very careful management, or both.
- Class V soils have little or no erosion hazard but has other limitations that are impractical to remove and limit their use largely to pasture, range, woodland, or wildlife food and cover.
- Class VI soils have severe limitations that make them generally unsuited to cultivation and limit their use largely to pasture or range, woodland, or wildlife food and cover.
- Class VII soils have very severe limitation that make them unsuited to cultivation and that restrict their use largely to grazing, woodland, or wildlife.
- Class VIII soils and landforms limitations that preclude their use for commercial plant production and restrict their use to recreation, wildlife, water supply, or to aesthetic purposes.

Capability Subclasses

Subclasses are groups of capability units within classes with the same kind of major limitations for agricultural use. The capability subclass is a grouping of capability units having similar kinds of limitations and hazards. Four general kinds of limitations or hazards are recognized:

1. Erosion hazard
2. Wetness
3. Rooting-zone limitations
4. Climate

The limitations recognized and the symbols used to identify them are risk of erosion designated by the symbol e; wetness, drainage, or overflow, w; rooting-zone limitations, s; and climatic limitations, c. There is a priority of use among the subclasses if the limitations are approximately of the same degree. The order of use is e, w, s, and c. For example, if a group of soils has both erosion and excess water hazard, e takes precedence over w. However, for some uses, it may be desirable to show two kinds of limitations. This combination is rarely used but when used, the higher priority one is shown first, i.e., IIew.

Capability Units

Capability units provide more specific and detailed information than class or subclass. They consist of soils that are nearly alike in suitability for plant growth and responses to the same kinds of soil management. However, they may have characteristics that place them in different soil series or soil map units. Soil map units in any capability unit are adapted to the same kinds of common cultivated and pasture plants and require similar alternative systems of management for these crops. Another assumption is that the longtime estimated yields of adapted crops for individual soil map units within the unit under comparable management do not vary more than about 25%.

Conclusions

The grouping of soil map units as used in the land capability classification is a technical classification. The basis for this classification is the natural soil classification system in which the soils and soil map units are defined based on their characteristics and interpretations made based on those properties that affect use and management. For conservation planning, environmental quality, and generation of interpretive maps, it is important to know the kind of soil, its location on the landscape, its extent, and its suitability for various uses.

References

1. Brinkman, R.; Smyth, A.J. *Land Evaluation for Rural Purposes*; Publication 17; International Institute for Land Reclamation and Improvement: Wageningen, the Netherlands, 1973; 116.
2. United States Department of Agriculture. *Soils and Men: Yearbook of Agriculture*; U.S. Govt. Printing Office: Washington, DC, 1938; 1232.
3. Jenny, H. *The Soil Resource: Ecological Studies*; Springer: New York, 1980; Vol. 37, 377.
4. Olson, G.W. *Land Classification*; Cornell University Agricultural Experimental Station Agronomy 4: Ithaca, New York, 1970.
5. National Resources Board Report. Part II. *Report of the Land Planning Committee, Section II*; U.S. Govt. Printing Office: Washington, DC, 1934; 108–152.
6. Norton, E.A. Classes of land according to use capability. Soil Sci. Soc. Am. Proc. **1939**, *4*, 378–381.
7. Norton, E.A. *Soil Conservation Survey Handbook*; Misc. Publication 352; U.S. Department Agricultural Soil Conservation Service: Washington, DC, 1939; 40.
8. Miller, M.F. *The Classification of Land*, Proceeding of First International Conference on Land Classification, Bulletin 421; Missouri Agricultural Experimental Station; 1940, 334.
9. Simonson, R.W. The National Conference on Land Classification. Soil Sci. Soc. Am. Proc. **1940**, 5, 324–326.
10. Orvedal, A.C.; Edwards, M.J. General principles of technical grouping of soils. Soil Sci. Soc. Am. Proc. **1941**, *6*, 386–391.
11. Simonson, R.W.; Englehorn, A.J. Interpretation and use of soil classification in the solution of soil management problems. Soil Sci. Soc. Am. Proc. **1942**, *7*, 419–426.
12. Hockensmith, R.D.; Steele, J.G. Recent trends in the use of the land-capability classification. Soil Sci. Soc. Am. Proc. **1949**, *14*, 383–387.
13. Hockensmith, R.D. Using soil survey information for farm planning. Soil Sci. Soc. Am. Proc. **1953**, *18*, 285–287.
14. Klingebiel, A.A. Soil survey interpretation-capability groupings. Soil Sci. Soc. Am. Proc. **1958**, *23*, 160–163.

15. Hockensmith, R.D. *Classification of Land According to its Capability as a Basis for a Soil Conservation Program*, Reprinted from Proceedings of the Inter-American Conference on Conservation of Renewable Natural Resources, Denver, CO, Sep 7–20, 1948.
16. Klingebiel, A.A.; Montgomery, P.H. Land capability classification. In *Agricultural Handbook 210*; U.S. Department of Agricultural Soil Conservation Service: Washington, DC, 1961; 21.
17. Hudson, N. Land use and soil conservation. In *Soil Conservation,* 3rd Ed.; Iowa State University Press: Ames, IA, 1995; 391.
18. U.S. Department of Agriculture, Natural Resources Conservation Service. *National Soil Survey Handbook, title 430-VI.* Available online at http://soils.usda.gov/technical/handbook/(accessed October 2012).

III

Soil

24

Pollution: Point Source

Mallavarapu Megharaj, Peter Dillon, Ravendra Naidu, Rai Kookana, and Ray Correll
Commonwealth Scientific and Industrial Research Organisation (CSIRO)

W. W. Wenzel
University of Natural Resources and Life Sciences

Introduction

Some naturally occurring pollutants are termed geogenic contaminants and these include fluorine, selenium, arsenic, lead, chromium, fluoride, and radionuclides in the soil and water environment. Significant adverse impacts of geogenic contaminants (e.g., As) on environmental and human health have been recorded in Bangladesh, West Bengal, India, Vietnam, and China. More recently reported is the presence of geogenic Cd and the implications to crop quality in Norwegian soils.[1]

The terms contamination and pollution are often used interchangeably but erroneously. Contamination denotes the presence of a particular substance at a higher concentration than would occur naturally and this may or may not have harmful effects on human or the environment. Pollution refers not only to the presence of a substance at higher level than would normally occur but is also associated with some kind of adverse effect.

Nature and Sources of Contaminants

The main activities contributing to PS pollution include industrial, mining, agricultural, and commercial activities as well as transport and services (Table 24.1). Uncontrolled mining, manufacturing, and disposal of wastes inevitably cause environmental pollution. Military land and land for recreational shooting are also important sites of PS contamination. The contaminants associated with such activities are listed in Table 24.1. Contamination at many of these sites appear to have resulted because of lax regulatory measures prior to the establishment of legislation protecting the environment.

TABLE 24.1 Industries, Land Uses, and Associated Chemicals Contributing to Points, Non-Point Source Pollution

Industry	Type of Chemical	Associated Chemicals
Airports	Hydrocarbons	Aviation fuels
	Metals	Particularly aluminum, magnesium, and chromium
Asbestos production and disposal	Asbestos	
Battery manufacture and recycling	Metals	Lead, manganese, zinc, cadmium, nickel, cobalt, mercury, silver, and antimony
	Acids	Sulfuric acid
Breweries/distilleries	Alcohol	Ethanol, methanol, and esters
Chemicals manufacture and use	Acid/alkali	Mercury (chlor/alkali), sulfuric, hydrochloric and nitric acids, sodium and calcium hydroxides
	Adhesives/resins	Polyvinyl acetate, phenols, formaldehyde, acrylates, and phthalates
	Dyes	Chromium, titanium, cobalt, sulfur and nitrogen organic compounds, sulfates, and solvents
	Explosives	Acetone, nitric acid, ammonium nitrate, pentachlorophenol, ammonia, sulfuric acid, nitroglycerine, calcium cyanamide, lead, ethylene glycol, methanol, copper, aluminum, *bis*(2-ethylhexyl) adipate, dibutyl phthalate, sodium hydroxide, mercury, and silver
	Fertilizer	Calcium phosphate, calcium sulfate, nitrates, ammonium sulfate, carbonates, potassium, copper, magnesium, molybdenum, boron, and cadmium
	Flocculants	Aluminum
	Foam production	Urethane, formaldehyde, and styrene
	Fungicides	Carbamates, copper sulfate, copper chloride, sulfur, and chromium
	Herbicides	Ammonium thiocyanate, carbanates, organochlorines, organophosphates, arsenic, and mercury
	Paints	
	Heavy metals	Arsenic, barium, cadmium, chromium, cobalt, lead, manganese, mercury, selenium, and zinc
	General	Titanium dioxide
	Solvent	Toluene, oils natural (e.g., pine oil) or synthetic
	Pesticides	Arsenic, lead, organochlorines, and organophosphates
	Active ingredients	Sodium, tetraborate, carbamates, sulfur, and synthetic pyrethroids
	Solvents	Xylene, kerosene, methyl isobutyl ketone, amyl acetate, and chlorinated solvents
	Pharmacy	Dextrose and starch
	General/solvents	Acetone, cyclohexane, methylene chloride, ethyl acetate, butyl acetate, methanol, ethanol, isopropanol, butanol, pyridine methyl ethyl ketone, methyl isobutyl ketone, and tetrahydrofuran
	Photography	Hydroquinone, pheidom, sodium carbonate, sodium sulfite, potassium bromide, monomethyl paraaminophenol sulfates, ferricyanide, chromium, silver, thiocyanate, ammonium compounds, sulfur compounds, phosphate, phenylene diamine, ethyl alcohol, thiosulfates, and formaldehyde
	Plastics	Sulfates, carbonates, cadmium, solvents, acrylates, phthalates, and styrene

(*Continued*)

TABLE 24.1 (*Continued*) Industries, Land Uses, and Associated Chemicals Contributing to Points, Non-Point Source Pollution

Industry	Type of Chemical	Associated Chemicals
	Rubber	Carbon black
	Soap/detergent	
	General	Potassium compounds, phosphates, ammonia, alcohols, esters, sodium hydroxide, surfactants (sodium lauryl sulfate), and silicate compounds
	Acids	Sulfuric acid and stearic acid
	Oils	Palm, coconut, pine, and tea tree
	Solvents	
	General	Ammonia
	Hydrocarbons	e.g., BTEX (benzene, toluene, ethylbenzene, xylene)
	Chlorinated organics	e.g., trichloroethane, carbon tetrachloride, and methylene chloride
Defense works		See "Explosives" under "Chemicals Manufacture and Use, Foundries, Engine Works, and Service Stations"
Drum reconditioning		See "Chemicals Manufacture and Use"
Dry cleaning		Trichlorethylene and ethane Carbon tetrachloride Perchlorethylene
Electrical		PCBs (transformers and capacitors), solvents, tin, lead, and copper
Engine works	Hydrocarbons Metals Solvents Acids/alkalis Refrigerants Antifreeze	Ethylene glycol, nitrates, phosphates, and silicates
Foundries	Metals	Particularly aluminum, manganese, iron, copper, nickel, chromium, zinc, cadmium and lead and oxides, chlorides, fluorides and sulfates of these metals
	Acids	Phenolics and amines Coke/graphite dust
Gas works	Inorganics	Ammonia, cyanide, nitrate, sulfide, and thiocyanate
	Metals	Aluminum, antimony, arsenic, barium, cadmium, chromium, copper, iron, lead, manganese, mercury, nickel, selenium, silver, vanadium, and zinc
	Semivolatiles	Benzene, ethylbenzene, toluene, total xylenes, coal tar, phenolics, and PAHs
Iron and steel works		Metals and oxides of iron, nickel, copper, chromium, magnesium and manganese, and graphite
Landfill sites Marinas	Antifouling paints	Methane, hydrogen sulfides, heavy metals, and complex acids Engine works, electroplating under metal treatment Copper, tributyltin (TBT)
Metal treatments	Electroplating metals Acids General	Nickel, chromium, zinc, aluminum, copper, lead, cadmium, and tin Sulfuric, hydrochloric, nitric, and phosphoric Sodium hydroxide, 1,1,1-trichloroethane, tetrachloroethylene, toluene, ethylene glycol, and cyanide compounds
	Liquid carburizing baths	Sodium, cyanide, barium, chloride, potassium chloride, sodium chloride, sodium carbonate, and sodium cyanate
	Mining and extracting industries	Arsenic, mercury, and cyanides and also refer to "Explosives" under "Chemicals Manufacture and Use"

(*Continued*)

TABLE 24.1 (*Continued*) Industries, Land Uses, and Associated Chemicals Contributing to Points, Non-Point Source Pollution

Industry	Type of Chemical	Associated Chemicals
	Power stations	Asbestos, PCBs, fly ash, and metals
	Printing shops	Acids, alkalis, solvents, chromium (see "Photography" under "Chemicals Manufacture and Use")
Scrap yards	Service stations and fuel storage facilities	Hydrocarbons, metals, and solvents Aliphatic hydrocarbons
		BTEX (i.e., benzene, toluene, ethylbenzene, xylene) PAHs (e.g., benzo(a) pyrene)
		Phenols
		Lead
Sheep and cattle dips		Arsenic, organochlorines and organophosphates, carbamates, and synthetic pyrethroids
Smelting and refining		Metals and the fluorides, chlorides and oxides of copper, tin, silver, gold, selenium, lead, and aluminum
Tanning and associated trades	Metals	Chromium, manganese, and aluminum
	General	Ammonium sulfate, ammonia, ammonium nitrate, phenolics (creosote), formaldehyde, and tannic acid
Wood preservation	Metals	Chromium, copper, and arsenic
	General	Naphthalene, ammonia, pentachlorophenol, dibenzofuran, anthracene, biphenyl, ammonium sulfate, quinoline, boron, creosote, and organochlorine pesticides

Source: Adapted from Barzi et al.[11]

Contaminant Interactions in Soil and Water

Inorganic Chemicals

Inorganic contaminant interactions with colloid particulates include: adsorption–desorption at surface sites, precipitation, exchange with clay minerals, binding by organically coated particulate matter or organic colloidal material, or adsorption of contaminant ligand complexes. Depending on the nature of contaminants, these interactions are controlled by solution pH and ionic strength of soil solution, nature of the species, dominant cation, and inorganic and organic ligands present in the soil solution.[2]

Organic Chemicals

The fate and behavior of organic compounds depend on a variety of processes including sorption-desorption, volatilization, chemical and biological degradation, plant uptake, surface runoff, and leaching. Sorption–desorption and degradation (both biotic and abiotic) are perhaps the two most important processes as the bulk of the chemicals is either sorbed by organic and inorganic soil constituents, and chemically or microbially transformed/degraded. The degradation is not always a detoxification process. This is because in some cases the transformation or degradation process leads to intermediate products that are more mobile, more persistent, or more toxic to non-target organisms. The relative importance of these processes is determined by the chemical nature of the compound.

Implications to Soil and Environmental Quality

A considerable amount of literature is available on the effects of contaminants on soil microorganisms and their functions in soil. The negative impacts of contaminants on microbial processes are important from the ecosystem point of view and any such effects could potentially result in a major ecological perturbance. Hence, it is most relevant to examine the effects of contaminants on microbial processes in combination with communities. The most commonly used indicators of metal effects on microflora in soil are: (1) soil respiration, (2) soil nitrification, (3) soil microbial biomass, and (4) soil enzymes.

Contaminants can reach the food chain by way of water, soil, plants, and animals. In addition to the food chain transfer, pollutants may also enter via direct consumption or dust inhalation of soil by children or animals. Accumulation of these pollutants can take place in certain target tissues of the organism depending on the solubility and nature of the compound. For example, DDT and PCBs accumulate in human adipose tissue. Consequently, several of these pollutants have the potential to cause serious abnormalities including cancer and reproductive impairments in animal and human systems.

Sampling for PS Pollution

The aims of the sampling system must be clearly defined before it can be optimized.[3] The type of decision may be to determine land use, how much of an area is to be remediated, or what type of remediation process is required. Because sampling and the associated chemical and statistical analyses are expensive, careful planning of the sampling scheme is therefore a good investment. One of the best ways to achieve this is to use any ancillary data that are available. These data could be in the form of emission history from a stack, old photographs that give details of previous land uses, or agricultural records. Such data can at least give qualitative information.

As discussed before, PS pollution will typically be airborne from a stack, or waterborne from some effluent such as tannery waste, cattle dips, or mine waste. In many cases, the industry will have modified its emissions (e.g., cleaner production) or point of release (increased stack height), hence the current pattern of emission may not be closely related to the historic pattern of pollution. For example, liquid effluent may have been discharged previously into a bay, but that effluent may now be treated and perhaps discharged at some other point. Typically, the aim of a sampling scheme in these situations is to assess the maximum concentrations, the extent of the pollution, and the rate of decline in concentration from the PS. Often the sampling scheme will be used to produce maps of concentration isopleths of the pollutant.

The location of the sampling points would normally be concentrated towards the source of the pollution. A good scheme is to have sufficient samples to accurately assess the maximum pollution, and then space additional samples at increasing intervals. In most cases, the distribution of the pollutant will be asymmetric, with the maximum spread down the slope or down the prevailing wind. In such cases more samples should be placed in the direction of the expected gradient. This is a clear case of when ancillary data can be used effectively. A graph of concentration of the pollutant against the reciprocal of distance from the source is often informative.[4] Sampling depths will depend on both the nature of the pollution and the reason for the investigation. If the pollution is from dust and it is unlikely to be leached, only surface sampling will be required. An example of this is pollution from silver smelting in Wales.[5] In contrast, contamination from organic or mobile inorganic pollutants such as F compounds may migrate well down to the profile and deep sampling may be required.[6,7]

Assessment

In order to assess the impacts of pollution, reliable and effective monitoring techniques are important. Pollution can be assessed and monitored by chemical analyses, toxicity tests, and field surveys. Comparison of contaminant data with an uncontaminated reference site and available databases for baseline concentrations can be useful in establishing the extent of contamination. However, this may not always be possible in the field. Chemical analyses must be used in conjunction with biological assays to reveal site contamination and associated adverse effects. Toxicological assays can also reveal information about synergistic interactions of two or more contaminants present as mixtures in soil, which cannot be measured by chemical assays alone.

Microorganisms serve as rapid detectors of environmental pollution and are thus of importance as pollution indicators. The presence of pollutants can induce alteration of microbial communities and reduction of species diversity, inhibition of certain microbial processes (organic matter breakdown, mineralization of carbon and nitrogen, enzymatic activities, etc.). A measure of the functional diversity of the bacterial flora can be assessed using ecoplates (see http://www.biolog.com/section_4.html). It has been shown that algae are especially sensitive to various organic and inorganic pollutants and thus may serve as a good indicator of pollution.[8] A variety of toxicity tests involving microorganisms, invertebrates, vertebrates, and plants may be used with soil or water samples.[9]

Management and/or Remediation of PS Pollution

The major objective of any remediation process is to:

(1) reduce the actual or potential environmental threat; and (2) reduce unacceptable risks to man, animals, and the environment to acceptable levels.[10] Therefore, strategies to either manage and/or remediate contaminated sites have been developed largely from application of stringent regulatory measures set up to safeguard ecosystem function as well as to minimize the potential adverse effects of toxic substances on animal and human health.

The available remediation technologies may be grouped into two categories: (1) ex situ techniques that require removal of the contaminated soil or groundwater for treatment either on-site or off-site; and (2) in situ techniques that attempt to remediate without excavation of contaminated soils. Generally, in situ techniques are favored over ex situ techniques because of: (1) reduced costs due to elimination or minimization of excavation, transportation to disposal sites, and sometimes treatment itself; (2) reduced health impacts on the public or the workers; and, (3) the potential for remediation of inaccessible sites, e.g., those located at greater depths or under buildings. Although in situ techniques have been successful with organic contaminated sites, the success of in situ strategies with metal contaminants has been limited. Given that organic and inorganic contaminants often occur as a mixture, a combination of more than one strategy is often required to either successfully remediate or manage metal contaminated soils.

Global Challenges and Responsibility

The last 100 years have seen massive industrialization. Indeed such developments were coupled with the rapid increase in world population and the desire to enhance economy and food productivity. While industrialization has led to increased economic activity and much benefit to the human race, the lack of regulatory measures and appropriate waste management strategies until the early 1980s (including the use of agrochemicals) has resulted in contamination of our biosphere. Continued pollution of the environment through industrial emissions is of global concern. There is, therefore, a need for politicians, regulatory organizations, and scientists to work together to minimize environmental contamination and to remediate contaminated sites. The responsibility to check this pollution lies with every individual and country although the majority of this pollution is due to the industrialized nations. There is a clear need for better coordination of efforts in dealing with numerous forms of PS pollution problems that are being faced globally.

References

1. Mehlum, H.K.; Arnesen, A.K.M.; Singh, B.R. Extractability and plant uptake of heavy metals in alum shale soils. Commun. Soil Sci. Plant Anal. **1998**, *29,* 183–198.
2. McBride, M.B. Reactions controlling heavy metal solubility in soils. Adv. Soil Sci. **1989**, *10,* 1–56.
3. Patil, G.P.; Gore, S.D.; Johnson, G.D. *EPA Observational Economy Series Volume 3: Manual on Statistical Design and Analysis with Composite Samples*; Technical Report No. 96-0501; Center for Statistical Ecology and Environmental Statistics: Pennsylvania State University, 1996.
4. Ward, T.J.; Correll, R.L. Estimating background concentrations of heavy metals in the marine environment, Proceedings of a Bioaccumulation Workshop: Assessment of the Distribution, Impacts and Bioaccumulation of Contaminants in Aquatic Environments, Sydney, 1990; Miskiewicz, A.G., Ed.; Water Board and Australian Marine Science Association: Sydney, **1992**; 133–139.
5. Jones, K.C.; Davies, B.E.; Peterson, PJ. Silver in welsh soils: physical and chemical distribution studies. Geoderma **1986**, *37,* 157–174.
6. Barber, C.; Bates, L.; Barron, R.; Allison, H. Assessment of the relative vulnerability of groundwater to pollution: A review and background paper for the conference workshop on vulnerability assessment. J. Aust. Geol. Geophys. **1993**, *14*(2–3), 147–154. 880 Pollution: Point Source (PS)
7. Wenzel, W.W.; Blum, W.E.H. Effects of fluorine deposition on the chemistry of acid luvisols. Int. J. Environ. Anal. Chem. **1992**, *46,* 223–231.
8. Megharaj, M.; Singleton, I.; McClure, N.C. Effect of penta-chlorophenol pollution towards microalgae and microbial activities in soil from a former timber processing facility. Bull. Environ. Contam. Toxicol. **1998**, *61,* 108–115.
9. Juhasz, A.L.; Megharaj, M.; Naidu, R. Bioavailability: the major challenge (constraint) to bioremediation of organically contaminated soils. In *Remediation Engineering of Contaminated Soils*; Wise, D., Trantolo, D.J., Cichon, E.J., Inyang, H.I., Stottmeister, U., Eds.; Marcel Dekker: New York, 2000; 217–241.
10. Wood, P.A. Remediation methods for contaminated sites. In *Contaminated Land and Its Reclamation*; Hester, R.E., Harrison, R.M., Eds.; Royal Society of Chemistry, Thomas Graham House: Cambridge, UK, 1997; 47–73.
11. Barzi, F.; Naidu, R.; McLaughlin, M.J. Contaminants and the Australian soil environment. In *Contaminants and the Soil Environment in the Australasia-Pacific Region*; Naidu, R., Kookana, R.S., Oliver, D., Rogers, S., McLaughlin, M.J., Eds.; Kluwer Academic Publishers: Dordrecht, the Netherlands, 1996; 451–484.

25

Soil Carbon and Nitrogen (C/N) Cycling

Sylvie M. Brouder
and Ronald F. Turco
Purdue University

Introduction

Carbon (C) and nitrogen (N) exist in soils in both inorganic and organic forms. Globally, only 780 petagrams (Pg; 10^{15}g) of C are in the atmosphere while the top one meter of soil is estimated to contain approximately 2700 Pg of C, 57% of which is soil organic matter (SOM).[1] In contrast, N in soil is less than one-hundredth of one percent of the N content of the atmosphere but, as with soil C, most soil N is in organic versus inorganic forms.[2] The inorganic C and N forms in soils include the gases carbon dioxide (CO_2) and dinitrogen (N_2) in the air space of soil pores and soluble forms in soil water such as carbonates from the chemical breakdown or weathering of rocks and minerals (e.g., silicates, limestones) and nitrate (NO_3^-) and ammonium (NH_4^+) from the breakdown or mineralization of SOM. C and N occur in all living tissues: C in sugars, starches, lipids, and a staggering array of complex molecules and N in nucleic acids (DNA and RNA) and in amino acids that are assembled into proteins. Both C and N are an integral component of SOM, which is formulated from plant and animal residues in various stages of decomposition including particulates of various size fractions, dissolved OM in soil water, and humus or stable OM as well as the living biomass of soil microorganisms (e.g., bacteria and fungi).[3] To a large degree, microorganisms control or drive global C and N (C/N) cycling. Even under relatively extreme environmental conditions, soil C and N are in a state of constant flux where microbiology transforms C and N among an array of molecules and drives C and N movement between the soil and the surrounding air and water. The goal of this entry is to summarize the state-of-the-art knowledge on the microbiology of soil C/N cycling and discuss how human activities including agriculture are causing large perturbations in soil C/N cycles for which outcomes are uncertain but anticipated to continue to contribute to global climate change.

Bacteria and Fungi Control Soil C/N Cycles

The proportion of soil C and N in the biomass of soil microorganisms is surprisingly small (1.2% and 2.6%;[4]), given the importance of soil microbiology in C/N cycling. Soil bacteria are typically 1 μm or less in length and live in tiny colonies (only 100 or so cells per colony) found inside of soil aggregates and are typically attached to clay and silt fractions. However, while the actual bacteria are small in physical size, soils are rich in cell number, microbial diversity, and, consequently, metabolic abilities. A surface soil can contain greater than 4000 different microorganism communities (i.e., genomes) and can support as many as 10^9 (one billion) bacteria in one gram.[5,6] Subsurface soils will tend to have lower population levels but will exceed 10^7 cells g^{-1}.[7] Because bacteria are small but soil surface area large, bacteria are found on less than 0.17% of the total surface area of SOM and less than 0.02% of the soil's mineral surface area.[8] Further, soil bacteria are largely sessile (not freely moving) and prefer to adhere to surfaces using a form of electrostatic interaction (London-van der Waals forces) followed by a hydrophobic interaction where the cells secrete a polymer that binds them to the surface.[9] This binding reaction attaches bacteria and soil particles together, helping to create the soil structure. Water is the critical connecting force in soil as it moves nutrients and oxygen from their sources to the sessile resident population (Figure 25.1, Process [P] 1). The exact locations within soil colonized by bacteria reflect the availability of nutrients (typically embedded organic carbon) and a preference by bacteria for the small soil pores (between 0.25 and 6 μm diameter) associated with the microaggregate fractions.[10,11] Analysis has shown that greater than 80% of soil bacteria are located within the soil's microaggregate (2 and 53 μm in size) fraction.[12]

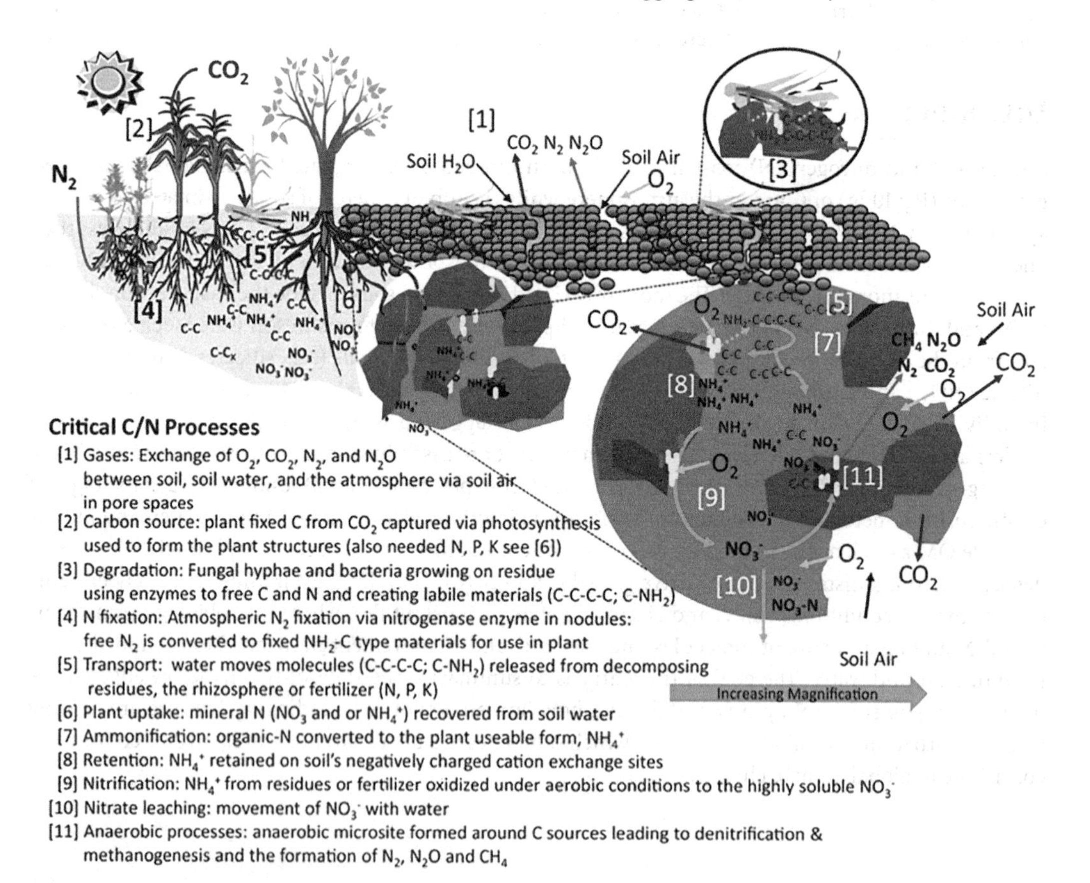

CO_2
[2]
N_2
[1]
CO_2 N_2 N_2O
Soil H_2O
Soil Air
O_2
[3]
[5]
[4]
[6]
[7]
[8]
[9]
[10]
[11]
Soil Air
CH_4 N_2O
N_2 CO_2
CO_2
O_2
Soil Air
Increasing Magnification
Critical C/N Processes
[1] Gases: Exchange of O_2, CO_2, N_2, and N_2O between soil, soil water, and the atmosphere via soil air in pore spaces
[2] Carbon source: plant fixed C from CO_2 captured via photosynthesis used to form the plant structures (also needed N, P, K see [6])
[3] Degradation: Fungal hyphae and bacteria growing on residue using enzymes to free C and N and creating labile materials (C-C-C-C; C-NH_2)
[4] N fixation: Atmospheric N_2 fixation via nitrogenase enzyme in nodules: free N_2 is converted to fixed NH_2-C type materials for use in plant
[5] Transport: water moves molecules (C-C-C-C; C-NH_2) released from decomposing residues, the rhizosphere or fertilizer (N, P, K)
[6] Plant uptake: mineral N (NO_3 and or NH_4^+) recovered from soil water
[7] Ammonification: organic-N converted to the plant useable form, NH_4^+
[8] Retention: NH_4^+ retained on soil's negatively charged cation exchange sites
[9] Nitrification: NH_4^+ from residues or fertilizer oxidized under aerobic conditions to the highly soluble NO_3^-
[10] Nitrate leaching: movement of NO_3^- with water
[11] Anaerobic processes: anaerobic microsite formed around C sources leading to denitrification & methanogenesis and the formation of N_2, N_2O and CH_4

FIGURE 25.1 **(See color insert.)** Diagram of the major C/N cycle soil processes. Left side shows C/N cycling at the terrestrial scale; right side is magnified to highlight cycling within the soil matrix.

Small pores offer an ideal location for bacteria as they allow for the passage of water, which carries nutrients and oxygen but blocks predators such as protozoa.

Fungi also contribute to soil C/N cycling, but fungi differ from bacteria in several important attributes. Soil fungi are resident on the outside of aggregates and are associated with larger organic materials and the coarse sand fractions.[13] More importantly, fungi are made mobile by extension of their hyphae and in doing so are the early colonizers and decomposers of fresh organic residues added to soil via natural and anthropogenic activities (Figure 25.1, P 3). Fungi function by secreting a wide array of enzymes, which start many important soil processes. The combined abilities of residue colonization and enzyme secretion give fungi a critical role in initiating the C/N cycling processes in fresh residue additions to soil.

Continuous Soil C/N Cycling Requires Diversity in Organisms and Substrates

Soils and fresh additions of plant and animal materials present the resident microorganisms with a compositionally diverse array of organic substrates as potential energy and nutrient sources; genetic and associated functional diversity of the soil microbiology ensures ongoing soil C/N and other nutrient cycling in the ever-changing physical and chemical environments.[14] Simplistically, nutrient cycling reflects two sets of processes by which microorganisms meet their need for (i) energy and (ii) essential nutrients (e.g., C, N, phosphorus [P], sulfur) to maintain or increase their biomass (cell numbers). Because both bacteria and fungi must assimilate nutrients across a membrane, organic substrates must first be solubilized into physically smaller fractions. In organic residues of plant origin, the cell walls composed of lignin and cellulose form a recalcitrant physical barrier to degradation. Fungi are one of a few kinds of organisms that can secrete the enzymes necessary to break down cellulose to glucose and the only known organism that can completely degrade lignin via in-place oxidation. Thus, initial decomposition typically involves hyphal extension and enzyme excretion into newly introduced organic material to release labile substances (Figure 25.1, P 3). This release process is general and creates a pool of soluble materials including physically smaller C and N molecules such as simple sugars, amino acids, and proteins that can be accessed by other bacteria, fungi, and growing plants. Labile materials not immediately assimilated by fungi may be washed out of decomposing tissue by rainfall; once these substrates reach the soil, they are transported into the soil matrix in soil water and diffuse to the resident microorganisms inside and outside of the soil aggregates (Figure 25.1, P 5).

Under most conditions, once soluble organic N molecules are transported into the soil matrix, they rapidly enter a multistep process that sequentially creates the inorganic N forms of NH_4^+ and NO_3^-. Proteins and other organics rich in N are acted upon by bacteria that produce ammonification enzymes that free ammonia (NH_3) from C; in water, NH_3 dissolves to form NH_4^+ (Figure 25.1, P 7). The NH_4-N can be assimilated by soil bacteria or plant roots (Figure 25.1, P 6). When soil water concentrations of NH_4^+ are high, excess NH_4^+ may adsorb onto negatively charged soil surfaces where it is held until soil water NH_4^+ concentrations dissipate and the NH_4^+ diffuses back into solution (Figure 25.1, P 8). Additionally, most soils are rich in chemoautotrophic bacteria. Whereas heterotrophic bacteria and fungi power their assimilation of organic C by the oxidation of organics, chemoautotrophic bacteria rely on the oxidation of inorganics (e.g., NH_3) and free CO_2 as their energy and C nutrition sources, respectively. Under aerobic conditions, chemoautotrophic (e.g., nitrifying) bacteria oxidize NH_4^+, consume CO_2, producing NO_3^- instead of CO_2 as an end product (Figure 25.1, P 9). Nitrate is both water soluble and non-reactive with the negatively charged soil surface; thus, once in the NO_3-N form, N is not only available for assimilation by plants and microorganisms but also highly susceptible to loss from the soil system via the physical process of leaching with excess rainfall (Figure 25.1, P 10).

It is important to note that different organisms and processes dominate when water from rainfall or flooding causes soils to become anoxic; these anaerobic processes have significant impacts on soil emission of greenhouse gases. When soil water content is high, macropores in soil are filled with water and O_2 consumption by aerobic bacteria, and plant root respiration (energy derived by electron transfer to O_2) exceeds the O_2 replacement rate from the soil atmosphere. As soil O_2 is depleted, the activity of anaerobic microorganisms that can use electron acceptors other than O_2 will increase. Provided adequate C substrate, many soil bacteria can use NO_3–N instead of O_2 as a terminal electron acceptor in the energy formation process with N_2O and N_2 as end products, a process called denitrification (Figure 25.1, P 11). Alternatively, in fermentation, bacteria use the organic substrate itself to replace O_2 as the oxidizing agent; an important fermentation end product is methane (CH_4). Not only is denitrification an important loss pathway for soil N but products of both denitrification and fermentation have also been identified as potent greenhouse gases. The global-warming potentials on the individual molecule basis are 321- and 12-fold that of CO_2 for N_2O and CH_4, respectively (100-year timeframe).[15]

An additional subset of microorganisms is critical to the soil C/N cycle. While most soil microorganisms and plants cannot use the wealth of N_2 in the soil atmosphere, some have the ability to convert or "fix" the inert N_2 gas into ammonia NH_3. This biological N fixation (BNF) has three forms: it can be independent (free living), associative, or symbiotic with respect to higher plants. Free-living N_2-fixing bacteria are widely distributed and ubiquitous in the soil, but their contributions to terrestrial N is comparatively minor (<1kg ha^{-1} yr^{-1}) as these organisms lack easy access to C substrates for energy to drive the extremely energy-intensive fixation process. These bacteria fix N for their own metabolic needs but, when they die and decompose, the fixed N becomes available to plants. Some of these free-living bacteria live on the soil surface and can photosynthesize and have been estimated to add 10–38kg N ha^{-1} yr^{-1}.[16] Much larger contributions come from soil microorganisms that can colonize root surfaces and invade plant tissues to gain more direct access to energy sources provided by plants. The largest contributions occur when soil microorganisms form true symbiotic relationships with plants such as occurs in legumes. Here, molecular dialogue between the soil microorganism and the plant root leads to the development of a specialized structure or nodule in the plant root to house the N_2 fixation process (Figure 25.1, P 4). In the resulting symbiosis, plants and microorganisms achieve a relatively direct exchange of C for N. Symbiotic N fixation can contribute 24 to 250kg N ha^{-1} in a growing season[17] and, when plant tissues die, this organic N is returned to the soil. Estimates of total annual contributions of BNF to fixed N in terrestrial ecosystems range from 107[18] to 195[19] teragrams (Tg; 10^{12}g) N yr^{-1}.

Natural versus Anthropogenic Controls of C/N Cycles

Among all the nutrient elements essential for soil microorganisms to grow, C and N are required in the largest quantities and are typically the most limiting in soils of natural systems; consequently, the general state of soil microbiological communities in their natural condition is one of repressed activity. Higher plants capture their C from atmospheric CO_2 with photosynthesis (Figure 25.1, P 2) but, as with soil microorganisms, N must be acquired from soils (Figure 25.1, P 6) and is required in higher quantities than any other soil-derived nutrient. Thus, N also constrains plant productivity in unmanaged ecosystems. Over the past several decades, human activities have increasingly perturbed pedologic C/N cycles by physically disturbing soils and/or by increasing quantities of C and/or N in labile form in the soil matrix thereby enhancing the overall activity of microorganisms and increasing opportunities for C and N loss from soils. Chief among anthropogenic activities known to disrupt previously stable cycles are (i) conversion of natural to managed systems with the goal of growing selected plant species (e.g., crops) with enhanced net primary productivity and (ii) the common agricultural practices of soil tillage and N fertilization.

In natural ecosystems, equilibrium exists between what happens when fresh organic residues are added and the genesis and decay of the resident soil humus. The rates of C/N cycling are modulated both by the physical environment and by the composition of added organic materials. Whereas labile substances such as simple sugars, amino acids, and proteins decompose rapidly, subsequent, slower rates of decomposition involve microorganisms attacking increasingly recalcitrant materials. Soil humus, the stable components of SOM that, by definition, bear no resemblance to the floral, faunal, or waste material of origin, is relatively resistant to biodegradation and includes materials derived from lignins, tannins, cutins, and so on.[20] In the absence of fresh residues, SOM serves as the major nutrient source for resident flora. However, as SOM is difficult to degrade and is N limited, the function of microorganisms is slow when they are constrained to this nutrient source. Without fresh additions, soil microorganisms are "starved" and waiting for the arrival of nutrient (C, N, P)-rich materials; as a general rule, new additions from an N-rich residue such as from an annual legume will decompose faster as compared to an equivalent mass of lignin-rich residues from tree branches and trunks.

Major physical determinants of cycling include the soil temperature and moisture regimes. Soil temperatures fluctuate daily and seasonally, especially at the soil surface. Cold or freezing temperatures can drastically slow or halt activity of microorganisms, often without killing them, while temperatures above 35°C not only halt activity but may also kill heat-sensitive organisms. Likewise, drought and low soil water status can restrict microbiology and desiccate microorganisms causing dormancy while soil flooding and anoxia slows growth of organisms that prefer aerobic conditions. However, niche differentiation is the hallmark of successful microbiological communities: heat-tolerant will proliferate at the expense of heat-sensitive soil microorganisms under high soil temperatures and, as discussed in the preceding text, anaerobic will take over from aerobic organisms with prolonged soil flooding.

Anthropogenic Factors Disrupting C/N Cycles: Land Conversion, Tillage, and N Fertilizers

Long-term stability in soil function is a direct result of centuries of buildup of SOM by soil microorganisms. However, the impacts of human activities related to agriculture can result in SOM degradation at rates that far outpace the comparatively slow but constant rates of SOM accumulation in natural systems. Ruddiman[21,22] dates anthropogenic soil losses of C as CO_2 to 8000 to 10,000 years ago with the beginnings of settled agriculture; soil C losses as CH_4 date to 5000 years ago and the introduction of paddy rice production. Recent estimates are that net anthropogenic emissions to the atmosphere total approximately 9.1 Pg C yr^{-1}, 82% of which comes from fossil fuel emissions; the remaining 1.6 Pg C yr^{-1} has been sourced to land use conversion (deforestation and biomass burning), agricultural cultivation, and associated erosion.[1] A meta-analysis of impacts of land use change on soil C showed that C stocks in the top 60 cm of soil decreased an average of 50% when soils were converted from native forests to row crop agriculture.[23] In contrast, converting forest to pasture did not negatively impact soil C stocks, but converting pastures into crops was as detrimental as direct conversion to crops from native forests. Conversion of cropland back to pasture or secondary forest restored some or all soil C, respectively.

Tillage, the mechanical disturbance of soil, is an agricultural practice that dates from the first settled agriculture.[24] It is a proven technology for conditioning the soil for agricultural crops including breaking up soil layers that are too hard and impede root exploration, mechanically controlling weeds that compete with crops for water and nutrients, incorporating nutrients, manure, and residue, controlling residue-borne pests and pathogens, and aerating the soil. Incorporation promotes rates of decomposition of residues of the soil surface by mechanically breaking down residues into small pieces, mixing residue fragments with soil, and enhancing the surface area of the residue directly in contact with the decomposer microorganisms in the soil. However, potential negative impacts are equally numerous.[25] Tilled soils that are not covered with a vigorously growing crop or even a thick mat of fallen residue are highly susceptible to erosion by wind and rainfall. Further, pulverization and compaction of the surface and underlying soil layers, respectively, can lead to reduced water infiltration and poor root development that reduce yields. The implementation of tillage on previously pastured and forested land typically results in dramatic losses of soil C.[23] The implementation of surface residue conservation and no-tillage crop management can restore some soil C but not under all conditions[26] and the value of these practices as a universal solution remains uncertain.[27]

Unlike tillage, the application of large amounts of fertilizer N is a relatively recent practice but one with strong linkages to global climate change. The objective of fertilization in agriculture is to ensure crop productivity and is not limited by competition with soil microorganisms over inadequate mineral N supplies. Early agriculture could rely on only modest amounts of additional N from manures or legumes. Dramatic accelerations of soil C/N cycling in intensively managed agricultural systems are linked to the 1908 discovery of the Haber–Bosch process for high temperature and pressure conversion of N_2 to NH_3. Subsequent commercialization of the process revolutionized the production and thus availability of N fertilizers. At present, estimates of global N inputs into agricultural land are 213 Mt (10^6t) N yr^{-1} of which 47% is fertilizer.[28] In contrast, livestock manures and BNF by leguminous crops account for only 16% and 28%, respectively, of annual global N inputs; cereals, primarily wheat, rice and maize, account for 55 Mt N yr^{-1} of fertilizer N. These fertilizer applications are associated with large increases in crop CO_2 capture but also with the acceleration of rates of C/N cycling including losses from soil. The negative impacts of N fertilizer on global N cycles are known to be substantial and range from eutrophication of aquatic systems to acid rain and ozone depletion.[29] However, implications of an altered N cycle for other biogeochemical cycles are much less certain and include a complex array of climate-relevant feedbacks.[30]' Galloway and Gruber[30] identify reducing uncertainty in our understanding of coupled C/N cycle perturbations as essential to improving global climate change projections.

Conclusion

With reference to ecological function, Becking and Beijerinck concluded more than a century ago "everything is everywhere, but the environment selects."[31] Anthropogenic activities change the environment, and any form of agriculture should be viewed as a selective pressure on soil that will cause the equilibrium of the natural C/N cycle to shift. Soil microbiology, with its great diversity of organisms, will respond with strategies optimizing the use of both quantities and forms of C and N introduced by soil management. For agriculture, the ongoing challenge for the production of major staple crops that underpin global food security remains the design of management systems that appropriately balance the trade-offs between yield objectives, soil biology and long-term soil stability and sustainability of ecosystem function. The challenge is particularly complex as the nature and magnitude of trade-offs are site and soil specific. As we are managing soils to maximize crop yield, the soil biology is constantly responding to our interventions. For example, when reduced forms of N are applied, a subset of soil microorganisms will increase cell numbers and activity to oxidize the material and create the soluble forms such as NO_3–N. These soluble forms can be either leached out of the soil causing water pollution or transported to reduced locations where denitrifying microorganisms consume them and emit greenhouse gases. Conversely, when residues high in C but low in N are applied, soil microorganisms will tie up and use all forms of available nutrients, limiting the amount of N and P available to growing plants. Food security goals that seek to double current agricultural yields on all arable lands will require enhanced quantities of plant-available nutrients. More efficient nutrient management strategies to increase crop production while minimizing impacts on air and water quality and global climate change have been frequently identified as critical to achieving global food security. At present, much of our research investment has gone toward exploring genetic aspects of crop yield and resource (e.g., N and water)-use efficiencies. Research on functional roles of soil microorganisms and their responses to management has lagged and must be elevated to the prominence of plant genetics as the soil's resident microorganisms are the systems-level control point for many key processes governing use efficiency of soil resources.

References

1. Lal, R. Sequestration of atmospheric CO_2 in global carbon pools. Energy Environ. Sci. **2008**, *1*, 86–100.
2. Tamm, C.O. *Nitrogen in Terrestrial Ecosystems: Questions of Productivity, Vegetational Changes, and Ecosystem Stability*; Springer-Verlag: New York, USA, 1991.

3. Stevenson, FJ. *Humus Chemistry: Genesis, Composition, Reactions,* 2nd Ed.; John Wiley & Sons: New York, USA. 1994.
4. Xu, X.; Thornton, P.E.; Post, W.M. A global analysis of soil microbial biomass carbon, nitrogen and phosphorus in terrestrial ecosystems. Global Ecol. Biogeogr. **2013**, *22*, 737–749.
5. Torsvik, V.; Goksoy, J.; Daae, F.L. High diversity in DNA of soil bacteria. Appl. Envir. Micro. **1990**, *56*, 782–787.
6. Torsvik, V.; Salte, K; Sorheim, R.; Goksoyr, J. Comparison of phenotypic diversity and DNA heterogeneity in a population of soil bacteria. Appl. Envir. Micro. **1990**, *56*, 776–781.
7. Konopka, A.E.; Turco, R.F. Biodegradation of organic compounds in vadose zone and aquifer sediments. Appl. Environ. Micro. **1991**, *57*, 2260–2268.
8. Hissett R.; Gray, T.R.G. Microsites and time changes in soil microbial ecology. In *The Role of Terrestrial and Aquatic Organisms in Decomposition Processes*; Anderson, J.M., MacFadyen, A. Eds.; Blackwell: Oxford, Great Britain, 1976; 23–39.
9. van Loosdrecht, M.C.M.; Lyklema, J.; Norde, W; Zehnder, A.J.B. Influence of interfaces on microbial activity. Microbiol. Rev. **1990**, *54*, 75–87.
10. Hattori, T.; Hattori, R. The physical environment in soil microbiology: An attempt to extend the principles of microbiology to soil microbiology. Crit. Rev. Microbiol. **1976**, *4*, 423–461.
11. Killham, K.; Amato, M.; Ladd, J.N. Effect of substrate location in soil and soil pore-water regime on carbon turnover. Soil Biol. Biochem. **1993**, *25*, 57–62.
12. Ranjard, L.; Poly, F.; Combrisson, J.; Richaume, A.; Gourbiere, F.; Thioulouse, J.; Nazaret, S. Heterogeneous cell density and genetic structure of bacterial pools associated with various soil microenvironments as determined by enumeration and DNA fingerprinting approach (RISA). Microbial. Ecol. **2000**, *39*, 263–272.
13. Kandeler, E.; Tscherko, D.; Bruce, K.D.; Stemmer, M.; Hobbs, P.J.; Bardgett, R.D.; Amelung, W. Structure and function of the soil microbial community in microhabitats of a heavy metal polluted soil. Biol. Fert. Soils **2000**, *32*, 390–400.
14. Zak, D.R.; Tilman, D.; Paramenter, R.; Fischer, F.M.; Rice, C.; Vose, J.; Milchunas, D.G.; Martin, C.W. Plant production and soil microorganisms in late successional ecosystems: A continental-scale study. Ecology **1994**, *75*, 2333–2347.
15. Intergovernmental Panel on Climate Change. *Climate Change: The Physical Science Basis*; Cambridge Univ. Press: Cambridge, UK, 2007; 31–35.
16. Witty, J.F.; Keay, P.J.; Frogatt, P.J.; Dart, PJ. Alagal nitrogen fixation on temperate arable fields: The Broadbalk experiment. Plant Soil **1979**, *52*, 151–164.
17. Werner, D. Production and biological nitrogen fixation of tropical legumes. In *Nitrogen Fixation in Agriculture, Forestry, Ecology, and the Environment*; Werner, D., Newton, W.E., Eds; Springer: Dordrecht, the Netherlands, 2005; 1–13.
18. Galloway, J.N.; Dentener, F.J.; Capone, D.G.; Boyer, E.W.; Howarth, R.W.; Seitzinger, S.P; Asner, G.P.; Cleveland C.C; Green, P.A.; Holland, E.A.; Karl, D.M. Michaels, A.F.; Porter, J.H.; Townsend, A.R; Vorosmarty, C.J. Nitrogen cycles: Past, present and future. Biogeochemistry **2004**, *70*, 153–226.
19. Cleveland, C.C.; Townsend, A.R.; Fisher, H.; Howarth, R.W.; Hedin, L.O.; Perakis, S.S.; Latty, E.F.; Von Fischer, J.C.; Elseroad, A.; Wasson, M.F. Global patterns of terrestrial biological nitrogen (N2) fixation in natural ecosystems. Global Biogeochem. Cycles **1999**, *13*, 623–645.
20. Derenne, S.; Largeau, C. A review of some important families of refractory macromolecules: Composition, origin, and fate in soils and sediments. Soil Sci. **2001**, *166*, 833–847.
21. Ruddiman, W.F. *Plows, Plagues and Petroleum: How Humans Took Control of Climate*; Princeton Univ. Press: Princeton, USA, 2003.
22. Ruddiman, W.F. The anthropogenic greenhouse gas era began thousands of years ago. Clim. Change **2005**, *61*, 262–292.

23. Guo, L.B.; Gifford, R.M. Soil carbon stocks and land use change: A meta analysis. Global Change Biol. **2002**, *8*, 345–360.
24. Lal, R. The plow and agricultural sustainability. J. Sust. Agric. **2009**, *33*, 66–84.
25. Hobbs, P.R.; Sayre, K; Gupta, R. The role of conservation agriculture in sustainable agriculture. Phil. Trans. R. Soc. B. **2008**, *363*, 543–555.
26. Six, J.; Ogle, S.M.; Breidt, F.J.; Conant, R.T.; Mosier, A.R.; Paustian, K. The potential to mitigate global warming with no-tillage management is only realized when practiced in the long term. Global Change Biol. **2004**, *10*, 155–160.
27. Giller, K.E.; Witter, E.; Corbeels, M.; Tittonell, P. Conservation agriculture and smallholder farming in Africa: The heretics' view. Field Crops Res. **2009**, *114*, 23–34.
28. Connor, D.J.; Loomis, R.S.; Cassman, K.G. *Crop Ecology: Productivity and Management in Agricultural Systems*, 2nd Ed.; Cambridge University Press: New York, 2011; 195–261.
29. Galloway, J.N.; Aber, J.D.; Erisman, J.W.; Seitzinger, S.P.; Howarth, R.W.; Cowling, E.B. The nitrogen cascade. Bioscience **2003**, *53*, 341–356.
30. Gruber, N.; Galloway, J.N. An earth-system perspective of the global nitrogen cycle. Nature **2008**, *451* (17), 293–296.
31. De Wit, R.; T. Bouvier. Everything is everywhere, but, the environment selects; What did Baas Becking and Beijerinck really say? Environ. Microbiol. **2006**, *8*, 755–758.

26

Soil Degradation: Food Security

Michael A. Zoebisch
German Agency for International Cooperation

Eddy De Pauw
International Center for Agricultural Research in the Dry Areas (ICARDA)

Introduction

Food security is commonly defined as the access by all people at all times to enough food for an active, healthy life. Extended concepts include food-quality aspects, cultural acceptability of the food, equitable distribution among the different social groups, and gender balance.[1–4] These concepts imply that, in addition to producing enough food, people must also have access to it. This opens the arena for the socioeconomic, cultural, policy, and political dimensions of food production and supply. Thus, food security—and consequently food insecurity—are multi-dimensional. It is not possible to separate clearly the biophysical and socioeconomic dimensions of food security, because they are closely interrelated and interdependent.[2] However, the most limiting of the natural resources needed for food production are soil and water. If these resources are depleting, land productivity—the key factor for crop production—declines, and the land is no longer capable of producing the biomass needed for direct and indirect human consumption. If soil degrades, its plant-life supporting functions are lost, together with its role in the hydrological cycle. Soil is a limited resource and, because of the long duration of soil-forming processes, can be considered as nonrenewable. The production of food and feed is directly linked to the productive capacity of the soil. The degree and extent of soil degradation and, consequently, its effect on food production are dependent on how the land is used.

Undernourishment—A Major Indicator of Food Insecurity

The proportion of chronically undernourished people is a manifestation of the degree of food insecurity. Worldwide, more than 830 million people are undernourished; 800 million of these people live in the developing countries, i.e., about 18% of the population of these countries.[1] Table 26.1 shows

TABLE 26.1 Undernourishment in Developing Countries

	Undernourished Population		
Region/Subregion	Proportion of Total Population 1979/1981 (%)	Proportion of Total Population 1995/1997 (%)	Number of People Affected 1995/1997 (millions)
Total Developing World	*29*	*18*	*791.5*
Asia and Pacific	*32*	*17*	*525.5*
East Asia	29	14	176.8
Oceania	31	24	1.1
Southeast Asia	27	13	63.7
South Asia	38	23	283.9
Latin America and Caribbean	*13*	*11*	*53.4*
North America	5	6	5.1
Caribbean	19	31	9.3
Central America	20	17	5.6
South America	14	10	33.3
Near East and North Africa	*9*	*9*	*32.9*
Near East	10	12	27.5
North Africa	8	4	5.4
Sub-Saharan Africa	*37*	*33*	*179.6*
Central Africa	36	48	35.6
East Africa	35	42	77.9
Southern Africa	32	44	35
West Africa	40	16	31.1

Source: Adapted from Lal & Singh.[10]

that the most seriously affected region is SubSaharan Africa, with 33% of the population chronically undernourished. In Asia and the Pacific chronic undernourishment has decreased significantly over the past two decades.

Yields of the major food crops in many countries of these regions have risen remarkably. But in most cases the yield increases have not been able to keep pace with the needs of growing populations. Subregional trends, especially from Africa, China, South Asia, and Central America, indicate large yield declines due to soil degradation.[2,5,6] Countries of the less degraded temperate regions can substitute for the losses in other regions. However, in the affected areas, food prices will increase and hence the incidence of malnutrition. Densely populated countries and regions that depend solely on agriculture will be most affected. As input requirements—and their costs—grow with increasing degradation, these countries will probably not be able to maintain adequate levels of soil productivity without special efforts to stabilize the system.

Vulnerable Areas

Only about one tenth of the world's arable areas are not endangered by degradation. Most soils are vulnerable to degradation and have a low resilience to fully recover from stresses. The soils in the tropics and subtropics are more vulnerable to degradation than those in temperate areas, mainly due to the climatic conditions.[7] The total land area available for agricultural production is limited. Different types of land have different capabilities and limitations. Not all land is suitable for cultivation. The major natural limiting factors for soil productivity are steep slopes, shallow soils, low levels of natural fertility, poor soil drainage, sandy or stony soil, salinity, and sodicity.

Different agro-ecosystems are typically susceptible to different types of soil degradation.[6–8] In mountainous areas and steeplands, water erosion is the dominant form of soil degradation. For arid areas, both water and wind erosion and the loss of soil organic matter are typical. In humid areas, soil acidification and fertility decline are of importance. In irrigated areas, the hazards of soil salinization and waterlogging are of special significance. Knowledge of these hazards can help understand the limitations of the land and introduce appropriate land-use practices that reduce soil degradation and loss of soil productivity.

Underlying Causes

Misuse of Land

The main causative factors of human-induced soil degradation, in a broad sense, are overuse and inappropriate management of agricultural land, deforestation and the removal of natural vegetation, and overgrazing as well as industrial activities that pollute the soil.[6] Worldwide, soil erosion by wind and water are the most important forms of soil degradation, accounting for between 70% and 90% of the total area affected by soil degradation.[5,6] High population pressure and the resultant increased need for higher land productivity is the main factor leading to inappropriate and exploitative land use.

In high-input systems, loss of soil quality due to degradation can be compensated to a certain extent by inputs, such as fertilizers and irrigation water. In low-input systems—usually practiced by resource-poor farmers (in poor countries)—soil quality contributes relatively more to agricultural productivity.[9] Therefore, soil degradation in these systems has more drastic and immediate effects on soil productivity and, hence, food production.

Pressure on the Land

When the land is not enough, i.e., when land scarcity becomes a limiting factor for food production, people usually resort to intensification of their land use and opening up new, often unsuitable, land for cultivation.[4,9] Both pathways will most certainly lead to soil degradation, if unsuitable land-use technologies are used. The majority of land users in the developing world do not have adequate access to appropriate land-use technologies and the financial means to invest in their land.[1] The area available for crop production therefore puts a definite limit to sustainable food production in many parts of the developing world. Alternative, i.e., nonagricultural means of income generation would enable the people to supplement their food requirements on the market and could relieve immediate pressure on the land. This would save soil and land resources.

Losses and Gains of Agricultural Land

Estimates show that more than 10 million hectares of agricultural land worldwide are lost per year, most of it due to soil degradation but also due to urbanization (i.e., between 2 and 4 million hectares).[4] Annual deforestation and clearing of savanna land accounts for about 20 million hectares, of which 16 million hectares are converted to cropland. Thus, there is an annual net gain of cropland of about 5–6 million hectares. Most of these converted lands can be considered marginal, i.e., less suitable for cropping. Loss of traditional sources of grazing and firewood, together with the ecological functions of forests, i.e., their role in regulating watershed hydrology, are factors inducing the degradation of soil.

Minimum Per-Capita Cropland Requirement

On a global scale, it is estimated that the minimum per-capita area of cropland required to produce an adequate quantity of food is around 0.1 ha.[2,5] This rough estimate is dependent on many different location-specific factors and circumstances, such as soil, terrain, climate, farming system, and land-use technology. However, it permits a suitable comparison at the overall global scale. Estimates predict

TABLE 26.2 Trends of Available Per-Capita Cropland and Average Yield Levels of Main Staple Crops for Selected Countries

	Per-Capita Cropland (ha)						Average Yield (kg/ha)			
Country	1961	1980	1990	2000	2015[a]	Crop	1961	1980	1990	2000[b]
Syria	1.40	0.62	0.46	0.34	0.17	Wheat	575	1,536	1,544	1,882
						Barley	461	1,312	310	381
Kenya	0.21	0.15	0.10	0.08	0.04	Wheat	1,090	2,156	1,864	1.400
						Maize	1,253	1,200	1,580	1,321
Pakistan	0.34	0.25	0.17	0.13	0.07	Wheat	822	1,568	1,825	2,492
						Rice	1.391	2.423	2,315	2.866
Bangladesh	0.17	0.10	0.09	0.07	0.05	Wheat	573	1,899	1,504	2.258
						Rice	1,700	2.019	2,566	2.851
Nepal	0.19	0.17	0.14	0.11	0.07	Wheat	1.227	1.199	1,415	1.820
						Rice	1.937	1.932	2,407	2.600
Zambia	1.52	0.92	0.65	0.57	0.28	Wheat	1.600	4.068	4,399	5.454
						Maize	882	1,688	1,432	1,431
Côte d'Ivoire	0.69	0.44	0.31	0.24	0.10	Rice	757	1.166	1,155	1.548
						Sorghum	667	583	575	352
Sudan	0.97	0.62	0.52	0.43	0.22	Rice	1.409	634	1,250	563
						Sorghum	970	712	428	634
Peru	0.20	0.20	0.17	0.14	0.10	Wheat	1.001	939	1,085	1.289
						Potato	5.287	7.196	7,881	10.818

[a] UN Projection, medium level.
[b] FAO Estimates.
Source: Adapted from Engelman & LeRoy,[2] and FAO.[12]

that a large number of countries will have reached the 0.1 ha limit of per-capita cropland by the year 2025.[1,2,7] In the absence of more land for cultivation, future food needs of the people in these countries will have to depend on increased soil productivity through intensified land use. Resource-poor farmers with no option to expand their cultivated area usually cannot invest in necessary soil-conservation and soil-fertility maintenance measures. Over time, this leads to nutrient depletion and other forms of soil degradation that affect soil productivity.[4,9,10]

Although some developing countries have indeed increased their yields on a per-unit area basis significantly, these increases cannot compensate for the increased demands by growing populations. Table 26.2 illustrates past and future trends in per-capita available cropland in selected countries with high population growth and gives examples of yield-level trends. The table shows that, to a large extent, the gains in yields are offset by the decrease in available cropland to grow the required total quantities of crops to feed the population. In these countries, food insecurity can be directly related to the pressure on the land, or land scarcity.

The growing pressure on the land is not only a reflection of population growth and the direct needs for food, but also of changed diets and increasing cash needs (by the land users) that have to be met from land cultivation, putting additional stress on the soil resources.

Estimating Soil Degradation on a Global Scale

Limitations of Methodology

The scientific relationships between different soil-degradation processes and soil productivity have been established mainly on field and plot level.[9,11] Although reliable data are available for a wide range of different agro-ecologies and farming systems, they cannot give a precise picture of degradation-dependent soil productivity for larger land units. The increasing complexity of the determining factors and processes on larger land units, the endless array of location-specific conditions, and the lack of adequate data make upscaling extremely difficult for country and regional levels. Estimates of the extent and severity of soil degradation—and their effects on soil productivity—on a country, regional, continental, and global level therefore bear a substantial degree of uncertainty. Most estimates use the methodology developed by the Global Assessment of Human-Induced Soil Degradation (GLASOD) Project.[5,6] The GLASOD estimates are based on expert assessment and are therefore largely subjective. However, the methodology has been widely adopted, and GLASOD estimates of soil degradation are accepted as the best estimates available. Global estimates are, therefore, only rough approximations and should not be taken literally.

Effects on Soil Productivity and Food Production

Table 26.3 shows estimates of degradation and cumulative losses in soil productivity on a continental level for different land-use types. The figures show that, overall, Africa and Central America are the most severely affected continents, both in terms of soil degradation and reduction of productivity. Sixty-five percent of African and 75% of Central American cropland are affected by soil degradation. The overall loss in productivity in these two regions over the last 50 years is estimated between 25% and 37%. On a continental basis, Africa still has land reserves, but large areas of the continent are marginal for crop production, and their food production is therefore most seriously affected by degradation. Europe also shows relatively high trends in degradation, but mainly of pastures and forests. Productivity losses generally are low, as for North America and Oceania (i.e., mainly Australia), and these can most probably be compensated for by improvements in technology and input supply.

A more detailed picture of trends in food productivity for the main food cereals in selected countries with high population growth rates and agriculture-based economies is given in Figure 26.1. The overall past trends in the different countries show a consolidation of per-capita food production. However, the projections until 2025 indicate clear decreases. This suggests that long-term food security in these countries is at stake.

TABLE 26.3 Estimates of Degradation and Losses in Soil Productivity for Different Land-Use Types in the Major Regions

	Degraded Areas									
	Agricultural Land		Pasture Land		Forests		Total Degraded area		Loss in Productivity[a]	
Region	(Mha)	(%)	(Mha)	(%)	(Mha)	(%)	(Mha)	(%)	Cropland (%)	Pasture (%)
Africa	121	65	243	31	130	19	494	30	25	6.6
Asia	206	38	197	20	344	27	747	27	12.8	3.6
South America	64	45	68	14	112	13	244	16	13.9	2.2
Central America	28	74	10	11	25	38	63	32	36.8	3.3
North America	63	26	29	11	4	1	96	9	8.8	1.8
Europe	72	25	54	35	92	26	218	27	7.9	5.6
Oceania	8	16	84	19	12	8	104	17	3.2	1.1
World	562	38	685	21	719	18	1,966	23	12.7	3.8

[a] Cumulative loss in productivity since 1945. Adjusted figures according to type and degree of degradation.

Source: Adapted from Scherr,[5] van Lynden & Oldeman,[6] and Oldeman.[13]

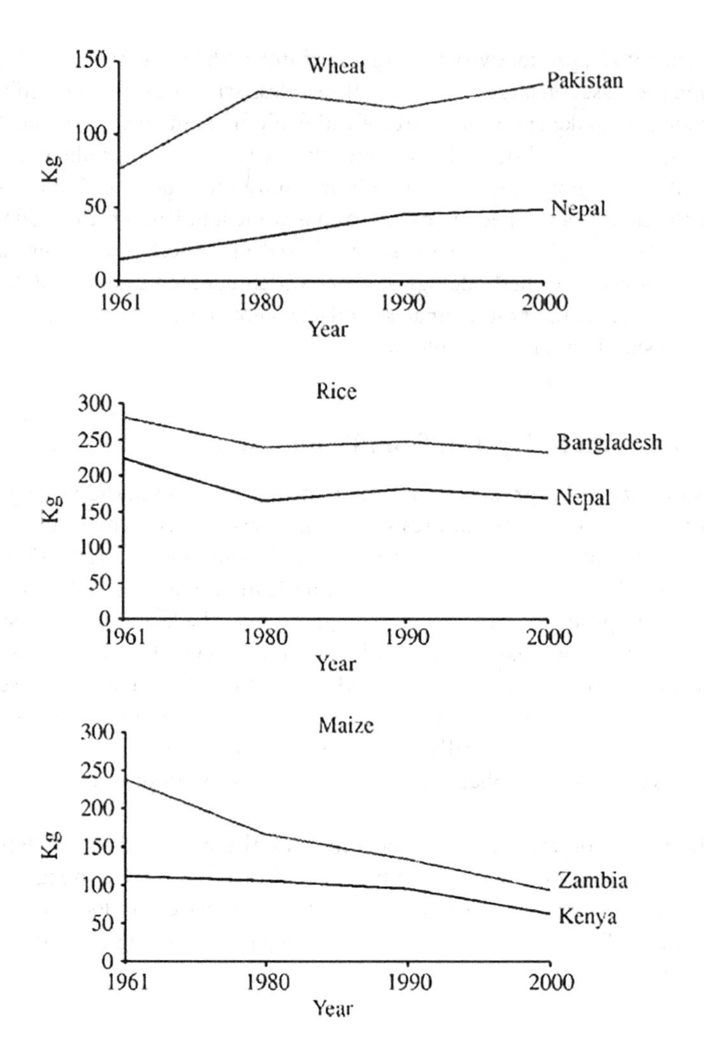

FIGURE 26.1 Per-capita production of selected crops in countries with high population pressure.
Source: Adapted from Engelman & LeRoy,[2] and FAO.[12]

Soil Degradation and Decline in Productivity

Over the past decades, the cumulative loss of productivity of cropland has been estimated at about 13%. However, aggregate global food security does not appear to be under a significant threat.[2,5] There is no conclusive evidence that soil degradation has in the past affected global food security. We cannot conclude to such relationship because 1) the database on the extent and magnitude of various types of soil degradation is inadequate and 2) the impact of soil degradation on land productivity is very site-specific, often anecdotal and difficult to quantify at regional and global levels. In fact, globally, per-capita food production has increased by about 25% since 1961 when the first production surveys were conducted by Food and Agriculture Organization of the United Nations (FAO).[12] The yields per unit area of the major cereals (i.e., wheat, rice, maize) have steadily increased and are still rising.[1,12] This, however, does not imply that sufficient food is—and will be—available in all countries and regions. Food is not necessarily produced where it is most needed. A top priority should therefore be to improve inventories of land degradation at regional, national, and subnational levels.

However, the evidence is conclusive that soil degradation affects food security at the subnational and national levels in the developing countries by the gradual decline of the land's productive capacity and that this trend will continue in the future. We therefore have to assume that, eventually, land degradation will constitute a serious threat to global food security by its particular impact on the developing countries. According to most scenarios, these countries are most vulnerable to degradation induced by increasing pressure on their land resources, the effects of climate change and their inability to finance programs to rehabilitate affected areas and prevent further degradation and decline of productivity.

References

1. FAO. *The State of Food Insecurity in the World*; Food and Agriculture Organization of the United Nations (FAO): Rome, Italy, 1999.
2. Engelman, R.; LeRoy, P. *Conserving Land: Population and Sustainable Food Production*; Population and Environment Program; Population Action International: Washington, DC, U.S.A., 1995.
3. Saad, M.B. *Food Security for the Food-Insecure: New Challenges and Renewed Commitments*; Position Paper: Sustainable Agriculture; Centre for Development Studies, University College Dublin: Dublin, Ireland, 1999.
4. Kindall, H.W.; Pimentel, D. Constraints on the expansion of the global food supply. Ambio **1994**, *23* (3), 198–205.
5. Scherr, S. J. *Soil Degradation—A Threat to Developing-Country Food Security by 2020*? Discussion Paper 27; International Food Policy Research Institute: Washington, DC, 1999.
6. van Lynden, G.W.J.; Oldeman, L.R. *The Assessment of the Status of Human-Induced Soil Degradation in South and Southeast Asia*; International Soil Reference and Information Centre: Wageningen, the Netherlands, 1997.
7. Greenland, D.; Bowen, G.; Eswaran, H.; Rhoades, R.; Valentin, C. *Soil, Water, and Nutrient Management Research—A New Agenda*; IBSRAM Position Paper; International Board for Soil Research and Management: Bangkok, Thailand, 1994.
8. Lal, R. Soil management in the developing countries. Soil Science **2000**, *165* (1), 57–72.
9. Stocking, M.; Murnaghan, N. *Land Degradation—Guidelines for Field Assessment*; UNU/UNEP/PLEC Working Paper; Overseas Development Group, University of East Anglia: Norwich, UK, 2000.
10. Lal, R.; Singh, B.R. Effects of soil degradation on crop productivity in East Africa. J. Sustain. Agricul. **1998**, *13* (1), 15–36.
11. Stocking, M.; Benites, J. *Erosion-Induced Loss in Soil Productivity: Preparatory Papers and Country Report Analyses*; Food and Agriculture Organization of the United Nations: Rome, Italy, 1996.
12. FAO. FAOSTAT. Statistical Database of the Food and Agriculture Organization of the United Nations. Internet Web-Source http://www.fao.org (accessed January 2001), Rome, Italy, 2000.
13. Oldeman, L.R. *Soil Degradation: A Threat to Food Security*? Report 98-01; International Soil Reference and Information Centre (ISRIC): Wageningen, the Netherlands, 1998.

27

Soil Degradation: Global Assessment

Selim Kapur
University of Qukurova

Erhan Akça
Adiyaman University

Introduction

One of the major challenges that humanity will face in the coming decades is the need to increase food production to cope with the ever-increasing population, which is indeed the greatest threat for food/soil security.[1] The worldwide loss of 12 million hectares (ha) of agricultural land per year was reflected in the average annual loss up to €38 billion for EU25,[2] in spite of the strict laws for environmental protection enacted there. This is the unpalatable but important indicator of a steady decline in agricultural production and increase in soil damage, reflecting the "impaired productivity" in a quarter of the world's agricultural land.[3–5]

Background

Soil can degrade without actually eroding. It can lose its nutrients and soil biota and can become damaged by waterlogging and compaction. Erosion is only the most visible part of degradation, where the forces of gravity, water flow, or wind actively remove soil particles.

Rather than taking the classical view that soil degradation was, is, and will remain an ongoing process, mainly found in countries of the developing world, this phenomenon should be seen as a worldwide process that occurs at different scales and different time frames in different regions. The causes of biophysical and chemical soil degradations are enhanced by socioeconomic interventions, which are the main anthropogenic components of this problem, together with agricultural mismanagement, overgrazing, deforestation, overexploitation, and pollution as reiterated by Lal,[6] UNEP/ISRIC,[7] Lal,[8] Eswaran & Reich,[9] Eswaran et al.,[10] Kapur et al.,[11] and Cangir et al.,[12] as the main reasons for erosion and chemical soil degradation.

Soil degradation, the threat to "soil security," is ubiquitous across the globe in its various forms and at varying magnitudes, depending on the specific demands of people and the inexorably increasing pressures on land. Europe provides many telling examples of the fragile nature of soil security and the destructive consequences of a wide range of soil degradation processes. Asia, Africa, and South and

North America are not only partly affected by the nonresilient impacts of soil degradation but also experiencing more subtle destruction of soils through political developments, which seek to provide temporary relief and welfare in response to the demands of local populations.

Europe

The major problems concerning the soils of Europe are the loss of such resources owing to erosion, sealing, flooding, large mass movements as well as local and diffuse soil contamination, especially in industrial and urban areas, and soil acidification.[13,14] Salinity is a minor problem in some parts of Western and Eastern Europe, but with severe effects at the northern and western parts of the Caspian Sea (with low salinity)[15] mainly because of the shift to irrigated agriculture and destruction of the natural vegetation (Figure 27.1).

Urbanization and construction of infrastructure at the expense of fertile land are widespread in Europe, particularly in the Benelux countries, France, Germany, and Switzerland, and such effects are most conspicuously destructive along the misused coasts of Spain, France, Italy, Greece, Turkey, Croatia, and Albania. The drastic increase in the rate of urbanization since the 1980s is now expected to follow the Blue Plan, which seeks to create beneficial relationships among populations, natural resources, the main elements of the environment, and the major sectors of development in the Mediterranean Basin and to work for sustainable development in the Mediterranean region. The very appropriate term "industrial desertification" remains valid for the once degraded soils of East Europe, under the pressure of mining and heavy industry, as in Ukraine where such lands occupy 3% of the total land area of the country.[22,23]

There are three broad zones of "natural" erosion across Europe, including Iceland: (1) the southern zone (the Mediterranean countries); (2) a northern loess zone comprising the Baltic States and part of Russia; and (3) the eastern zone of Slovenia, Croatia, Bosnia-Herzegovina, Rumania, Bulgaria, Poland, Hungary, Slovakia, the Czech Republic, and Ukraine (Figure 27.1). Seasonal rainfalls are responsible for severe erosion owing to overgrazing and the shift from traditional crops. Erosion in southern Europe is an ancient problem and still continues in many places, with marked on-site impacts and with significant decreases in soil productivity as a result of soil thinning. The northern zone of high-quality loess soils displays moderate effects of erosion with less intense precipitation on saturated soils. Local wind erosion on light textured soils is also responsible for the transportation of agricultural chemicals used in the intensive farming systems of the northern zone to adjacent water bodies, along with eroded sediments. The high erodibility of the soils of the eastern zone is exacerbated by the presence of large state-controlled farms that have introduced intensive agriculture at the expense of a decrease in the natural vegetation. Contaminated sediments are also present in this zone, particularly in the vicinity of former industrial operations/deserts, with high rates of erosion in Ukraine (33% of the total land area) and Russia (57% of the total land area), whose agricultural land has been subjected to strong water and wind erosion ever since the beginning of industrialization.

Localized zones of likely soil contamination through the activities of heavy industry are common in northwestern and central Europe as well as northern Italy, together with more scattered areas of known and likely soil contamination caused by the intensive use of agricultural chemicals. Sources of contamination are especially abundant in the "hot spots" associated with urban areas and industrial enclaves in the northwestern, southern, and central parts of the continent (Figure 27.1). Acidification through deposition of windborne industrial effluents and aerosols has been a longstanding problem for the whole of Europe; however, this is not expected to increase much further, especially in western Europe, as a result of the successful implementation of emission-control policies over the recent years.[3]

The desertification of parts of Europe has been evident for some decades, and the parameters of the problem are now becoming clear, with current emphasis on monitoring of the environmentally sensitive areas[22,20] on selected sites, seeking quality indicators for (1) soil (particularly organic carbon; (2) vegetation; (3) climate; and (4) human management throughout the Mediterranean basin. Apart from the human factor, these indicators are inherent.

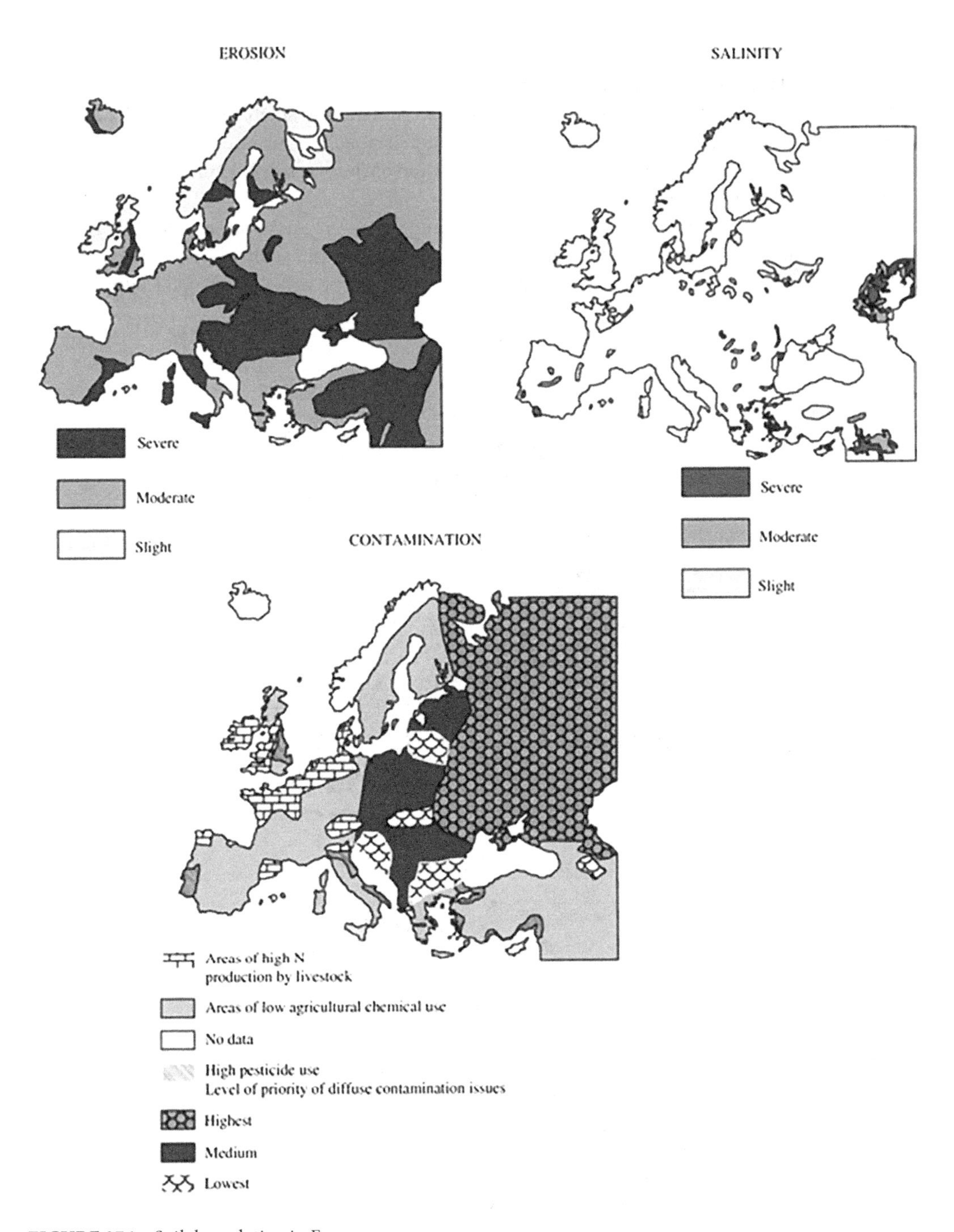

FIGURE 27.1 Soil degradation in Europe.
Source: Adapted from UNEP/ISRIC,[7] Kobza,[16] USDA-NRCS,[17,18] Erol,[19] Kosmas et al.[20] Kharin et al.[21]

Asia

The most severe aspect of soil degradation on Asian lands has been desertification owing to the historical, climatic, and topographic character of this region as well as the political and population pressures created by the conflicts of the past 500 years or so. Salinization caused by the rapid drop in the level of the Aral Sea and the water-logging of rangelands in Central Asia owing to the destruction of the vegetation cover by overgrazing and cultivation provide the most striking examples of an extreme version of degradation—desertification caused by misuse of land. Soil salinity, the second colossal threat to the Asian environment, has occurred through the accumulation of soluble salts, mainly deposited from saline irrigation water or through mismanagement of available water resources, as in the drying Aral Sea and the Turan lowlands as well as the deterioration of the oases in Turkmenistan, with excessive abstraction of water in Central Asia (Figure 27.2).[21]

The dry lands of the Middle East have been degrading, since the Sumerian epoch, with excessive irrigation causing severe salinity and erosion-siltation problems,[21] especially in Iraq, Syria, and Saudi Arabia. Iraq has been unique in the magnitude of the historically recorded build-up of salinity levels,

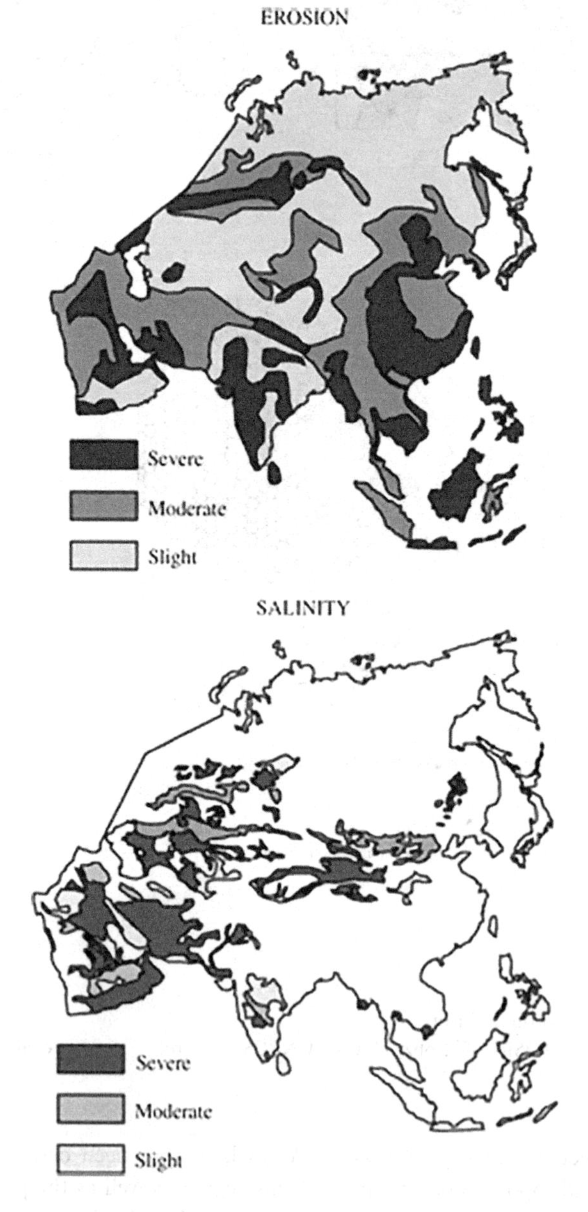

FIGURE 27.2 Soil degradation in Asia.
Source: Adapted from UNEP/ISRIC,[7] Erol,[19] Kosmas et al.[20] Lowdermilk.[24]

with 4.81 million ha saline land, which is 74% of the total arable land surface (i.e., 90% of the land in the southern part of that country). The historical lands of Iran, Pakistan, Afghanistan, India, and China are also subject to ancient and ongoing soil degradation processes, which are subtle in some areas but evident and drastic in others (Figure 27.2).

Africa

Africa's primary past and present concern has been the loss of soil security by nutrient depletion, that is, the decreasing NPK levels (kg/ha) along with micronutrients in cultivated soils following the exponential growth in population and the resulting starvation and migrations (Figure 27.3). Intensification

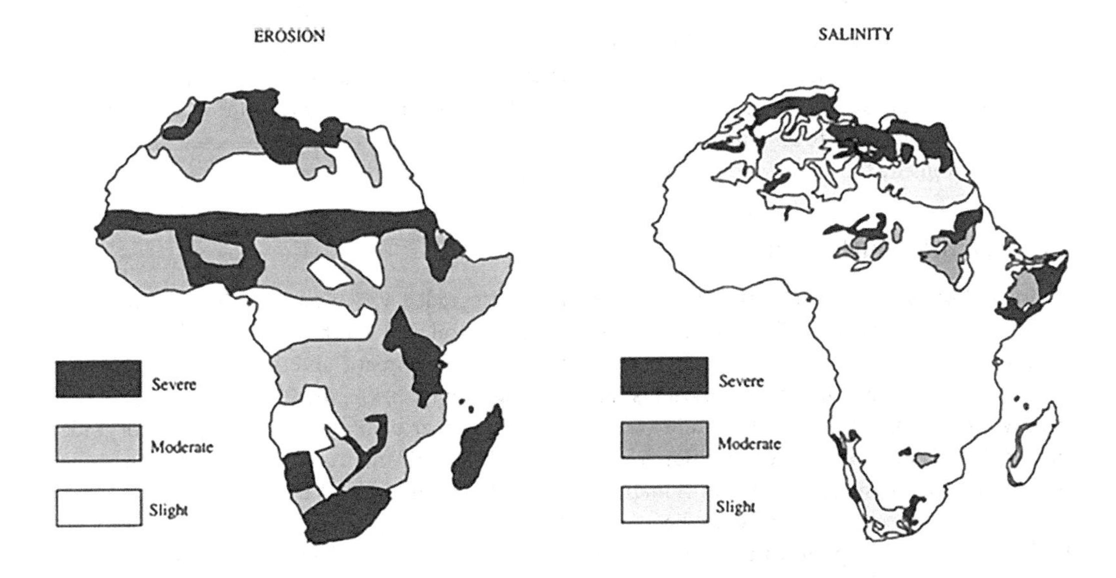

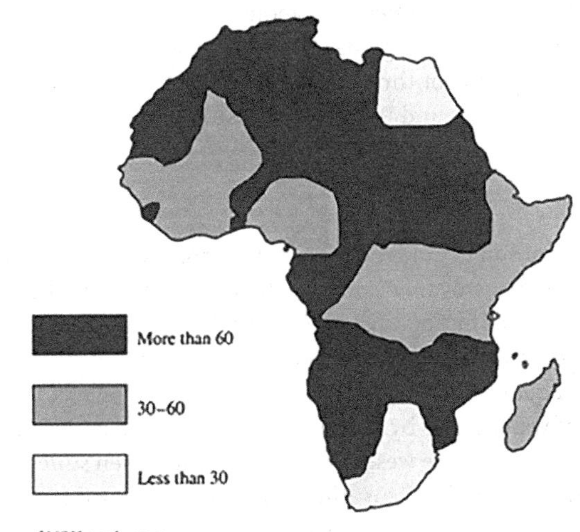

FIGURE 27.3 Soil degradation in Africa.
Source: Adapted from UNEP/ISRIC,[7] Lowdermilk.[24]

of land use to meet the increased food demands combined with the mismanagement of the land leads to the degradation of the continental soils. This poses the ultimate question of how the appropriate sustainable technologies that will permit increased productivity of soils can be identified. This problem is illustrated by the example of the Sudan, where nutrient depletion has steadily increased through more mechanized land preparation, planting, and threshing without the use of inorganic fertilizers, and legume rotations. Thus, aggregate yields have been falling as it became more difficult to expand the cultivated area without substantial public investments in infrastructure. This decline in yield has occurred at enormous rates in Vertisols of Sudan, with the mean annual sorghum yields decreasing from 1000 to 500 kg/ha and the wholesale price of sorghum increased 3.7% per month since 2007.[25] In Burkina Faso, the decreased infiltration and increased runoff causing erosion are further consequences of repeated cultivation. Thus, the technological measures to be identified for these two African examples must include development of water retention technologies in Burkina Faso, while polyculture/rotations with proper manuring and fertilization for cost-efficient provisions of N and P and preferably green manuring are all needed in Sudan to permit the balanced management of soil moisture, nutrients, and organic matter (and to enhance C-sequestration, a main goal for sustainability based on the earth sciences—to ensure the security of both the soil and global climate).[25,26]

Central and South America

Africa and Latin America have the highest proportion of degraded agricultural land. Water and wind erosion are the dominant soil degradation processes in Central and South America and have caused the loss of the topsoil at alarming rates because of the prevailing climatic and topographic conditions. Almost as important is the loss of nutrients from the Amazon basin (Figure 27.4).[27] These effects are mainly attributable to deforestation and overgrazing, the former being responsible for the degradation of 576 million ha out of 1,000 million ha potential agricultural land. Another important factor has been the ever-increasing introduction of inappropriate agricultural practices derived from the so-called imported technology, which have not been properly adapted to indigenous land-use procedures. The traditional methods of permitting the land to recover naturally have been almost totally abandoned and replaced by unsuitable technological measures designed to maintain production levels (temporarily) and to overcome the loss of soil resilience, thus increasing chemical inputs.

The rapid industrialization/urbanization of the limited land resources in the Caribbean region has been expelling agricultural communities to remote and marginal regions that are at present rich in biodiversity and biomass—a major global C sink. Moreover, large-scale livestock herding of Central and South America is also a major threat to soil security and has been responsible for degrading 1 million km^2 of Argentinean, Bolivian, and Paraguayan pasturelands.

North America

The most prominent outcome of soil degradation (or more correctly desertification) in United States is exemplified by the accelerated dust storm episodes of the 1930s—the Dust Bowl years, marked by the "Black Blizzards," which were caused by persistent strong winds, droughts, and overuse of the soils. These resulted in the destruction of large tracts of farmland in the south and central United States. Recently, salinization has become an equally severe problem in the western part of the country (Figure 27.5) through artificial elevation of water tables by extensive irrigation, with associated acute drainage problems. An area of about 10 million ha in the west of United States has been suffering from salinity-related reductions in yields, coupled with very high costs in both the Colorado River basin and the San Joaquin Valley.[28] Unfortunately, new irrigation technologies, such as the center pivot irrigation system (developed as an alternative to the conventional irrigation systems causing the salinity problems), have caused a decline of the water-table levels in areas north of Lubbock, Texas, by around 30–50 m, leading to a dramatic decrease in the thickness of the well-known Ogallala aquifer by 11% decrease between 2003

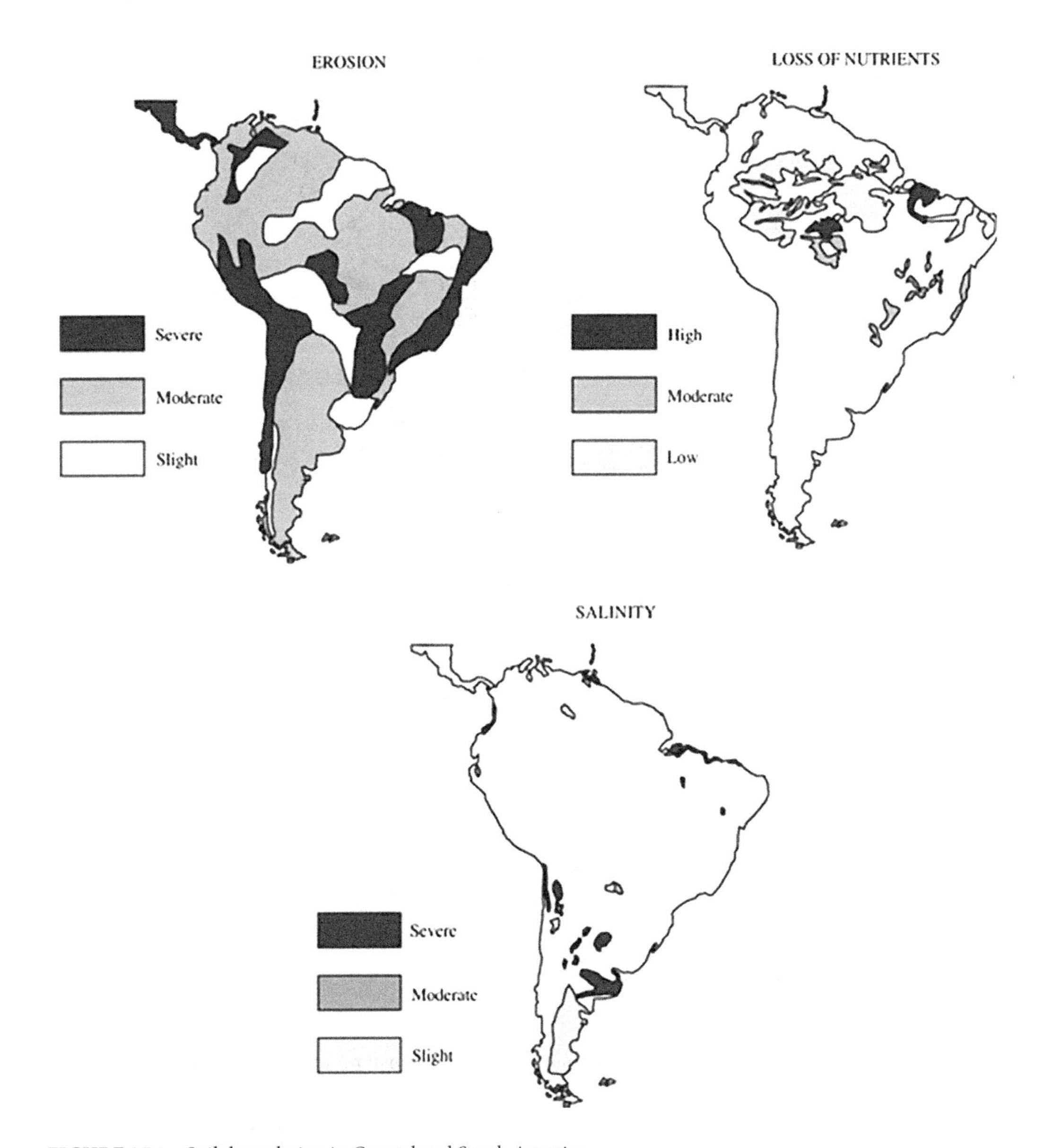

FIGURE 27.4 Soil degradation in Central and South America.
Source: Adapted from UNEP/ISRIC,[7] Kharin et al.[21]

and 2011 only. In some areas, this has been followed by ground subsidence, which is an extreme form of soil structure degradation, that is, loss of the physical integrity of the soil.

Loss of topsoil, as a result of more than 200 years of intensive farming in United States, is estimated to vary between 25% and 75% and exceeds the upper limit in some parts of the country.[18,19] United States provides good examples of the difficulties involved in erosion control, with its large-scale intensive agriculture—deteriorating soil structure and increasing erosion of its susceptible soils. This problem could be overcome primarily by the strict introduction of the no-till system. No-till areas have increased from 4 million ha in 1989 to 25.3 million ha in 2004, and they are forecast to follow a linear extrapolation until 48 million ha out of the total 81 million ha of cultivated land will be attained.[29] Conservation farming is practiced in only about half of all U.S. agricultural land and on less than half of the country's most erodible cropland. Conservation farmers are encouraged to use only the basic types of organic

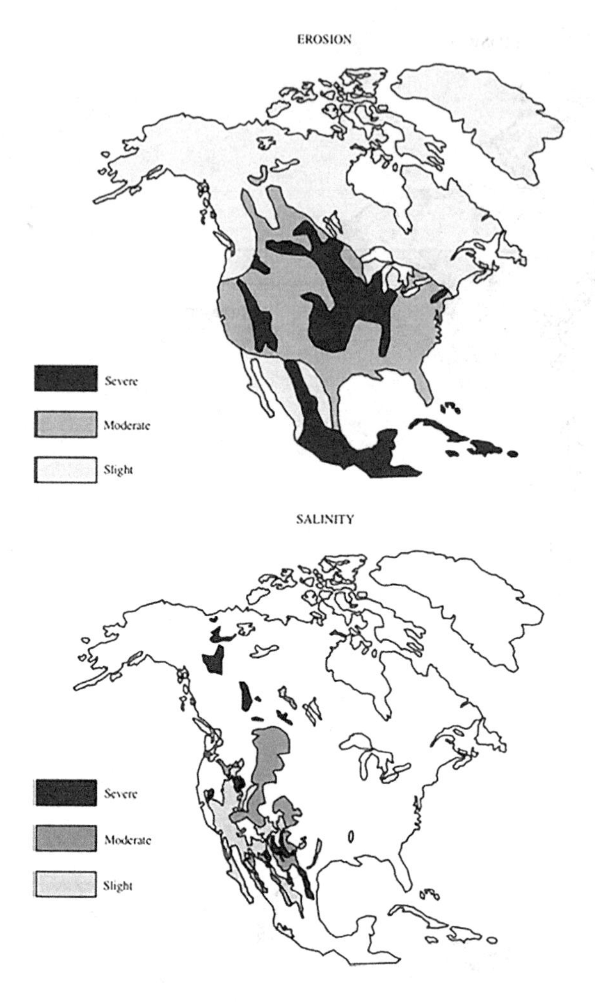

FIGURE 27.5 Soil degradation in North America.
Source: Adapted from UNEP/ISRIC,[7] Kharin et al.[21]

fertilizers, such as animal and green manure together with compost, mulch farming, improved pasture management, and crop rotation, to conserve soil nutrients.

Canada is a large country where 68 million ha of available land is cultivated, which is only the 7.4% of the territorial of the country with an average farm size of 450 ha. It is reported that Canada has experienced annual soil losses on the prairies, through wind and water erosions, that are similar to the Asian steppes, amounting, respectively, to 60 and 117 million tons. These annual rates are much higher than the rate of soil formation, resulting in an annual potential grain production loss of 4.6 million tons of wheat. With regard to primary soil salinity, during historic times, the prairies have experienced steady increases related partially to increasing groundwater levels. Major problems of secondary salinity are estimated to affect 2.2 million ha of land in Alberta, Saskatchewan, and parts of Manitoba, with an immense economic impact each year.

Australia

The Australian agricultural/soil resource base has been endangered, as the "business as usual" concept was adopted on the continent to achieve temporary economic betterment. Identification of different types of soil degradation in Australia reveals that erosion has been the main component, primarily via dust storms, which are still a serious problem, especially where cropping practices do not include retention of cover and minimum tillage methods. Water erosion effects are also particularly severe in areas of summer rainfall and topographic extremities (Figure 27.6). Although water is a scarce resource in Australia, about 14 million tons of soil is lost in Australia by water erosion. Remedial actions for this include the well-known measures of maintaining adequate cover and changing prevailing attitudes toward stock management, storage feed, redesign of watering sites, and management of riparian areas.

Part of the excess salinity in Australia is of primary origin and was retained in the subsoil by trees, which have now been cleared to create soil surfaces for cropping and pastures, allowing penetration of water to the saline subsoil, then followed by abstraction from the water table, thus leading to the ultimate disaster. About 30% of Australia's agricultural land is sodic, creating poor physical conditions and impeded productivity. This problem can only be alleviated by massive revegetation programs and by taking extra care of the water table and plant cover. Despite the introduction of costly conventional measures for reclamation, salinity levels continue to increase across Australia in the dry and irrigated

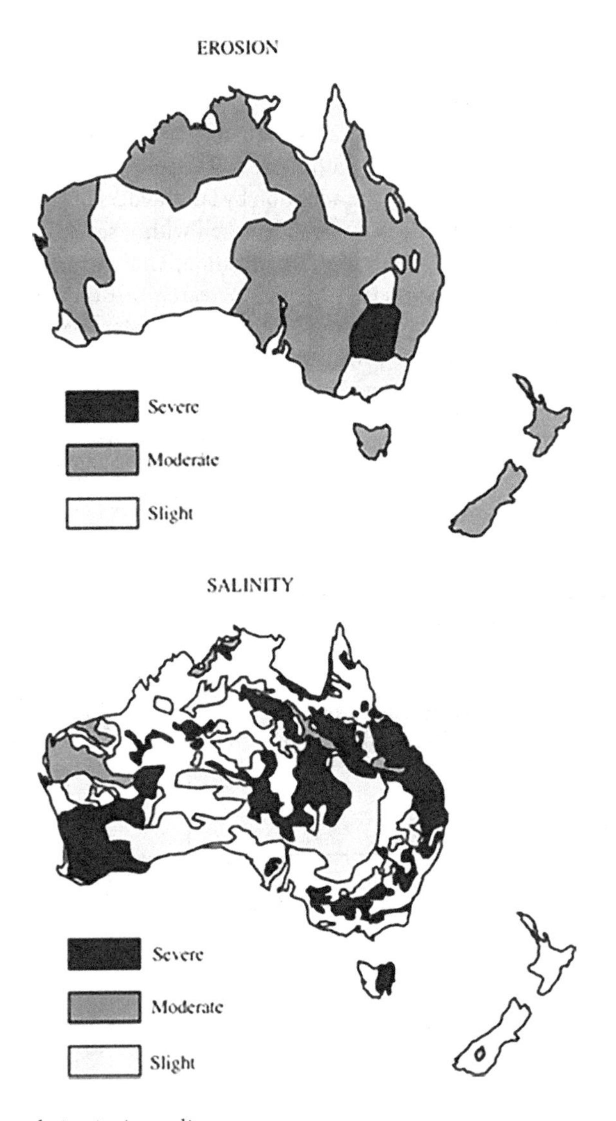

FIGURE 27.6 Soil degradation in Australia.
Source: Adapted from UNEP/ISRIC,[7] Kharin et al.[21]

soils. The dryland salinity in the continent affects about 5 million ha of farmland and is expanding at a rate of 3–5% per year.[30]

The retardation of organic matter levels also requires remediation measures, with economically justified fertilizer use strategies to be utilized throughout the continent. Moreover, overgrazing has resulted in the impoverishment of plant communities and loss of habitats as well as the decline in the chemical fertility of the soil by progressive depletion of organic matter in the topsoil, followed by deterioration in the soil structure.

Acidification caused by legume-based mixed farming plus use of ammonia-based fertilizers threatens 55 million ha of Australian land. Liming seems to be the most effective present remedy, but is costly, does not lead to rapid recovery, and is impractical for subsoil acidity. Thus, the precise remedies are yet to be developed for the conditions on this continent, utilizing careful, long-term monitoring and the experience of farmers to devise specific treatment and conservation procedures.

Conclusions

The state of soil degradation and its remedies as a multi-function-multi-impact approach have been identified through a Driving Force-Pressure-State-Impact-Response matrix by the European Environmental Agency[2] (Figure 27.7) leading to sustainable land management (SLM)[31–33] measures to be taken for the future. SLM is concerned with more soil-friendly farming practices that minimize the erosion potential of soils, together with the adoption by landholders of property management planning procedures that involve community actions undertaken within several concerted action projects of the European Union, Food and Agriculture Organization, Global Environment Facility, United Nations Environment Programme, International Crops Research Institute for the Semi-Arid Tropics, Canadian International Development Agency, and World Bank.[34] Moreover, as Smyth & Dumanski[35] have stated, these combine socioeconomic principles with environmental concerns so that the production is enhanced, together with the reduction of its level of risk with the protection of natural resources,

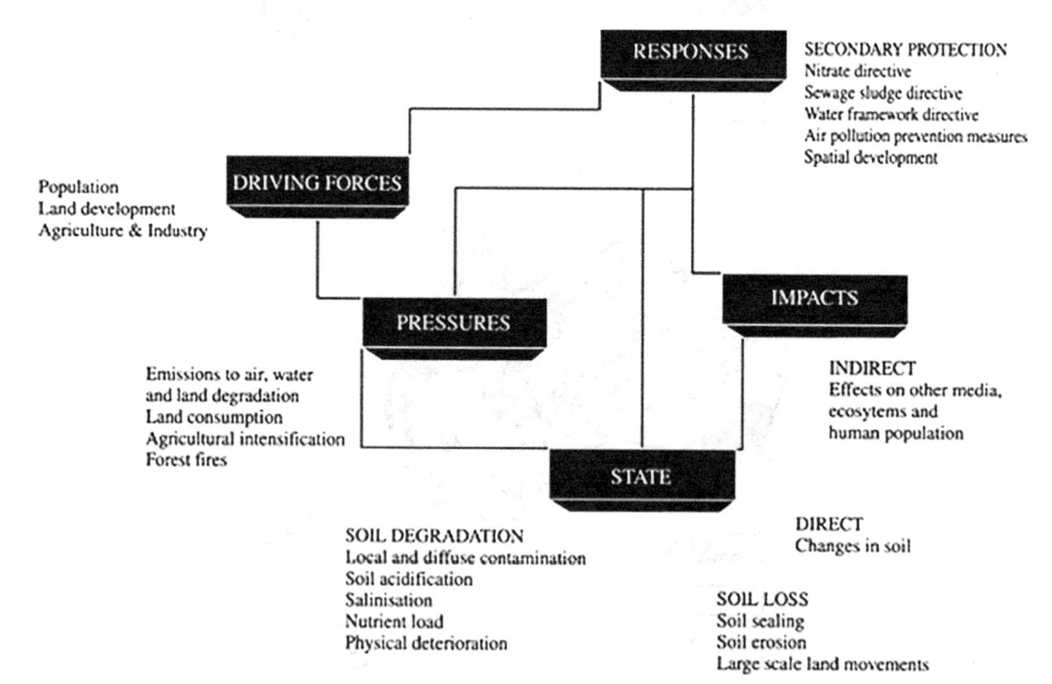

FIGURE 27.7 Driving Force-Pressure-State-Impact-Response framework applied to soil.
Source: Adapted from World Bank.[31]

which would prevent degradation of soil and water quality to be successfully accepted by the farmer. The methods to be adopted for SLM via community actions include contour farming, terracing, vegetative barriers, and other land-use practices amalgamated with indigenous (traditional) technical knowledge (ITK)[33] as applied to farming and landscape preservation. The impetus to the use of ITK by scientists and local communities in creating new strategies for sustainable resource management was initiated in the United Nations Conference on Environment and Development held in Rio de Janeiro (Brazil) in 1992 and the need for raising awareness and combatting land degradation is man's priority issue for global social security.

References

1. International Food Policy Research Institute. The Economics of Desertification, Land Degradation, and Drought. Toward an Integrated Global Assessment. IFPRI Discussion Paper 01086, May 2011. Bonn, 188 p.
2. Toth, G.; Montanarella, L.; Rusco, E. *Threats to Soil Quality in Europe*; Institute for Environment and Sustainability, Land Management and Natural Hazards Unit: EUR 23438 EN. Milan, 2008; 162 p.
3. Amman, M.; Bertok, I.; Cabala, R.; Cofala, J.; Heyes, C.; Gyarfas, F.; Klimont, Z.; Schöpp, W.; Wagner, F. *A Further Emission Control Scenario for the Clean Air for Europe (CAFE) Programme*; International Institute for Applied Systems Analysis: Laxenburg, Austria, 2005.
4. Oldeman, L.R. The global extent of soil degradation. In *Soil Resilience and Sustainable Land Use*; CAB International: Oxon, U.K., 1994; 115 pp.
5. Steiner, K.G. *Causes of Soil Degradation and Development Approaches to Sustainable Soil Management*; Steiner, K.G., Ed.; GTZ. Weikersheim. Margraf Verlag: Filderstadt, Germany, 1996: 50 pp.
6. Lal, R. Forest soils and carbon sequestration. Forest Ecol. Manag. **2005**, *220*, 242–258 pp.
7. UNEP/ISRIC. In *GLASOD World Map of the Status of Human-Induced Soil Degradation*, 1:10 M. 3 Sheets; 2nd Revised Ed.; ISRIC: Wageningen, 1991.
8. Lal, R. Methods and guidelines for assessing sustainable use of soil and water resources in the tropics. In *N.21, SMSS Technical Monograph*; U.S. Agency for International Development: Washington, DC, 1994, 79 pp.
9. Eswaran, H.; Reich, P. Desertification: a global assessment and risks to sustainability. In *Proceedings of the 16th Int. Cong. Soil Science*; Montpellier, France, 1998; CD-ROM.
10. Eswaran, H.; Beinroth, F.H.; Reich, P. Global land resources and population supporting capacity. Am. J. Alternative Agric. **1999**, *14*, 129–136.
11. Kapur, S.; Atalay, P. İ.; Ernst, F.; Akça, E.; Yetiş, C.; İşler, F.; Öcal, A.D.; Üzel, I.; Şafak, Ü. A review of the late quaternary history of Anatolia. In *1st International Symposium on Cilician Archaeology*; Durugönül, S., Ed.; Mersin University: Turkey, 1999; 253–272.
12. Cangir, C.; Kapur, S.; Boyraz, D.; Akça, E.; Eswaran, H. Land resource consumption in Turkey. J. Soil Water Conserv. **2000**, *3*, 253–259.
13. Millennium Ecosystem Assessment. *Ecosystems and Human Well-being: Desertification Synthesis*; World Resources Institute: Washington, DC, 2005.
14. Oldeman, L.R.; Hakkeling, R.T.A.; Sombroek, W.G. *World Map of the Status of Human-Induced Soil Degradation: An Explanatory Note*; International Soil Reference Center: Wageningen: the Netherlands, 1992; 34 pp.
15. Karpinsky, M.G. Aspects of the Caspian Sea benthic ecosystem. Mar. Pollut. Bull. **1992**, *24*, 3849–3862.
16. Kobza, J. Monitoring and contamination of soils in Slovakia in relation to their degradation. Proceedings of the 1st International Conference on Land Degradation. Kapur, S.; Akça, E.; Eswaran, H.; Kelling, G.; Vita-Finzi, C.; Mermut, A.R.; Öcal, A.D., Eds.; Adana, Turkey, 1996; 10–14.

17. USDA-NRCS. *Risk of Human Induced Desertification, NRSC*; Soil Survey Division, World Soil Resources: 2003.
18. USDA-NRCS. *Global Desertification Vulnerability, NRSC*; Soil Survey Division, World Soil Resources: 2003.
19. Erol, O. Misuse of the Mediterranean coastal area in Turkey. In *Proceedings of the 1st International Conference on Land Degradation,* Kapur, S.; Akça, E.; Eswaran, H.; Kelling, G.; Vita-Finzi, C.; Mermut, A.R.; Öcal, A.D., Eds.; Adana: Turkey, 1999; 195–203.
20. Kosmas, C.; Ferrara, A.; Briassouli, H.; Imeson, A. Methodology for mapping environmentally sensitive areas (ESAs) to desertification. In *The MEDALUS Project. Mediterranean Desertification and Land Use Report*; Kosmas, C.; Kirkby, M.; Geeson, N., Eds.; European Commission: Belgium, 199; 31–47.
21. Kharin, N.G.; Tateishi, R.; Harahsheh, H. *Degradation of the Drylands of Asia*; Center for Environmental Remote Sensing Chiba Univ.: Japan, 1999; 81 pp.
22. Mnatsakanian, R. *Environmental Legacy of the Former Soviet Republics*; Center for Human Ecology, University of Edinburgh: Edinburgh, 1992.
23. Manfred, K.; Christian, P.; Markus, M. Strategic Environmental Assessment. Environmental Report. Central Europe Programme 2007–2013. Austrian Institute of Ecology: Wien. 44.
24. Lowdermilk, W.C. *Conquest of the Land through Seven Thousand Years. No. 99*; USDA, Soil Conservation Service, U.S. Government Printing Office: Washington, DC, 1997, 1–30.
25. Sanders, J.H.; Southgate, D.D.; Lee, J.G. *The Economics of Soil Degradation: Technological Change and Policy Alternatives, no. 22*; SMSS Technical Monograph; Department of Agricultural Economics, Purdue University: 1995; 8–11.
26. Croitoru, L.; Sarraf, M. *The Cost of Environmental Degradation*; Case Studies from the Middle East and North Africa. The World Bank Washington: DC 2010.
27. Lal, R. Monitoring soil erosion's impact on crop productivity. In *Soil Erosion Research Methods*; Lal, R., Ed.; Soil and Water Conservation Society: Ankeny, IA, 1988, 187–200.
28. Barrow, C.J. *Land Degradation*; Cambridge University Press: Cambridge, 1994.
29. Derpsch, R.; Friedrich, T.; Kassam, A.; Hongwen, L. Current status of adoption of no-till farming in the world and some of its main benefits. Int J Agric and Biol Eng. 1–25 pp.
30. State of the Environment 2011 Committee. *Australia State of the Environment 2011*; Independent report to the Australian Government Minister for Sustainability, Environment, Water, Population and Communities. Canberra: DSEWPaC, 2011.
31. World Bank. Sustainable Land Management. Challenges, Opportunities, and Trade-offs. World Bank: Washington DC. 2006; 112 P.
32. Eswaran, H.; Beinroth, F.; Reich, P. Biophysical considerations in developing resource management domains. In *Proceedings of the IBSRAM International Workshop on Resource Management Domains*, Kuala Lumpur, Aug. 26–29, 1996; Leslie, R.N., Ed.; The International Board for Soil Research and Management (IBSRAM): Thailand, 1996; 61–78.
33. Eswaran, H.; Berberoglu, S.; Cangir, C.; Boyraz, D.; Zucca, C.; Özevren, E.; Yazici, E.; Zdruli, P.; Dingil, M.; Dönmez, C.; Akça, E.; Çelik, I.; Watanabe, T.; Koca, Y.K.; Montanarella, L.; Cherlet, M.; Kapur, S. The Anthroscape Approach in Sustainable Land Use. In *Sustainable Land Management: Learning from the Past for the Future.* Kapur, S.; Eswaran, H.; Blum, W.E.H. Eds.; Heidelberg, Springer: 2011; 1–50.
34. Zdruli, P.; Steduto, P.; Kapur, S.; Akça, E., Eds. Ecosystem-based assessment of soil degradation to facilitate land users' and land owners' prompt actions. Medcoastland Project Workshop Proceedings, June 2–7, 2003; Adana, Turkey, 2006.
35. Smyth, A.J.; Dumanski, J. A frame work for evaluating sustainable land management. Canadian J. soil Sci. **1995**, *75* (4), 401–406.

28

Soil: Erosion Assessment

John Boardman
University of Oxford

Introduction

Soil erosion is the loss of soil from the surface of the Earth (Figure 28.1). It is a two-stage process with the detachment of soil particles preceding transport by the agency of water or wind. It is a natural process occurring at relatively low rates, but which may be accelerated by human actions. Most erosion is associated with the formation of rills and gullies. Erosion has serious impacts on the fertility of soils, on water quality, and on reservoir storage capacity. Assessment of actual and potential erosion risk is necessary so that measures can be taken to prevent the loss of soil.

Agents of erosion are wind, water, and gravity (causing mass movement). Movement of people and animals may contribute, and soils will be more susceptible if weakened by weathering (wetting and drying, frost, etc.).

Erosion is an important contributor to global soil degradation. It is also a major component of desertification. Under semiarid conditions, vegetation damage typically precedes erosion. If climate variability is a major influence, then degradation may be cyclic, interspersed with periods of improved land quality. Frequently, because of population pressure, degraded land does not recover.

FIGURE 28.1 Severely eroded winter wheat fields, South Downs, UK, October 1987 (J. Boardman).

The major impact of soil erosion is on agricultural land, both arable and grazing. Other areas of concern are construction sites and protected natural areas (e.g., footpath erosion in National Parks).

This entry is mainly concerned with erosion by water which globally appears to affect twice the area of wind erosion.[1] Overgrazing as a cause of erosion is very widespread but is difficult to classify as it often leads to erosion by water and, in some areas, by wind.

The causes of erosion may be regarded as (i) proximal and (ii) ultimate. Proximal causes relate to physical factors such as rainfall, runoff, wind velocity, gradient, and soil type. Ultimate causes relate to socioeconomic conditions that influence farmer decision-making, and therefore land use and farming practices. Important farmer decisions concern what crops to plant, where to plant them, and how to plant them. Such decisions are strongly influenced by farmer attitudes and traditions, community views, and, especially in the Developed World, by government support for agriculture. Some such support may be regarded as a "perverse subsidy" in that it leads to unwanted and unforeseen environmental impacts including erosion and pollution; examples include European Union support for almonds in Spain,[2] land leveling in Norway,[3] and olive oil production in Spain.[4] A major socioeconomic impact on eastern European landscapes was the collectivization of agriculture under Communism.[5,6]

The relationship between population growth, food production, and exploitation of the soil is suggested by Montgomery (p. 84):[7]

> A fundamental model of agricultural development in which prosperity increases the capacity of the land to support people, allowing the population to expand to use the available land. Then, having eroded soils from marginal land, the population contracts rapidly before soil rebuilds in a period of low population density.

Major texts on erosion, its causes, and consequences include those by Bennett,[8] Blaikie[9] Morgan,[10] Boardman and Poesen,[11] and Montgomery.[7]

Scale and Extent of Erosion

Rates of erosion on grassed or wooded areas are usually below 1 t ha^{-1} yr^{-1}. Rates of soil formation are very low – similar to natural rates of erosion. The oft-quoted USDA figure is 1 inch per 500 years which is about 0.6 t ha^{-1} yr^{-1}.[7] Erosion in excess of this figure means the loss of, or decline in, a finite resource. Erosion rates on farmland in temperate areas occasionally exceed 100 t ha^{-1} yr^{-1}. On the Loess Plateau of China, 7.2 M ha was reported to have erosion rates in excess of 100 t ha^{-1} yr^{-1}. [12]

Hotspots of global erosion are difficult to define and rank due to a lack of or unreliable data. However, we may suggest the following:[13]

- Loess plateau, the Yangtze basin, and the southern hill country of China
- Ethiopia
- Swaziland and Lesotho
- The Andes
- South and East Asia
- Madagascar
- The Himalayas
- The West African Sahel
- Caribbean and Central America.

There are very limited data for some of these areas particularly the Andes and Madagascar. If areas that have been severely affected in the past and to some extent are still vulnerable are to be included, then the list would be extended:

FIGURE 28.2 Dust storm approaching Stratford, Texas. Dust bowl surveying in Texas, April 18, 1935. Image ID: thebl365, Historic C&GS Collection, NOAA George E. Marsh Album.

- The Dust Bowl in the 1930s in the United States (parts of Colorado, Kansas, New Mexico, Texas, and Oklahoma)[14] (Figure 28.2)
- The Virgin Lands Project of the former Soviet Union[15]
- The Mediterranean basin[16]
- Iceland.

Impact of Erosion

Impacts of erosion are the following:

1. On-site, which includes loss of crop, seeds, and fertilizer and thinning of the soil layer, thus directly affecting present and future productivity. The amount of yield reduction is disputed with recent work suggesting a figure of 4% per 10 cm of soil loss.[17] In extreme cases, land may become unworkable or unprofitable because of soil loss and nutrient depletion, as in the case of badlands (Figure 28.3). Montgomery discusses many cases where land is abandoned and farmers move on, e.g., the 18th-century tobacco growing in Virginia, Georgia, and the Carolinas or the Amazon lowlands today (p. 118).[7] This process may result in exploitation of less suitable land or land at greater risk of erosion, e.g., Rome (p. 62).[7]

FIGURE 28.3 Badlands, northern Spain (J. Boardman).

2. Off-site, where people, water/air, or property not directly responsible for erosion are affected. Runoff from agricultural land generally transports sediment particles and attached phosphorous and pesticides. Impacts include dam sedimentation,[18] pollution of water courses,[19] and property damage by muddy floods.[20]

Costs of erosion are under-researched. This is partly because it is difficult to put a value on soil. As soils thin, crop yields decline or they are artificially enhanced by the use of fertilizers. This is a cost which less-developed societies cannot afford. Serious soil depletion results in abandonment and migration (see above). Costs resulting from off-site impacts are easier to estimate; for example, water cleaning costs in England and Wales are estimated at several 100 million pounds per year for the removal of nitrates, phosphates, and pesticides, some of which are associated with runoff and erosion events.[21] The costs of muddy floods in central Belgium range from 16 to 172 million euros per year.[22] Fine-particle sediments (mainly of agricultural origin) affect fish spawning in gravel-bedded rivers.[23–25] In Europe, much of the drive to improve the ecological condition of fresh water systems is due to the influence of the EU Water Framework Directive.[26] Sedimentation of ports is recorded from the Mediterranean and from the eastern states of the United States (p. 139).[7]

Impacts may also be either immediate or long term. The former includes the formation of gullies, the burial of crops, and the flooding of houses; the latter refers to the gradual thinning of soils or sedimentation of reservoirs. Unfortunately, the long-term effects are easy to overlook. Societies in the Middle East, Rome, Greece, Mesoamerica, etc. have all been seriously affected by soil loss.[7]

Erosion Assessment

Erosion assessment may describe the "actual" risk of erosion: the Global Assessment of Soil Degradation (GLASOD) project is an example of this type of assessment applied globally.[1] Similarly, Canada has published maps of the risk of wind and water erosion at the provincial level.[27] New Zealand has combined, on single maps at 1: 250,000 scale, the actual and "potential" risk of erosion.[28] Potential in this case refers to a change of land use to pastoral or arable. Maps from the CORINE project show actual and potential soil erosion in southern Europe; the latter is defined as the worst possible situation that might be reached.[29] The National Soil Maps for England and Wales show similar soils grouped as "soil associations."[30] The risk of soil erosion on each soil association has been assessed by Evans based on empirical evidence.[31]

Erosion assessment may focus on changes through time in the same area. Trimble[32] has measured and modeled changes in erosion and deposition rates in the Coon Creek catchment in Wisconsin since the beginning of European settler farming in the 1850s. This study relates erosion to land use change, management, and conservation techniques, as well as the capacity of rivers to transport eroded materials. It therefore links on-site and downstream effects through time and space. Erosion rates since Neolithic clearance have been modeled on the loessic soils of the South Downs, UK. This assessment is based on changing land use, climate, soils, and farming practices. It is noteworthy that most erosion occurred during periods of intensive farming during the Bronze Age and Iron Age 4000–2000 BP (Figure 28.4).[33] Assessment of erosion and its impacts may be combined in complex scenarios of change involving population pressure, e.g., Hurni for Ethiopia.[34]

Modeling constitutes an important approach to erosion assessment. This is for several reasons: many areas lack the data that allow direct estimates of erosion; modeling may be based on scenarios of future land use or climate, and therefore, estimates may be made of erosion under assumed conditions; and application of a model may be far less expensive and time-consuming than field surveys.

The spatial scale of modeling may vary from a few square meters to thousands of square kilometers. The global change and terrestrial ecosystems (GCTEs) model evaluation exercise divides models into field, catchment, and landscape, based on spatial scale.[35] Resolution of erosion models is determined by grid size, that is, km2 grid size used in CORINE for southern Europe and the 250 m grid used in the maps of seasonal erosion risk in France.

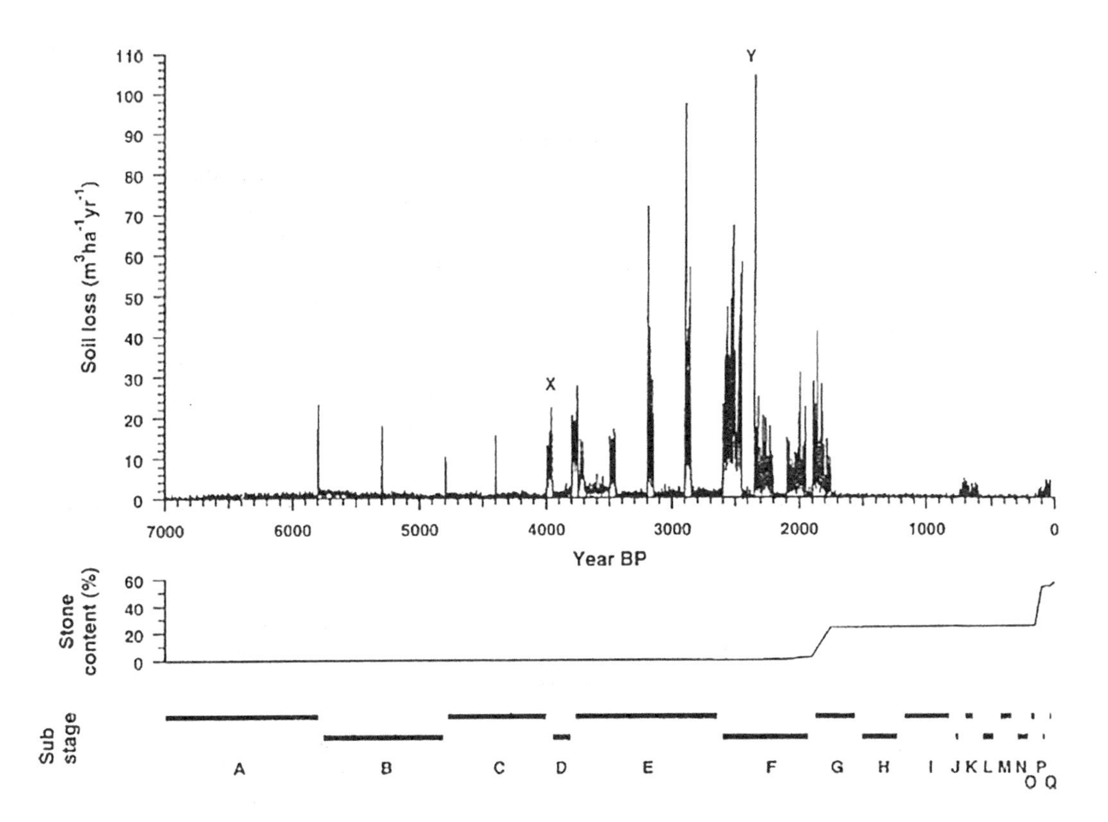

FIGURE 28.4 Simulated annual erosion rates and surface stone content for a "thin" soil profile (1.22 m depth to chalk) on the South Downs, southern England. (Adapted from Favis-Mortlock et al.[33])

Assessments are frequently expressed in map form.[36] Landscape features indicative of erosion may be mapped as a record of an actual event or series of events. In arable landscapes features may be temporary; thus, rills, ephemeral gullies, and fans are usually plowed out annually, e.g., erosional features produced as a result of a snowmelt event in May 1993 in Slovakia.[37] The ephemeral gully in Figure 28.5 was plowed out a few weeks later. In semiarid, grazed landscapes erosional features may become permanent, e.g., gullies and badlands. Assessments should distinguish features active at the present time from those that are inactive and represent erosion under former conditions.

FIGURE 28.5 Ephemeral gully, winter 2007, near Midhurst, West Sussex, UK (J. Boardman).

Several methods of erosion assessment are commonly used as shown in Table 28.1[38] and will be reviewed briefly.

1. Small experimental plots in the laboratory or in the field are typically 22 m × 2 m in size. Experiments on plots may be conducted with natural or simulated rainfall[39] (Figures 28.6 and 28.7). Plots are assumed to represent larger parts of the landscape. Results are frequently extrapolated to fields or catchments, and are used as input to models. Extrapolation from small plots to regional, national, or continental size areas is not acceptable.[40] Plots are also unsatisfactory in that the key erosion process of gullying is not reproducible at this scale. Plots are best used to investigate small scale processes such as rill/interrill relations[41] or rill/land use relations.[42]
2. Sediment yield of rivers represents the net loss of soil from a catchment.[43] Erosion rates on fields may be very different owing to storage of soils between the field and the river. The highest rates of sediment yields from a large river are from the Yellow River, China, with an average silt content

TABLE 28.1 Assessment Methods: Their Merits and Limitations

Assessment Method	Merits	Limitations
Experimental plots in lab or field	Control over erosion factors; ease of measurement of runoff and soil	Small size: difficult to extrapolate to larger areas
Sediment yield of rivers	Ease of measurement; may cover >100 years if reservoir sediments used	Measures total soil lost from watershed, not from specific areas
Field monitoring	Records soil lost from rills and gullies; covers large areas; inexpensive	Gives very approximate amounts of soil lost
Remote sensing	Covers large areas and may cover long time-span	Limited availability for many areas; may not be suitable for small scale features
Cesium-137	Gives 65-year average	Many unknowns: unreliable without careful calibration; expensive and time-consuming
Stratigraphy and pedology	Deposited soil may contain artifacts; suitable for long-time-period studies	Scarcity of suitable sites; low precision of some dating methods
Expert opinion	May reveal unrecorded information	Subjectivity: difficult to compare different areas
Models	Objectivity; ease of use even by "non-experts"; numerical output	Availability of data; complexity of process descriptions; frequent lack of validation
Sediment source fingerprinting	Assessment of source of eroded material, not available by other methods	Analytical time and expense; possible ambiguous results

FIGURE 28.6 Experimental plot (flume), University of Toronto (J. Boardman).

FIGURE 28.7 Experimental plots, South Limbourg, the Netherlands (J. Boardman).

of 38 kg m^{3}.[12] Rivers of east and south Asia deliver about 67% of the sediment reaching the world's oceans; rates are the highest in this region owing to tectonic uplift and steep slopes, high rainfall, deforestation, and population pressure resulting in intensive farming activities.[44] Alternatively, the sediment yield from river basins may be estimated by measurement of the volume of sediment trapped in a still water body such as a lake, reservoir, or pond. A date for the construction of a reservoir is often known. Sediment including bedload will be trapped. Losses in water passing through the dam may be estimated (trap efficiency). The advantage of this method is the possibility of medium-term records of >100 years.[45] Sediment yield to three small dams in the Karoo, South Africa, is estimated at 115, 357 and 654 t km^{-2} yr^{-1} for a period of 65 years. These rates are adjusted to consider trap efficiency, and the contrasts are due to differences in land use within the catchments.[46] Similar approaches have been used in Australia[47] and in Ethiopia.[48] Estimates of soil losses for the same river basin derived from river yields and reservoir sedimentation may differ substantially probably because of the influence of basin scale on sediment yield.[49]

3. Field monitoring is defined as "field-based measurement of erosional and/or depositional forms over a significant area (e.g., >10 km^{2}) and for a period of >2 years."[50] Regular measurements of volumes of soil loss from rills and gullies or deposited in fans are made. Notable examples are the monitoring of 17 localities in England and Wales during 1982–1986,[51] a 10-year program on the South Downs, England, during 1982–1991,[52] monitoring of ephemeral gullies in Belgium during 1982–1993,[53] and a long-term monitoring scheme in the Swiss midlands.[54,55]

   Simple techniques of measurement and assessment of erosion suitable for Less-Developed Countries include measurement of pedestal height, depth of armor layer, plant/ tree root and fence post-exposure, tree mounds, and buildup of sediment behind barriers and in drains.[56] Many studies have used erosion pins to assess rates of erosion over time (Figure 28.8). These are particularly suited to bare degraded lands such as badlands. Rates may be related to topographic position, presence of vegetation, weathering, and rainfall/wind factors. Measurement errors should be estimated and taken into account[57] (Figure 28.9).

4. Remote sensing is used increasingly in erosion assessment. For rill and gully erosion, it may be used as a locational technique to allow detailed field measurements to be made at selected sites. Gullies and badlands may be identified on black and white air photographs of 1:20,000 scale. In the Karoo of South Africa, comparisons have been made between the extent of gullying in 1945 and 1980.[58] Also in South Africa, in a pioneering study, Talbot mapped gullying in the Swartberg resulting from land conversion to wheat farming[59] (Figure 28.10). More recently, the extent of Icelandic land degradation has been assessed using Landsat imagery.[60] The value of Google Earth images for mapping erosional and degradational features should not be overlooked. The main disadvantage is the irregular, rather random time sequencing of the images.[61]

FIGURE 28.8 Erosion pin site in the Karoo, South Africa (J. Boardman).

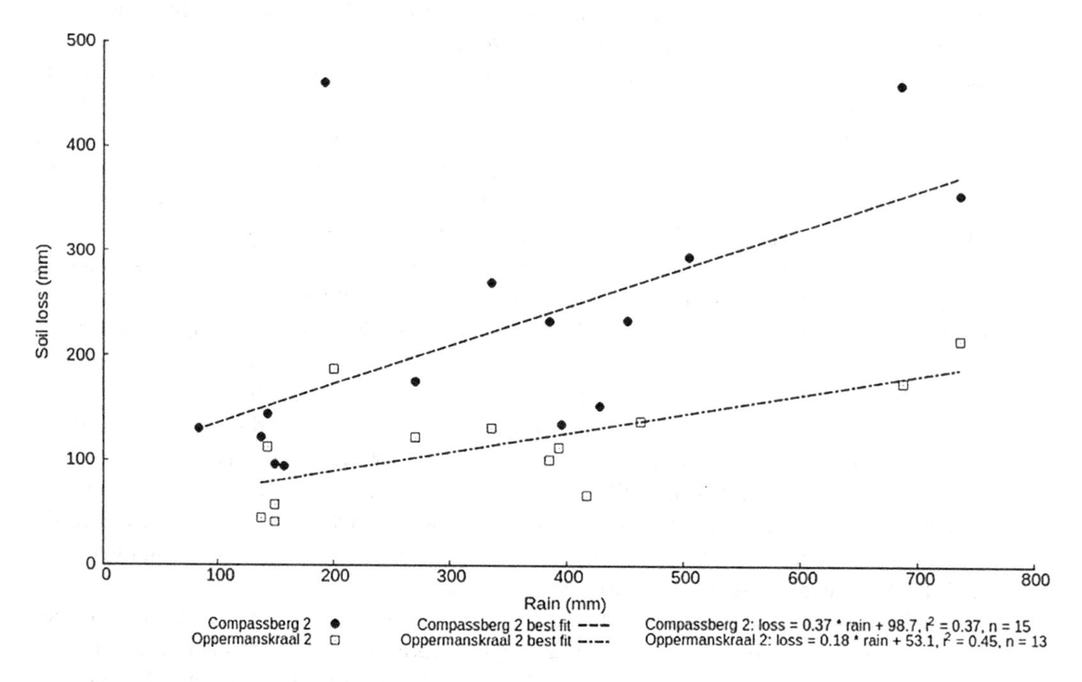

FIGURE 28.9 Average rates of erosion for eroding pins at two sites in the Karoo, South Africa, plotted against rainfall amount for each measurement period (J. Boardman).

5. Cesium-137, derived from weapons testing since 1954, is used as a tracer to assess amounts of erosion and deposition. Annual average rates for the last ca. 65 years are estimated by measurement of the amount and distribution of cesium in comparison with an uneroded reference site.[62] Studies have been carried out in Australia, Canada, China, the Netherlands, Poland, Thailand, the United Kingdom, and the United States. However, results are controversial with issues regarding the original assumed regular distribution of cesium across the landscape via the assumed homogenous spatial distribution of rainfall, and the assumption that cesium "sticks" to fine soil particles in a predictable way. These assumptions have recently been strongly challenged, and erosion rates obtained using this technique now appear unreliable.[63] Comparison of field-measured erosion data with cesium-derived predictions shows gross over-estimation using the latter method.[64]

FIGURE 28.10 Gullied hill country ca. 8 km northwest of Durbanville, South Africa. Gullies are on ploughed land. Photography March 1938: the land had been abandoned by 1944. (Adapted from Talbot.[59])

6. Total amounts of soil loss since a historical baseline such as woodland clearance or the beginning of European settlement have been assessed by comparison of existing soil profiles with uneroded ones. Evans estimates historical losses of topsoil of 150–250 mm in lowland England and Wales, much of which is now stored as alluvium and colluvium.[56] Sedimentation above marker horizons in reservoirs may be used to compute sediment yield for periods of time.[46] Pedological and stratigraphic approaches have also been used to assess the effect of past extreme events, for example, the major storms and floods in Germany in the early 14th century.[65]
7. Expert opinion has been used to assess erosion on a regional or global scale. GLASOD assesses erosion at a global scale (10 M).[1] A total of 12 degradation types are recognized under the broad headings of water erosion, wind erosion, chemical deterioration, and physical deterioration. The "degree" of degradation is assessed for each mapping unit, as is its "relative extent." A combination of degree and relative extent gives an assessment of "severity" which is expressed in "severity classes" and mapped in four colors. However, there is evidence that GLASOD is a very imperfect predictive tool. Van-Camp et al.[66] are particularly critical and suggest that the GLASOD approach should be abandoned: comparison with maps and data for Spain show that the assessment of water erosion was poor.[67] GLASOD has also been assessed by other authors.[68] Comparison of GLASOD with other approaches to assessing the area of global degraded land gives a range of under 1 billion ha to over 6 billion ha, depended on the approach adopted.[69] At a more local and detailed level, expert opinion appears more reliable: see, for example, a recent South African assessment.[70]
8. Assessment of erosion has frequently been based on the application of models. The Universal Soil Loss Equation (USLE) is most widely used because of its simplicity and low-level data requirements.[71] Recent developments in modeling have concentrated on the simulation of erosional processes and computerization. Models may be used to estimate average, long-term erosion rates under specified conditions (USLE) or to assess the effect of a particular rainfall/runoff event, e.g., EUROSEM.[72] Erosion models have also been used to predict wind erosion, and to assess chemical losses in solution or attached to soil particles (for reviews, see Morgan).[73] However, problems with models are widely recognized, for example, because of lack of validation at the field scale,[74] inappropriate calibration,[75] and the quality of field data.[76] Regional mapping using the Revised Universal Soil Loss Equation (RUSLE) is shown to give unreliable results.[77,78]

9. Sediment source fingerprinting is used in investigating, for example, the source of sediments in a lake: are they from hillslope, topsoil, subsoil, or river bank sources?[79] The method depends on a variety of distinct sources within a catchment that can be matched with the lake sample. Ongoing geomorphological controversies such as the contribution of river bank sediment to total erosion in a catchment may be investigated using fingerprinting.[80]

Controlling Soil Erosion and Off-Site Impacts

The control of erosion on the field may or may not coincide with a desire to prevent off-site impacts such as down-valley flooding. With different aims, different approaches may be needed. In Western Europe, most recent concern has been about off-site impacts because of their high social and economic costs rather than the long-term loss of the soil resource associated with erosion on the field. Off-site impacts must be dealt with at a catchment scale. The challenge is to find mechanisms of incentivizing land managers in parts of the catchment where runoff and erosion are occurring, to invest in mitigation measures for the benefit of others. Thus, the emphasis has shifted to the problem of connectivity. This is especially apparent in arable-dominated landscapes where both anthropogenic features such as tracks, roads, and ditches enhance long-distance flow routes and large areas of the same crop give rise to the risk of runoff at the same time.[81]

Controlling soil erosion is based on two principles: protecting the soil from runoff or wind by providing vegetative protection, and reducing the slope and thereby decreasing water velocity. Erosion generally takes place on bare soil (Figure 28.1), and therefore, the protection of a crop or crop debris at vulnerable times of the year is important. Most radical solutions such as minimum tillage and conservation tillage aim to have some degree of vegetation cover throughout the year and therefore drill the new crop through the remnants of the old, thus avoiding the bare ground inherent in tillage systems involving the plough.

The break-up of large fields into strips of different crops can also reduce the risk. Valley-bottom zones of concentrated flow may be grassed (grass waterways) to prevent erosion. Buffer strips of grass are widely used around arable fields in order to reduce runoff volumes and loss of soil from fields. Evans and Boardman show how return to grass of a small area interrupts valley-bottom runoff from large areas of winter cereals, thus preventing flooding of houses.[82,83]

The desire to reduce slope gradients has traditionally been addressed by the construction of terraces. An added advantage is that a terrace allows for the cultivation of slopes that were too steep for agriculture. Terrace systems are known from the Middle Bronze Age on Crete[84] and from the Incas of Peru. Many systems around the Mediterranean are now abandoned, and erosion associated with their post-abandonment state has been much researched.[85] Maintenance of terraces is therefore vital if they are to prevent erosion (Figure 28.11).

It is doubtful if any one conservation technique will be effective on its own. Most successful schemes involve several approaches. For example, the Melsterbeek catchment soil and water conservation scheme in Flanders uses grass waterways, buffer strips, and retention ponds (Figure 28.12). Successful schemes, such as this, must involve the local population of land owners and managers which generally requires financial support and technical assistance.[86,87] However, the scheme is successful in that it reduces the amount of damage caused by muddy flooding and the costs of mitigation measures are far less than the costs of damage.[87]

In the Palouse wheat-growing area of Washington State, high erosion rates have persisted for over 100 years and are particularly associated with the practice of summer fallow. The costs of reducing erosion by conservation practices are well illustrated:

> Applying various measures of conservation treatment to the land will affect the economy of the basin. But erosion rates can be reduced by 40% in the low and high precipitation zones and 60% in the intermediate precipitation zone—without decreasing farm income. The erosion rate can be

FIGURE 28.11 (**A**) Old abandoned terraces and recently rebuilt terraces, Mallorca, Spain (J. Boardman). (**B**) Abandoned terraces, central Corsica, France (J. Boardman). (**C**) Abandoned and degraded terraces, Eritrea (J. Boardman).

FIGURE 28.12 Mitigation measures: grass buffers along field edges, the Melsterbeek catchment, Flanders (Belgium) (Karel Vandaele, Samenwerking Land & Water).

reduced 80% through maximum levels of conservation practices and retirement of 35,000 acres of steep, erodible land. Achieving this level of reduction would cost more than $29 million in reduced productivity and increased operating costs.[89]

Soil erosion is often best addressed in a wide framework of natural resource improvements. The issue of sustaining the livelihoods of the farmers as well as introducing conservation measures is critical. Examples of successful approaches from West Africa, New Zealand, southern Brazil, northwest India, northern Cameroon, and Kenya are discussed by El Swaify et al.[88] There is also the well-known case of the Machakos District, Kenya.[89] Hudson discusses reasons for the failure of soil conservation projects.[90]

Previously quoted extremely high erosion rates for the Loess Plateau, China, and associated high sediment yields for the Yellow River, appear to have been substantially reduced in recent years. Water–soil conservation practices are credited with 40% reduction in sediment loads and sediment trapping by major reservoirs with 30%, and most of the remaining is due to precipitation decrease.[91]

We are left with the challenge of how to better organize agriculture and food production in a world of increasing population, climate change, and inequality. Soil erosion and its assessment play a part in this conundrum. Some form of changed or non-conventional agriculture is clearly needed. Among many suggestions, "smart intensification" offers some answers, but putting it into practice remains a major challenge.[76]

Conclusion

The assessment of erosion is necessary in order to understand the scale and extent of the problem. Many assessment methods have been used including experimental plots, sediment yield in rivers, field monitoring, remote sensing, Cesium-137, historical analysis, expert opinion, and modeling. All have advantages and disadvantages and may not be appropriate for particular circumstances. Assessment will include an evaluation of the type of erosion processes and their relative importance. The design of mitigation measures will depend on the outcome of the assessment. Mitigation measures must consider the driving forces of erosion particularly its socioeconomic context and the need to support local peoples in recommending appropriate solutions.

References

1. Oldeman, L.R.; Hakkeling, R.T.A.; Sombroek, W.G. World Map of the status of human-induced soil degradation: An explanatory note; ed. ISCRIC and UNEP, 1990; 32.
2. Faulkner, H. Gully erosion associated with the expansion of unterraced almond cultivation in the coastal Sierra de Lujar, S. Spain. *Land Degrad. Rehabil.* 1995, *6* (1), 115–127.
3. Lundekvam, H.E.; Romstad, E.; Oygarden, L. Agricultural policies in Norway and effects on soil erosion. *Environ. Sci. Policy* 2003, *6* (1), 57–68.
4. de Graaff, J.; Eppink, L.A.A.J. Olive oil production and soil conservation in southern Spain, in relation to EU subsidy policies. *Land Use Policy* 1999, *40*, 259–267.
5. Stankoviansky, M. Geomorphic effect of surface runoff in the Myjava Hills, Slovakia. *Zietshrift fur Geomorphologie* 1997, *110*, 207–217.
6. Boardman, J.; Poesen, J.; Evans, R. Socio-economic factors in soil erosion and conservation. *Environ. Sci. Policy* 2003, *6* (1), 1–6.
7. Montgomery, D.R. *Dirt: The Erosion of Civilizations*; University of California Press: Berkeley, 2007.
8. Bennett, H.H. *Soil Conservation*; McGraw Hill: New York, 1939.
9. Blaikie, P. *The Political Economy of Soil Erosion in Developing Countries: Longman*; Routledge: London, 1985.
10. Morgan, R.P.C. *Soil Erosion and Conservation*; Third Edition, Blackwell: Oxford, 2005.

11. Boardman, J.; Poesen, J. *Soil Erosion in Europe*; John Wiley: Chichester, 2006.
12. Wen, D. Soil erosion and conservation in China. In *World Soil Erosion and Conservation*, Pimentel, D., Ed.; Cambridge University Press: Cambridge, 1993, 63–85.
13. Boardman, J. Soil erosion science: Reflections on the limitations of current approaches. *Catena* 2006, *68*, 73–86.
14. Worster, D. *Dust Bowl: The Southern Plains in the 1930s*; Oxford University Press: New York, 1979.
15. McCauley, M. *Khrushchev and the Development of Soviet Agriculture; The Virgin Lands Program 1953–1964*; Holmes & Meier Publishers, Inc.: New York, 1976.
16. Faulkner, H.; Hill, A. Forests, soils and the threat of desertification. In *The Mediterranean: Environment and Society*; King, R., Proudfoot, L., Smith, B. Eds.; Arnold: London, 1997; 252–272.
17. Bakker, M.M.; Govers, G.; Rounsevell, M.D.A. The crop productivity-erosion relationship: An analysis based on experimental work. *Catena* 2004, *57*, 55–76.
18. Renwick, W.H. Continental-scale reservoir sedimentation patterns in the United States. In *Erosion and Sediment Yield: Global and Regional Perspectives*; Walling, D.E., Webb, B.W. Eds., Proceedings of the Exeter Symposium, July 1996. IAHS Publication: Wallingford, *236*, 1996, 513–522.
19. Evans, R. Pesticide run off into English rivers – a big problem for farmers. *Pesticides News* 2009, *85*, 12–15.
20. Boardman, J.; Verstraeten, G.; Bielders, C. Muddy floods. In *Soil Erosion in Europe*; Boardman, J., Poesen, J. Eds.; John Wiley: Chichester, 2006, 743–755.
21. Evans, R. Soil erosion and land use: Towards a sustainable policy. In *Soil Erosion and Land Use: Towards a Sustainable Policy*; Cambridge Environmental Initiative, Professional Seminar Series 7, White Horse Press: Cambridge, 1995; 14–26.
22. Evrard, O.; Bielders, C.; Vandaele, K.; van Wesemael, B. Effectiveness of erosion mitigation measures to prevent muddy floods in central Belgium, off-site impacts and potential control measures. *Catena* 2007, *70*, 443–454.
23. Theurer, F.D.; Harrod, T.R.; Theurer, M. *Sedimentation and Salmonids in England and Wales*; Technical Report, Environment Agency: England and Wales, 1998.
24. Birt, K. Runoff from potato farms blamed for fish kills on Canadian Island. *J. Soil Water Conserv.* 2007, *62* (6), 136A.
25. Collins, A. L.; Davison, P.S. Mitigating sediment delivery to watercourses during the salmonid spawning season: Potential effects of delayed wheelings and cover crops in a chalk catchment, southern England. *Int. J. River Basin Manag.* 2009, *7* (3), 209–220.
26. European Parliament. Establishing a framework for community action in the field of water policy. Directive EC/2000/60, EU, Brussels, 2000.
27. Manitoba. Water erosion risk Map 1:1,000,000, Agriculture Canada; Publication 5259/B: Ottawa, Canada, 1988.
28. Prickett, R.C. Sheet 23 Oamaru: Erosion map of New Zealand, 1:250,000, National Water and Soil Organization, Wellington, New Zealand, 1984.
29. Corine. Soil erosion risk and important land resources in the Southern Regions of the European Community; EUR 1323: Luxembourg, 1992.
30. Soil Survey of England and Wales. National soil map, soil survey of England and Wales: Harpenden, UK; 1984.
31. Evans, R. Soils at risk of accelerated erosion in England and Wales. *Soil Use and Manag.* 1990, *6*, 125–131.
32. Trimble S.W. Decreased rates of alluvial sediment storage in the Coon Creek Basin, Wisconsin, 1975–1993. *Science* 1999, *285*, 1244–1246.
33. Favis-Mortlock, D.T., Boardman, J., Bell, M. Modelling long-term anthropogenic erosion of a loess cover, South Downs, UK. *Holocene* 1997, *7* (1), 79–89.
34. Hurni, H. Land degradation, famine, and land resource scenarios in Ethiopia. In *World Soil Erosion and Conservation*; Pimentel, D., Ed.; Cambridge University Press: Cambridge, 1993; 27–61.

35. Boardman, J.; Favis-Mortlock, D.T. Modelling soil erosion by water. In *Modelling Soil Erosion by Water*; Boardman, J., Favis-Mortlock, D.T. Eds.; NATO ASI Series; Springer: Berlin, 1998; 3–6.
36. Morgan, R.P.C. Chapter 4, Erosion hazard assessment. In *Soil Erosion and Conservation*; Third Edition, Blackwell: Oxford, 2005.
37. Stankoviansky, M. Geomorphic effect of surface runoff in the Myjava Hills, Slovakia. *Zeitshrift fur Geomorphologie* 1997, *110*, 207–217.
38. Boardman, J. Soil erosion: The challenge of assessing variation through space and time. In *Geomorphological Variations*; Goudie, A.S., Kalvoda, J. Eds., Nakladatelsti P3K: Prague, 2007; 205–220.
39. Mutchler, C.K.; Murphree, C.E.; McGregor, K.C. Laboratory and field plots for erosion studies. In *Soil Erosion Research Methods*; Lal, R., Ed., Soil and Water Conservation Society: Ankeny, Iowa, 1988.
40. Boardman, J. An average soil erosion rate for Europe: Myth or reality? *J. Soil Water Conserv.* 1998, *53* (1), 46–50.
41. Govers, G.; Poesen, J. Assessment of the interrill and rill contributions to a total soil loss from and upland field plot. *Geomorphology* 1988, *1*, 343–354.
42. Cerdan, O.; Govers, G.; Le Bissonnais, Y.; Van Oost, K.; Poesen, J.; Saby, N.; Gobin, A.; Vacca, A.; Quinton, J.; Auerswald, K.; Klik, A., Kwaad, K.J.P.M.; Raclot, D.; Ionita, I., Rejman, J.; Rousseva, S.; Muxart, T.; Roxo, M.J.; Dostal, T. Rates and spatial variations of soil erosion in Europe: A study based on erosion plot data. *Geomorphology* 2010, *122*, 167–177.
43. Walling, D.E. Measuring sediment yield from river basins. In *Soil Erosion Research Methods*; Lal, R. Ed., Soil and Water Conservation Society: Ankeny, IA 1988, 9–73.
44. Milliman, J.D.; Meade, R.H. World-wide delivery of river sediment to the oceans. *J. Geol.* 1983, *91*, 751–762.
45. Verstraeten, G.; Bazzoffi, P.; Lajczak, A.; Radoane, M.; Rey, F.; Poesen J.; de Vente, J. Reservoir and pond sedimentation in Europe. In *Soil Erosion in Europe*; Boardman, J., Poesen, J. Eds.; John Wiley: Chichester, 2006, 759–774.
46. Boardman, J.; Foster, I.; Rowntree, K.; Mighall, T.; Gates, J. Environmental stress and landscape recovery in a semi-arid area, the Karoo, South Africa. *Scott. Geogr. J.* 2010, *126* (2), 64–75.
47. Erskine, W.D.; Mahmoudzadeh, A.; Myers, C. Land use effects on sediment yields and soil loss rates in small basins of Triassic sandstone near Sydney, NSW, Australia. *Catena*, 2002, *49* (4), 271–287.
48. Haregeweyn, N.; Poesen, J.; Nyssen, J.; De Wit, J.; Haile, M.; Govers, G.; Deckers, S. Reservoirs in Tigray (northern Ethiopia): Characteristics and sediment deposition problems. *Land Degrad. Dev.* 2006, *17*, 211–230.
49. Foster, I.D.L.; Boardman, J. Monitoring and assessing land degradation: New approaches. In *Southern African Landscapes and Environmental Change*; Holmes, P.J. and Boardman, J. Eds.; Routeledge, Abingdon, UK, 2018; 249–274.
50. Boardman, J. Soil erosion by water: Problems and prospects for research. In *Advances in Hillslope Processes*; Anderson, M.G. Brooks, S.M. Eds.; Volume 1, Wiley: Chichester, 1996; 489–505.
51. Evans, R. Extent, frequency and rates of rilling of arable land in localities in England and Wales. In *Farm Land Erosion: In Temperate Plains Environment and Hills*; Wicherek, S. Ed., Elsevier: Amsterdam, 1993; 177–190.
52. Boardman, J. Soil erosion and flooding on the eastern South Downs, southern England, 1976–2001. *Trans. Inst. Br. Geogr.* 2003, *28*, 176–196.
53. Poesen, J. Gully typology and gully control measures in the European loess belt. In *Farm Land Erosion: In Temperate Plains Environment and Hills*; Wicherek, S. Ed.; Elsevier: Amsterdam, 1993, 221–239.
54. Prasuhn, V. Soil erosion in the Swiss midlands: Results of a 10-year field survey. *Geomorphology* 2011, *126*, 32–41.

55. Prasuhn, V. On-farm effects of tillage and crops on soil erosion measured over 10 years in Switzerland. *Soil & Till. Res.* 2012, *120*, 137–146.
56. Stocking M.A.; Murnaghan, N. *Handbook for the Field Assessment of Land Degradation*; Earthscan: London, 2001.
57. Boardman, J.; Favis-Mortlock, D. The use of erosion pins in geomorphology. In *Geomorphological Techniques (Online Edition)*; Cook, S.J., Clarke, L.E., Nield, J.M. Eds., Chap. 3, Sec. 5.1, British Society for Geomorphology: London, UK, 2016; ISN: 2047-0371.
58. Keay-Bright, J.; Boardman, J. Evidence from field based studies of rates of erosion on degraded land in the central Karoo, South Africa. *Geomorphology* 2009, *103*, 455–465.
59. Talbot, W.J. *Swartland and Sandveld*; Oxford University Press: Cape Town, 1947.
60. Arnalds, O.; Borarinsdottir, E.F.; Metusalemsson, S.; Jonsson, A.; Gretarsson, E.; Arnason, A. *Soil Erosion in Iceland*; Soil Conservation Service and the Agricultural Research Institute: Iceland, 2001.
61. Boardman, J. 2016. The value of Google Earth for erosion mapping. *Catena 143*, 123–127.
62. Quine, T.A.; Walling, D.E. Assessing recent rates of soil loss from areas of arable cultivation in the UK. In *Farm Land Erosion: In Temperate Plains Environment and Hills*; Wicherek, S., Ed., Elsevier: Amsterdam, 1993; 357–371.
63. Parsons, A.J.; Foster, I.D.L. What can we learn about soil erosion from the use of ^{137}Cs? *Earth-Sci. Rev.* 2011, *108*, 101–113.
64. Evans, R. Soil erosion: its impact on the English and Welsh landscape since woodland clearance. In *Soil Erosion on Agricultural Land*; Boardman, J., Foster, I.D.L., Dearing, J.A., Eds., Wiley: Chichester, 1990; 231–254.
65. Bork, H-R. Soil erosion during the past millennium in Central Europe and its significance within the geomorpho-dynamics of the Holocene. *Catena* 1989, *15*, 121–131.
66. Van-Camp, L.; Bujarrabal, B.; Gentile, A-R.; Jones, R.J.A.; Montarnarella, L.; Olazabal, C.; Selvaradjou, S-K. Reports of the technical working groups established under the thematic strategy for soil protection. EUR 21319 EN/2, 2004, Office for Official Publications of the European Communities, Luxembourg.
67. Sanchez, J.; Recatala, L.; Colomer, J.C.; Ano, C. Assessment of soil erosion at national level: a comparative analysis for Spain using several existing maps. In *Ecosystems and Sustainable Development III*; Villacampa, Y., Brevia, C.A., Uso, J.L. Eds.; Advances in Ecological Sciences 10, WITT Press: Southampton, UK, 2001; 249–258.
68. Sonneveld, B.G.J.S.; Dent, D.L. How good is GLASOD? *J. Environ. Manag.* 2009, *90*, 274–283.
69. Gibbs, H.K.; Salmon, J.M. Mapping the world's degraded lands. *Appl. Geogr.* 2015, *57*, 12–21.
70. Hoffman, T.; Ashwell, A. *Nature Divided: Land Degradation in South Africa*; University of Cape Town Press: Cape Town, 2001.
71. Wischmeier, W.H.; Smith, D.D. *Predicting Rainfall Erosion Losses, Agricultural Research Service Handbook 537*; U.S. Department of Agriculture: Washington, DC, 1978.
72. Morgan, R.P.C.; Quinton, J.N.; Smith, R.E.; Govers, G.; Poesen, J.W.A.; Auerswald, K.; Chischi, G.; Torri, D.; Styczen, M.E. The European Soil Erosion Model (EUROSEM): A dynamic approach for predicting sediment transport from fields and small catchments. *Earth Surf. Process. Landforms* 1998, *23*, 527–544.
73. Morgan, R.P.C. *Soil Erosion and Conservation*; Third Edition, Blackwell: Oxford, 2005; ch. 6.
74. Boardman, J., Evans, R. 2019. The measurement, estimation and monitoring of soil erosion by runoff at the field scale: Challenge and possibilities with particular reference to Britain. *Prog. Phys. Geog.* DOI:10.1177/0309133319861833
75. Govers, G.; Merckx, R.; van Wesemael, B.V.; Van Oost, K. Soil conservation in the 21st century: Why we need smart agricultural intensification. *SOIL* 2017, *3*, 45–59.
76. Garcia-Ruiz, J.M.; Begueria, S.; Lana-Renault, N.; Nadal-Romero, E.; Cerda, A. Ongoing and emerging questions in water erosion studies. *Land Degrad. Dev.* 2017, *28*, 5–21.

77. Evans, R.; Boardman, J. A reply to Panagos et al., 2016 (*Environmental Science & Policy* 59, 53–57). *Environ. Sci. Policy* 2016, *60*, 63–68.
78. Evans, R.; Boardman, J. 2016. The new assessment of soil loss by water erosion in Europe. Panagos P. et al. 2015 Environmental Science & Policy 54, 438-447 – a response. *Environ. Sci. Policy* 2016, (58), 11–15.
79. Walling, D.; Foster, I. Using environmental radionuclides, mineral magnetism and sediment geochemistry for tracing and dating fine fluvial sediments. In *Tools in Fluvial Geomorphology*; Second Edition. Kondolf, M., Piegay, H., Eds., Wiley: Chichester, UK, 2016, 183–209.
80. Pulley, S.; Foster, I.; Antunes, P. The uncertainties associated with sediment fingerprinting suspended and recently deposited fluvial sediment in the Nene river basin. *Geomorphology* 2015, *228*, 303–319.
81. Boardman, J.; Vandaele, K.; Evans, R.; Foster, I.D.L. Off-site impacts of soil erosion and runoff: Why connectivity is more important than erosion rates. *Soil Use and Manag.* 2019, DOI:10.1111/sum.12496
82. Evans, R., Collins, A.L., Zhang, Y., Foster, I.D.L., Boardman, J., Sint, H., Lee, M.R.F., Griffith, B.A. 2017. A comparison of conventional and ^{137}Cs–based estimates of soil erosion rates on arable and grassland across lowland England and Wales. *Earth-Sci. Rev. 173*, 49–64.
83. Evans, R.; Boardman, J. The curtailment of muddy floods in the Sompting catchment, South Downs, West Sussex, southern England. *Soil Use Manag.* 2003, *19*, 223–231.
84. Grove, A.T.; Rackham, O. The Nature of Mediterranean Europe: An Ecological History. Yale University Press: New Haven, 2001; ch. 6 Cultivation terraces.
85. Lesschen, J.P.; Cammeraat, L.H.; Nieman, T. Erosion and terrace failure due to agricultural land abandonment in a semi-arid environment. *Earth Surf. Process. Landforms* 2008, *33*, 1574–1584.
86. Evrard, O.; Bielders, C.L.; Vandaele, K.; van Wesemael, B. Spatial and temporal variation of muddy floods in central Belgium, off-site impacts and potential control measures. *Catena* 2007, *70*, 443–454.
87. Boardman, J.; Vandaele, K. Managing muddy floods: Balancing engineed and alternative approaches. *J. Flood Risk Manag.* 2019. DOI:10.1111/jfr3.12578
88. El-Swaify, S.A. with contributors. *Sustaining the Global Farm – Strategic Issues, Principles, and Approaches*; International Soil Conservation Organisation (ISCO), and the Department of Agronomy and Soil Science, University of Hawaii at Manoa: Hololulu, 1999.
89. Tiffin, M.; Mortimore, M.; Gichuki, F. *More People, Less Erosion: Environmental Recovery in Kenya*; Wiley: Chichester, 1994.
90. Hudson, N.W. A study of the reasons for success or failure of soil conservation projects. FAO Soil Bull. FAO, Rome, 1991; 64.
91. Peng, P.; Chen, S.; Dong, P. Temporal variation of sediment load in the Yellow River basin, China, and its impacts on the lower reaches and the river delta. *Catena* 2010, *83*, 135–147.

29

Soil: Evaporation

William P. Kustas
U.S. Department of Agriculture (USDA-ARS)

Introduction

Soil evaporation not only determines partitioning of available energy between sensible and latent heat flux for bare soil surfaces but can also significantly influence energy flux partitioning of partially vegetated surfaces. This latter effect occurs via the impact of soil evaporation on the resulting surface soil moisture and temperature. These, in turn, strongly influence the microclimate in partially vegetated canopies, indirectly affecting plant transpiration.[1] Over a growing season, soil evaporation can be a significant fraction of total water loss for agricultural crops.[2] On a seasonal basis in semiarid and arid regions, soil evaporation can significantly alter the relative fraction of runoff to rainfall, which in turn has a major impact on the available water for plants.[3] In deserts, in spite of its small magnitude, soil evaporation can introduce significant errors in meteorological forecasting if neglected.[4]

The measurement of soil evaporation at field scale is typically obtained using standard micrometeorological techniques, namely Bowen ratio and eddy covariance methods. Traditionally, due to fetch and measurement requirements, under partial canopy cover conditions, these techniques are not able to partition the total evapotranspiration into its soil evaporation and plant transpiration components. Recently, a novel procedure for partitioning evapotranspiration through utilizing the measured high-frequency time series of carbon dioxide and water vapor concentrations has been developed and tested.[5] This approach relies upon the simple assumption that contributions to the time series of carbon dioxide and water vapor concentrations derived from stomatal processes (i.e., photosynthesis and transpiration), and nonstomatal processes (i.e., respiration and direct evaporation) separately conform to flux-variance similarity. Vegetation water-use efficiency is the only parameter needed to perform the partitioning. Further work is needed to evaluate the utility of this technique with eddy covariance data collected over a variety of land cover and climate conditions.

Soil evaporation in partial canopy cover conditions varies spatially depending primarily on soil water distribution, canopy shading, and under-canopy wind patterns. These effects are magnified in row crops and under various irrigation techniques (e.g., drip irrigation). Soil evaporation can be measured using microlysimeters,[6] chambers,[7] time-domain reflectometers (TDRs),[8] a combination of microlysimetry and TDR,[9] micro-Bowen ratio systems,[3,10] or heat pulse probes.[11] Given the high spatial variability in the driving forces under partial canopy cover conditions, these point-based measurements are difficult to extrapolate to the field scale. Therefore, models have been developed to estimate the contribution of soil evaporation to the total evapotranspiration process.

Measurement methods are described, and models of varying degrees of complexity are reviewed, focusing primarily on relatively simple analytical models, some of which provide daily estimates and can be implemented operationally. The potential application of models using remote sensing data for large-scale estimation is also briefly discussed.

Methodologies

Measurement Methods

Microlysimeters

Microlysimeters have been widely used to measure evaporation from the soil surface of irrigated crops.[8,12,13] Typically, an undisturbed soil sample (a representative vertical section of the soil profile) is inserted into a small cylinder open at the top. The microlysimeter is inserted back into the soil with its upper edge level with the soil surface and weighed either periodically or continuously. Changes in weight reflect an evaporative flux. To eliminate vertical heat conduction through the microlysimeter cylinder and minimize horizontal heat flux in the deeper layers of the sample, poly(vinyl chloride) (PVC) has been found to be the most suitable material. The microlysimeter's dimensions are typically a diameter of ~8 cm and a depth of 7–10 cm.[14–16] Theoretically, the microlysimeters provide absolute reference for soil evaporation, as long as their soil and the heat balance are similar to the surrounding area.

Chambers

Chambers are used to directly measure the flux of gases between the soil surface and the atmosphere by enclosing a volume and measuring all flux into and out of the volume.[17] Reviews of chamber designs and calculations of fluxes based on chamber methods can be found in Livingston and Hutchinson[18] and Hutchinson and Livingston.[19]

With infrared gas analyzers (IRGAs) becoming increasingly common, they are widely considered to be the method of choice today for chamber-based soil respiration and evaporation measurements.[20] Chambers can be used in either of two modes to calculate fluxes:[18] (1) in steady-state mode, the flux is calculated from the concentration difference between the air flowing at a known rate through the chamber inlet and outlet after the chamber headspace air has come to equilibrium concentration of carbon dioxide; (2) in the non-steady-state mode, the flux is calculated from the rate of increasing concentration in the chamber headspace of known volume shortly after the chamber is put over the soil.

In both modes, air is circulated between a small chamber that is placed on the soil and an IRGA. Typically, a soil chamber of ca. 1 L volume is placed on a PVC collar of about 80 cm^2 area. This collar is inserted about 2 cm into the soil and secured to prevent movement when the chamber is placed on it. When the chamber is placed on the collar, circulation of air between the chamber and the external IRGA is induced by a pump, and the water vapor concentration is measured.[21]

Soil Water Balance

Soil evaporation (E) can be extracted from the water balance equation, provided that all other components are known:

$$E = I + P - R + F - \Delta S \tag{29.1}$$

where I is irrigation, P is precipitation, R is runon or runoff, F is deep soil water flux (percolation), and ΔS is change in soil water storage. For an experimental field site, irrigation and precipitation can be easily monitored, and runoff and runon may be controlled to near-zero amounts by diking. Deep soil water flux errors can be reliably estimated in several ways, with the most important being the monitoring or measuring of the soil water content well below the root zone. The change in soil water storage can be determined fairly accurately with profile measurements of soil water content over multiple depths at the beginning and end of a defined time period.

There are many soil water content sensors, all of which work by measuring a surrogate property that is empirically or theoretically related to the soil water content. A recent comparative review by Evett et al.[22] concluded that soil water content is best determined using the neutron probe, gravimetric sampling, and conventional TDR[23] methods as compared to bore hole capacitance methods. Of the three optimal methods in the study, TDR is the only methodology capable of providing automated continuous measurements. However, other continuous measurement sensors such as frequency-domain reflectometers (FDRs),[24] and time-domain transmission (TDT)[25] sensors are also emerging as options for long-term installations, given appropriate calibration.

Micro-Bowen Ratio Systems

The Bowen ratio energy balance (BREB) approach is one of the simplest and most practical methods of estimating water vapor flux[10,26] and has thus been used extensively under a wide range of conditions providing robust estimates.[27] Use of the BREB concept[28] enables solving the energy balance equation by measuring simple gradients of air temperature and vapor pressure in the near-surface layer above the evaporating surface. The BREB equation is

$$LE = \frac{R_n - G}{1 + B_o} \tag{29.2}$$

in which LE is the latent heat flux, R_n is the net radiation, G is the soil heat flux, and B_o the Bowen ratio, which is found from measurements of temperature and vapor pressure at two heights within the constant flux layer.[29] Assuming equal transfer coefficients for heat and vapor, the Bowen ratio is defined as

$$B_o = \frac{H}{LE} = \gamma \frac{\partial T}{\partial e} \tag{29.3}$$

where H is the sensible heat flux, γ is the psychrometric constant, T is the air temperature, and e is the water vapor pressure.

Application of the Bowen ratio concept to measure bare soil surface evaporation was suggested[3,10] by measuring temperature and vapor pressure close to the soil surface (e.g., at 1 and 6 cm). Compared to microlysimeters, the micro-Bowen ratio (MBR) system yielded good results over bare soil.[10] The potential of the MBR approach was demonstrated by successfully measuring soil evaporation within a maize field.[30] To date, testing of the MBR is very limited, and further research is required to examine the performance of the technique under various environmental and agronomic conditions.

Heat Pulse Probes

A novel approach for measuring soil evaporation has been recently proposed, based on the soil sensible heat balance.[11,31] In this approach, a sensible heat balance is used to determine the amount of latent heat involved in the vaporization of soil water following Gardner and Hanks:[32]

$$LE = (H_o - H_1) - \Delta S \tag{29.4}$$

where H_0 and H_1 are soil sensible heat fluxes at depths 0 and 1, respectively; ΔS is the change in soil sensible heat storage between depths 0 and 1; L is the latent heat of vaporization; and E is evaporation.

Typically, three-needle sensors like those described by Ren et al.[33] are used, which are spaced 6 mm apart, in parallel. Temperature is measured by all three needles, and the central needle also contains a resistance heater for producing a slight pulse of heat required for the heat pulse method. At a given time interval (2–4 hours), a heat pulse is executed, and the corresponding rise in temperature at the outer sensor needles is recorded. Soil volumetric heat capacity and thermal diffusivity are then determined from the heat input and temperature response following the procedures described by Knight and Kluitenberg[34] and Bristow et al.,[35] respectively. Having measurements of soil temperature, thermal conductivity, and volumetric heat, evaporation is estimated from Equation 29.4.

Numerical Models R

Numerous mechanistic/numerical models of heat and mass flows exist and are primarily based on the theory of Philip and de Vries.[36] However, they continue to be refined through improved parameterization of the moisture and heat transport through the soil profile.[37–39] Some of these mechanistic models have been used recently to explore the utility of bulk transfer approaches used in weather forecasting models[40] and in soil–vegetation–atmosphere models,[41] computing field to regional-scale fluxes. These bulk transport approaches are commonly called the "alpha" and "beta" methods defined by the following expressions

$$LE_s = \frac{\rho C_P}{\gamma}\left(\frac{\alpha e_*(T_s) - e_A}{R_A}\right) \tag{29.5}$$

and

$$LE_s = \frac{\rho C_P}{\gamma}\beta\left(\frac{e_*(T_s) - e_A}{R_A}\right) \tag{29.6}$$

where ρ is the air density (~1 kg/m^3), C_p the heat capacity of air (~1000 J/kg/K), γ the psychrometric constant (~65 Pa), $e_*(T_S)$ the saturated vapor pressure (Pa) at soil temperature T_S(K), e_A the vapor pressure (Pa) at some reference level in the atmosphere, and R_A is the resistance (s/m) to vapor transport from the surface usually defined from surface layer similarity theory.[42] For α, several different formulations exist[43,44] with one of the first relating α to the thermodynamic relationship for relative humidity in the soil pore space, h_R[45]

$$h_R = \exp\left(\frac{\psi g}{R_V T_S}\right) \tag{29.7}$$

where ψ is the soil matric potential (m), g the acceleration of gravity (9.8 m/s^2), and R_V the gas constant for water vapor (461.5 J/kg/K). From Equation 29.6, β can be defined as a ratio of aerodynamic and soil resistance to vapor transport from the soil layer to the surface, R_S, namely,

$$\beta = \frac{R_A}{R_A + R_S} \tag{29.8}$$

Both modeling[44] and observational results[46,47] indicate that more reliable results are obtained using the beta method. In fact, Ye and Pielke[44] formulate an expression similar to Camillo and Gurney,[48] which combines Equations 29.5 and (29.6) and provides more reliable evaporation rates for a wider range of conditions,

$$LE_s = \frac{\rho C_P}{\gamma}\beta\left(\frac{\alpha e_*(T_S) - e_A}{R_A}\right) \tag{29.9}$$

Unfortunately, there is no consensus concerning the depth in the soil profile to consider in defining the α and β terms. In particular, studies evaluating the soil resistance term R_S use a range of soil moisture depths: 0–1/2 cm,[48] 0–1 cm,[7] 0–2 cm,[39,46] and 0–5 cm.[39] The study conducted by Chanzy and Bruckler[38] appears to be one of the few studies that attempts to determine the most useful soil moisture depth for modeling soil evaporation by considering the penetration depth of passive microwave sensors of varying wavelengths. Using field data and numerical simulations with a mechanistic model, they find that the 0–5 cm depth, which can be provided by the L-band microwave frequency, appears to be the most adequate frequency for evaluating soil evaporation.

Besides the depth of the soil layer to consider, the equations relating soil moisture to R_S have ranged from linear to exponential (Table 29.1). Furthermore, observations and numerical models have shown that varies significantly throughout the day and that its magnitude is also affected by climatic conditions.[38,39]

These are not the only complicating factors that make the use of such a bulk resistance approach somewhat tenuous. From detailed observations of soil moisture changes and water movement, Jackson et al.[49] found the soil water flux in the 0–9 cm depth to be very dynamic with fluxes at all depths continually changing in magnitude and sometimes direction over the course of a day. These phenomena observed by Jackson et al.[49] are owing in part to a process that occurs during soil drying where dry surface soil layer forms, significantly affecting the vapor transport through the profile.[50] Yamanaka et al.[51] recently developed and verified, using wind tunnel data, a simple energy balance model in which the soil moisture available for evaporation is defined using the depth of the evaporating/drying front in the soil.

TABLE 29.1 Bulk Soil Resistance Formulations, R_S, from Previous Studies

RS Formula (s/m)	Value of Coefficients	Soil Type	Depth (cm)	References
$R_s = a\left(\frac{\theta_s}{\theta}\right)^n$	$a = 3.5$	Loam[a]	0–1/2	[21]
	$b = 33.5$			
	$n = 2.3$			
$R_S = a(\theta_S - \theta) + b$	$a = 4{,}140$	Loam[b]	0–1/2	[48]
	$b = -805$			
$R_s = R_{SMIN}\exp[a(\theta_{MIN} - \theta)]$	$R_{SMIN} = 10$ s/m	Fine sandy loam	0–1	[7]
	$\theta_{MIN} = 15\%$			
	$a = 0.3563$			
$R_S = \frac{a(\theta_S - \theta)}{2.3\times10^{-4}(T_s/273.16)^{1.75}}$	$a = 2.16 \times 10^2$ $n = 10$ $\theta_S = 0.49$	Loam	0–2	[46]
	$a = 8.32 \times 10^5$	Sand	0–2	[46]
	$n = 16.16$			
	$\theta_S = 0.392$			
$R_S = \alpha\theta + b$	$a = -73{,}420 - -51{,}650$	Sand	0–2	[39]
	$b = 1{,}940 - 3{,}900$			
	$a = 4.3$	Silty clay loam[a]	0–5	[41]
$R_S = \exp\left(b - a\left(\frac{\theta}{\theta_S}\right)\right)$				
	$b = 8.2$			
	$a = 5.9$	Gravelly sandy loam	0–5	[66]
	$b = 8.5$			

[a] Soil type for the data from Jackson et al.[49] was determined from texture-dependent soil hydraulic conductivity and matric potential equations of Clapp and Hornberger evaluated by Camillo and Gurney.[48]

[b] Soil type was determined from texture-dependent soil hydraulic conductivity and matric potential equations of Clapp and Hornberger evaluated by Sellers et al.[41]

This approach removes the ambiguity of defining the thickness of the soil layer and resulting moisture available for evaporation. However, the depth of the evaporating surface is not generally known *a priori*, nor can it be measured in field conditions; hence, this approach at present is limited to exploring the effects of evaporating front on R_S type formulations.

Analytical Models

To reduce the effect of temporal varying, R_S, Chanzy and Bruckler[38] developed a simple analytically based daily LE_S (E_D) model using simulations from their mechanistic model for different soil texture, moisture, and climatic conditions as quantified by potential evaporation (E_{PD}), as given by Penman.[52] The analytical daily model requires midday 0–5 cm soil moisture θ, daily potential evaporation, and daily average wind speed (U_D). The simple model has the following form:

$$\frac{E_D}{E_{PD}}=\left[\frac{\exp(A\theta+B)}{1+\exp(A\theta+B)}\right]C+(1+C) \tag{29.10a}$$

$$A=a+5\max(3-E_{PD},0) \tag{29.10b}$$

$$B=b-5(-0.025b-0.05)\max(3-E_{PD},0)+\alpha(U_D-3) \tag{29.10c}$$

$$C=0.90-0.05c(U_D-3) \tag{29.10d}$$

where the coefficients *a*, *b*, and *c* depend on soil texture (Table 29.1) and were derived from their detailed mechanistic model simulations for loam, silty clay loam, and clay soils.[38] In Figure 29.1, a plot of Equation 29.10a is given for two soil types, loam (Figure 29.1a) and silty clay loam (Figure 29.1b) under two climatic conditions, namely, a relatively low evaporative demand condition with $U_D = 1$ m/s and $E_{PD} = 2$ mm/d and high demand $U_D = 5$ m/s and $E_{PD} = 10$ mm/d. Notice the transition from $E_D/E_{PD} \sim 1$ to $E_D/E_{PD} < 1$ as a function of θ varies not only with the soil texture, but also with the evaporative demand. The simplicity of such a scheme outlined in Equation 29.10 needs further testing for different soil textures and under a wider range of climatic conditions.

The ratio E_D/E_{PD} as a function of θ illustrated in Figure 29.1 also depicts the effect of the two "drying stages" typically used to describe soil evaporation.[42] The "first stage" (S_1) of drying is under the condition where water is available in the near-surface soil to meet atmospheric demand, i.e., $E_D/E_{PD} \sim 1$. In the "second stage" (S_2) of drying, the water availability or θ falls below a certain threshold where the soil evaporation is no longer controlled by the evaporative demand, namely, $E_D/E_{PD} < 1$. Under S_2, several studies find that a simple formulation can be derived by assuming that the time change in θ is governed by desorption, namely, as isothermal diffusion with negligible gravity effects from a semi-infinite uniform medium. This leads to the rate of evaporation for S_2 being approximated by[42,53]

$$E_D=0.5D_E\,t^{-1/2} \tag{29.11}$$

where the desorptivity D_E (mm/d$^{1/2}$) is assumed to be a constant for a particular soil type and t is the time (in days) from the start of S_2. Although both numerical models and observations indicate that the soil evaporation is certainly a more complicated process than the simple analytical expression given by Equation 29.11, a number of field studies[3,54–58] have shown that for S_2 conditions, reliable daily values can be obtained using Equation 29.11. In many of these studies for determining D_E, the integral of Equation 29.11 is used, which yields the cumulative evaporation as a function of $t^{1/2}$

$$\sum E_D=D_E(t-t_O)^{1/2} \tag{29.12}$$

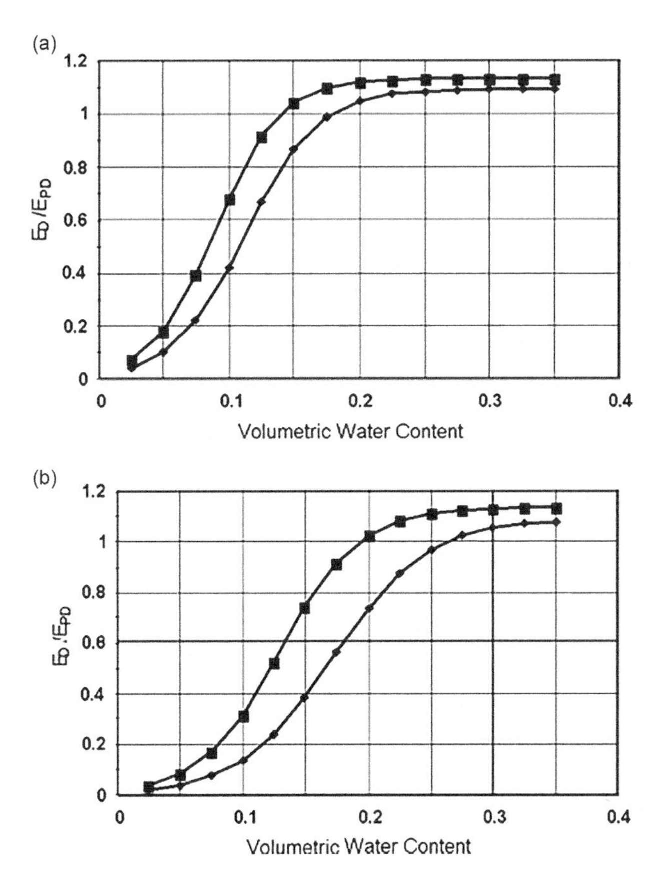

FIGURE 29.1 A plot of E_D/E_{PD}, estimated using Equation 29.10 from Chanzy and Bluckler[38] vs. volumetric water content for (a) loam and (b) silty clay loam soil under two evaporative demand conditions: $U_D = 1$ m/s and $E_{PD} = 2$ mm/d (squares) and $U_D = 5$ m/s and $E_{PD} = 10$ mm/d (diamonds).

where t_O is the number of days where $E_D/E_{PD} \sim 1$ or is in S_1. In practice, observations of ΣE_D are plotted vs. $(t - t_O)^{1/2}$ and in many cases the choice of the starting point of S_2 is $t_O \approx 0$ or immediately after the soil is saturated. As shown by Campbell and Norman,[58] the course of evaporation rate for three drying experiments (see Figure 9.6 in Campbell and Norman[58]) indicates that for a loam soil, t_O depends on the evaporative demand or E_{PD} with $t_O \sim 2$ days when E_{PD} is high vs. $t_O \sim 5$ days when E_{PD} is low. On the other hand, for a sandy soil, there is almost an immediate change from S_1 to S_2 conditions with $t_O \approx 1$ day. As suggested by the analysis of Jackson et al.[56] and as stated more explicitly by Brutsaert and Chen,[59] the value of t_O can significantly influence the value computed for D_E. Jackson et al.[56] also show that for the same soil type, the value of D_E has a seasonal dependency (ranging from 4 to 8 mm/d$^{1/2}$) most likely related to the evaporative demand, which they correlate to daily average soil temperature (see Figure 29.2 in Jackson, Idso, et al.[56]). Values of D_E from the various studies are listed in Table 29.2. Brutsaert and Chen[59] modified Equation 29.11 for deriving D_E by rewriting in terms of a "time-shifted" variable $T = t - t_O$ and expressing it in the form

$$E_D^{-2} = \left(\frac{2}{D_E}\right)^2 T \tag{29.13}$$

where D_E and t_O will come from the slope and intercept (see Figure 29.1 in Brutsaert and Chen[59]). It follows that ΣE_D under S_2 will start at $T = T_O$ and not at $T = 0$, so that Equation 29.12 is rewritten as

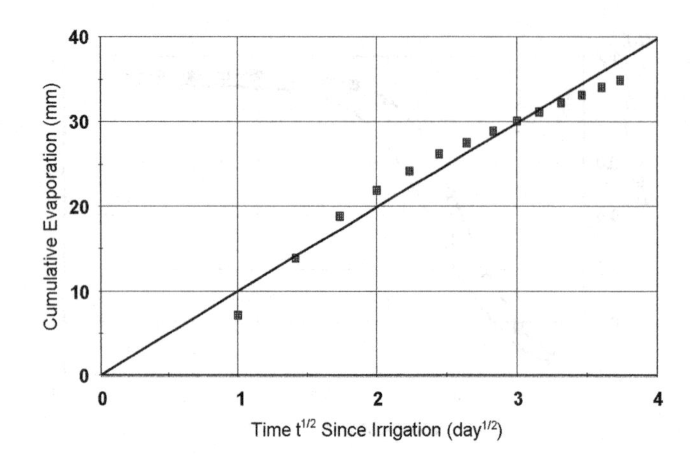

FIGURE 29.2 The desorptivity D_E (mm/d$^{1/2}$) estimated from a least squares linear fit to the data from Jackson et al.[56] assuming $t_O = 0$ (i.e., stage-two drying occurs immediately after irrigation/precipitation).

TABLE 29.2 Values of Desorptivity, D_E, Evaluated from Various Experimental Sites

Desorptivity D_E (mm/d$^{1/2}$)	Soil Type	References
4.96–4.30	Sand	[54]
5.08	Clay loam	[55]
4.04	Loam	[55]
3.5	Clay	[55]
~4 to ~8[a]	Loam[b]	[56]
5.8	Clay loam	[57]
4.95[c]	Silty clay loam[d]	[59]
2.11	Gravelly sandy loam	[3]

[a] The magnitude of D_E was found to have a seasonal dependency.
[b] Soil type was determined from texture-dependent soil hydraulic conductivity and matric potential equations of Clapp and Hornberger evaluated by Camillo and Gurney.[48]
[c] This value was evaluated for a vegetated surface.
[d] Soil type was determined from texture-dependent soil hydraulic conductivity and matric potential equations of Clapp and Hornberger evaluated by Sellers et al.[41]

$$\sum E_D = D_E \left(T^{1/2} - T_O{}^{1/2}\right) \tag{29.14}$$

They evaluated the effect on the derived D_E using this technique with the data from Black et al.[54] The value of D_E using Equations 29.13 and 29.14 was estimated to be approximately 3.3 mm/d$^{1/2}$, which is smaller than D_E values reported by Black et al.,[54] namely, 4.3–5 mm/d$^{1/2}$. However, Brutsaert and Chen[59] show that this technique yields a better linear fit to the data points that were actually under S_2 conditions.

Equations 29.13 and 29.14 were used with the September 1973 data set from Jackson et al.[56] and compared to using Equation 29.12 with $t_O = 0$. The plot of Equation 29.12 with the regression line in Figure 29.2 yields $D_E \sim 10$ mm/d$^{1/2}$, which is significantly larger than any previous estimates (Table 29.2). Moreover, it is obvious from the figure that Equation 29.12 should not be applied with $t_O = 0$, as this relationship is not linear over the whole drying processes. With Equation 29.13, applied to the data, t_O is estimated to be approximately 4.3 days, and thus a linear relationship should start at the shifted time scale $T = t - 4.3$; this means ΣE_D should start on day 5 or $T_O \sim 5 - 4.3$ (Figure 29.3a). With Equation 29.14,

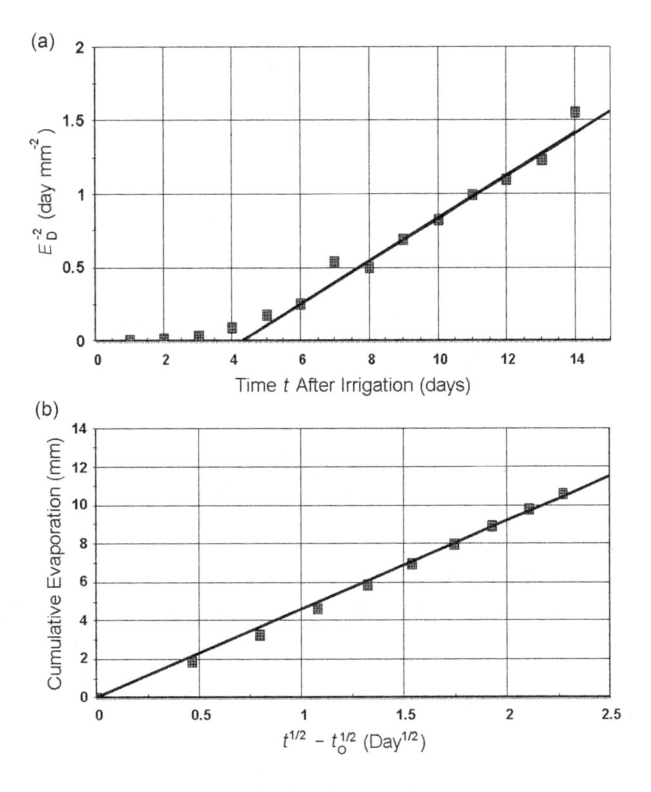

FIGURE 29.3 Estimation of (a) D_E and t_O with the data from Figure 29.2 using Equation 29.13 from Brutsaert and Chen[59] and (b) the resulting cumulative evaporation ΣE_D curve under second stage drying using Equation 29.14.

a more realistic $D_E \sim 4.6$ mm/d$^{1/2}$ is estimated for the linear portion of daily evaporation following the S_2 condition (Figure 29.3b).

While this approach is relatively easy to implement operationally, D_E will likely depend on climatic factors as well as soil textural properties. However, it might be feasible to describe the main climate/seasonal effect on D_E from soil temperature observations.[56] These might come from weather station observations or possibly from multitemporal remote sensing observations of surface temperature.

The difficulty in developing a formulation for R_S, which correctly describes the water vapor transfer process in the soil, was recognized much earlier by Fuchs and Tanner[60] and Tanner and Fuchs.[61] They proposed a combination method instead that involves atmospheric surface layer observations and remotely sensed surface temperature, T_{RS}. Starting with the energy balance equation

$$R_N = H + G + \text{LE} \tag{29.15}$$

where R_N is the net radiation, H the sensible heat flux, LE the latent heat flux and G the soil heat flux all in W/m^2, and assuming the resistance to heat and water vapor transfer are similar yielding,

$$H = \rho C_P \left(\frac{T_{RS} - T_A}{R_A} \right) \tag{29.16}$$

$$\text{LE} = \frac{\rho C_P}{\gamma} \left(\frac{e_{RS} - e_A}{R_A} \right) \tag{29.17}$$

an equation of the following form can be derived

$$\mathrm{LE}=\left(\frac{\gamma+\Delta}{\Delta}\right)\mathrm{LE}_{\mathrm{P}}-\frac{\rho C_{\mathrm{P}}}{\Delta}\left(\frac{e_{*}(T_{\mathrm{RS}})-e_{\mathrm{A}}}{R_{\mathrm{A}}}\right) \quad (29.18)$$

where

$$\mathrm{LE}_{\mathrm{P}}=\rho C_{\mathrm{P}}\left(\frac{e_{*}(T_{\mathrm{A}})-e_{\mathrm{A}}}{R_{\mathrm{A}}(\Delta+\gamma)}\right)+\Delta\left(\frac{R_{\mathrm{N}}-G}{\Delta+\gamma}\right) \quad (29.19)$$

The difference $e_{*}(T_{\mathrm{A}}) - e_{\mathrm{A}}$ is commonly called the saturation vapor pressure deficit, and the value of soil surface vapor pressure e_{RS} is equal to $h_{\mathrm{RS}}e_{*}(T_{\mathrm{RS}})$ where h_{RS} is the soil surface relative humidity. Substituting Equation 29.19 into Equation 29.18 yields

$$\mathrm{LE}=R_{\mathrm{N}}-G-\frac{\rho C_{\mathrm{P}}}{\Delta}\left(\frac{e_{*}(T_{\mathrm{RS}})-e_{*}(T_{\mathrm{A}})}{R_{\mathrm{A}}}\right) \quad (29.20)$$

This equation has the advantage over the above bulk resistance formulations using R_{S} in that there are no assumptions made concerning the saturation deficit at or near the soil surface or how to define h_{RS}. Instead, this effect is accounted for by T_{RS} because as the soil dries, T_{RS} increases and hence $e_{*}(T_{\mathrm{RS}})$, which generally results in the last term on the right-hand side of Equation 29.20 to increase, thus causing LE to decrease. In a related approach,[62] the magnitude of LE is simply computed as a residual in the energy balance equation, Equation 29.19, namely,

$$\mathrm{LE}=R_{\mathrm{N}}-G-\rho C_{\mathrm{P}}\left(\frac{T_{\mathrm{RS}}-T_{\mathrm{A}}}{R_{\mathrm{A}}}\right) \quad (29.21)$$

Particularly crucial in the application of either Equation 29.20 or 29.21 is a reliable estimate of R_{A} and T_{RS}. Issues involved in correcting radiometric temperature observations for surface emissivity, viewing angle effects, and other factors are summarized in Norman and Becker.[63] The aerodynamic resistance R_{A} is typically expressed in terms of Monin–Obukhov similarity theory[42]

$$R_{\mathrm{A}}=\frac{\left\{\ln\left[\frac{z-d_{\mathrm{O}}}{z_{\mathrm{OM}}}\right]-\psi_{\mathrm{M}}\right\}\left\{\ln\left[\frac{z-d_{\mathrm{O}}}{z_{\mathrm{OS}}}\right]-\psi_{\mathrm{S}}\right\}}{k^{2}U} \quad (29.22)$$

where z is the observation height in the surface layer (typically 2–10 m), d_{O} the displacement height, z_{OM} the momentum roughness length, z_{OS} the roughness length for scalars (i.e., heat and water vapor), k (~0.4) von Karman's constant, ψ_{M} the stability correction function for momentum, and ψ_{S} the stability correction function for scalars. Both d_{O} and z_{OM} are dependent on the height and density of the roughness obstacles and can be considered a constant for a given surface, while the magnitude of z_{OS} can vary for a given bare soil surface as it is also a function of the surface friction velocity.[42] Experimental evidence suggests that existing theory with possible modification to some of the "constants" can still be used to determine z_{OS} providing acceptable estimates of H for bare soil surfaces. However, application of Equation 29.21 in partial canopy cover conditions has not been successful in general because z_{OS} is not well defined in Equation 29.22, exhibiting large scatter with the existing theory.[64]

For this reason, estimating soil evaporation from partially vegetated surfaces using T_{RS} invariably has to involve "two-source" approaches whereby the energy exchanges from the soil and vegetated components are explicitly treated.[65] Similarly, when using remotely sensed soil moisture for vegetated surfaces, a two-source modeling framework needs to be applied.[66] In these two-source approaches, there is the added complication of determining aerodynamic resistances between soil and vegetated surfaces

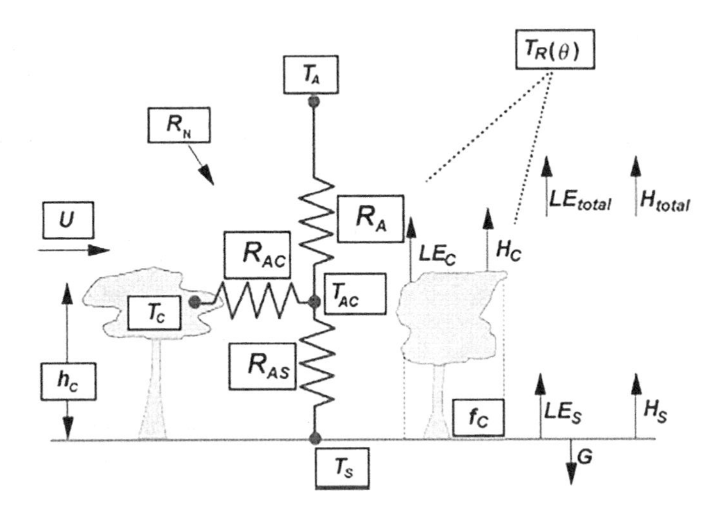

FIGURE 29.4 A schematic diagram illustrating the resistance network for the two-source approach where the subscript *c* refers to the vegetated canopy and s refers to the soil surface. The symbol $T_R(\theta)$ refers to a radiometric temperature observation at a viewing angle θ, T_{AC} is the model-derived within-canopy air space temperature, h_C is the canopy height, f_C is the fractional vegetation/canopy cover, and R_{AC} and R_{AS} are the aerodynamic resistances to sensible heat flux from the canopy and soil surface, respectively. The main meteorological inputs required for the model are also illustrated, namely wind speed, *U*, air temperature, T_A, and the net radiation, R_N.
Source: Adapted from Norman et al.[65]

and the canopy air space. Schematically, the resistance network and corresponding flux components for two-source models is shown in Figure 29.4. An advantage with the two-source formulation of Norman et al.[65] is that R_S is not actually needed for computing LE_S as it is solved as a residual. Nevertheless, the formulations in such parameterizations that are used (such as the aerodynamic resistance formulations) are likely to strongly influence LE_S values.[1] Yet this two-source formulation is found to be fairly robust in separating soil and canopy contributions to evapotranspiration.[67]

Conclusions

Techniques for measuring soil evaporation, separately from plant transpiration, exist and continually improve. Given the high variability in soil water content often exhibited in the field, the exact position of the instrumentation may have a large effect on the measurement. This is magnified in row crops, where the variability is structured, rather than random. In row crops, soil evaporation is not only dependent on the soil properties, but also local climatic conditions, and the timing of irrigation/precipitation. Under low-to-moderate wind speed conditions (~1 to 4 m s^{-1}), row orientation relative to the wind direction also affects the amount of soil evaporation.[68] This large variability requires that there be careful thought to the design of soil evaporation measurements in the field. The methodology proposed by Scanlon and Kustas[5] using eddy covariance measurements suffers less from the sampling issue but requires *a priori* knowledge of the vegetation water-use efficiency, which will vary with vegetation type and condition.

For many landscapes having partial vegetative cover, the contribution of soil evaporation to the total evapotranspiration flux cannot be ignored, particularly with regard to the influence of surface soil moisture and temperature on the microclimate in the canopy air space (Figure 29.4). Numerical models for simulating soil evaporation have been developed and are shown to be fairly robust. However, the required inputs for defining model parameters often limit their application to field sites having detailed soil profile information and that are well instrumented with ancillary weather data.

For many operational applications where detailed soil and ancillary weather data are unavailable or where daily evaporation values are only needed, some of the analytical models described may provide the necessary level of accuracy. Moreover, in the application of weather forecast and hydrologic models, the use of simplified approaches is necessitated by the computational requirements or the lack of adequate data or both for defining more complex numerical model parameters and variables.

For large area estimation, the use of remotely sensed soil moisture and surface temperature offer the greatest potential for operational applications. Development of modeling schemes that can incorporate this remote sensing information and readily apply it on a regional scale basis has been proposed and shows promise.[66,69–71]

References

1. Kustas, W.P.; Norman, J.M. Evaluation of soil and vegetation heat flux predictions using a simple two-source model with radiometric temperatures for partial canopy cover. Agric. Forest Meteorol. **1999,** *94,* 13–25.
2. Tanner, C.B.; Jury, W.A. Estimating evaporation and transpiration from a row crop during incomplete cover. Agron. J. **1976,** *68,* 239–242.
3. Wallace, J.S.; Holwill, C.J. Soil evaporation from tiger-bush in south-west Niger. J. Hydrol. **1997,** *188–189,* 426–442.
4. Agam, N.; Berliner, P. R.; Zangvil, A.; Ben-Dor, E. Soil water evaporation during the dry season in an arid zone. J. Geophys. Res. Atmos. **2004,** *109,* D16103, doi:16110.11029/12004JD004802.
5. Scanlon, T.M.; Kustas, W.P. Partitioning carbon dioxide and water vapor fluxes using correlation analysis. Agric. Forest Meteorol. **2010,** *150,* 89–99.
6. Allen, S.J. Measurement and estimation of evaporation from soil under sparse barley crops in northern Syria. Agric. Forest Meteorol. **1990,** *49,* 291–309.
7. Van de griend, A.A.; Owe, M. Bare soil surface resistance to evaporation by vapor diffusion under semiarid conditions. Water Resour. Res. **1994,** *30,* 181–188.
8. Plauborg, F. Evaporation from bare soil in a temperate humid climate – Measurement using micro-lysimeters and time domain reflectometry. Agric. Forest Meteorol. **1995,** *76,* 1–17.
9. Baker, J.M.; Spaans, E.J. Measuring water exchange between soil and atmosphere with TDR-Microlysimetry. Soil Sci. **1994,** *158,* 22–30.
10. Ashktorab, H.; Pruitt, W.O.; Pawu, K.T.; George, W.V. Energy-balance determinations close to the soil surface using a micro-Bowen ratio system. Agric. Forest Meteorol. **1989,** *46,* 259–274.
11. Heitman, J.L.; Horton, R.; Sauer, T.J.; Desutter, T.M. Sensible heat observations reveal soil-water evaporation dynamics. J. Hydrometeorol. **2008a,** *9,* 165–171.
12. Shawcroft, R.W.; Gardner, H.R. Direct evaporation from soil under a row crop canopy. Agric. Meteorol. **1983,** *28,* 229–238.
13. Lascano, R.J.; Van bavel, C.H.M. Simulation and measurement of evaporation from a bare soil. Soil Sci. Soc. Am. J. **1986,** *50,* 1127–1133.
14. Boast, C.W.; Robertson, T.M. A "micro-lysimeter" method for determining evaporation from bare-soil: description and laboratory evaluation. Soil Sci. Soc. Am. J. **1982,** *46,* 689–696.
15. Walker, G.K. Measurement of evaporation from soil beneath crop canopies. Can. J. Soil Sci. **1983,** *63,* 137–141.
16. Evett, S.R.; Warrick, A. W.; Matthias, A.D. Wall material and capping effects on microlysimeter temperatures and evaporation. Soil Sci. Soc. Am. J. **1995,** *59,* 329–336.
17. Denmead, O.T.; Dunin, F.X.; Wong, S.C.; Greenwood, E.A.N. Measuring water use efficiency of Eucalyptus trees with chambers and micrometeorological techniques. J. Hydrol. **1993,** *150,* 649–664.
18. Livingston, G.P.; Hutchinson, G.L. Enclosure-based measurement of trace gas exchange: Applications and sources of error. In *Biogenic Trace Gases: Measuring Emissions from Soil and Water;* Matson, P.A., Harriss, R.C., Eds.; Blackwell Scientific Publications: Oxford, 1995; pp. 14–51.

19. Hutchinson, G.L.; Livingston, G.P. Gas flux. In *Methods of Soil Analysis: Part 1, Physical Methods;* Dane, J.H., Topp, G.C., Eds.; Soil Science Society of America: Madison, WI: 2002.
20. Davidson, E.A.; Savage, K.; Verchot, L.V.; Navarro, R. Minimizing artifacts and biases in chamber-based measurements of soil respiration. Agric. Forest Meteorol. **2002,** *113,* 21–37.
21. Norman, J.M.; Kucharik, C.J.; Gower, S.T.; Baldocchi, D.D.; Crill, P.M.; Rayment, M.; Savage, K.; Striegl, R.G. A comparison of six methods for measuring soil-surface carbon dioxide fluxes. J. Geophys. Res.-Atmospheres **1997,** *102,* 28771–28777.
22. Evett, S.R.; Schwartz, R.C.; Casanova, J.J.; Heng, L.K. Soil water sensing for water balance, ET and WUE. Agric. Water Manag. **2012,** *104,* 1–9.
23. Topp, G.C.; Davis, H.L.; Annan, A.P. Electromagnetic determination of soil water content: measurements in coaxial transmission lines. Water Resour. Res. **1980,** *16,* 574–582.
24. Logsdon, S.D.; Green, T.R.; Seyfried, M.S.; Evett, S.R.; Bonta, J. Hydra probe and twelve-wire probe comparisons in fluids and soil cores. Soil Sci. Soc. Am. J. **2010,** *74,* 5–12.
25. Miralles-Crespo, J.; Van Lersel, M.W. A calibrated time domain transmissometry soil moisture sensor can be used for precise automated irrigation of container-grown plants. Hort. Science **2011,** *46,* 889–894.
26. Allen, R.G.; Pereira, L.S.; Howell, T.A.; Jensen, M.E. Evapotranspiration information reporting: I. Factors governing measurement accuracy. Agric. Water Manag. **2011,** *98,* 899–920.
27. Farahani, H.J.; Howell, T.A.; Shuttleworth, W.J.; Bausch, W.C. Evapotranspiration: Progress in measurement and modeling in agriculture. Trans. Am. Soc. Agric. Biol. Engineers **2007,** *50,* 1627–1638.
28. Bowen, I.S. The ratio of heat losses by conduction and by evaporation from any water surface. Phys. Rev. **1926,** *27,* 779–787.
29. Monteith, J.L.; Unsworth, M.H. *Principles of Environmental Physics,* 3rd ed.; Elsevier/Academic Press: 2008.
30. Zeggaf, A.T.; Takeuchi, S.; Dehghanisanij, H.; Anyoji, H.; Yano, T. A Bowen ratio technique for partitioning energy fluxes between maize transpiration and soil surface evaporation. Agronomy J. **2008,** *100,* 988–996.
31. Heitman, J.L.; Xiao, X.; Horton, R.; Sauer, T.J. Sensible heat measurements indicating depth and magnitude of subsurface soil water evaporation. Water Resour. Res. **2008b,** *44.*
32. Gardner, H.R.; Hanks, R.J. Evaluation of the evaporation zone in soil by measurement of heat flux. Soil Sci. Soc. Am. Proc. **1966,** *32,* 326–328.
33. Ren, T.S.; Ochsner, T.E.; Horton, R. Development of thermo-time domain reflectometry for vadose zone measurements. Vadose Zone J. **2003,** *2,* 544–551.
34. Knight, J.H.; Kluitenberg, G.J. Simplified computational approach for dual-probe heat-pulse method. Soil Sci. Soc. Am. Journal. *68,* 447–449.
35. Bristow, K.L.; Kluitenberg, G.J.; Horton, R. Measurement of soil thermal-properties with a dual-probe heat-pulse technique. Soil Sci. Soc. Am. J. **1994,** *58,* 1288–1294.
36. Philip, J.R.; De vries, D.A. Moisture movement in porous materials under temperature gradients. Trans. Am. Geophys. Union **1957,** *38* (2), 222–232.
37. Camillo, P.J.; Gurney, R.J.; Schmugge, T.J. A soil and atmospheric boundary layer model for evapotranspiration and soil moisture studies. Water Resour. Res. **1983,** *19,* 371–380.
38. Chanzy, A.; Bruckler, L. Significance of soil surface moisture with respect to daily bare soil evaporation. Water Resour. Res. **1993,** *29,* 1113–1125.
39. Daamen, C.C.; Simmonds, L.P. Measurement of evaporation from bare soil and its estimation using surface resistance. Water Resour. Res. **1996,** *32,* 1393–1402.
40. Noilhan, J.; Planton, S. A simple parameterization of land surface processes for meteorological models. Mon. Weather Rev. **1989,** *117,* 536–549.
41. Sellers, P.J.; Heiser, M.D.; Hall, F.G. Relations between surface conductance and spectral vegetation indices at intermediate (100 m^2 to 15 km^2) length scales. J. Geophys. Res. **1992,** *97,* 19033–19059.

42. Brutsaert, W. *Evaporation into the Atmosphere: Theory, History and Applications;* D. Reidel Publishing Co.: Boston, 1982.
43. Mahfouf, J.F.; Noilhan, J. Comparative study of various formulations of evaporation from bare soil using in situ data. J. Appl. Meteorol. **1991,** *30,* 1354–1365.
44. Ye, Z.; Pielke, R.A. Atmospheric parameterization of evaporation from non-plant-covered surfaces. J. Appl. Meteorol. **199,** *32,* 1248–1258.
45. Philip, J.R. 1957, Evaporation, and moisture and heat fields in the soil. Journal of Meteorology. *14,* 354–366.
46. Kondo, J.; Saigusa, N.; Sato, T. A parameterization of evaporation from bare soil surfaces. J. Appl. Meteorol. **1990,** *29,* 385–389.
47. Cahill, A.T.; Parlange, M.B.; Jackson, T.J.; O'Neill, P.; Schmugge, T.J. Evaporation from nonvegetated surfaces: Surface aridity methods and passive microwave remote sensing. J. Appl. Meteorol. **1999,** *38.*
48. Camillo, P.J.; Gurney, R.J. A resistance parameter for bare-soil evaporation models. Soil Sci. **1986,** *141,* 95–104.
49. Jackson, R.D.; Kimball, B.A.; Reginato, R.J.; Nakayama, F.S. Diurnal soil water evaporation: Time-depth-flux patterns. Soil Sci. Soc. Am. Proc. **1973,** *37,* 505–509.
50. Hillel, D. *Applications of Soil Physics;* Academic Press: San Diego, 1980.
51. Yamanaka, T.; Takeda, A.; Sugita, F. A modified surface-resistance approach for representing bare-soil evaporation: Wind tunnel experiments under various atmospheric conditions. Water Resour. Res. **1997,** *33,* 2117–2128.
52. Penman, H.L. Natural evaporation from open water, bare soil and grass, Proc. R. Soc. Lond. **1948,** *A193,* 120–146.
53. Gardner, W.R. Solution of the flow equation for the drying of soils and other porous media. Soil Sci. Soc. Am. Proc. **1959,** *23,* 183–187.
54. Black, R.A.; Gardner, W.R.; Thurtell, G.W. The prediction of evaporation, drainage, and soil water storage for a bare soil. Soil Sci. Soc. Am. Proc. **1969,** *33,* 655–660.
55. Ritchie, J.T. Model for predicting evaporation from a row crop with incomplete cover. Water Resour. Res. **1972,** *8,* 1204–1213.
56. Jackson, R.D.; Idso, S.B.; Reginato, R.J. Calculation of evaporation rates during the transition from energy-limiting to soil-limiting phases using albedo data. Water Resour. Res. **1976,** *12,* 23–26.
57. Parlange, M.B.; Katul, G.G.; Cuenca, R.H.; Kavvas, M.L.; Nielsen, D.R.; Mata, M. Physical basis for a time series model of soil water content. Water Resour. Res. **1992,** *28,* 2437–2446.
58. Campbell, G.S.; Norman, J.M. *An Introduction to Environmental Biophysics;* Springer-Verlag: New York, 1998.
59. Brutsaert, W.; Chen, D. Desorption and the two stages of drying of natural tallgrass prairie. Water Resour. Res. **1995,** *31,* 1305–1313.
60. Fuchs, M.; Tanner, C.B. Evaporation from a drying soil. J. Appl. Meteorol. **1967,** *6,* 852–857.
61. Tanner, C.B.; Fuchs, M. Evaporation from unsaturated surfaces: a generalized combination method. J. Geophys. Res. **1968,** *73,* 1299–1304.
62. Jackson, R.D.; Reginato, R.J.; Idso, S.B. Wheat canopy temperature: A practical tool for evaluating water requirements. Water Resour. Res. **1977,** *13,* 651–656.
63. Norman, J.M.; Becker, F. Terminology in thermal infrared remote sensing of natural surfaces. Agric. Forest Meteorol. **1995,** *77,* 153–166.
64. Verhoef, A.; De bruin, H.A.R.; Van den hurk, B.J.J.M. Some practical notes on the parameter kB-1 for sparse vegetation. J. Appl. Meteorol. **1997,** *36,* 560–572.
65. Norman, J.M.; Kustas, W.P.; Humes, K.S. A two-source approach for estimating soil and vegetation energy fluxes in observations of directional radiometric surface temperature. Agric. Forest Meteorol. **1995,** *77,* 263–293.

66. Kustas, W.P.; Zhan, X.; Schmugge, T.J. Combining optical and microwave remote sensing for mapping energy fluxes in a semiarid watershed. Remote Sensing Environ. **1998,** *64,* 116–131.
67. Kustas, W.P.; Anderson, M.C. Advances in thermal infrared remote sensing for land surface modeling. Agric. Forest Meteorol. **2009,** *149,* 2071–2081.
68. Agam, N.; Evett, S.R.; Tolk, J.A.; Kustas, W.P.; Colaizzi, P.D.; Alfieri, J.G.; Mckee, L.G.; Copeland, K.S.; Howell, T.A.; Chávez, J.L. Evaporative loss from irrigated interrows in a highly advective semi-arid agricultural area. Adv. Water Resour. **2012,** http://dx.doi.org/10.1016Zj.advwatres.2012.07.010.
69. Mecikalski, J.R.; Diak, G.R.; Anderson, M.C.; Norman, J.M. Estimating fluxes on continental scales using remotely sensed data in an atmospheric-land exchange model. J. Appl. Meteorol. **1999,** *38,* 1352–1369.
70. Kustas, W.P.; Jackson, T.J.; French, A.N.; Macpherson, J.I. Verification of patch- and regional-scale energy balance estimates derived from microwave and optical remote sensing during SGP97. J. Hydrometeorol. **2001,** *2,* 254–273.
71. Anderson, M.C.; Kustas, W.P.; Norman, J.M.; Hain, C.R.; Mecikalski, J.R.; Schultz, L.; Gonzalez-Dugo, M.P.; Cammalleri, C.; D'urso, G.; Pimstein, A.; Gao, F. Mapping daily evapotranspiration at field to continental scales using geostationary and polar orbiting satellite imagery. Hydrol. Earth Syst. Sci. **2011,** *15,* 223–239.

30

Soil: Fauna

Mary C. Savin
University of Arkansas

Introduction

Soil fauna are diverse, abundant, and critical to ecosystem functions. Soil fauna account for almost 25% of living species.[1] Although they compose a group of organisms that has been ignored historically when it comes to land management, there is growing recognition and understanding of the roles soil fauna play in ecosystems. Appreciation for which organisms are present in soil, their habitats, community diversity, and contributions to ecosystem functions, stability and sustainability is increasing with accumulating evidence from scientific research. Further development of our understanding of soil ecology is essential because soil fauna contribute directly and indirectly to properties and processes at the scale of the organism and its habitat with outcomes observed up to field and landscape levels. Future research efforts that build upon past discoveries should enhance our understanding of how soil fauna connect aboveground and belowground ecology for the proper management and sustainability of ecosystems and ecosystem services.[2]

Soil Fauna—Organisms

Soil fauna, protists, and animals include organisms that span orders of magnitude in size (from μm to cm to m). Generally, soil fauna are eukaryotic, aerobic chemoheterotrophs, but they occupy many niches in soil food webs.[3] Organisms are commonly distributed into groupings which can range from low to high taxonomic resolution. One common differentiation is classification by size. Sized-based groups are microfauna, mesofauna, macrofauna, and megafauna. (See soil microbiology and ecology textbooks, e.g., references,[4–7] for more detailed descriptions of the groups.)

Because categorization based on size is an empirical separation, the upper size cut-off value may differ depending on the reference in the literature that one is searching. Microfauna may be considered up to 100–200 μm in length and include protozoa and nematodes. Nematodes (phylum Nematoda) may also be designated as mesofauna in the literature. Protozoa (kingdom Protista) are often grouped into flagellates, amoeba, and ciliates. Flagellates and amoebae tend to be smaller and numerous while ciliates are larger relatively and less numerous. Protozoa consume bacteria, protozoa, and other organisms. Abundances range from $10–10^5$/g soil.[8] (The diverse range of numbers for soil faunal abundances is related to the heterogeneity of soil conditions.) Nematodes, roundworms, are often grouped by into trophic groups: bacterivores, fungivores, omnivores, predators, and plant parasites. Microfauna, and also the mesofauna enchytraeids, are considered to reside in water films and water-filled pores inside aggregates.[9,10]

Mesofauna range from 100 (or 200) to 2000 (sometimes considered up to 10,000) μm. At this larger size range, mesofauna are considered to be in the interaggregate pore space, which is likely to be air-filled, rather than the intra-aggregate pore space expected to be occupied by microfauna.[11] This category includes microarthropods such as springtails (subclass Collembola) and mites (subclass Acari), and the enchytraeids (family Enchytraeidae) (potworms). Microarthropod abundances may range from $10–10^5/m^2$ (reviewed in Coleman et al.).[5] Enchytraeidae are particularly important in temperate and boreal forests (reviewed in Huhta et al.)[12] at abundances ranging from 10^3 to greater than $10^5/m^2$. Enchytraeidae consume fine plant litter covered in fungi and bacteria. Collembola are likely omnivorous, although they are often considered fungivorous, and mites feed on fungi, plant detritus, or are predaceous.[5] Mesofauna are generally regarded as important consumers in litter transformation, especially important given that about 90% of terrestrial organic inputs are processed through detrital food webs.[5]

Macrofauna are generally considered to be greater than 2 mm, up to a few cm, and some species of earthworms are meters long. Macrofauna are often cited as including ants (widely distributed, family Formicidae), earthworms (Phylum Annelida, class Oligochaeta), termites (tropical and some temperate, Termitidae in the Order Isoptera), and macroarthropods (phylum Arthropoda). Earthworm abundances range from less than 10 to greater than $1000/m^2$ in temperate and tropical ecosystems.[13] When present, ants and termites can be so abundant that researchers often do not enumerate abundances. Megafauna sometimes include earthworms and include vertebrates such as moles, gophers, and prairie dogs which can be important in many ecosystems.[14] Macrofauna and megafauna can be ecosystem engineers. These organisms are too large to exist within the existing pores; in soil, they move soil particles, modify soil pore structure, and create structures such as aggregates.[15,16] In general, ecosystem engineers modify their physical environment and resources for others.[14–17] Further, they alter the environment distinctively from abiotic processes.[14,17]

Much of the biological activity in soil occurs in a small volume relative to whole which poses logistical challenges in representative sampling. However, these "hotspots" or small volumes of soil where conditions are conducive for high activity can be placed in important functional domains.[16] These domains, or spheres of influence regulating microbial activity, are often associated with the activity of larger organisms such as plant roots and macrofauna, and include examples such as the rhizosphere (zone of influence around a root), drilosphere (zone of influence around an earthworm burrow), and termitosphere (zone of influence under termite activity). Macrofauna create conditions conducive for high rates of activity by directly changing the "architecture" of soil at scales relevant to smaller organisms,[18] and the accumulation of their direct and indirect activities changes properties and processes in fields. Macrofauna may reside in the surface litter layer (e.g., epigeic earthworms), soil (e.g., endogeic earthworms), and/or move across soil-litter boundaries (e.g., anecic earthworms). Macrofauna are also involved in important symbiotic relationships in soil, such as protozoan gut communities in termites, fungus gardens of termites, and microbes in the external rumen of earthworms.[7]

Soil Fauna—Diversity and Biomass

The enormous diversity of fauna means that not all groups have been identified and described. Fauna grouped by size may be placed into broad taxonomic categories that use trophic status for differentiation. Beyond taxonomic classifications, soil fauna are often placed into functional groups. Characterizing communities based on functional groups or higher level taxonomic groups may provide stronger statistical power[19] than trying to identify organisms at finer resolution. However, species may feed across trophic levels,[20] and thus taxonomic trophic groups may not be most appropriate level of resolution for understanding the functional importance of soil fauna. Salamon et al.,[21] for example, found that different species of Collembola respond to increases in fungal compared to bacterial biomass. Although there is one food web aboveground, there are both bacterial-based and fungal-based food webs operating belowground.

Despite the difficulty of determining the appropriate level of resolution to understand soil faunal diversity, diversity is an important consideration for processes, ecosystem functions, and productivity. There remains much that is not well known about the diversity of belowground communities. Regardless, it is safe to state that both local and global diversity of soil fauna is huge (reviewed in Coleman).[3] Estimates for numbers of species can span hundreds in a few hundred square meters.[22] However, relatively few taxa may dominate soil communities within a land use. In the southeastern United States, rank-abundance relationships suggest that a few taxa dominated in four land uses, with most abundant taxa belonging to different types of beetles in cultivated and pasture sites, centipedes in hardwood forest, and millipedes in pine stand forest.[23] A large portion of biomass in tundra, boreal forest, and temperate forest soil (high latitude ecosystems) was accounted for in enchytraeids, while earthworms dominated in tropical forests and temperate grassland soils.[24] Mites, Collembola, and Enchytraeidae are abundant in acidic forests with thick organic layers.[20] Despite the diversity of organisms, there are a few predominant organisms in ecosystems in many studies, that is, many rare, and few cosmopolitan taxa.

A recent survey of soil animals (i.e., not including protozoa) in 11 locations around the global using molecular analysis of 18S rDNA sequences resulted in a dominance of nematodes and microarthropods.[25] Almost 96% of operational taxonomic units (OUTs) (surrogate for species defined by 99% sequence similarity) were detected in one location only, indicating high endemism and a lack of cosmopolitan species.[25] Predominance of nematodes (grasslands) and microarthropods (forests) was related to pH with nematodes positively correlated and microarthropods negatively correlated with pH. In fact, several environmental variables were related to abundances of different groups; microarthropods predominate in forests with lower pH, root biomass, mean annual temperature, inorganic N, and higher C:N, litter, and moisture.[25] Further, on a global scale the soils from ecosystems with higher aboveground plant diversity had lower belowground faunal diversity.[25] Diversity matrices have been less informative than multivariate analyses.[25,26] Thus, it may be that diversity per se is not the best approach to assessing soil faunal communities, rather multivariate approaches relating abundance and functional structure to environmental variables may be informative.

While it is generally postulated that high biodiversity levels are positive for ecosystem health, stability, and function, there may be much functional redundancy among soil fauna. Impacts of fauna depend in part on characteristics of the ecosystem. In a field experiment investigating the contributions of soil fauna to litter decomposition, fauna of all size classes increased decomposition of litter in its home environment when the plant community was in late succession and contained more recalcitrant compounds.[27] In early succession, when litter contained easily degraded compounds, the presence of an indigenous community was less important. Often there are changes in taxonomic groups within a system, but little change in diversity. Thus, it may be that functional structure of soil fauna is more important than taxonomic structure in ecosystems. Responses to elevated CO_2 concentrations were related to trophic structure among soil biota in a meta-analysis.[28] Abundances of detritivores increased while there were no changes at higher trophic levels. So, change in the functional characteristics of species composition is important. In a microcosm study, Cole et al.[29] found that an increasing density of microarthropods increased nitrate concentrations, but increased richness decreased N mineralization and nitrification. The authors suggest that increased predation pressure with increased richness decreased the microbivore grazing of remaining microarthropods. Thus, they concluded it was not a change in richness per se that was important but a change that altered functional group status that was important for functioning.

The study by Cole et al.[29] also brings to light another question in understanding the impacts of soil fauna on ecosystems. In addition to the extent of diversity, be it functional or taxonomic, among ecosystems, there is also a question of whether total abundance of soil fauna or diversity is more important. In another global meta-analysis of seven biomes, the authors estimated biomass of five major faunal groups. Despite the challenges in computing estimates for biomass that resulted in low to medium confidence for several of the estimates from several biomes, faunal biomass was found to be 40–80% of animal biomass in different biomes except desert.[24] However, soil faunal biomass (with the exception of deserts at <0.02% of microbial biomass) averaged 2% (1.5–3.6%) of microbial biomass.[24] Microbial biomass is 2–5% soil organic carbon, which is usually 1–5% of soil by volume. Despite composing a minor amount of the biomass in soil, soil fauna are major players in many important functions in terrestrial ecosystems.

Soil Fauna—Roles in Ecosystem Functions, Stability, and Sustainability

Lavelle et al.[2] argue that invertebrates are a "resource that needs to be properly managed to enhance ecosystem services" since fauna mediate soil formation, nutrient cycling, primary production, and regulation services. Soil organisms alter soil aggregation through the production of particular compounds, the physical breakdown of organic materials, and the movement of soil particles. Changing soil pore and aggregate structure impacts water infiltration or run-off, and distribution and water storage within the soil profile. Water content affects the habitat, diffusion of gases and compounds, and activity and access of soil organisms, which has implications for the rates and extent of nutrient cycling. Grazing consumes microbes, changes community structure and activity, and releases nutrients. Soil fauna may consume microorganisms, other fauna, and plant detritus. Thus, soil fauna impact nutrient cycling directly through incorporation and comminution of plant detrital materials, and indirectly through consumption, movement, selection, and activation of microbial communities. These processes impact sequestration of carbon (and other nutrients) in soil and gaseous releases which affect climate. The activities of soil fauna impact root production, and enhanced nutrient cycling can increase aboveground plant biomass.[30]

While the negative impacts of pests and pathogens are well studied and appreciated because of the economic damage they can cause, the positive impacts of the activities of soil fauna on plant growth and other ecosystem functions are often less appreciated and intentionally utilized. Ingham et al.[30] demonstrated the positive influence of free-living nematodes on plant biomass by inoculating soil with bacteria and bacterial-feeding nematodes. About 30% N taken up by plants can come from mineralization as a result of grazing of fauna on soil microorganisms.[31] The presence of earthworms has been shown to decrease plant parasitic nematode infection of rice by more than 80%, allowing plants to maintain biomass production similar to uninfected plants.[32] In a recent study as part of the biodiversity experiment in Jena, Germany, use of the insecticide chlorpyrifos decreased biomass of two functional plant groups (forbs and grass), indicating that negative effects from reductions in Collembola abundance, whose grazing of microorganisms contributes to nutrient cycling, outweighed benefits that might have been gained from reductions in herbivore populations.[33]

Soil fauna are an integral component of belowground ecology which can alter aboveground ecology, but these fauna also respond to aboveground management. Fauna may respond to or emit plant signaling compounds (reviewed in Kardol and Wardle).[34] Re-establishing mixed forests in coniferous stands changed *Collembola* species richness and diversity, highlighting how aboveground management can affect belowground ecology.[35] In a study comparing conventional practices to organic wheat farming, organic management promoted both above-and below-ground interactions with positive effects on soil fauna compared to a conventional system using mineral fertilizers and herbicides.[36] Organic management improved soil quality by enhancing generalist predators (top-down control); however, the system was also enhanced through the addition of resources (bottom-up control).[36]

Subject to much debate is whether terrestrial systems and soil food webs are controlled by bottom-up or top-down forces. Invertebrate numbers and populations respond to or covary with soil properties, such as soil organic matter (OM), temperature, and electrical conductivity.[26] However, there are conflicting data about which properties and the degree of importance of properties that control soil fauna. Mites correlated positively with high soil OM.[27] Soil pH is an important controlling variable. For example, enchytraeids are abundant in low pH soil.[20] Soils are often considered to be controlled through bottom-up controls. However, research results do not necessarily clarify the debate. Although microorganisms are the primary agents of decomposition in soil, microbes do not always respond to resource inputs. This may be in part because fauna are consuming biomass, and thus preventing measurable increases in microbial production. In a temperate deciduous forest in Germany, while particular organisms responded to resource input, the structure of the food web was not clearly related to bottom-up control.[21] Earthworms changed habitat for other organisms, reducing effects of resource inputs.[21] Increased predation pressure reduces N mineralization, presumably through cascading effects that reduce grazing of microorganisms.[29,36] These results indicate that top-down control of food web dynamics is also important in understanding how functions emerge in systems. Taken together, there remain many gaps in the data for many groups of fauna to adequately assess what conditions lead to the greatest diversity. It is often hypothesized that with little disturbance and much competition, diversity will be limited, and under too much stress, diversity will be limited. However, the heterogeneity of soil, patchy community structures, and the activities of ecosystem engineers may all complicate these relationships and lead to high diversity of fauna in soil ecosystems.[37]

In addition to diverse community composition, inherent spatial and temporal variability in soil properties and activities and interactions of organisms poses challenges in accurate assessments. In a study investigating the literature from 1940 to 1992, spatial variability of mesofauna and macrofauna was analyzed.[19] While common species showed lower variation than rare species, the coefficient of variation for common species was often 100% with much of the variability attributed to random error.[19] An important step in making cross-study comparisons is to standardize field collection methods to assess soil faunal communities.[38] Abundances and diversity are often greatest at the soil surface. While organisms can be found at depths below 30 cm, Callaham et al.[23] discontinued sampling in four land uses in the southeastern United States at the 15–30 cm depths for their faunal community assessments because fauna were found at the 0–15 cm soil depth. Abundances often need to be transformed to fit parametric statistical analyses. Distributions can be difficult to sample representatively because they are patchy at best within an ecosystem and variable in response to soil properties and to disturbances.[39,40] Further, distributions and activity are temporally variable. For example, enchytraeid activity is closely related to rainfall.[41]

Soil fauna are responsive to changes in their environment. This responsiveness, which occurs in part because the soil is their habitat, makes soil fauna useful indicator organisms of disturbance, recovery, and soil quality.[42] Nematodes, for example, have been used as indicators of ecosystem recovery and function, maturity, and the soil food web.[43,44] Earthworms are utilized as indicators of pollution, including heavy metals. However, their diversity in niche separation may make it difficult to interpret responses. Currently, we remain limited not only by scale of resolution in sampling and identification, but by what we do not know about the biology and ecology of unidentified and even identified organisms. Barbercheck et al.[26] suggested separating Collembola by habitat type for effectiveness of using Collembola as indicators. However, relying on one indicator or group of indicators to describe effects of disturbance among ecosystems is not advised.[19,26]

In restoration ecology, it has been proposed that to restore ecosystems, one must integrate the belowground with the aboveground.[34] Organisms are useful as indicators because they integrate cumulative effects of chemistry and physics. However, there are different types of disturbances which occur at different frequencies, intensities, durations, and extent of spatial coverage. Greater diversity of fauna was measured in the less disturbed hardwood forest site, followed by the pine stand, followed by pastures, and the least diversity occurred in cultivated soil in four land uses in southeastern United States.[23] Native earthworms were present in fields and forests, but most commonly collected from forests.[23] Introduced earthworms were most often collected from cultivated fields and pastures. Introduced species are well known for drastically altering characteristics and functions of ecosystems, especially if they become invasive. Introduction of non-native earthworms has resulted in change in location and cycling of OM in forests, such that ecosystems are no longer sustainable (northern U.S. forests).[45] It is frequently expected that introduced organisms will be more common in disturbed areas and native species more common in undisturbed locales.[46] However, although reduced in abundance, native species are not always eliminated from an ecosystem as a result of a disturbance. To what extent native and introduced organisms co-occur and alter ecosystem function still needs research.

Conclusion

Although there is growing recognition of their integral effects on soil functions, soil fauna have been largely ignored in development of land management plans. There is an accumulating body of knowledge of the importance of these organisms in driving ecosystem functions. Beyond the small size of many organisms, soil fauna may not have received deserved levels of research attention because there is general agreement that most important drivers of nutrient cycling are microorganisms. However, soil fauna interact with microbes, influence community structure and dispersal, change the physical environment, and facilitate decomposition and nutrient cycling. Soil fauna may be the link that is necessary to connect microorganisms to fields and landscapes, that is, the scale at which processes are observed.

References

1. Decaëns, T.; Jiménez, J.J.; Gioia, C.; Measey, G.J; Lavelle, P. The values of soil animals for conservation biology. Eur. J. Soil Biol. **2006**, *42*, S23–S38.
2. Lavelle, P.; Decaens, T.; Aubert, M.; Barot, S.; Blouin, M.; Bureau, F.; Margerie, P.; Mora, P.; Rossi, J.P. Soil invertebrates and ecosystem services. Eur. J. Soil Biol. **2006**, *42*, S3–S15.
3. Coleman, D.C. From peds to paradoxes: Linkages between soil biota and their influences on ecological processes. Soil Biol. Biochem. **2008**, *40*, 271–289.
4. Amador, J.A.; Gërres, J.H. Chapter 8 Fauna. In *Principles and Applications of Soil Microbiology 2nd Edition;* Sylvia, D.M.; Fuhrmann, J.J.; Hartel, P.G.; Zuberer, D.A., Eds.; Pearson Education Inc.: Upper Saddle River, NJ, 2005; 181–200.
5. Coleman, D.C.; Crossley, Jr., D.A.; Hendrix, P.F. *Fundamentals of Soil Ecology 2nd Edition*; Elsevier Academic Press: Burlington, MA, 2004.
6. Coleman, D.C.; Wall, D.H. Fauna: The engine for microbial activity and transport. In *Soil Microbiology, Ecology, and Biochemistry 3rd Edition;* Paul, E.A., Ed.; Elsevier Academic Press: Burlington, MA. 2007; 163–191.
7. Lavelle, P.; Spain, A.V. *Soil Ecology;* Kluwer Academic Publishers: Dordrecht, the Netherlands, 2001.
8. Clarholm, M. Protozoan grazing of bacteria in soil impact and importance. Microb. Ecol. **1981**, *7*, 343–350.
9. Bouwman, L.A.; Zwart, K.B. The ecology of bacterivorous protozoans and nematodes in arable soil. Agric. Ecosyst. Environ. **1994**, *51*, 145–160.
10. Görres, J.H.; Savin, M.C.; Neher, D.A.; Weicht, T.H.; Amador. J.A. Grazing in a porous environment: 1. The effect of pore structure on C and N mineralization. Plant Soil **1999**, *212*, 75–83.
11. Lavelle, P. Faunal activities and soil processes: Adaptive strategies that determine ecosystem function. Adv. Ecol. Res. **1997**, *27*, 95–132.
12. Huhta, V.; Persson, T.; Setälä, H. Functional implications of soil fauna diversity in boreal forests. Appl. Soil Ecol. **1998**, *10*, 277–288.
13. Curry, J.P. Factors affecting earthworm abundances in soils. In *Earthworm Ecology;* Edwards, C.A. Ed; CRC Press: Boca Raton, 2000; 37–64.
14. Reichman, O.J.; Seabloom, E.W. The roles of pocket gophers as subterranean ecosystem engineers. Trends Ecol. Evol. **2002**, *17*, 44–49.

15. Lavelle, P.; Barois, I.; Martion, A.; Zaidi, Z.; Schaefer, R. Management of earthworm populations in agro-ecosystems: A possible way to maintain soil quality? In *Ecology of Arable Land;* Clarholm, M.; Bergstrom, L. Eds.; Kluwer Academic Publishers: Dordrecht, the Netherlands, 1989; 109–122.
16. Brown, G.G.; Barois, I.; Lavelle, P. Regulations of soil organic matter dynamics and microbial activity in the drilosphere and the role of interactions with other edaphic functional domains. Eur. J. Soil Biol. **2000**, *36,* 177–198.
17. Jones, C.G.; Lawton, J.H; Shachak, M. Organisms as ecosystem engineers. Oikos **1994,** *69,* 373–386.
18. Görres, J.H.; Savin, M.C.; Amador, J.A. Soil micropore structure and carbon mineralization in burrows and casts of anecic earthworms (*Lumbricus terrestris*). Soil Biol. Bio-chem. **2001**, *33,* 1881–1887.
19. Ekschmitt, K. Population assessments of soil fauna: General criteria for the planning of sampling schemes. Appl. Soil Ecol. **1998**, *9,* 439–445.
20. Scheu, S.; Falca, M. The soil food web of two beech forests (*Fagus sylvatica*) of contrasting humus type: stable isotope analysis of a macro-and a mesofauna-dominated community. Oecologia **2000**, *123,* 285–286.
21. Salamon, J.A.; Alphei, J.; Ruf, A.; Schaefer, M.; Scheu, S.; Schnieder, K.; Sührig, A.; Maraun, M. Transitory dynamic effects in the soil invertebrate community in a temperate deciduous forest: Effects of resource quality. Soil Biol. Bio-chem. **2006**, *38,* 209–221.
22. Hansen, R.A. Effects of habitat complexity and composition on a diverse litter microarthropods assemblage. Ecology **2000**, *81,* 1120–1132.
23. Callaham, Jr., M.A.; Richter, Jr, D.D.; Coleman, D.C.; Hofmockel, M. Long-term land-use effects on soil invertebrate communities in Southern Piedmont soils, USA. Eur. J. Soil Biol. **2006**, *42,* S150–S156.
24. Fierer, N.; Strickland, M.S.; Liptzin, D.; Bradford, M.A.; Cleveland, C.C. Global patterns in below-ground communities. Ecol. Lett. **2009**, *12,* 1238–1249.
25. Wu, T.; Ayres, E.; Bardgett, R.D; Wall, D. H.; Garey, J.R. Molecular study of worldwide distribution and diversity of soil animals. PLOS **2011**, doi: 10.1073/pnas.1103824108.
26. Barbercheck, M.E.; Neher, D.A.; Anas, O.; El-Allaf, S.M.; Weicht, T.R. Response of soil invertebrates to disturbance across three resource regions in North Carolina. Environ. Monit. Assess. **2009**, *152,* 283–298.
27. Milcu, A.; Manning, P. All size classes of soil fauna and litter quality control the acceleration of litter decay in its home environment. Oikos **2011**, *120,* 1366–1370.
28. Blankinship, J.C.; Niklaus, P.A.; Hungate, B.A. A metaanalysis of responses of soil biota to global change. Oecologia **2011**, *165,* 553–565.
29. Cole, L.; Dromph, K.M.; Boaglio, V.; Bardgett, R.D. Effect of density and species richness of soil mesofauna on nutrient mineralisation and plant growth. Biol. Fertil. Soil **2004**, *39,* 337–343.
30. Ingham, R.E.; Trofymow, J.A.; Ingham, E.R.; Coleman, D.C. Interactions of bacteria, fungi, and their nematode grazers: Effects on nutrient cycling and plant growth. Ecol. Monogr. **1985**, *55,* 119–140.
31. Verhoef, H.A.; Brussaard, L. Decomposition and nitrogen mineralization in natural and agro-ecosystems: The contribution of soil animals. Biogeochemistry **1990**, *11,* 175–211.
32. Blouin, M.; Zuily-Fodil, Y.; Pham-Thi, A.-T.; Laffray, D.; Reversat, G.; Pando, A.; Tondoh, J.; Lavelle, P. Below-ground organism activities affect plant aboveground phenotype, inducing plant tolerance to parasites. Ecol. Lett. **2005**, *8,* 202–208.
33. Eisenhauer, N.; Sabais, A.C.W.; Schonert, F.; Scheu, S. Soil arthropods beneficially rather than detrimentally impact plant performance in experimental grasslands systems of different diversity. Soil Biol. Biochem. **2010**, *42,* 1418–1424.
34. Kardol, P.; Wardle, D.A. How understanding aboveground-belowground linkages can assist restoration ecology. Trends Ecol. Evol. **2010**, *25,* 670–679.

35. Chauvet, M.; Titsch, D.; Zaytsev, A.S.; Wolters, V. Changes in soil faunal assemblages during conversion from pure to mixed forest stands. For. Ecol. Manag. **2011**, *262,* 317–324.
36. Birkhofer, K.; Bezemer, T.M.; Bloem, J.; Bonkowski, M.; Christensen, S.; Dubois, D.; Ekelund, F.; FlieBbach, A.; Gunst, L. Hedlund, K.; Mader, P.; Mikola, J.; Robin, C.; Setala, H.; Tatin-Froux, F.; Van der Putten, W.H.; Scheu, S. Long-term organic farming fosters below and aboveground biota: Implications for soil quality, biological control and productivity. Soil Biol. Biochem. **2008**, *40,* 2297–2308.
37. Decaëns, T. Macroecological patterns in soil communities. Global Ecol. Biogeograph. **2010**, *19,* 287–302.
38. Römbke, J.; Sousa, J.P.; Schouten, T.; Riepert, F. Monitoring of soil organisms: A set of standardized field methods proposed by ISO. Eur. J. Soil Biol. **2006**, *42,* S61–S64.
39. Jiménez, J.-J.; Decaëns, T.; Amézquita, E.; Rao, I.; Thomas, R.J. and Lavelle, P. Short-range spatial variability of soil physico-chemical variables related to earthworm clustering in a neotropical gallery forest. Soil Biol. Biochem. **2011**, *43,* 1071–1080.
40. Rossi, J.P.; Huerta, E.; Fragoso, C.; Lavelle, P. Soil properties inside earthworm patches and gaps in a tropical grassland (la Mancha, Veracruz, Mexico). Eur. J. Soil Biol. **2006**, *42,* S284–S288.
41. Nieminen, J.K. Enchytraeid population dynamics: Resource limitation and size-dependent mortality. Ecol. Monit. **2009**, *220,* 1425–1430.
42. Wardle, D.A.; Yeates, G.W.; Watson, R.N.; Nicholson, K.S. The detritus food-web and the diversity of soil fauna as indicators of disturbance regimes in agro-ecosystems. Plant Soil **1995**, *170,* 35–43.
43. Bongers, T. The maturity index: An ecological measure of environmental disturbance based on nematode species composition. Oecologia **1990**, *83,* 14–19.
44. Neher, D.A. Role of nematodes in soil health and their use as indicators. J. Nematol. **2001**, *33,* 161–168.
45. Bohlen, P.J.; Groffman, P.M.; Fahey, T.J.; Fisk, M.C.; Suarez, E.; Pelletier, D.M.; Fahey, R.T. Ecosystem consequences of exotic earthworm invasion of north temperate forests. Ecosystems **2004**, *7,* 1–12.
46. Hendrix, P.F.; Baker, G.H.; Callaham, Jr., M.A.; Damoff, G.A.; Fragoso, C.; Gonzalez, G.; James, S.W.; Lachnicht, S.L.; Winsome, T.; Zou, X. Invasion of exotic earthworms into ecosystems inhabited by native earthworms. Biol. Invasions **2006**, *8,* 1287–1300.

31

Soil: Fertility and Nutrient Management

John L. Havlin
North Carolina State University

Introduction

Mineral nutrients essential to support life are supplied by the soil. As plants are removed from a field or soil sediments are transported off-site, nutrients in the soil are depleted. Increasing demand for food and fiber production with increasing population growth will require increased plant productivity and subsequently soil nutrient supply. Since nearly all of the global land areas suitable for agricultural production are currently under cultivation, future food and fiber demand must be met through increased plant yield per unit of land area. Therefore, to optimize nutrient supply to crops and minimize environmental risk of nutrient use, it is essential to understand nutrient reactions and processes in soils (*soil fertility*), and to efficiently manage inorganic and organic nutrient inputs (*nutrient management*) to ensure adequate soil nutrient supply. Agricultural producers must take advantage of soil and plant management technologies that increase plant productivity, and minimize soil productivity loss through erosion and nutrient depletion.

Soil Fertility

Soil fertility represents the diverse interactions between biological, chemical, and physical properties and processes in soil that influence plant nutrient supply. Understanding these processes and interactions is essential to optimize plant nutrient availability and minimize nutrient losses to the environment. Maximum plant production depends on the growing season environment and the producers' skill to minimize yield-limiting factors using appropriate management practices. Plant productivity can be maximized only when the *most* limiting factor to yield potential is corrected first, the second most limiting factor next, and so on (Figure 31.1). For example, optimum plant growth or yield potential cannot be

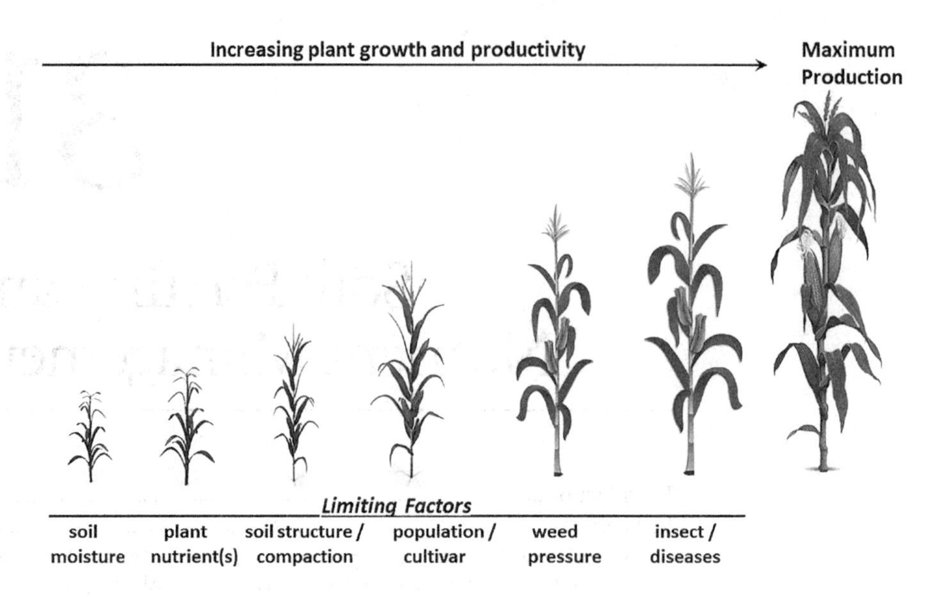

FIGURE 31.1 Principle of *law of the minimum* where achieving maximum plant growth requires identifying the most limiting factors to yield potential and adjusting soil and crop management practices to minimize their effect on yield.

obtained if plant-available water is the *most* limiting factor. Thus, soil water, as a yield-limiting factor, must be addressed before the crop can fully respond to management of any other factor (Figure 31.1). Once plant-available water is no longer a limiting factor to plant growth, some other factor is the most limiting and must then be addressed.

Essential Plant Nutrients

Seventeen essential plant nutrients are involved in metabolic functions such that the plant cannot complete its life cycle without these nutrients (Table 31.1). The most abundant non-mineral nutrients (C, H, and O) are obtained from CO_2 and H_2O and converted into carbohydrates, amino acids, proteins, nucleic acids, and other compounds through photosynthesis. Since 1960, atmospheric CO_2 has increased from ~310 to 410 ppm, increasing growth in some plants. The supply of H_2O rarely limits photosynthesis, but reduces plant growth under limited plant-available H_2O. The remaining 14 macronutrients and micronutrients are classified on their relative abundance in plants. Nutrient content varies greatly between plant species; but in most plants, N concentration is greater than the other nutrients, although some plants require substantial amounts of Ca and K (Table 31.1). Plants also absorb many nonessential elements. For example, Al^{3+} accumulates in plants when soils contain relatively large amounts of soluble Al (low pH soil). In some severely acid soils, Al toxicity can reduce plant growth and yield. In high pH soils, Na^+ can be elevated in soil solution and/or on the exchange complex, increasing Na^+ in the plant, which can be detrimental to plant growth.

Many soil, climate, and management factors influence plant nutrient availability. When nutrient concentration is *deficient* enough to severely reduce plant growth, distinct visual deficiency symptoms appear (Figure 31.2). Although rare, extreme deficiencies can result in plant death. Visual symptoms may not appear when a nutrient is marginally deficient, but plant yield may be reduced. The *critical nutrient deficiency range* represents the nutrient concentration below which plant growth is reduced. Critical nutrient deficiency ranges vary among plants but always represent the transition between nutrient deficiency and sufficiency (Figure 31.2). Plant nutrient *sufficiency* represents the concentration range where increasing nutrient supply does not increase plant growth, but increases plant nutrient

TABLE 31.1 Essential Plant Nutrients and Their Relative and Average Concentration in Plants

	Nutrient		Concentration[a]	
Classification	Name	Symbol	Relative	Average
Macronutrients	Hydrogen	H	60,000,000	6%
	Carbon	C	40,000,000	45%
	Oxygen	O	30,000,000	45%
	Nitrogen	N	1,000,000	1.5%
	Potassium	K	250,000	1.0%
	Calcium	Ca	125,000	0.5%
	Magnesium	Mg	80,000	0.2%
	Phosphorus	P	60,000	0.2%
	Sulfur	S	30,000	0.2%
Micronutrients	Chloride	Cl	3,000	100 ppm
	Iron	Fe	2,000	100 ppm
	Boron	B	2,000	20 ppm
	Manganese	Mn	1,000	50 ppm
	Zinc	Zn	300	20 ppm
	Copper	Cu	100	6 ppm
	Nickel	Ni	2	<1 ppm
	Molybdenum	Mo	1	<1 ppm

[a] Concentration expressed on a dry matter weight basis.

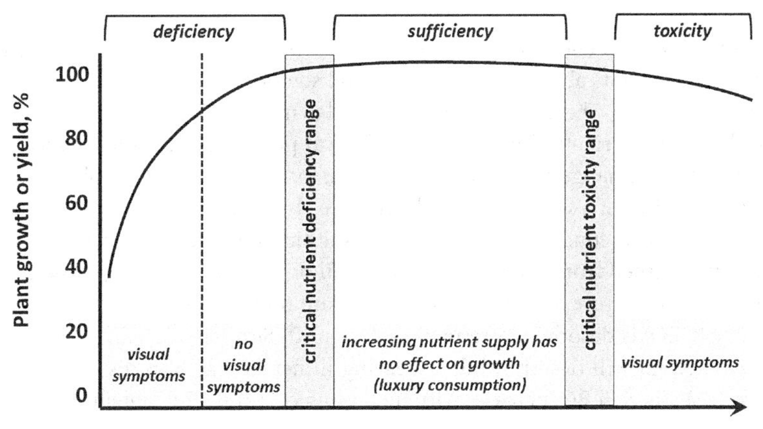

FIGURE 31.2 Relationship between plant nutrient concentration and plant growth. The *critical nutrient deficiency range* represents the nutrient concentration below which nutrients should be added; however, nutrients added beyond this level increases plant nutrient concentration without a response in plant growth or yield.

concentration. Similarly, *critical nutrient toxicity range* represents the nutrient concentration above which plant growth is reduced.

Soil Processes Supplying Plant-Available Nutrients in Soil

Plant nutrients absorbed from soil solution are exported in harvested plant materials or returned to soil in plant or animal residues. These nutrients are subject to the same reactions as nutrients are added as fertilizers and organic wastes. Although specific cycling processes vary between nutrients, understanding

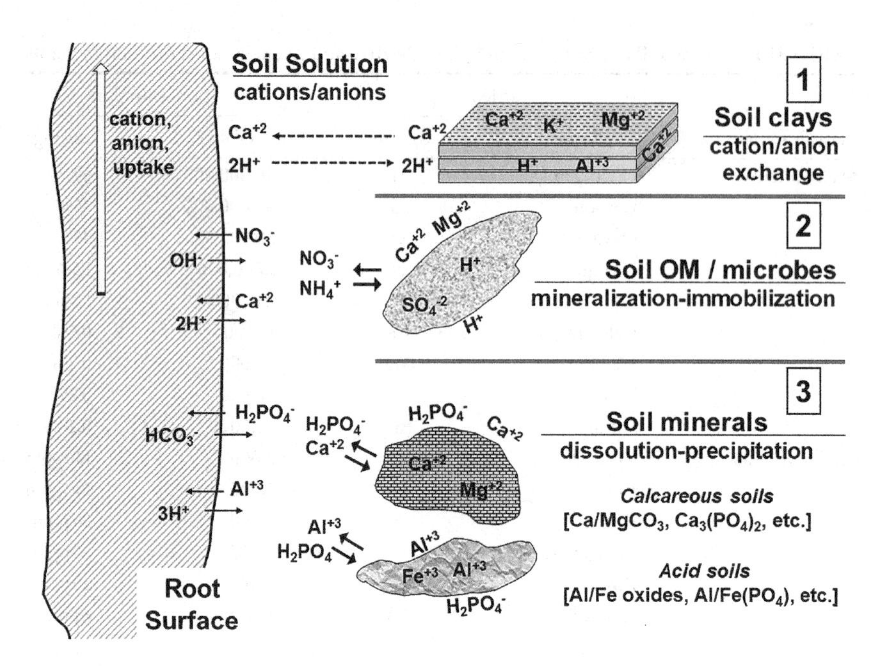

FIGURE 31.3 Primary processes of buffering nutrients in soil solution and absorbed by plant roots. Clay minerals (cation/anion exchange),[1] OM (mineralization),[2] and mineral compounds (dissolution)[3–5] resupply nutrients to soil solution as they are removed from solution by plant uptake.

nutrient transformations in the soil–plant–atmosphere system is essential for maximizing plant recovery of applied nutrients. As plants absorb nutrients from soil solution, reactions occur to resupply or *buffer* the soil solution (Figure 31.3). These reactions are essential to nutrient availability; however, the dominant reaction varies between nutrients. For example, biological processes are more important for N and S availability, and surface exchange reactions are important for Ca, Mg, and K supply to roots.

The buffer capacity (BC) of a soil represents the relative ability of the soil to resupply the soil solution with nutrients. As nutrient concentration in the soil solution decreases by plant uptake, nutrients in soil solution are replenished from exchange surfaces, soil organic matter (OM) mineralization, or dissolution of soil minerals (Figure 31.3). For example, as plant roots *absorb* K^+, adsorbed K^+ on the *Cation exchange capacity* (CEC) is desorbed to resupply solution K^+. Similarly, $H_2PO_4^-$ adsorbed to the *anion exchange capacity* (AEC) will desorb, while P-bearing minerals may also dissolve to supply solution $H_2PO_4^-$ (Figure 31.3). The soil BC increases with increasing clay and OM content.

Cation and Anion Exchange

Exchange of cations and anions on surfaces of clay minerals and OM is one of the most important soil chemical properties and greatly influences nutrient availability (Figure 31.3). Ions adsorbed on mineral and organic surfaces are reversibly exchanged with other ions in the solution. CEC represents the quantity of negative (−) charge available to attract cations in solution, and AEC represents positive (+) charge attracting anions in solution. In most soils, CEC > AEC.

The source of (+) or (−) surface charges is comprised of permanent charge unaffected by solution pH, and pH-dependent charge located on clay mineral edges and soil OM (Figure 31.4). At low pH, more (+) charge exists due to higher H^+ on mineral edges and OM; as pH increases, H^+ concentration decreases, which increases (−) edge or surface charge. At pH > 7, H^+ on edges are neutralized, which maximizes (−) pH-dependent charge (Figure 31.4). Soils with greater clay and OM contents have a higher CEC than sandy, low OM soils (Table 31.2).

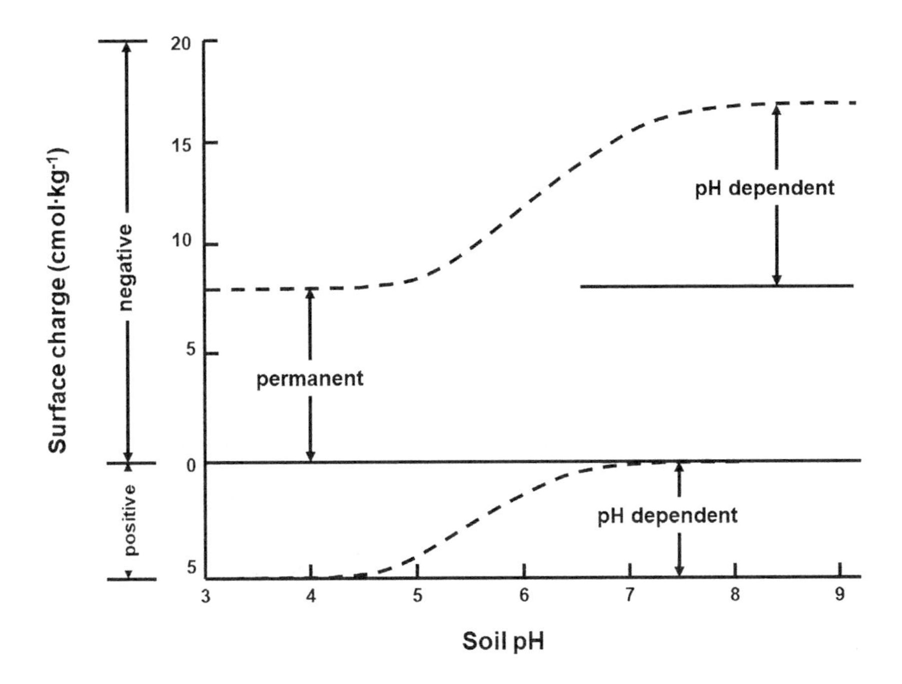

FIGURE 31.4 Effect of soil pH on CEC and AEC.

TABLE 31.2 Typical Range in Cation Exchange Capacity for Moderately Weathered (Mollisol) and Highly Weathered (Ultisol) Soils

Soil Textural Class	Mollisol (cmol kg^{-1})	Ultisol (cmol kg^{-1})
Sands (light colored)	3–5	~1
Sands (dark colored)	10–20	1–3
Loams	10–15	1.5–5
Silt loams	15–25	2–6
Clay and clay loams	20–50	3–5
Organic soils	50–100	20–40

Most adsorbed cations on CEC are essential plant nutrients, except for Al^{3+} in acid soils and Na^+ in neutral and alkaline soils. Cations are adsorbed to CEC with different strengths, which influences the relative ease of desorption. The *lyotropic series,* or relative strength of adsorption, is Al^{3+} $(H^+)>Ca^{2+}>Mg^{2+}>K^+ = NH_4^+>Na^+$. Cations with greater charge are more strongly adsorbed. For cations of similar charge (Ca^{2+} vs Mg^{2+}), adsorption strength is determined by the size of the hydrated cation; greater hydrated cation radii (Mg^{2+}) reduce adsorption because the cation cannot get as close to the exchange surface as cations with smaller hydrated radii (Ca^{2+}).

Base saturation (BS) represents the percentage of CEC occupied by basic cations [Ca^{2+}, Mg^{2+}, K^+, Na^+], which increases with increasing soil pH (Figure 31.5). The availability of Ca^{2+}, Mg^{2+}, and K^+ increases with increasing %BS. For example, 80% BS provides essential nutrients (cations) to plants easier than 40% BS. At pH 5 and 6, most soils have ~50% and 80% BS, respectively (Figure 31.5).

Anions are adsorbed to (+) charged sites on clay mineral and soil OM surfaces, where AEC increases with decreasing soil pH (Figure 31.4). The adsorption strength is $H_2PO_4^- > SO_4^{2-} \gg NO_3^- = Cl^-$. In most soils, $H_2PO_4^-$ is the primary anion adsorbed, although in severely acid soils adsorbed SO_4^{2-} represents a major source of plant-available S.

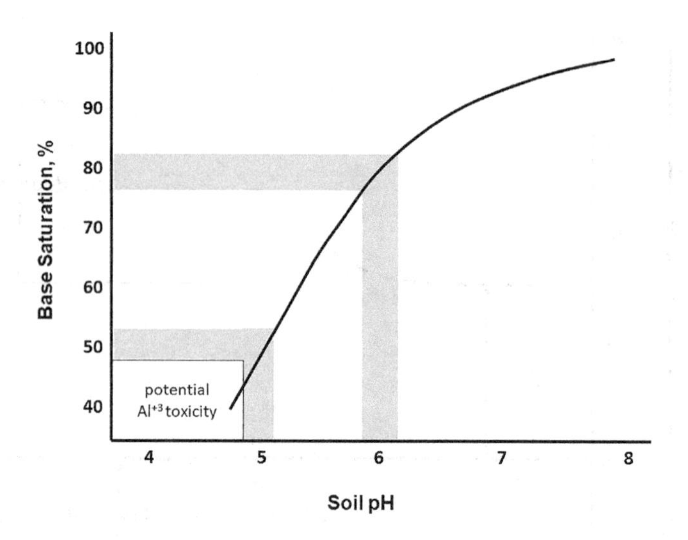

FIGURE 31.5 Influence of soil pH on base saturation % (BS%) in mineral soils.

Soil Acidity and Alkalinity

Acid soils usually occur in regions where annual precipitation is >800 mm. Natural soil acidification is enhanced with increasing rainfall since rainfall pH is ≤ 5.7. Management factors increasing soil acidity include (i) crop removal and/or leaching of cations decreasing %BS, (ii) use of acid-forming fertilizers, and (iii) decomposition of acidic organic residues. Optimum soil pH depends on the crop and ranges between 4.5 and 7.5. In many agricultural crops, productivity can be reduced by Al^{3+} toxicity commonly observed in soils with pH < 5 (low BS%). Neutralizing soil acidity by liming to pH ≥ 5.5 will reduce exchangeable Al^{3+} to <20% of the CEC and will usually prevent Al toxicity-related yield loss (Figure 31.5).

Calcareous soils contain the solid mineral $CaCO_3$, exhibit soil pH ≥ 7.2, and occur in regions of <600 mm annual precipitation. In areas where annual rainfall is >800 mm, the acidifying effect of rain increases depth to $CaCO_3$ to below the root zone. If $CaCO_3$ exists, soil pH is buffered at >7.2, and all the $CaCO_3$ must be dissolved before soil pH could decrease. In most situations, reducing soil pH by neutralizing $CaCO_3$ is costly and unnecessary, although high soil pH in calcareous soils reduces the availability of several micronutrients (e.g., Fe^{3+} and Zn^{2+}).

Excessive salts in alkaline soils can reduce plant growth and yield depending on crop sensitivity. In arid and semiarid regions as water evaporates, salts are deposited near the soil surface to form saline and/or sodic soils, and are characterized by their electrical conductivity (EC_{se}), soil pH, and exchangeable Na% (ESP) (Table 31.3). In saline soils, soluble salts interfere with plant growth, although salt tolerance varies with plant species. In sodic soils (ESP > 15%), soil aggregates disperse reducing infiltration and hydraulic conductivity, while excess Na^+ can also create nutritional disorders. Saline-sodic soils exhibit both high salt and Na^+ content.

TABLE 31.3 Classification and Properties of Salt-Affected Soils

Classification	EC_{se} (dS m^{-1})[a]	Soil pH	ESP (%)[a]	Physical Condition
Saline	>4	<8.5	<15	Normal
Sodic	<4	>8.5	>15	Poor
Saline–sodic	>4	<8.5	>15	Normal

[a] EC_{se} represents electrical conductivity of soil water removed from a sample of saturated soil. Increasing salt concentration increases EC_{se}. ESP% (exchangeable Na%) represents % of CEC occupied with Na^+.

Soil Organic Matter

Soil OM represents organic materials in various stages of decay (Figure 31.6). The most important fractions to soil fertility are the active and stable OM comprised of large molecular weight organic compounds collectively referred to as *humus* (e.g., humic and fulvic acids) produced from microbial digestion of plant and animal residues. These fractions (soil humus) are essential for maintaining optimum soil physical conditions (aggregation, infiltration, aeration, etc.) and nutrient cycling important for plant growth. Fresh plant and animal residues added to soil undergo rapid decomposition depending on C:N ratio of the residue. Low C:N residues (e.g., legumes) degrade rapidly compared to high C:N residues (non-legumes) because of greater N content and availability to microbes. The microbial biomass represents soil microorganisms responsible for mineralization and immobilization processes that influence plant-available N, plant-available P, plant-available S, and other nutrients. Residue decomposition and the extent of soil humus formation depend on climate, soil type, and soil and crop management practices. Soil OM affects various soil processes and properties that influence nutrient availability (Table 31.4). For example, increasing soil OM increases soil aggregation and decreases bulk density, which improves water infiltration, air exchange, root proliferation, and crop productivity. Increasing crop yield results in greater above ground and root residues returned to soil, further increasing soil OM. Soil OM levels strongly depend on soil and crop management practices. If management practices increase C input relative to C loss, soil OM slowly increases. Soil and crop management systems should be adopted that sustain or increase soil OM, which will enhance soil health and plant productivity. Nutrient management practices that enhance crop productivity will increase the quantity of residue returned and ultimately soil OM content.

Mineral Solubility in Soils

Mineral solubility refers to the solution ion concentration in equilibrium with solid minerals. There are many soil minerals influencing nutrient concentrations in soil solution. For example, in acid soils, $FePO_4 \cdot H_2O$ (strengite) influences P availability by

$$FePO_4 \cdot 2H_2O + H_2O \leftrightarrows H_2PO_4^- + H^+ + Fe(OH)_3$$

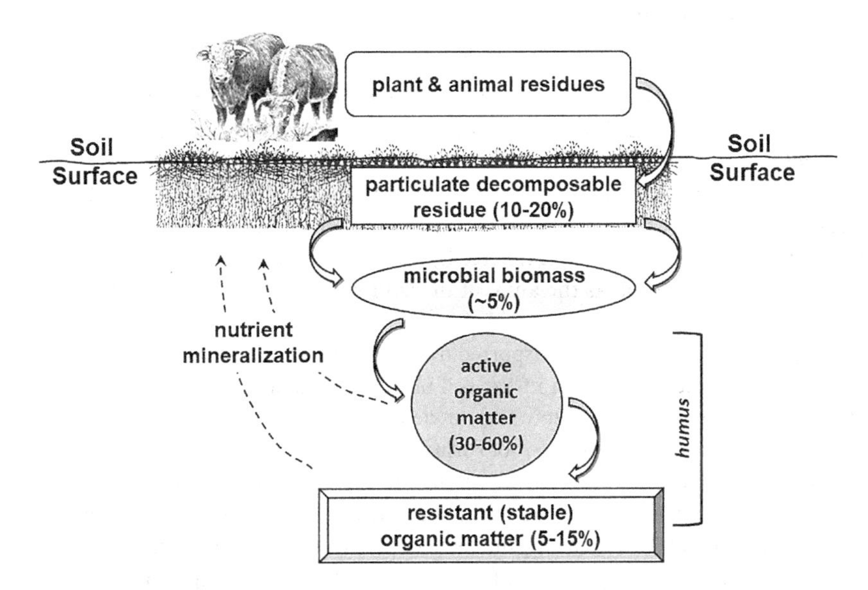

FIGURE 31.6 Pathway of degradation of plant and animal residues to produce soil humus. Mineralization of organic compounds in soil produces plant-available nutrients.

TABLE 31.4 Characteristics of Soil OM and Associated Effects on Soil and Plants

Property	Effect on Soil	Effect on Plant
Soil color	Imparts dark color	Soils to warm faster; > surface soil temperature enhances germination and seedling growth (depending on surface residue cover)
H_2O holding capacity	Holds ~20 times weight in H_2O	> H_2O holding capacity, > plant-available H_2O especially in sandy soils
OM-clay interaction	Cements clay particles into aggregates	Improves soil structure and porosity, enhances gas exchange, infiltration, root growth
Ion exchange/buffering	Increases CEC and AEC 20%–70%	> nutrient retention and availability
Mineralization/ immobilization	Increases nutrient cycling	Increases nutrient availability, retains/ conserves nutrients
Chelation	OM forms stable metal (Fe^{+3}, Zn^{+2}, etc.) complexes	Increases micronutrient availability
Solubility	Insoluble humus–clay complexes, many soluble low MW organic compounds	OM-bound nutrients less likely to leach

AEC, anion exchange capacity; CEC, cation exchange capacity; OM, organic matter.

The K_{sp} for this reaction is

$$K_{sp} = \left(H_2PO_4^-\right)\left(H^+\right)$$

As $H_2PO_4^-$ decreases with P uptake, strengite dissolves to resupply or maintain solution $H_2PO_4^-$ concentration. The reaction also shows that as H^+ increases (decreasing pH), $H_2PO_4^-$decreases. Therefore, specific soil P minerals and the solution P concentration supported by these minerals are dependent on soil pH. Solubility relationships are particularly important for plant availability of P and many micronutrients.

Nutrient Transport from Soil to Roots

Nutrient absorption by plant roots requires contact between the nutrient ion and the root surface (Figure 31.2). Nutrients reach the root surface by *root interception*, *mass flow*, and *diffusion*. Although each is important, mass flow and diffusion dominate nutrient transport (Table 31.5).

Root interception represents ion exchange through physical contact between the root and exchange surfaces. Ions adsorbed to root surfaces (e.g. H^+) may exchange with ions adsorbed to mineral and OM surfaces through direct contact of both surfaces. As roots and associated mycorrhiza develop, more ions adsorbed to soil surfaces are contacted by increasing root volume. The quantity of nutrients in direct contact with plant roots is the amount in a soil volume equal to the root volume (~2% soil volume).

Mass flow represents solution ions transported to the root by H_2O transpired by the plant, evaporation at the soil surface, and percolation in the soil profile. Nutrients transported to the root by mass flow is determined by the rate of H_2O flow and nutrient concentration in the solution. Decreasing soil moisture decreases water transport to the root surface.

Diffusion occurs when an ion moves from an area of high to low concentration. As roots absorb nutrients from soil solution, concentration at the root surface is lower than that in solution not influenced by the root. Therefore, increasing nutrient concentration gradient increases diffusion rate toward the root. A high plant requirement for a nutrient results in a large concentration gradient, favoring a high diffusion rate. Most solution P, K, and micronutrients move to the root by diffusion (Table 31.5).

TABLE 31.5 Significance of Root Interception, Mass Flow, and Diffusion in Nutrient Transport to Roots

Nutrient	Nutrients Required (kg ha^{-1}) for 10 t ha^{-1} of Corn	Percentage Supplied By:		
		Root Interception	Mass Flow	Diffusion
N	200	1	99	0
P	40	2	4	94
K	180	2	20	78
Ca	45	120	440	0
Mg	50	27	280	0
S	22	4	94	2
Cu	0.11	8	400	0
Zn	0.36	25	30	45
B	0.22	8	350	0
Fe	2.2	8	40	52
Mn	9.3	25	130	0
Mo	0.01	8	200	0

For some nutrients, the quantity transported can exceed that required by the crop.

Note: Contribution of diffusion was estimated by the difference between total nutrient needs and amounts supplied by interception and mass flow. If interception + mass flow is >100%, then diffusion = 0.

The importance of diffusion and mass flow in supplying ions to the root surface depends on the ability of the solid phase of the soil to replenish or buffer the soil solution. Ion concentrations are influenced by the types of clay minerals and OM and the distribution of cations and anions on CEC or AEC.

Mass flow and diffusion processes are also important in nutrient management. Soils that exhibit low diffusion rates because of high BC, low soil moisture, or high clay content may require application of immobile nutrients near roots to maximize nutrient availability and plant uptake.

Nutrient Mobility in Soil

Nutrient mobility in soil also influences ion transport to plant roots, evaluation of nutrient availability to plants, and ultimately nutrient management decisions. Nutrient mobility varies between ions, where NO_3^-, Cl^-, and $H_3BO_3^{\circ}$ are highly soluble and not strongly attracted to exchange sites and readily move throughout the root zone. As a result, mobile nutrients within the root volume are available for transport to the root in percolating and transpirational water (Figure 31.7). The relative mobility of each nutrient will depend on soil pH, temperature, moisture, soil texture, and OM content. For example, Ca^{2+} and Mg^{2+} are dominantly adsorbed to the CEC; however, their solution concentrations are also relatively high and thus are relatively mobile in the soil and transported to root surface by mass flow. Similarly, SO_4^{2-} can be strongly adsorbed to the AEC in acid subsoils; however, high concentrations of adsorbed $H_2PO_4^-$ can limit SO_4^{2-} adsorption which would increase its' mobility.

Immobile nutrients interact with mineral and OM surfaces, are less soluble, and do not readily move throughout the root zone (Figure 31.7). As these nutrients are relatively immobile in soil, plant roots access them from a small soil volume surrounding individual roots. Plants create a small zone or soil volume around the root with very low concentration of these immobile nutrients due to plant uptake, causing a concentration gradient encouraging nutrient diffusion. If the soil has a high BC for an immobile nutrient, then the solution can be replenished and diffusion continues. With a low BC, solution concentration (and diffusion) ultimately decreases, causing a nutrient deficiency.

Understanding nutrient mobility in soils is essential to managing nutrient applications to maximize plant growth and recovery of applied nutrients by the plant. For example, mobile nutrients can be broadcast or band applied with relatively similar results because of their mobility in soil. However, P is generally placed in concentrated bands because it is relatively immobile in soil.

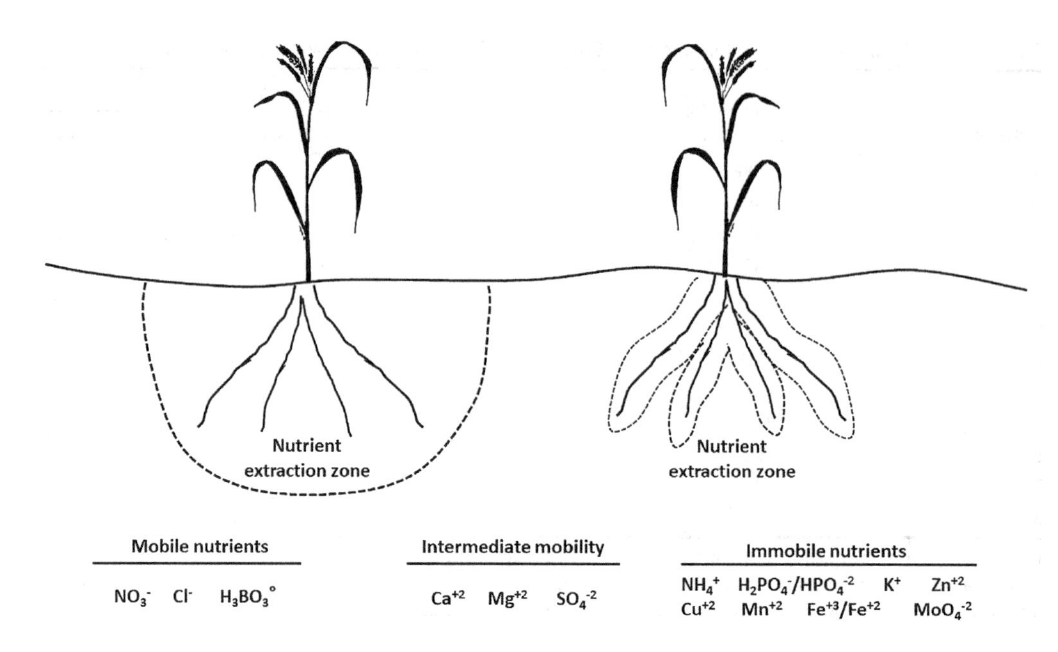

FIGURE 31.7 Relationship between nutrient mobility and nutrient extraction zones. Plants obtain mobile nutrients from the whole soil volume occupied by plants roots. In contrast, plants obtain immobile nutrients from the small soil volume immediately surrounding the plant root.

Nutrient Management

Effective nutrient management programs supply sufficient plant nutrients to sustain maximum plant growth and yield while minimizing off-site transport of applied nutrients. Substantial economic and environmental consequences occur when nutrients limit plant productivity or when nutrients are applied in excess of plant requirement. The essential requirements of an effective nutrient management plan include (i) assessment of crop nutrient requirement, (ii) evaluation of nutrient supplying capacity of the soil, (iii) quantify the optimum nutrient rate, (iv) identify appropriate nutrient source(s), (v) determine most efficient nutrient placement method, and (vi) schedule nutrient applications to meet plant nutrient demand. Management decisions will vary depending on the specific nutrient and crop.

Assessing Plant Nutrient Requirement

The quantity of nutrients required by plants depends on the nutrient (Table 31.1), plant characteristics (specific crop, yield level, variety, or hybrid), growing season environmental factors (moisture, temperature, etc.), soil characteristics (soil properties, soil fertility, and landscape position), and soil and crop management (previous crop and nutrient applications). Management practices that enhance crop productivity (yield) will increase nutrient requirements.

Evaluation of Nutrient Supplying Capacity of the Soil

The *soil testing–nutrient recommendation system* is comprised of four consecutive steps: (i) collect a representative soil sample from the field, (ii) determine the quantity of plant-available nutrient in the soil sample (soil test), (iii) interpret the soil test results, and (iv) estimate the nutrient application rate. The greatest potential for error is in collecting a soil sample that accurately represents the field sampling area. Great care is required in identifying the sampling area from which a composited sample is sent to a laboratory for analysis. Selected parameters used in separating sampling areas within a field include differences in topography, soil type, crop productivity, and past management (Figure 31.8).

Soil testing is essential for determining the relative availability of plant nutrients in soil. Soil tests extract a portion of the total nutrient in soil that is related to (but not equal to) the quantity of nutrients removed by plants. Thus, the soil test level represents an *index* of nutrient availability or relative nutrient sufficiency in a soil (Figure 31.9). Soil test extracting solutions vary with specific nutrients, and selected examples are shown in Table 31.6.

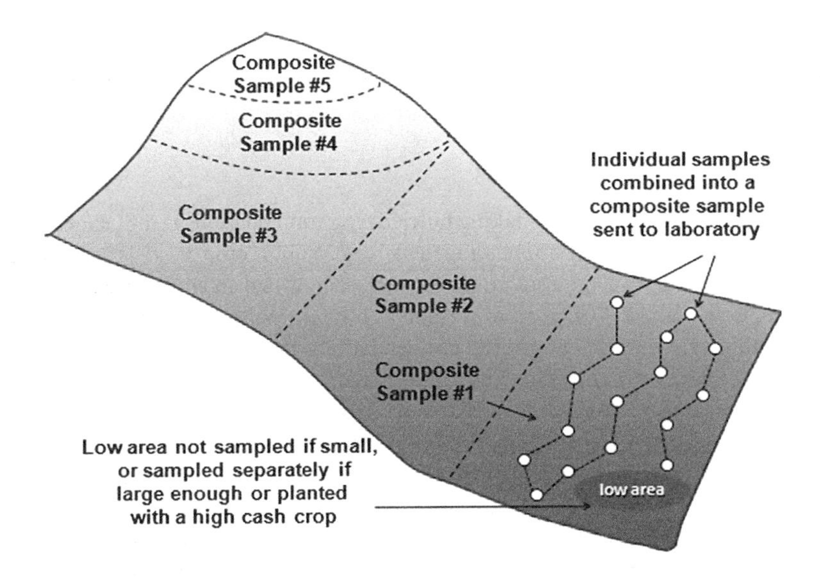

FIGURE 31.8 Composite soil sampling areas determined by variation in elevation within a field. Soil samples are collected within uniform elevation areas, composited into a single sample, and sent to the laboratory for analysis.

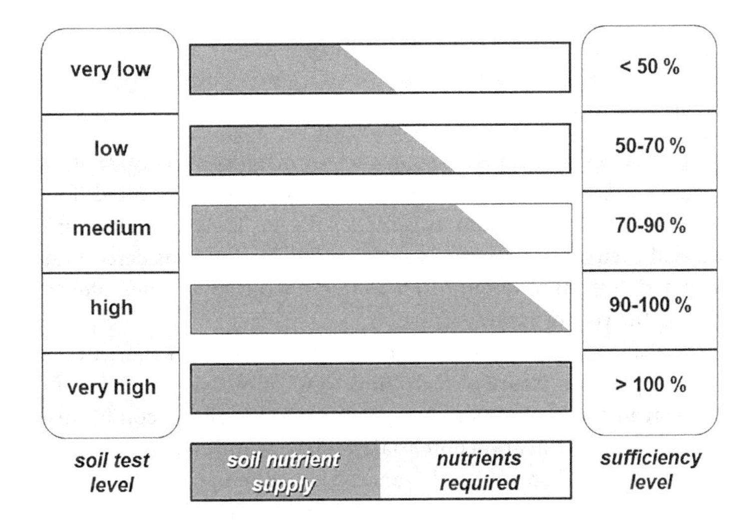

FIGURE 31.9 As soil test levels increase, the capacity of the soil to provide sufficient plant nutrients increases, reducing the quantity of nutrients added to meet crop requirement.

TABLE 31.6 Common Soil Test Solutions Used to Extract Plant Nutrients from Soil

Plant Nutrient	Common Soil Test Extractant	Primary Nutrient Source in Soil
NO_3^-; NH_4^+	KCl, $CaCl_2$	Solution and CEC (NH_4^+)
$H_2PO_4^-$; HPO_4^{-2}	NH_4F-HCl (Bray-P)	Fe/Al-P mineral solubility
	NH_4F-CH_3COOH/HNO_3 (Mehlich-P)	Fe/Al-P mineral solubility
	$NaHCO_3$ (Olsen-P)	Ca-P mineral solubility
K^+	NH_4OAc	CEC
SO_4^{-2}	$Ca(H_2PO_4)$; $CaCl_2$	Solution; AEC
Zn^{+2}, Fe^{+3}, Cu^{+2}, Mn^{+2}	DTPA, EDTA (chelates)	Zn, Fe, Cu, Mn mineral solubility
$H_3BO_3^{\circ}$	Hot water	Solution
Cl^-	Water	Solution

Quantify Optimum Nutrient Rate

Soil test interpretation for purposes of making nutrient recommendations is influenced by the mobility of the nutrient (Figure 31.7). With mobile nutrients (e.g. NO_3^-), crop yield is proportional to total quantity of nutrient present in the root zone, because of minimal interaction with soil constituents (Figure 31.2). For example, a 0.3–0.6 m profile sample depth is important for accurately assessing mobile nutrient availability (Figure 31.7). For N, recommendations are often based on yield goal, where N required to produce each unit of yield is known (e.g. 0.3 kg N kg^{-1} grain). This concept is also evident when additional in-season N is recommended because better-than-average growing conditions increase yield potential above initial estimates provided before planting. Additional factors affecting optimum N rates include estimates of N mineralized during the growing season from previous legume crops and organic N amendments (manures, biosolids, etc.). With immobile nutrients (e.g., $H_2PO_4^-$, K^+, Zn^{2+}), crop yield potential is limited by the quantity of nutrient available at the soil–root interface. Generally, solution concentrations of immobile nutrients are low, and replenishment occurs through exchange, mineralization, and mineral solubility reactions (Figure 31.2). If nutrient uptake demand exceeds the soil's capacity to replenish solution nutrients, then plant growth and yield will be limited. Immobile nutrient recommendations are based on sufficiency levels determined through soil testing (Figures 31.9 and 31.10). Soil tests for immobile nutrients provide an index of nutrient availability that is generally independent of environment.

Identify Nutrient Source(s)

The primary criteria for selecting a nutrient source are availability and cost. For example, if organic nutrient sources (manures, biosolids, etc.) are available they should be utilized if the cost per unit of plant-available nutrients is comparable to inorganic fertilizers. There are important benefits of using organic sources beyond their nutrient value, which should also be considered (Table 31.4). Nutrient content of organic materials must be quantified to determine the application rate needed to meet crop nutrient requirement in the year of application.

In most cases, there are few differences in crop response between inorganic fertilizer sources. Although most plants prefer a mixture of NH_4^+ and NO_3^-, most NH_4^+ or NH_4^+-forming N sources (urea-based products) readily dissolve and nitrify to form NO_3^-. Under conditions of high denitrification, volatilization, and/or nitrification potential, numerous additives or coatings (solid fertilizers) are available to reduce losses of applied NH_4^+ and NO_3^-. When applying nutrients with the seed, nutrient sources with a low salt index are important in reducing salt damage to germinating seeds and seedlings. If fertilizer rates are generally $<$10–15 kg N+K_2O ha^{-1}, the probability of seedling damage is reduced.

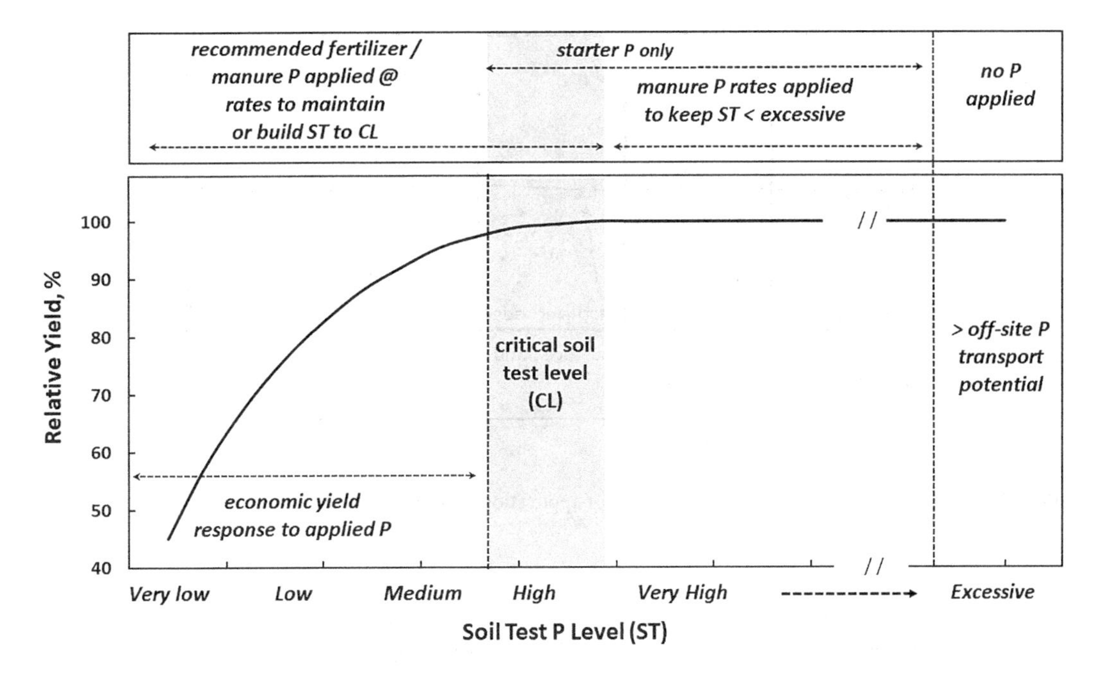

FIGURE 31.10 Example of the relationship between soil test P level and nutrient management options. When soil test P is below the critical level, P should be added at rates that will increase soil P to the critical level. Above the critical level, low P rates are used to maintain soil test P at the critical level. Nutrients should not be applied when soil test levels are considered excessive.

Nutrient Placement Methods

Nutrient placement decisions involve knowledge of crop and soil characteristics, whose interactions determine nutrient availability. Although numerous placement methods have been developed, the primary goal is to ensure optimum nutrient availability from plant emergence to maturity. Vigorous seedling growth (i.e., no early growth stress) is essential to optimize plant growth and yield. Merely applying nutrients does not ensure that they will be absorbed by the plant. Fertilizer placement options generally involve surface or subsurface applications before, at, or after planting. Placement practices depend on the crop and crop rotation, degree of deficiency or soil test level, nutrient mobility in the soil, degree of acceptable soil disturbance, and equipment availability.

Preplant Application

Preplant applications are important to provide adequate nutrient supply during germination and early plant growth stages. Broadcast nutrients are applied uniformly on the soil surface over the entire area before planting and can be incorporated by tillage (Figure 31.11). In no-till, there is limited opportunity for incorporation; thus, broadcast N applications may reduce N recovery by the crop due to enhanced immobilization, denitrification, and volatilization of applied N. Crop recovery of nutrients can be increased by placement below the soil surface where soil moisture might be more favorable for nutrient uptake (Figure 31.11). Surface band applied fertilizers can be effective before planting; however, if not incorporated, dry surface soil conditions can reduce nutrient uptake, especially with immobile nutrients. Surface band N applications can improve N availability compared with broadcast application in some soils and cropping systems due to reduced interaction with surface plant residues.

Animal waste and other organic nutrient sources are commonly preplant-applied, usually as a broadcast or subsurface band placement. Broadcast applications, especially unincorporated, are subject to greater denitrification and volatilization N losses, and surface runoff losses of all nutrients.

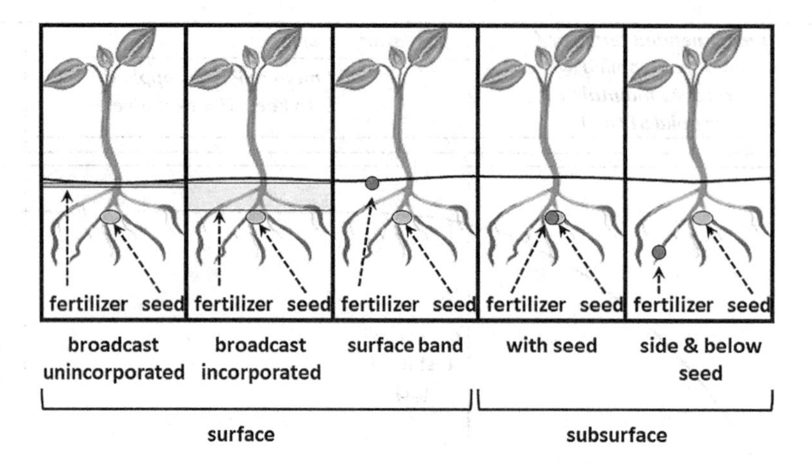

FIGURE 31.11 Plant nutrient placement options for application of nutrients on or below the soil surface.

At Planting Application

Fertilizer placement can occur at numerous locations near the seed (2–8 cm below and/or to side of seed), depending on equipment and crop (Figure 31.11). Fertilizer application with the seed is a subsurface band used to enhance early seedling vigor, especially in cold, wet soils. Usually, low nutrient rates are applied to avoid germination or seedling damage. Fertilizers can be surface band applied at planting directly over or to the side of the row (Figure 31.11).

After Planting Application

Broadcast application after planting is referred to as topdressing. Topdress N applications are common on permanent or close-seeded crops (e.g., turf, small grains, and pastures); however, N immobilization with high surface residue reduces recovery of topdress N. Topdressed P and K are generally not as effective as preplant applications. Sidedress N application is common in row crops (e.g., corn and grain sorghum) as either a surface or a subsurface band. Sidedress applications increase flexibility since applications can be made almost any time equipment is operated without damage to the crop. Sidedress placement is particularly suited for spatially variable N application using remote sensors mounted on the applicator to guide application rate. Subsurface sidedress applications too close to the plant can cause damage by either root pruning or nutrient toxicity. Sidedress application of immobile nutrients (e.g., P and K) is not recommended because most crops need P and K early in the season.

Timing of Nutrient Applications

Nutrients are needed in the greatest quantities during periods of maximum plant growth. Thus, nutrient management plans are designed to ensure adequate nutrient supply before the exponential growth period. In this case, N should be applied preplant or split applied before the stem extension phase. All of the immobile nutrients, including P and K, should be applied before planting. Knowledge of crop-uptake patterns facilitates improved management for maximum productivity and recovery of applied nutrients.

Variable Nutrient Management

Variable or *site-specific* nutrient management can improve nutrient use efficiency by distributing nutrients based on spatial variation in yield potential, soil test levels, and other spatially variable factors that influence soil nutrient availability and crop nutrient demand. Figure 31.12 illustrates how areas within a

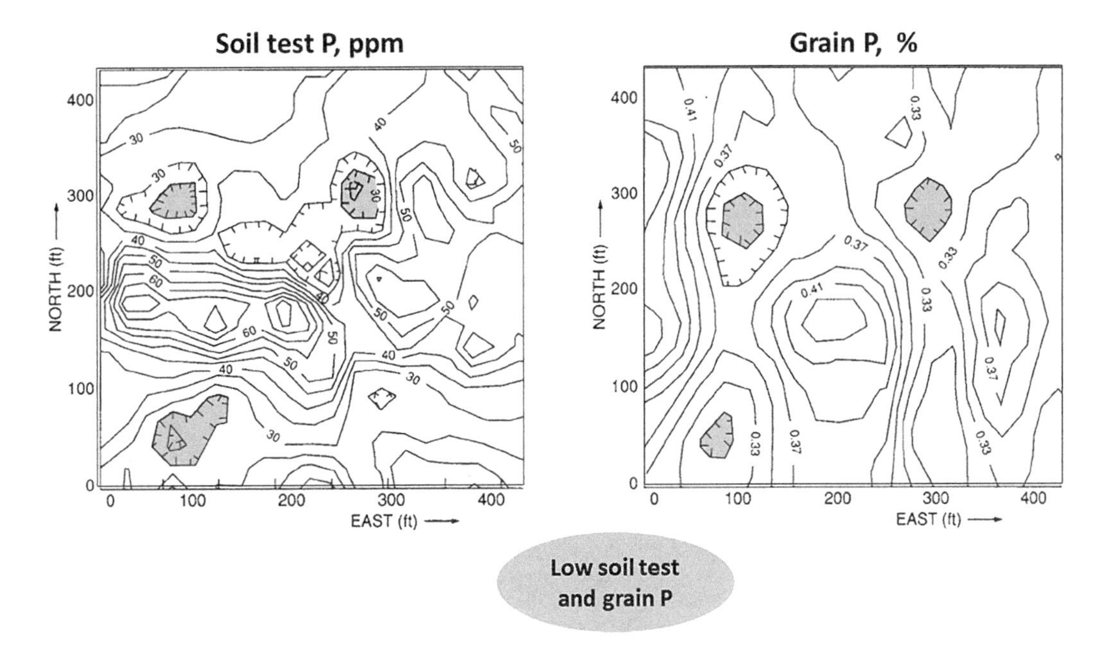

FIGURE 31.12 Spatial distribution of soil test P and wheat grain P over a 3-ha field. Low soil test P correlated with low plant P.

field exhibiting low grain P (P deficiency) are spatially similar to areas of low soil test P. Preseason and in-season assessments of spatial information are two variable nutrient management approaches used to guide nutrient application decisions.

Conclusion

Meeting food needs for a growing population requires increased food production on the same or less agricultural land area. Land managers must adopt economically viable technologies that maintain, enhance, or protect the productive capacity of our soil resources to ensure future food and fiber supplies. While organic nutrient sources are important to meeting the nutritional needs of higher-yielding cropping systems, inorganic fertilizer nutrients will remain the predominant nutrient source. The challenge to the agricultural community is to ensure maximum recovery of applied nutrients, regardless of source, through the use of diverse soil, crop, water, nutrient, and other input management technologies to maximize plant productivity. Accomplishing this will significantly reduce nutrient losses to the environment.

References

1. Barker, A.V.; Pilbeam, D.J. *Handbook of Plant Nutrition*; CRC Press: Boca Raton, FL, 2007.
2. Grant, C.A.; Milkha, M.S. *Integrated Nutrient Management for Sustainable Crop Production*; The Hanworth Press: New York, 2008.
3. Havlin, J.L.; Tisdale, S.L.; Nelson, W.L.; Beaton, J.D. *Soil Fertility and Fertilizers: An Introduction to Nutrient Management*, 8th Edition; Pearson Inc.: Upper Saddle River, NJ, 2014, p. 516.
4. Rengel, Z. Ed. *Handbook of Soil Acidity*; Marcel Dekker: New York, 2003.
5. Sparks, D.L. *Environmental Soil Chemistry*; Academic Press: Cambridge, MA, 2003.

32

Soil: Organic Matter

R. Lal
The Ohio State University

Introduction

The origin and importance of soil organic matter (SOM) were appropriately summarized in 1973 by Allison:[1] "Soil organic matter has over the centuries been considered by many as an elixir of life. Ever since the dawn of history, some eight thousand or more years ago, man has appreciated the fact that dark soils, commonly found in the river valleys and broad level plains, are usually productive soils. He also realized at a very early date that color and productivity are commonly associated with organic matter derived chiefly from decaying plant materials." The SOM comprises the sum of all organic substances in soil, and primarily consists of heterogeneous substances of plant, animal, and microbial origin at various stages of decomposition. Its decomposition products include (1) soluble organic compounds such as sugars, proteins, and other metabolites; (2) amorphous organic compounds such as humic acids, fats, waxes, oils, lignin, and polyuronides; and (3) organomineral complexes, which involve hybrid compounds of organic molecules attached to clay particles through polyvalent cations such as Al^{3+}, Ca^{2+}, and Mg^{2+}.[2]

Two principal components of SOM are highly active humus and relatively inert charcoal. Humus is a dark brown or black amorphous material. Being a colloidal substance, it is characterized by a large surface area, high charge density, and high affinity for clay. Charcoal is widely present in soils of the fire-dependent ecosystems (i.e., savanna, steppe, grasslands). The charcoal, also called black carbon (C), is of pyrogenic origin. It comprises of lightly charred organic matter, soot particles rich in graphite, and polycondensed aromatic groups. Highly fertile soils of the Amazon, *terra preta do Indios,* are enriched by addition of charcoal.

Soil Organic Matter Dynamics

In natural ecosystems, SOM is in dynamic equilibrium with its environment. Its amount in the soil is stabilized when the rate of input equals that of the output. The input of SOM is mainly through residues of plants grown in situ and from substances brought in by alluvial, aeolian, and colluvial processes. The output of SOM is through decomposition or oxidation, erosion, and leaching. The rate of decomposition depends on temperature and moisture regimes, and is more in warmer and wetter than in cooler and drier climates. Soil erosion, by water and wind, preferentially removes SOM because it is lighter (lower density) than mineral fraction, and is concentrated in the vicinity of the soil surface where erosional processes are the most effective. The soluble and colloidal components of SOM are leached into the subsoil and eventually into aquatic ecosystems along with the percolating water. Removal of crop residues for alternative uses, such as traditional or modern biofuels (cellulosic ethanol) and livestock feed, reduces input of biomass and decreases the soil organic C pool. Thus, there is no such thing as a free biofuel from crop residues.

Natural factors affecting the SOM concentration include climate, soil type, vegetation, hydrology, landscape position, and soil biodiversity. Principal climatic parameters are precipitation amount and distribution, temperature and its seasonal and diurnal variation, annual and seasonal water budget, and the duration of the growing season as determined by the degree days and hydrologic balance. Predominant soil properties that affect SOM include texture, clay minerals, pH, ionic composition and concentration, and soil depth. Internal drainage of the soil profile, affecting soil moisture regime, is also important. Landscape affects SOM content through the position, drainage density, and slope shape, and aspect. Slope aspect determines soil temperature, slope position determines moisture regime, and slope shape determines the sedimentation and depositional processes. Analyses of these factors indicate that once SOM has been severely depleted by natural or anthropogenic factors, it is difficult to restore in tropical agroecosystems, coarse-textured soils, drought-prone environments, and nutrient-deficit (N, P, S) conditions. Vegetation, and its residues, influences SOM content through the relative concentration of lignin, cellulose, polyphenols, and other recalcitrant compounds. Root system, its depth distribution and turnover are also important to SOM stock in the profile, especially in the subsoil horizons. Roots contribute more to refractory SOM (relatively stable soil C pool) than the same amount of above-ground residue-derived C.[3] Presence of stabilizing elements (especially N, P, S) is important to converting biomass C into humus.[4] In drier climates, SOM stock is favorably influenced by the hydraulic lift of deep-rooted shrubs.[5]

Assessment of Soil Organic Matter

There is a long history of determination of SOM concentration in the laboratories.[6] Concentration of SOM in the laboratory is determined by either the wet combustion[7] or dry combustion method.[8,9] There are numerous modern techniques of measuring SOM stock in situ,[10] and rapidly and in a cost-effective manner.[11,12] There are also molecular-level methods for monitoring SOM concentration in relation to climate change,[13] including nuclear magnetic resonance techniques and SOM biomarker techniques[14–19] Several practical indicators of changes in SOM concentration have been developed.[20,21] While concentration of SOM in the root zone (measured as % or g/kg) has been the unit of assessment for agricultural and forestry land uses, the assessment at regional, watershed, national, and global scales is done in units of Mg/ha, Tg, or Pg.

Management of Soil Organic Matter

The quantity and quality of SOM affect numerous soil processes and ecosystem services of significance at local, regional, and global scales (Figure 32.1). The SOM concentration of 2–3% dry soil weight is needed to improving the use of efficiency of inputs and enhancing agronomic production.[22] It is essential to advance global food security[23] and create a positive C budget by reducing losses and increasing inputs of biomass-C. Maintenance of SOM at above the critical level enhances physical, chemical, and biological processes, which are strong determinants of the overall soil quality (Table 32.1, Figure 32.2). The processes are enhanced by adoption of soil-specific practices. Some recommended management practices in agroecosystems include conservation tillage, mulch farming, cover cropping, integrated nutrient management including use of manures, biofertilizers, and biochar; agroforestry; controlled grazing; and improved pasture. Restoration of wetlands, erosion control, and afforestation of degraded and desertified lands are important options of enhancing the SOM concentration and pool. In 1938, William Albrecht[24] emphasized that "soil organic matter is one of our most precious resources; its

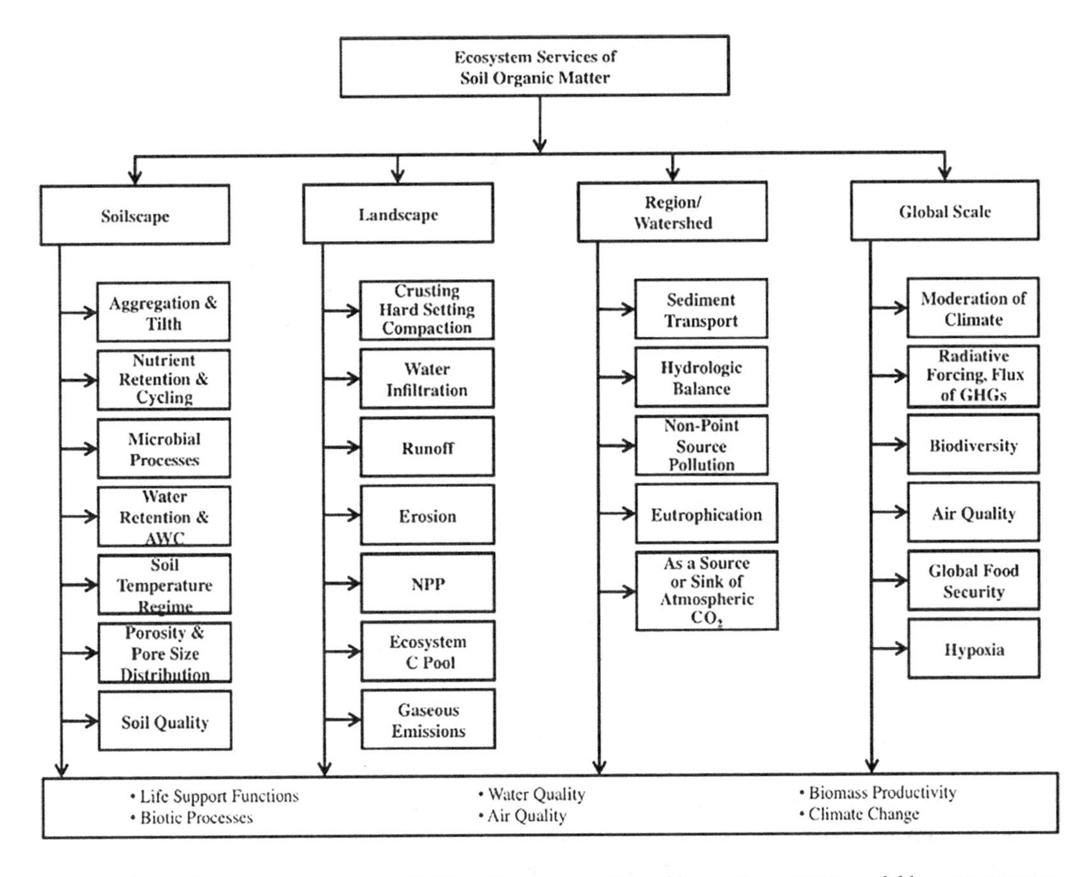

FIGURE 32.1 Ecosystem services provided by soil organic matter. Abbreviations: AWC, available water capacity; GHGs, greenhouse gases (e.g., CO_2, CH_4, N_2O); NPP, net primary productivity.

TABLE 32.1 Soil Organic Matter as a Determinant of Specific Components of Soil Quality

Soil Physical Quality	Soil Chemical Quality	Soil Biological Quality
1. Soil structure and tilth	1. Soil reaction (pH)	1. Food for soil organisms
2. Porosity and port size distribution; biopores	2. Cation exchange capacity	2. Habitat for biota
3. Plant's available water holding capacity	3. Elemental toxicity (Al3+, Fe3+, Mn3+)	3. Microbial biomass carbon
4. Soil erodibility	4. Electrical conductance, salt concentration	4. Products of decomposition
5. Infiltration, percolation, run off, and leaching	5. Plant available nutrients (i.e., N, P, K, Cu, Zn)	5. Soil respiration rate
6. Crusting, compaction, and hard setting	6. Nutrient/elemental cycling	6. Methanogenesis
7. Soil temperature regime and thermal conductivity	7. Chemical transformations	7. Nitrification/denitrification
8. Gaseous diffusion (diffusion coefficient)	8. Nutrient diffusion	8. Soil C sink capacity
9. Surface area	9. Organomineral complexes	9. Gaseous composition of soil air
10. Soil strength	10. Redox potential	10. Activity and species diversity of soil flora and fauna

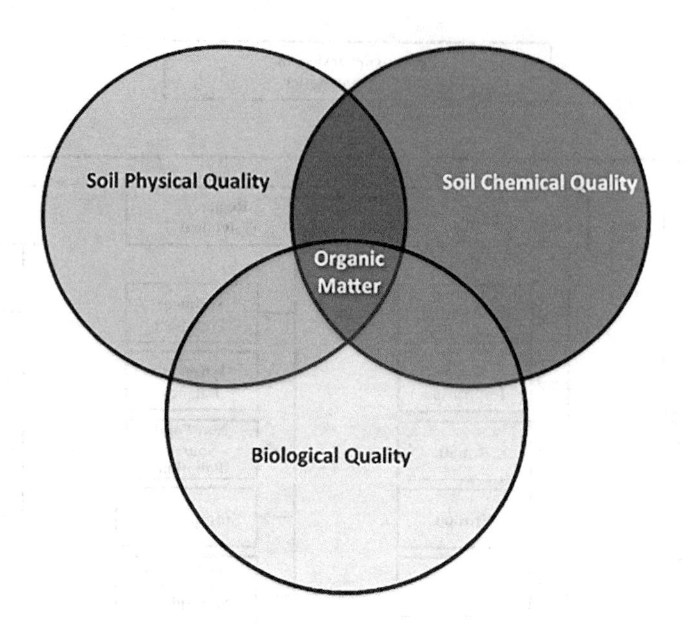

FIGURE 32.2 Interactive physical, chemical, and biological processes moderated by soil organic matter as a strong determinant of soil quality.

unwise exploitation has been devastating; and it must be given its proper rank in any conservation policy as one of the major factors affecting the level of crop production in the future." It was in a similar context, in view of the rising energy costs in late 1970s and the suggestion about the use of crop residues for biofuels, when Hans Jenny[25] stated in 1980 that "I am arguing against indiscriminate conversion of biomass and organic wastes to fuels. The humus capital, which is substantial, deserves being maintained because good soils are a national asset."

Conclusion

Quantity and quality of SOM and its maintenance above the threshold/critical level have been the foundation of past civilizations, and are the bases on which the future of mankind depends. Its concentration in the root zone determines numerous ecosystem services including food security, climate change mitigation and adaptation, biodiversity, and water resources. It is the global public good, whose importance to human well-being and ecosystem functions can never be overstated. While strengthening and promoting long-term basic and applied research on processes and practices of its management and assessment at different scales, policy instruments must be in place at national and international levels for its enhancement and sustainable management.

References

1. Allison, F.E. *Soil Organic Matter and Its Role in Crop Production*; Amsterdam: Elsevier, 1973.
2. Schnitzer, M. Soil organic matter – the next 75 years. Soil Sci. **1991**, *151*, 41–58.
3. Kätterer, T.; Bolinder, M.A; Andrén, O.; Kirchmann, H.; Menichetti, L. Roots contribute more to refractory soil organic matter than above-ground crop residues, as revealed by a long-term field experiment. Agr. Ecosyst. Environ. **2011**, *141*, 184–192.
4. Kirkby, C.A.; Kirkegaard, J.A.; Richardson, A.E.; Wade, L.J.; Blanchard, C.; Batten, G. Stable soil organic matter: A comparison of C:N:P:S ratios in Australian and other world soils. Geoderma **2011**, *163*, 197–208.

5. Armas, C.; Kim J.H.; Bleby, T.M.; Jackson, R.B. The effect of hydraulic lift on organic matter decomposition, soil nitrogen cycling, and nitrogen acquisition by a grass species. Oecologia **2012**, *168* (1), 11–22. doi:10.1007/s00442-01102065-2 (accessed August 2011).
6. Manlay, R.J.; Feller, C.; Swift M.J. Historical evolution of soil organic matter concepts and their relationships with the fertility and sustainability of cropping systems. Agric. Ecosyst. Env. **2007**, *119*, 217–233.
7. Allsion, F.E. Wet combustion apparatus and procedure for organic and inorganic carbon in soil. Soil Sci. Soc. Am. Proc. **1960**, *24*, 36–40.
8. Nelson, D.W.; Sommers, L.E. Total carbon, organic carbon and organic matter. In *Methods of Soil Analysis, Part 2. Chemical and Microbial Properties.* Amer. Soc. Agron. No. 9 (Part 2). Agronomy Series. Page, R.L.; Miller, R.H.; Keeney, D.R.; Eds.; ASA/SSSA: Madison WI, 1982; 539–580.
9. Tiesson, H.; Moir, J.O. Total and organic carbon. In *Soil Sampling and Methods of Analysis*; Carter, M.R. Ed; Lewis Publishers: Boca Raton, FL, 1993; 187–199.
10. Wielopolski, L.; Chatterjee, A.; Mitra, S.; Lal, R. In-situ determination of soil carbon pool by inelastic neutron scattering. Geoderma **2011**, *160* (3–4), 394–399, doi:1016/j/ Geoderma.2010.10.009 (accessed August 2011).
11. Chatterjee, A.; Lal, R.; Wielopolski, L.; Martin, M.Z.; Ebinger, M.H. Evaluation of different soil carbon determination methods. Crit. Rev. Plant Sci. **2009**, *28*, 164–178.
12. Lal, R; Cerri, C.; Bernoux, M; Etchevers, J.; Cerri, E. *The Potential of Soils of Latin America to Sequester Carbon and Mitigate Climate Change*; The Hawthorn Press: West Hazelton, PA 2006; 554.
13. Feng, X.; Simpson, M.J. Molecular-level methods for monitoring soil organic matter responses to global climate change. J. Environ. Monit. **2011**, *13*, 1246–1254.
14. Braun, S.; Agren, G.I.; Christensen, B.T.; Jensen, L.S. Measuring and modeling continuous quality distributions of soil organic matter. Biogeosciences **2010**, *7*, 27–41.
15. Parton, W.J.; Schimel, D.S.; Cole, C.V.; Ojima, D.S. Analysis of factors controlling soil organic matter levels in Great Plains grasslands. Soil Sci. Soc. Am. J. **1987**, *51*, 1173–1179.
16. Nye, P.H.; Greenland, D.J. The soil under shifting cultivation. Commonwealth Bureau Soils Tech. Commun. **1960**, *51*, 156.
17. Viaud, V.; Angers, D.A.; Walter, C. Toward landscape-scale modeling of soil organic matter dynamics in agroecosystems. Soil Sci. Soc. Amer. J. **2010**, *74*, 1847–1860.
18. Neill, C. Impacts of crop residue management on soil organic matter stocks: a modeling study. Ecol. Model. **2011**, *222*, 2751–2760.
19. Kutsch, W.L.; Bahn. M.; Heinemeyer, A. *Soil Carbon Dynamics: An Integrated Methodology;* Cambridge University Press: Cambridge, U.K. 2009; 286.
20. Sohi, S.P.; Yates, H.C.; Gaunt, J.L. Testing a practical indicator for changing oil organic matter. Soil Use Manag. **2010**, *26*, 108–117.
21. Plante, A.F.; Fernandez, J.M.; Haddix, M.L.; Steinweg, J.M.; Conant, R.T. Biological, chemical and thermal indices of soil organic matter stability in four grassland soils. Soil Biol. Biochem. **2011**, *43*, 1051–1058.
22. Lal, R. Beyond Copenhagen: Mitigating climate change and achieving food security through soil carbon sequestration. Food Secur. **2010**, 2, 169–177.
23. Magdoff, F.; Van Es, H. *Building Soils for Better Crops: A Sustainable Soil Management;* SARE/ USDA: Beltsville, MD, 2009; 294.
24. Albrecht, W. Loss of soil organic matter and its restoration. In *Soils and Men*; *USDA Yearbook of Agriculture*, U.S. Government Printing Office: Washington, D.C., 1938; 347–360.
25. Jenny, H. Alcohol or humus. Science **1980**, *209*, 444.

FIGURE 1.1 Examples of threats to protected lands: (**a**) Vandalism (destruction of signage); (**b**) illegal dumping of refuse; (**c**) all-terrain vehicle track damage to wetland habitat.
Source: Photo courtesy, The Rhode Island Chapter of The Nature Conservancy.

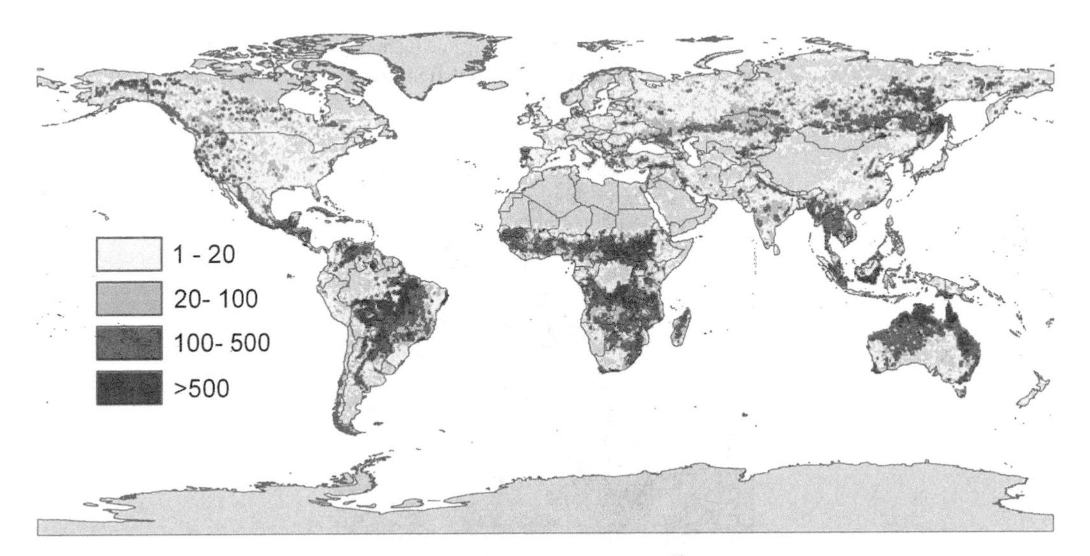

FIGURE 3.1 Spatial distribution of fire occurrence density (the number of active fire counts detected from the satellite using the Along Track Scanning Radiometers (ATSR) instrument between 1996 and 2006 within a 0.5°×0.5° grid cell). (Data adapted from http://due.esrin.esa.int/.)

FIGURE 3.3 Stand-replacement wildfire in a Chinese boreal forest.

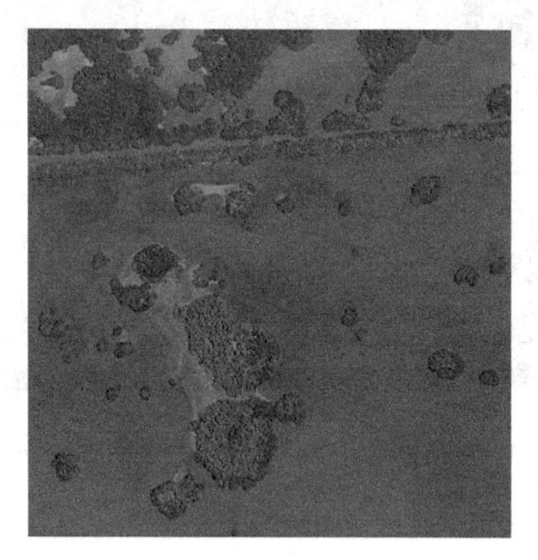

FIGURE 4.2 Example of metapopulation patch dynamics analogy using mangrove islands in Everglades National Park, Florida, USA. These island patches illustrate how focal populations occur within a network of local populations and may interact via migration within a landscape or region.

FIGURE 4.5 Example of intact naturally connected landscape and human-induced highly fragmented and isolated landscape.

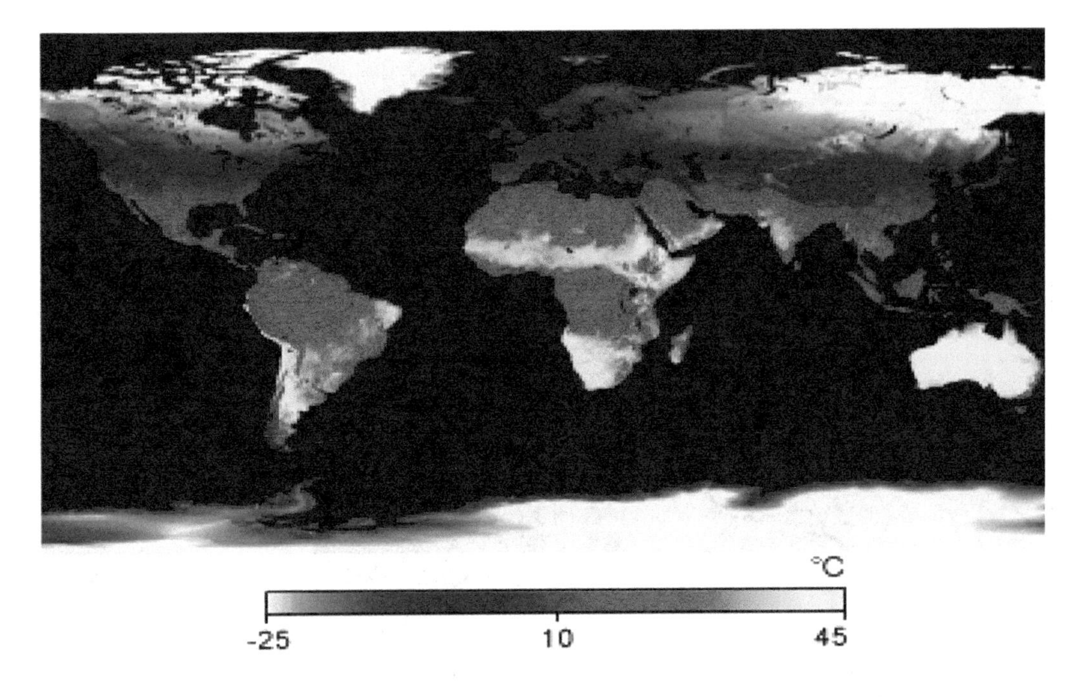

FIGURE 5.3 Average global LST map obtained from the NASA Terra satellite MODIS (Moderate Resolution Imaging Spectroradiometer) sensor.
Source: From http://modis.gsfc.nasa.gov/gallery/individual.php?db_date=2009-04-30.

FIGURE 10.1 Climate change and human-induced disturbances impose uncertainty on the habitat conditions, so as the future of biodiversity. Field photos illustrate critically endangered Siberian cranes (*Leucogeranus leucogeranus*) wintering in Poyang Lake area (2017), China (**a**); endangered Hawaiian monk seals (*Neomonachus schauinslandi*) in Kauai (2018), endemic to the Hawaiian Islands (**b**); and polar bears in the Hudson Bay area (2016), Canada (**c**). (Photos: Yeqiao Wang.)

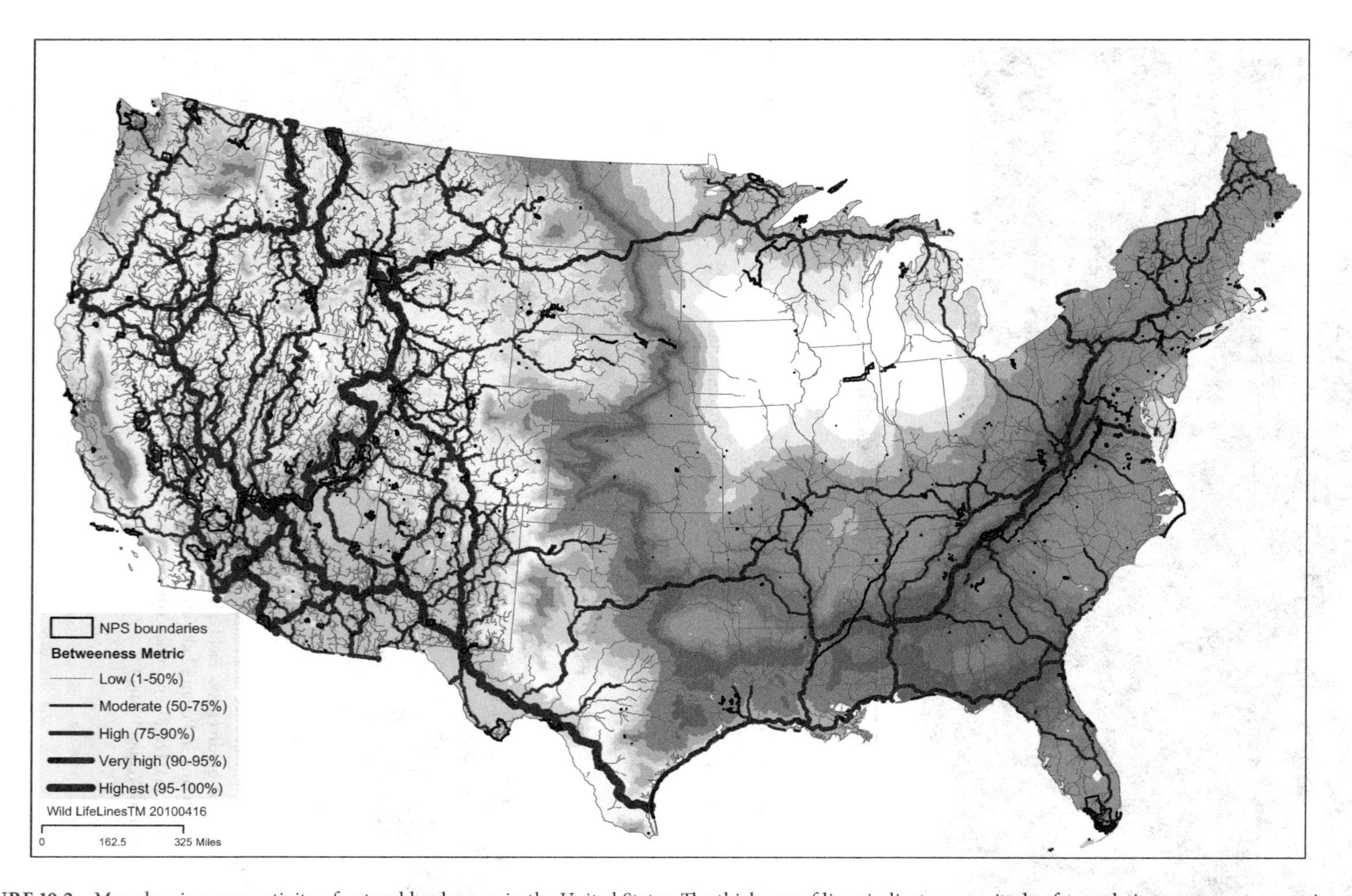

FIGURE 10.2 Map showing connectivity of natural landscapes in the United States. The thickness of lines indicates magnitude of cumulative movement, assuming that animals avoid human-modified areas. The surface underneath the pathways depicts the averaged cost-distance surfaces, or the overall landscape connectivity surface. NPS units are outlined in black. (Gross et al.[14])

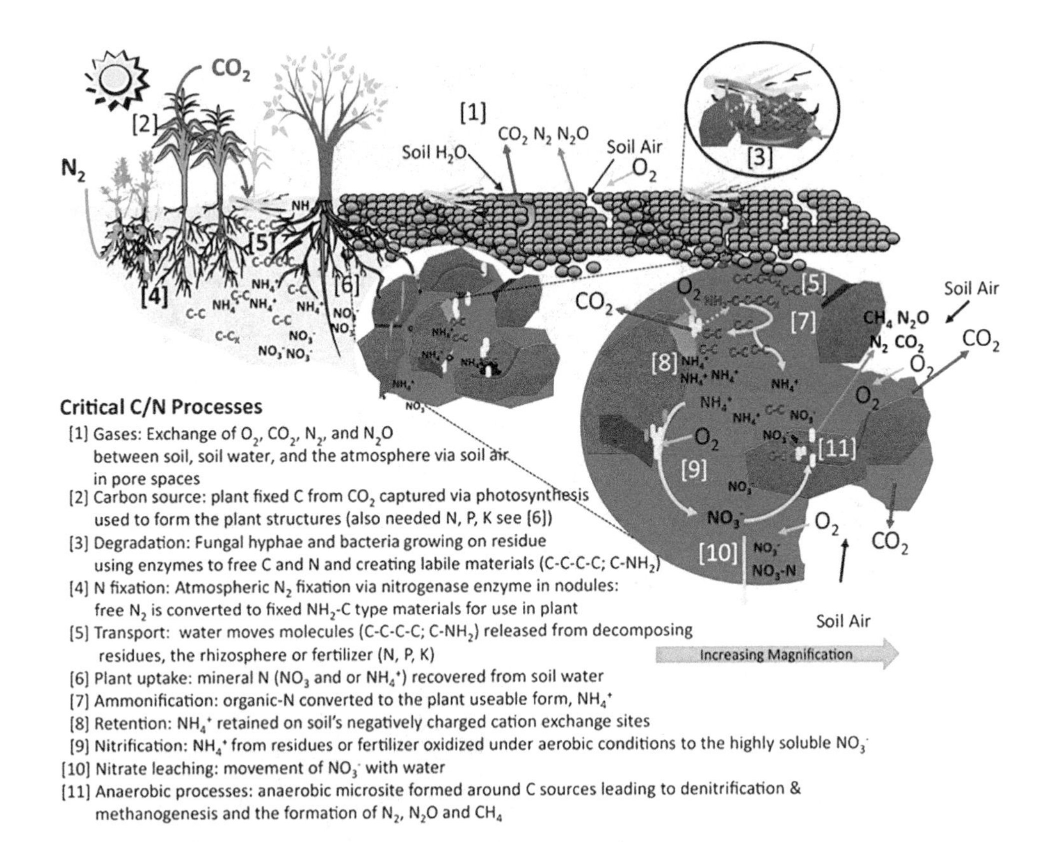

CO2
[2]
N2
[1]
Soil H2O
CO2 N2 N2O
Soil Air
O2
[3]
[5]
[4]
[6]
CO2
[7]
[8]
Soil Air
CH4 N2O
N2 CO2
O2
CO2
O2
[9]
[11]
NO3-
[10]
NO3-N
CO2
Soil Air
Increasing Magnification
Critical C/N Processes
[1] Gases: Exchange of O2, CO2, N2, and N2O between soil, soil water, and the atmosphere via soil air in pore spaces
[2] Carbon source: plant fixed C from CO2 captured via photosynthesis used to form the plant structures (also needed N, P, K see [6])
[3] Degradation: Fungal hyphae and bacteria growing on residue using enzymes to free C and N and creating labile materials (C-C-C-C; C-NH2)
[4] N fixation: Atmospheric N2 fixation via nitrogenase enzyme in nodules: free N2 is converted to fixed NH2-C type materials for use in plant
[5] Transport: water moves molecules (C-C-C-C; C-NH2) released from decomposing residues, the rhizosphere or fertilizer (N, P, K)
[6] Plant uptake: mineral N (NO3 and or NH4+) recovered from soil water
[7] Ammonification: organic-N converted to the plant useable form, NH4+
[8] Retention: NH4+ retained on soil's negatively charged cation exchange sites
[9] Nitrification: NH4+ from residues or fertilizer oxidized under aerobic conditions to the highly soluble NO3-
[10] Nitrate leaching: movement of NO3- with water
[11] Anaerobic processes: anaerobic microsite formed around C sources leading to denitrification & methanogenesis and the formation of N2, N2O and CH4

FIGURE 25.1 Diagram of the major C/N cycle soil processes. Left side shows C/N cycling at the terrestrial scale; right side is magnified to highlight cycling within the soil matrix.

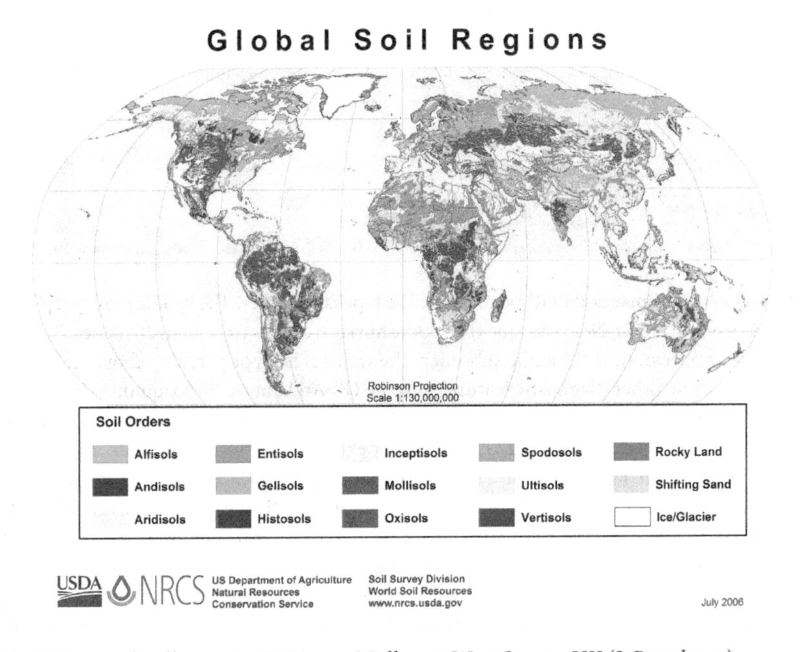

Global Soil Regions
Robinson Projection
Scale 1:130,000,000
Soil Orders
Alfisols
Entisols
Inceptisols
Spodosols
Rocky Land
Andisols
Gelisols
Mollisols
Ultisols
Shifting Sand
Aridisols
Histosols
Oxisols
Vertisols
Ice/Glacier
USDA NRCS
US Department of Agriculture
Natural Resources
Conservation Service
Soil Survey Division
World Soil Resources
www.nrcs.usda.gov
July 2006

FIGURE 34.1 Ephemeral gully, winter 2007, near Midhurst, West Sussex, UK (J. Boardman).

FIGURE 35.1 **(A)** A Typic Epiaquod in Rhode Island. The tape is in inches. The spodic diagnostic subsurface horizon varies in thickness and depth. Near the tape the spodic horizon occurs from 10 to 20 inches. The soil is saturated to the surface for parts of year, thus the aquic suborder. The water is held, or perched, above a restrictive layer (epi) starting at 36″. There are no other diagnostic features (Typic). **(B)** An Umbric Endoaquult in Pennsylvania. The soil has an umbric epipedon, an argillic diagnostic subsurface horizon, and a base saturation <35%. The soil is saturated throughout the profile and to the soil surface (Endoaqu) for extended periods of the year. The small marks on the tape are 10 cm increments. The large marks are in feet. **(C)** Fluvaquentic Eutrudept on a Pennsylvania floodplain. The tape is in cm. Fluvaquentic indicates the soil is saturated during the year for a considerable amount of time above 60 cm and there is buried soil organic carbon (note the buried A horizon between 80 and 90 cm). The soil has a cambic diagnostic subsurface horizon (20 to 80 cm) with a base saturation >60% (Eutr). The climate is humid, thus a Udept. **(D)** Calcic Haplustalf in Texas. The small increments on the tape are 10 cm apart. The larger marks are one-foot apart. The soil has calcic and argillic diagnostic subsurface horizons. The base saturation is greater than 35% and the moisture regime is semiarid (ustic).

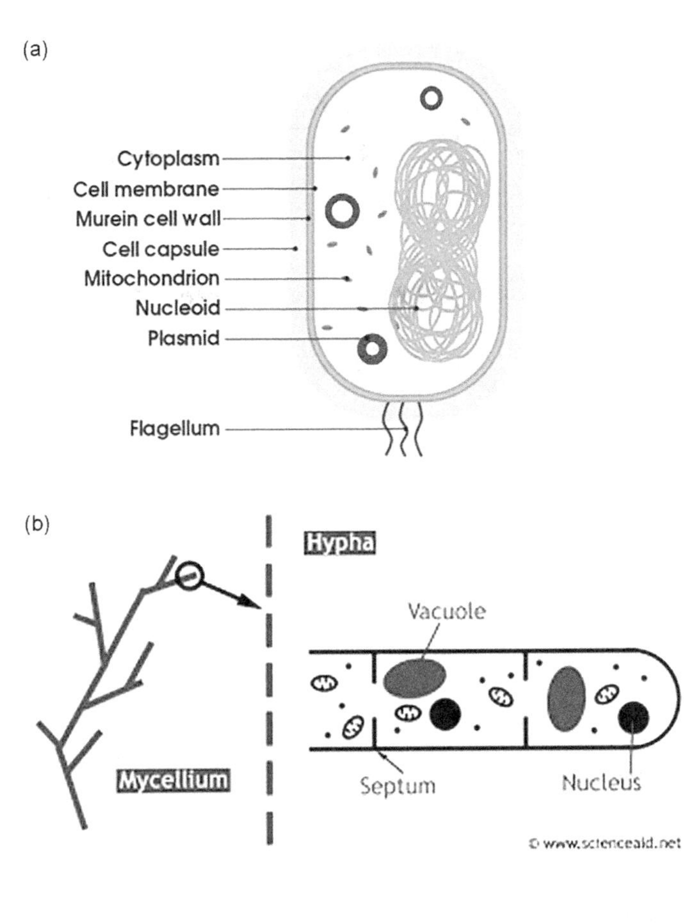

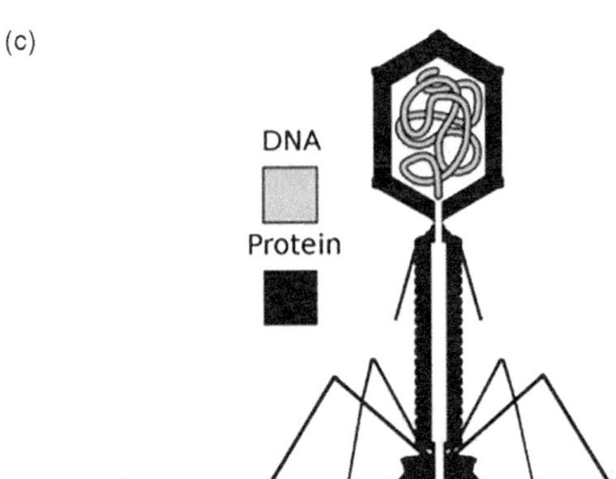

FIGURE 36.1 Morphology of bacteria (**a**), fungal hyphae (**b**), and bacteriophage (**c**). (Picture credit: (a) Image by Domdomegg, distributed under a CC-BY 4.0 license. (b) Image from ScienceAid.net. (c) Image by Mike Jones, distributed under a CC BY-SA 2.5 license.)

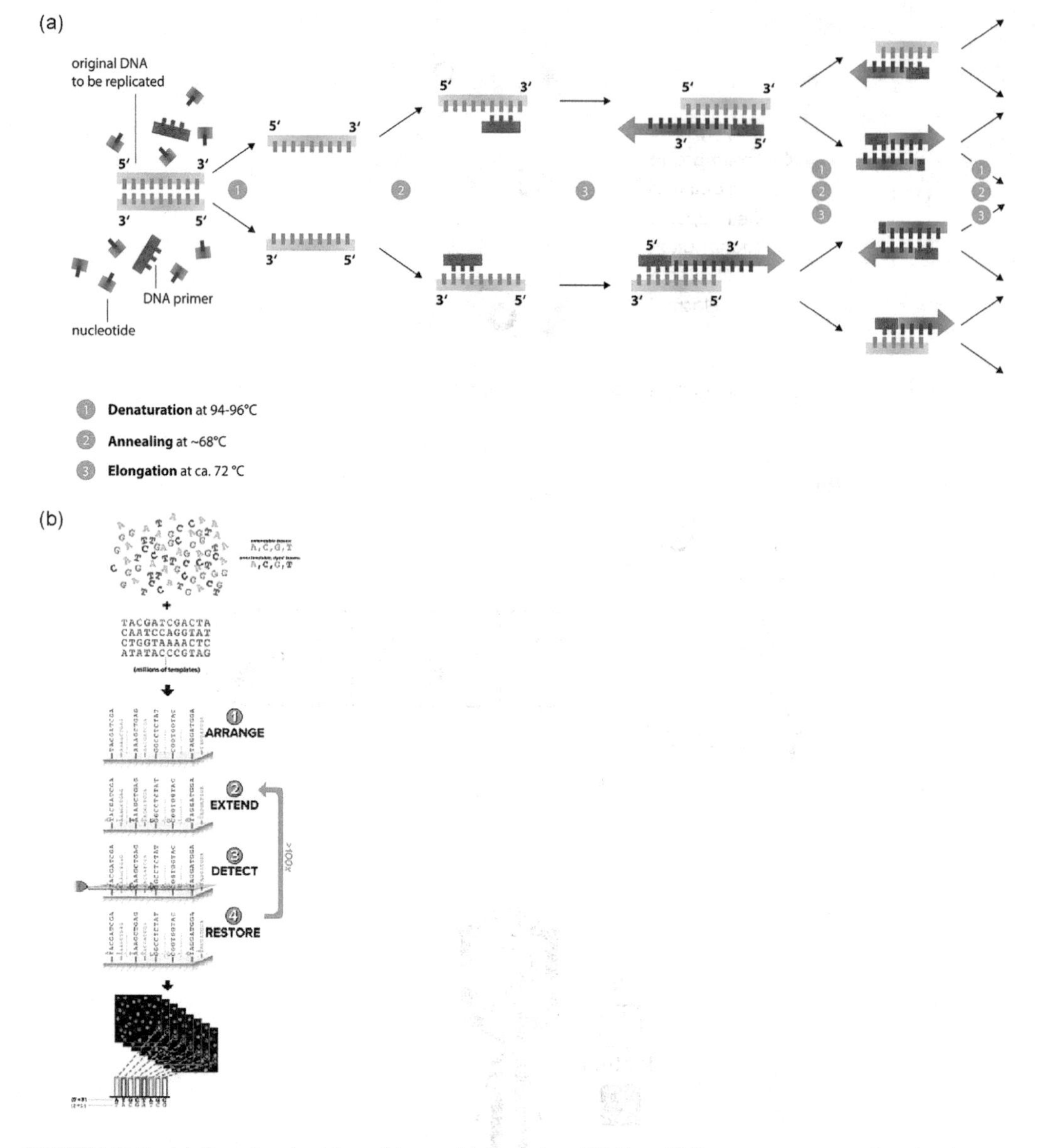

FIGURE 36.3 (**a**): Steps involved in making multiple copies of DNA or RNA sequences using the PCR. Primers have a sequence that is complementary to the target sequence in either RNA or DNA. (**b**): High-throughput sequencing process, which is used to sequence PCR products quickly and inexpensively. (Image credit: (a): Image by Enzoklop, distributed under a CC BY-SA 3.0 license. (b): This image was reproduced from Figure 36.1 in Price et al (2018). It is licensed under Creative Commons Attribution 4.0 International License, http://creativecommons.org/licenses/by/4.0/.)

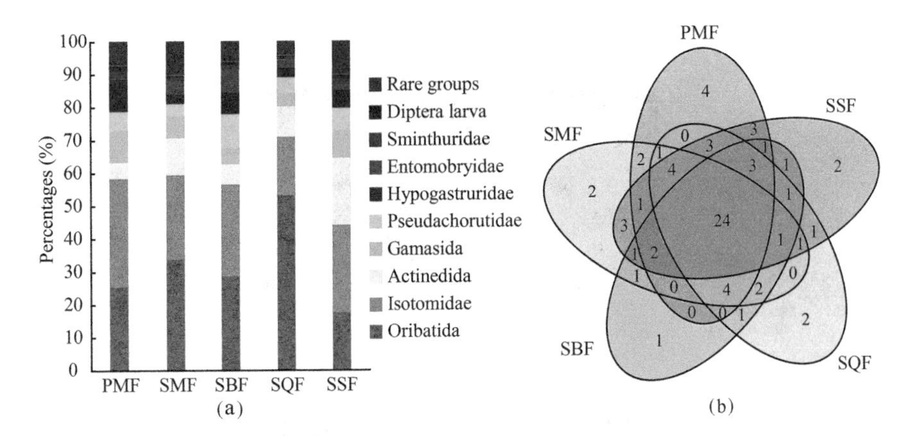

FIGURE 37.1 Characteristics of soil invertebrates' communities. (**a**) The relative abundance was defined as the percentage of soil invertebrates' community in different forest types of Changbai Mountains. (**b**) Venn diagram of the number of shared and unique soil invertebrates' taxa in different forest types of Changbai Mountains. The numbers in the circles indicate either the unique number of taxa in a forest or the number of shared taxa in the overlapping regions; PMFs, primeval coniferous broad-leaved mixed forests; SMFs, secondary coniferous broad-leaved mixed forests; SBFs, secondary broad-leaved forests; SQF, secondary *Q. mongolica* forests; SSFs, secondary shrub forests.

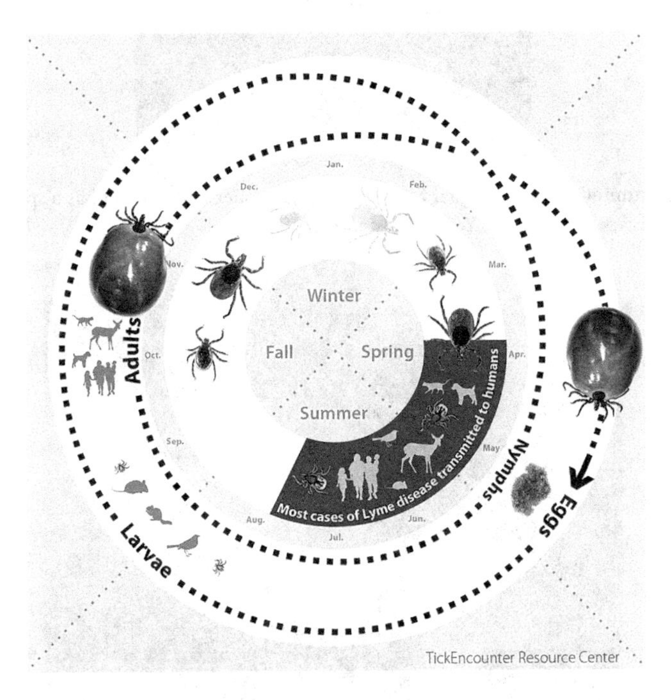

FIGURE 39.2 Life cycle of *Ixodes scapularis* (a.k.a. blacklegged or deer tick) in the northeast/mid-Atlantic/upper mid-western United States as an example of a typical three-host ixodid tick life stage. Larval deer ticks are active in August and September, but these ticks are pathogen free. Ticks become infected with pathogens when larvae (or nymphs) take a blood meal from infectious animal hosts. Engorged larvae molt over winter and emerge in May as poppy-seed-sized nymphal deer ticks. Most cases of Lyme disease are transmitted from May through July, when nymphal-stage ticks are active. Adult-stage deer ticks become active in October and remain active throughout the winter whenever the ground is not frozen. Blood-engorged females survive the winter in the forest leaf litter and begin laying their 1,500 or more eggs around Memorial Day (late May). These eggs hatch in July, and the life cycle starts again when larvae become active in August.

FIGURE 40.2 The summit of the Changbai Mountain cups a crater lake (**a**) and has a spectacular volcanic and subalpine tundra landscape (**b**). (Photos: Yeqiao Wang.)

FIGURE 40.3 The climate and terrain conditions support distinctive vertically distributed vegetation zones within the CMNR. Field photos illustrate the Alpine tundra zone between 2,100 and 2,400 meters with representative species such as short Rhododendron shrubs (**a**), the zone of subalpine dwarf birch (*Betula ermanii*) forest between 1,800 and 2,100 meters (**b**), the evergreen coniferous forest zone between 1,600 and 1,800 meters with dominant species of spruce (*Pieca jezoensis, Pieca koreana*) and fir (*Abies nephrolepis*) (**c**), and the needle- and broad-leaf mixed forest zone between 600 and 1,600 meters (**d**). (Photos: Yeqiao Wang.)

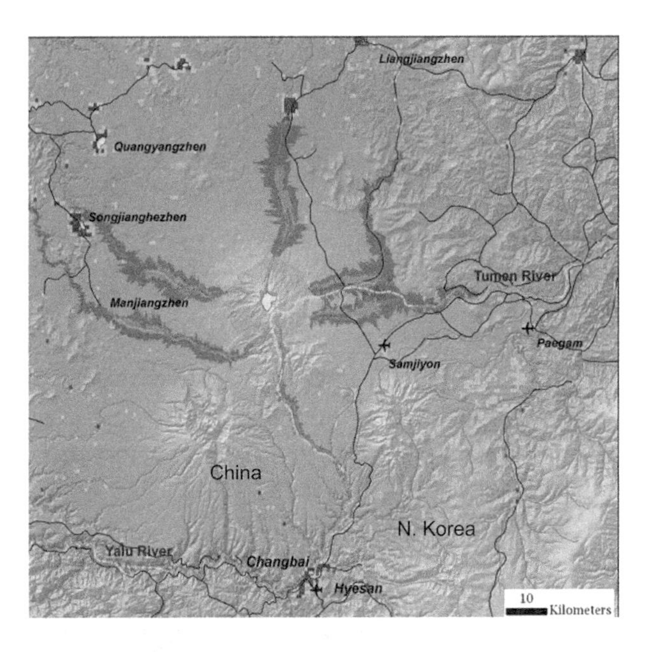

FIGURE 40.4 Distal hazard zones based on 3 scenarios of lahar volumes of 10 million cubic meters, 100 million cubic meters, and 1 billion cubic meters were mapped for the Changbai Mountain volcano with indication of nearby population centers of larger than 100 per km^2 and 1,000 per km^2. (Adapted from Lu et al., 2012.)

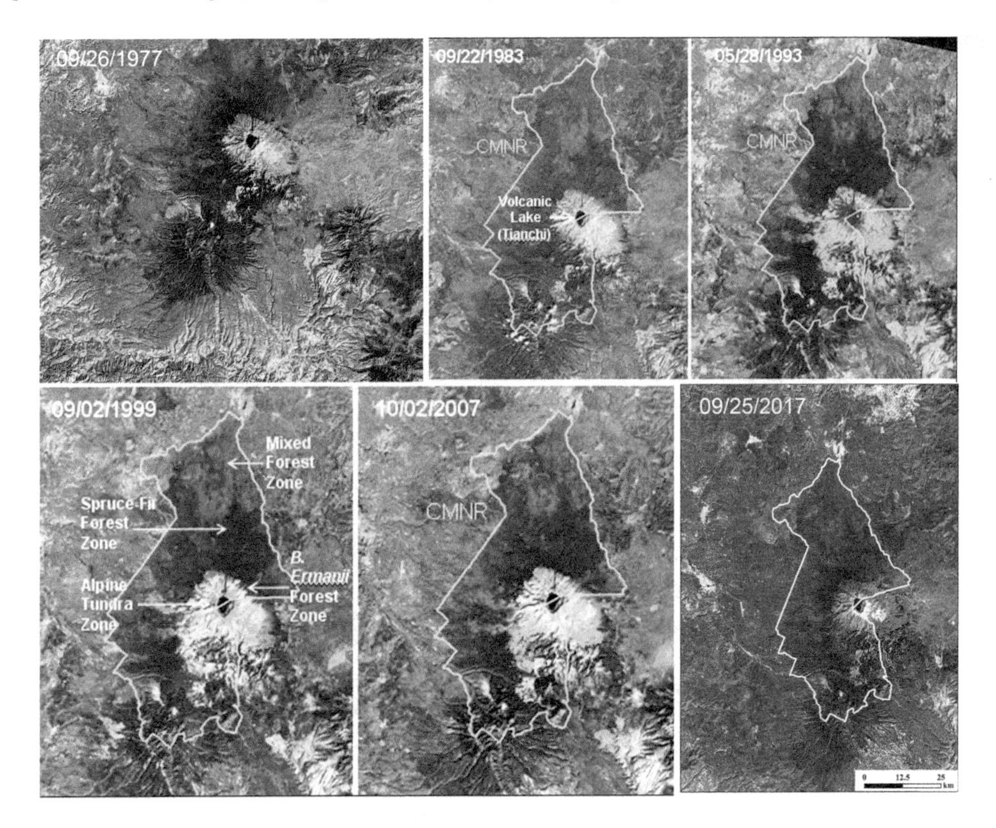

FIGURE 42.1 Landsat and Sentinel-2 satellite imagery data revealed vegetation patterns and change from 1977 to 2017 within and adjacent to the protected CMNR.

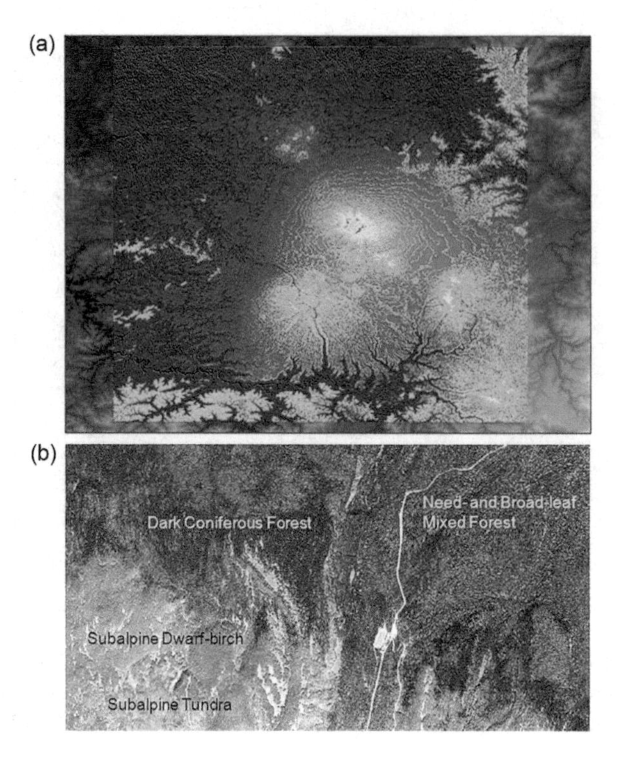

FIGURE 42.2 SRTM and derivative topographic information (elevation, slope, aspects) provide zonal guidance and reference (**a**) to differentiate vegetation distribution from high-resolution IKONOS data (**b**).

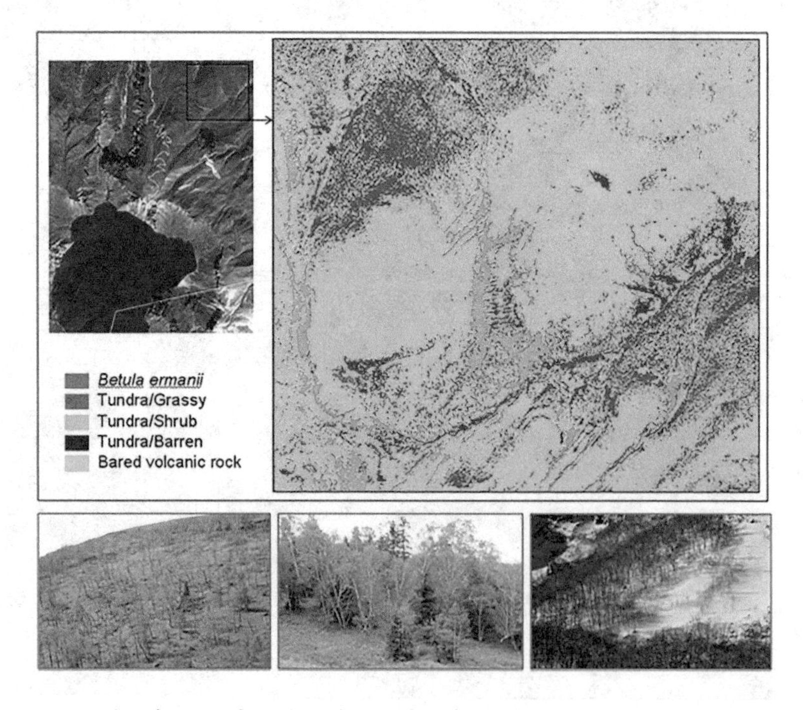

FIGURE 42.3 Image classification of IKONOS data to identify representative species in transition areas between vertical vegetation zones. Field survey confirmed the colonization of pioneer species of subalpine *B. ermanii* and *L. olgensis* into the tundra zone in high altitudes of the northern slope.

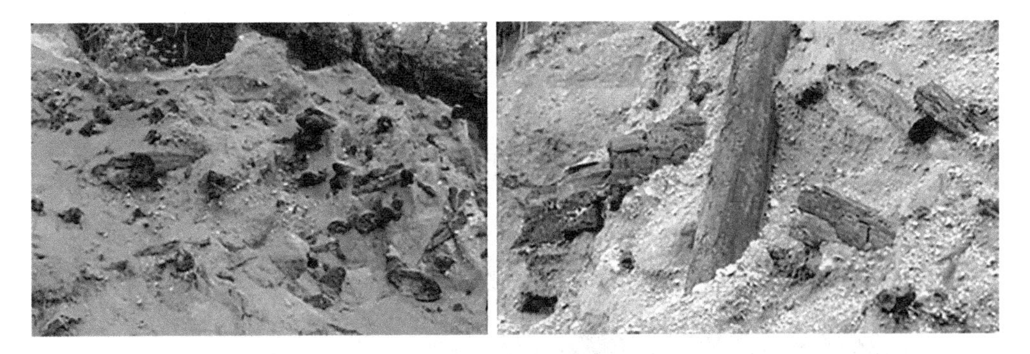

FIGURE 42.4 Samples of carbonized woods in species and abundance at different altitudes and slope/aspects of the area surrounding the volcanic summit were referenced to reconstruct pre-eruption vegetation structure and distribution.

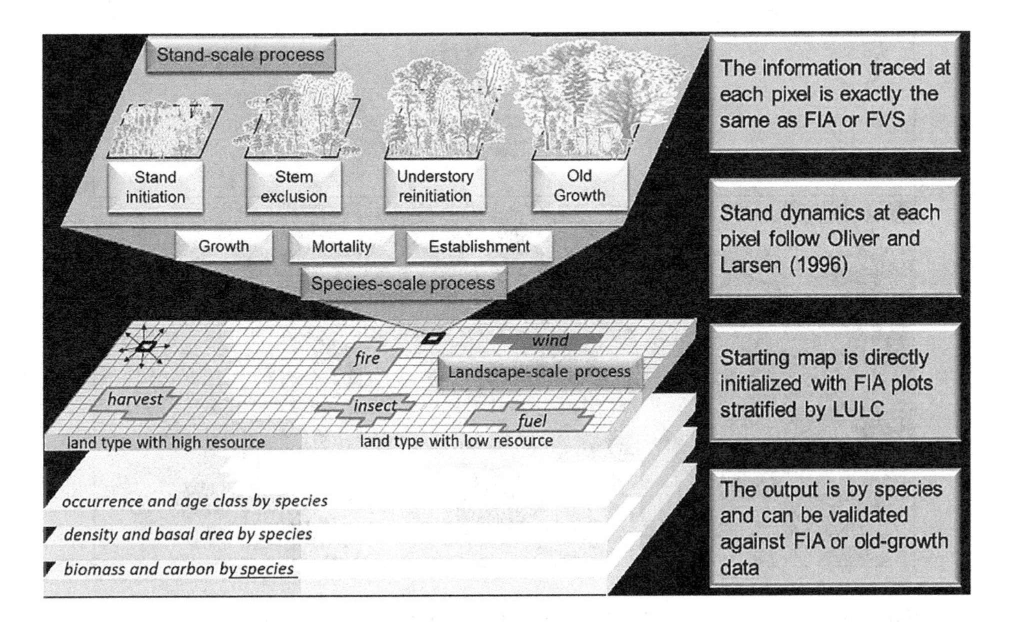

FIGURE 43.2 The conceptual design of LANDIS PRO that simulates forest change incorporating species-, stand-, and landscape-scale processes. (Modified from Wang et al. 2014a.)

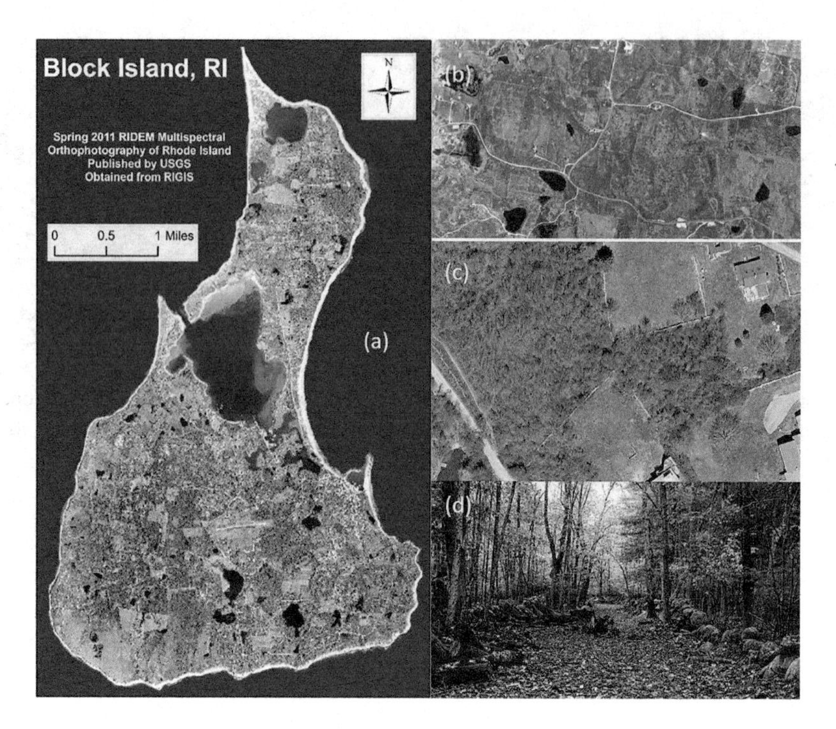

FIGURE 44.1 True color orthophotography data of the study area were collected in the spring of 2011 at a pixel resolution of 0.1524 m (**a**). The historical aerial photo of 1951 (**b**), recent high-spatial-resolution orthophotography of 2008 (**c**), and a field photo illustrate stone wall features on the landscape (**d**).

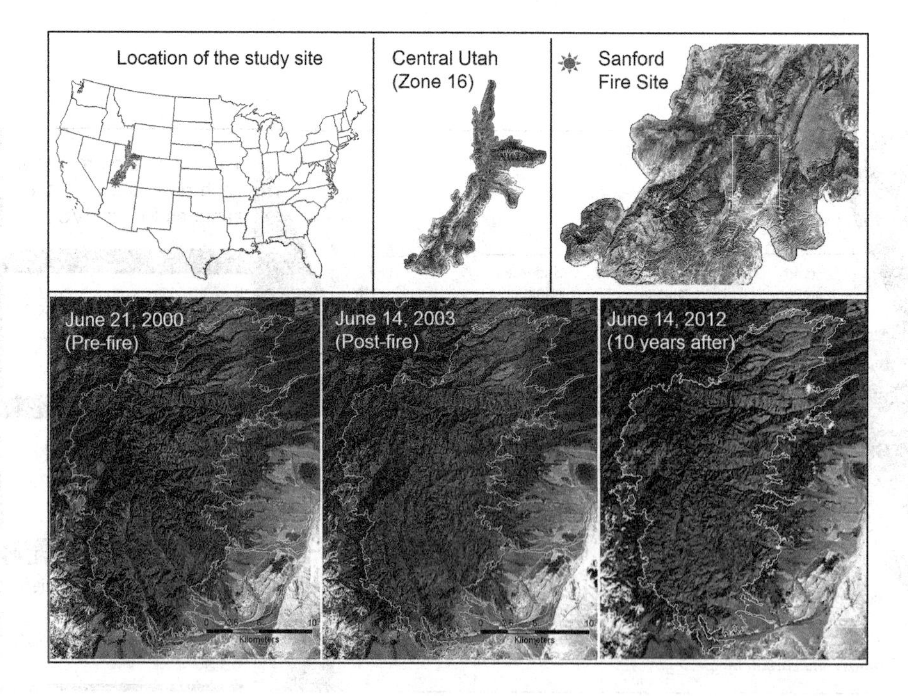

FIGURE 46.1 The Sanford Fire site is located in southern section of the Central Utah Valley (Zone 16) of the LANDFIRE data products. A comparison of pre- and post-fire Landsat TM reflectance images and Landsat-7 ETM+ image 10 years after the fire illustrates the impacts of the affected areas and vegetation recovery.

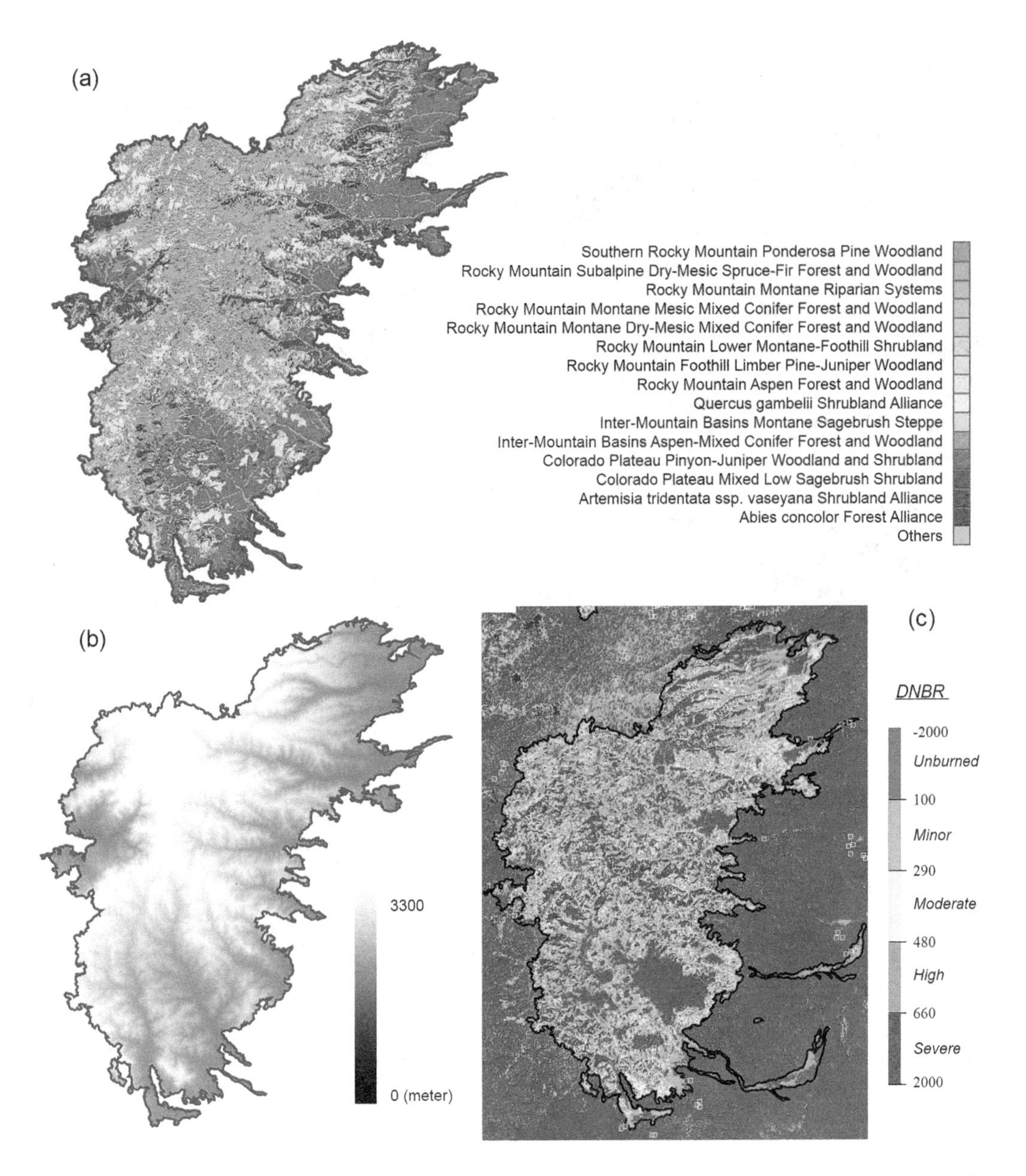

FIGURE 46.2 The EVT data (**a**) and DEM (**b**) provided pre-fire ecosystem data and environmental setting for the Sanford Fire site. The DNBR data (**c**) provided scales of burn severity on each pixel location as estimated measurements of fire impacts. For Plot 66: Pre-fire vegetation was identified as Rocky Mountain Mesic Mixed Conifer Forest and Woodland, canopy >50% and <60%; canopy height (10, 25 m); DNBR=996; CBI total=2.79. The 2005 site visit observed 50% vegetation cover, predominantly herbaceous species, creeping barberry, with evident of dense resprout of aspen. There was no evidence of recovery of fir or coniferous. For Plot 32: Pre-fire vegetation was identified as Colorado Plateau Mixed Low Sagebrush Shrubland, canopy >30% and <40%; canopy height (0, 0.5 m); DNBR=193; CBI total=2.72. The 2005 site visit observed good recovery 3 years after the fire with 60%–70% vegetation cover, predominantly herbaceous species.

FIGURE 46.3 An example of plot locations displayed on top of the post-fire Landsat TM image and DNBR map.
Source: Wang et al. 2011.

33

Soil: Organic Matter and Available Water Capacity

T. G. Huntington
U.S. Geological Survey (USGS)

Introduction

Available water capacity (AWC) is defined as the amount of water (cm^3 water/100 cm^3 soil) retained in the soil between "field capacity" (FC) and the "permanent wilting point" (PWP).[1] FC and PWP are defined as the volumetric fraction of water in the soil at soil water potentials of 10 to 33 kPa and 1500 kPa, respectively. One of the paradigms of soil science is that AWC is positively related to soil organic matter (SOM) because SOM raises FC more than PWP[2–4] (Figure 33.1). SOM enhances soil water retention because of its hydrophilic nature and its positive influence on soil structure.[5,6] Increasing SOM increases soil aggregate formation and aggregate stability[7] (Figure 33.2), thereby increasing porosity in the range of pore sizes that retain plant-available water and enhancing infiltration and water retention throughout the rooting zone. When SOM decreases, soil aggregation and aggregate stability decrease and bulk density increases.[8] These changes in physical properties result in lower infiltration rates and higher susceptibility to erosion.[9,10] This entry reviews the literature on the sensitivity of AWC to soil organic carbon (SOC) and discusses the environmental implications of changes in SOM and AWC. This entry supersedes and updates a 2003 review[11] with a description of the recent literature on the sensitivity of AWC to SOM and discusses the environmental implications of changes in SOM and AWC.

Sensitivity of AWC to SOM

Many studies have determined that AWC decreases as SOM decreases, but considerable variation in this relation has also been reported. A majority of studies have used regression analysis to quantify the sensitivity of AWC to SOM. Other estimates of the relation between SOM and AWC have been obtained from changes in water retention characteristics over time for a specific soil where different management, fertilization, or erosion histories have resulted in changes in SOM. To facilitate comparison among estimates of the relation between AWC and SOM in this review, sensitivity was defined as the average change in AWC (cm^3 water/100 cm^3 soil) predicted for a 5% relative decrease in SOM where the initial SOM concentration was in the range of 2% to 4% by weight. If texture was included in the published regression, it was held constant at 65% silt, 20% clay, and 15% sand. If bulk density was included in the

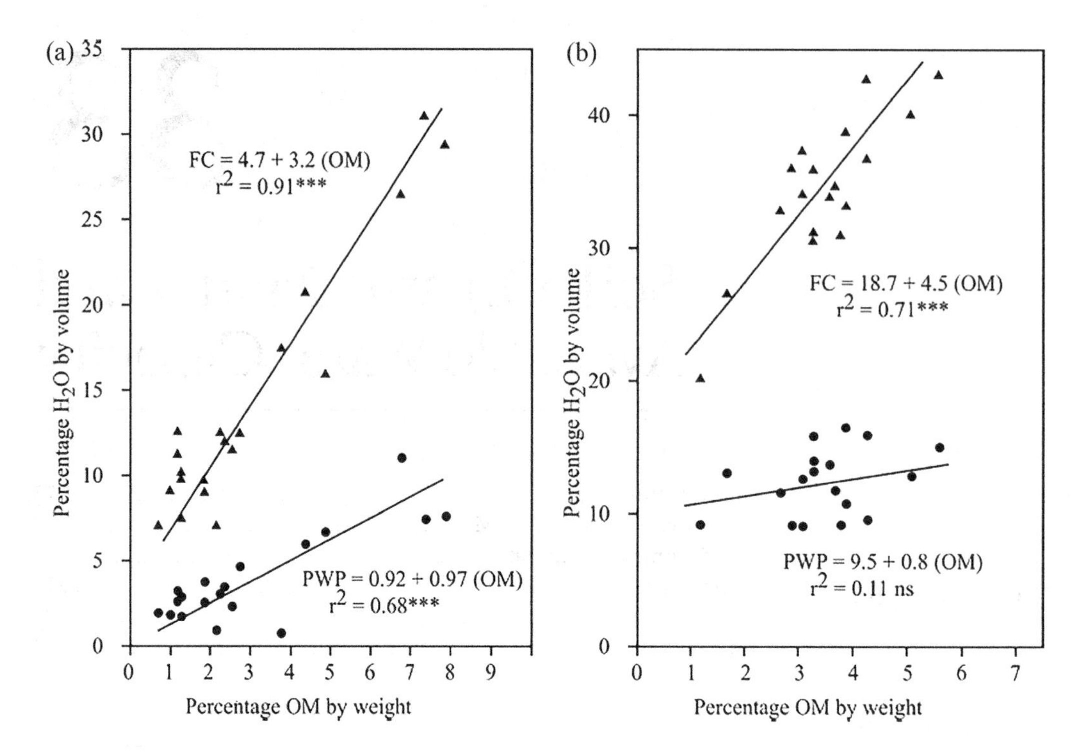

FIGURE 33.1 Water content at field capacity (FC) and permanent wilting point (PWP) for sandy (**a**) and silt loam (**b**) textured soils. *** indicates that the relationship is statistically significant with p-value <0.0001 and ns indicates that the relationship was not statistically significant with p-value >0.05.
Source: Reproduced from the original in Hudson[3] with permission.

regression, it was assumed that bulk density decreased with increasing SOM following the regression equation:

$$\text{Bulk density} = 1.723 - 0.212(\text{OC})^{0.5} - 0.0006(\text{WC15})^2$$

where OC = percent soil organic carbon and WC15 = water content at 1500 kPa.[12] WC15 for a soil with the texture of 65% silt, 20% clay, and 15% sand was assumed to be 11.3 cm^3 water/100 cm^3 soil. Where data or regression analysis required transformation of soil organic carbon (SOC) to SOM, it was assumed that SOM = 1.724 * SOC.

Reported sensitivities fell in the following ranges (0 to +0.27),[13–15] (0 to −1),[16–21] (−1 to −2),[22–25] and (<−2)[26–28] (cm^3 water/100 cm^3 soil) per 5% relative decrease in SOM concentration. The average value for the sensitivity was −1.1 (+/− 0.34 SE) cm^3 water/100 cm^3 soil per 5% relative decrease in SOM concentration. Out of the 15 studies, 3 found a negative or no relation between SOM and AWC. Regression models that indicated no relationship between SOM and AWC typically found that SOM resulted in equivalent increases in water retention at both FC and PWP with the result that AWC did not change.

In addition, two large studies have used the U.S. National Soil Inventory database to study the relationship between SOM and AWC where AWC was based on reported values for FC and PWP.[29,30] One study[29] showed a positive relationship between SOM and plant AWC that was dependent on soil texture and reported that FC was more sensitive to SOM than PWP. In another study,[30] a complex relation was reported such that a 1% increase in SOM causes a 2% to >5% increase in soil AWC depending on the soil texture. Both studies concluded that coarse-textured soils had a higher sensitivity to SOM than fine-textured soils and that sensitivity decreased as SOM content increased. These two studies[29,30] are consistent with the findings of most of the other cited studies that have found that increases in SOM result in proportionately larger increases in AWC in medium- to coarse-textured soils compared with finer-textured soils. Most studies found that aggregate stability has been shown to increase with increasing SOM at similar rates across a wide range of clay contents[7] (Figure 33.2). However, in that study[7] it was also shown that with increasing clay content, soils require a higher SOM content to maintain a given aggregate stability value (Figure 33.2). As SOM levels decrease, as occurs following chronic high rates of erosion, the effect of incremental decreases in SOM on AWC becomes progressively smaller. In heavily eroded soils, clay contents become increasingly more important than SOM in influencing hydraulic properties.[10]

The physical and chemical properties of SOM could influence how SOM affects soil water retention. Differences in SOM quality may arise from differences in inputs, such as natural plant resides, manure, or sludge, or they may result from different soil drainage conditions. These and other factors such as mineralogy, exchangeable cations, and surface chemical properties that can influence the formation and stability of aggregates and mesopores complicate the simple relation between AWC and SOM.[6]

There are several other possible explanations for the high variability in the reported sensitivity of AWC to SOM. Differences in methods used to measure water retention at both FC and PWP could introduce bias. Both *in situ* and laboratory methods have been used for the determination of water retention at FC and PWP. Laboratory methods can involve everything from intact cores to air-dried, sieved, rewetted, and recompacted soil samples.[1] Any pretreatment that changes soil porosity and soil aggregation will affect estimated water retention. There is no agreed-upon standard water potential for the estimation of field capacity in the laboratory. Different methods have been used to determine SOC and SOM. Differences in texture, mineralogy, bulk density, coarse fragment volume, and quality of the SOM of the soils used in the regression analyses could mask the effects of SOM. Some reports involve controlled regressions where variables such as texture are held constant. It may not be surprising that in uncontrolled regressions, the effects of SOM are obscured.[3,6]

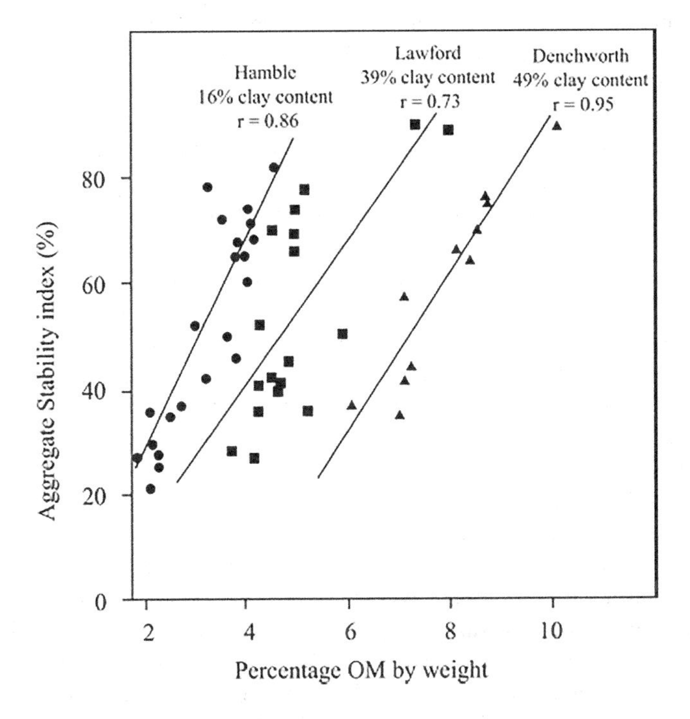

FIGURE 33.2 Relationship between aggregate stability and soil organic matter content for samples from the Hamble, Lawford, and Denchworth soil series from southern England.
Source: Redrawn from Haynes and Beare.[7]

Soil water retention data from long-term plots, such as classic crop rotations, fertility experiments, and no-till versus conventional tillage experiments, may be more appropriate for this type of sensitivity analysis. In these types of experiments, SOM changes measurably over time, but other variables are essentially held constant. However, there have been few reports of this kind. At the Sanborn Field in Missouri, continuous cropping for over a century resulted in large decreases in SOM and soil productivity.[31] Losses of SOM at the Sanborn Field could have increased erosion and loss of topsoil that likely decreased AWC.[31] However, in another study at the Sanborn Field,[15] no significant differences in water retention among treatments were observed for soil cores collected between 2.5 and 10 cm depth. A study in Denmark reported that after 90 years of manure additions, AWC was substantially higher on the amended plots than on unfertilized reference plots.[28] In a 32-year continuous corn experiment, cropping treatments that resulted in lower SOM (removal of corn stover and tillage) reduced SOM by 8% and 25% and decreased AWC by 8% and 13%, respectively.[32] In a shorter-term study, the addition of manures to grasses, cereals, and sugar beets for 7 or more years increased AWC at four sites with different soils in the United Kingdom.[33]

Environmental Implications

Environmental changes such as land-use conversions, changes in agricultural management practices, climate change, and changes in acidic deposition could affect SOM contents and, consequently, AWC. Land-use conversions from grasslands or forests to cultivated agriculture[34] or intensification of agriculture[35] are likely to result in net losses of SOM, but improvements in agricultural management practices have the potential to substantially enhance SOM contents and increase AWC.[36] Historical increases in acidic deposition followed by decreases in recent decades have been associated with concomitant increases and decreases in SOM in forest soils.[37,38]

It is now generally acknowledged that global mean annual temperature will likely increase in the range 1.4°C to 5.8°C during the 21st century as atmospheric carbon dioxide (CO_2) concentrations increase.[39] However, there is substantial uncertainty in how SOM levels will respond to the associated soil warming and elevated CO_2 concentrations. Some modeling studies[40–42] and empirical studies where land use and management were held constant[43–48] are consistent with declines in SOM following increases in temperature, but other environmental factors cannot be ruled out. Other studies suggest only no or small declines in SOM with increasing temperature or increases in SOM if precipitation increases along with increasing temperature.[49–52] In one study[53] it was concluded that temperature affects SOM polymerization and adsorption and desorption to mineral surfaces as well as the production and conformation of microbial enzymes. They argued that the net effect of these processes on the availability and degradability of SOM will determine how SOM will be affected by increasing soil temperatures. In another recent study,[54] it was reported that warming accelerates decomposition of decades-old carbon in forest soils. In yet another recent study,[55] elevated CO_2 was shown to have a priming effect on the decomposition of older SOM and to accelerate the decomposition of new detrital inputs. Complex relations have been reported between SOM and temperature and CO_2 that suggest both increases and decreases in SOM,[53,56,57] with the result that there is no consensus on the overall effect of these environmental changes on SOM in soils.

In areas where climate change results in a hotter and drier environment with more variability and extreme events of drought and heavy rainfall, as has been predicted for some regions,[58,59] soils will be subjected to more influence from a number of processes that are conducive to structural degradation.[60] Soil aggregate stability and soil structural integrity are likely to be adversely affected by losses in SOM that would accompany drying conditions. Structural degradation is associated with a loss in AWC and crop water use efficiency and increasing susceptibility to erosion. Decreases in SOM content would have adverse effects on soil chemical and physical properties because of the importance of SOM to soil structure and hydrologic properties. SOM is important for the maintenance of good soil structure through its positive influence on soil aggregate formation and stability[7] (Figure 33.2). Aggregation increases soil

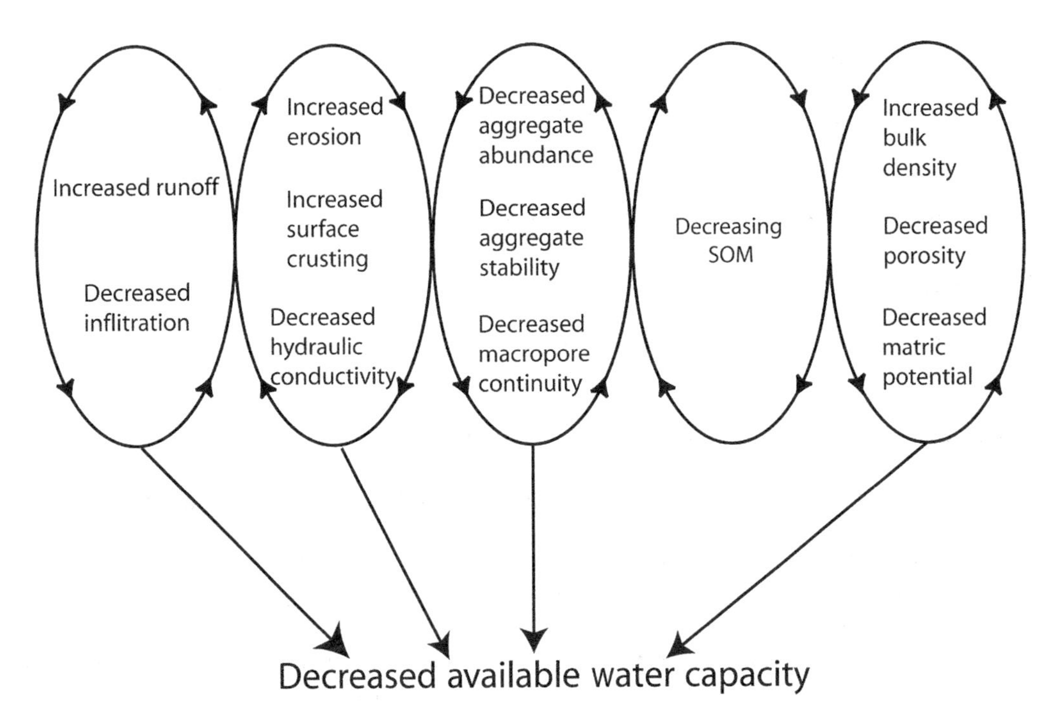

FIGURE 33.3 A conceptual model of the major effects of decreasing soil organic matter on soil physical properties that interact through the loops shown to decrease available water capacity in cultivated, eroding agricultural fields.

porosity and provides for improved water retention within the range of soil water potential where water is available to plants. It has been recognized that decreases in SOC could reduce AWC and soil fertility with resulting adverse consequences on soil health and productivity.[61,62]

Reductions in SOM in the surface soil result in increased bulk density, decreased porosity, and increases in matric potential that, in turn, result in decreased AWC (Figure 33.3). Decreases in SOM also result in decreases in soil aggregation, aggregate stability,[63,7] and macropore continuity, thus causing increases in surface crusting and decreases in hydraulic conductivity (Figure 33.3). These changes in physical properties lead to reduced rates of water infiltration and increases in runoff and erosion.[10] A climate warming-induced intensification of the hydrologic cycle could increase runoff and erosion in some regions.[64] Decreased infiltration and increased runoff result in less precipitation retained throughout the soil rooting zone so that less precipitation becomes available for plant growth. SOM exerts a strong influence on AWC through these interacting feedback loops that affect many soil physical properties (Figure 33.3).

Conclusions

Despite considerable uncertainty in the relation between AWC and SOM, most studies are consistent with the classical conceptual understanding that these soil properties are positively related. SOM promotes soil aggregate formation and the development of soil porosity that enhances infiltration and retention of water in a range that is available to plants. Environmental changes that result in increases in SOM will increase AWC. On the other hand, environmental changes that result in decrease in SOM will decrease AWC with resulting adverse consequences for the sustainability of agricultural productivity in some regions. There is a need to improve the quantitative understanding of the sensitivity of AWC to SOM to reduce the current level of uncertainty and to provide resource managers with better decision support systems.

Acknowledgments

Rattan Lal, Harry Vereecken, Alan Olness, and Jeffery Kern provided useful insights and references on the sensitivity of AWC to SOM. Miko Kirschbaum and Gregory Lawrence provided valuable insights that substantially improved this revised entry. Funding for this entry was provided by U.S. Geological Survey Climate and Land Use Change Research and Development Program.

References

1. Romano, N.; Santini, A. Measurement of water retention: Field methods. In *Methods of Soil Analysis, Part 4*; Dane, J. H., Topp, G. C., Eds., Soil Sci. Soc. Am. Madison, WI, **2002**, 721–738.
2. Jenny, H. *The Soil Resource;* Springer-Verlag: New York, 1980.
3. Hudson, B. D. Soil organic matter and available water capacity. J. Soil Water Conserv. **1994**, *49,* 189–194.
4. Emerson, W.W. Water retention, organic carbon and soil texture. Aust. J. Soil Res. **1995**, *33,* 241–251.
5. Stevenson, F.J. *Humus Chemistry;* John Wiley and Sons: New York, N.Y., 1982.
6. Kay, B.D. Soil structure and organic carbon: A review. In *Soil Processes and the Carbon Cycle;* Lal, R., Ed.; CRC Press: Boca Raton, 1997; 169–197.
7. Haynes, R.J.; Beare, M.H. Aggregation and organic matter storage in meso-thermal, humid soils. In *Structure and Organic Matter in Agricultural Soils;* Carter, M.R., Stewart, B.A., Eds.; CRC Lewis: Boca Raton, 1996.
8. Haynes, R.J.; Naidu, R. Influence of lime, fertilizer and manure applications on soil organic matter content and soil physical conditions: A review. Nutrient Cycling Agro-ecosyst. **1998**, *51,* 139–153.
9. Ascough, J.C. II; Baffaut, C.; Nearing, M.A.; Liu, B.Y. The WEPP watershed model: I. Hydrology and erosion. Trans. ASAE **1997**, *40,* 921–933.
10. Rhoton, F.E.; Lindbo, D.L. A soil depth approach to soil quality assessment. J. Soil Water Cons. **1997**, *52,* 66–72.
11. Huntington, T.G. Available water capacity and soil organic matter. In *Encyclopedia of Soil Science;* Lal, R., Ed.; Marcel Dekker: New York, 2003; 1–5 pp, DOI: 10.1081/E-ESS 120018496.
12. Manrique, L.A.; Jones, C. Bulk density of soils in relation to soil physical and chemical properties. Soil Sci. Soc. Am. J. **1991**, *55,* 476–481.
13. Veereecken, H.; Maes, J.; Feyen, J.; Darius, P. Estimating the soil moisture retention characteristics from texture, bulk density and carbon content. Soil Sci. **1989**, *148,* 389–403.
14. Lal, R. Physical properties and moisture retention characteristics of some Nigerian soils. Geoderma **1979**, *21,* 209–223.
15. Anderson, S.H.; Gantzer, C.J.; Brown, J.R. Soil physical properties after 100 years of continuous cultivation. J. Soil Water Cons. **1990**, *45,* 117–121
16. Gupta, S.C.; Larson, W.E. Estimating soil water retention characteristics from particle size distribution, organic matter content and bulk density. Water Resour. Res. **1979**, *15,* 1633–1635.
17. Thomasson, A.J.; Carter, A.D. Current and future uses of the UK soil water retention dataset. In *Indirect Methods for Estimating the Hydraulic Properties of Unsaturated Soils,* van Genuchten, T.M., Leij, F.J., Lund, L.J., Eds.; Proceedings of the International Workshop on Indirect Methods for Estimating the Hydraulic Properties of Unsaturated Soil, Riverside, California, USA, 1992; 355–358.
18. Salter, P.J.; Berry, G.; Williams, J.B. The influence of texture on the moisture characteristics of soils, III: Quantitative relationships between particle size, composition, and available water capacity. J. Soil Sci. **1966**, *17,* 93–98.
19. Pidgeon, J.D. The measurement and prediction of available water capacity of ferrallitic soils in Uganda. J. Soil Sci. **1972**, *23,* 431–444.

20. Schroo, H. Soil productivity factors of the soils of the Zandry formation in Surinam. De Surinaamse Landbouw **1976**, *24*, 68–84.
21. Karlen, D.L.; Wollenhaupt, N.C.; Erbach, D.C.; Berry, N.C.; Swain, J.B.; Eash, N.S.; Jordahl, J.L. Crop residue effects on soil quality following 10 years of no-till corn. Soil Tillage Res. **1994**, *31*, 149–167.
22. Rawls, W.L.; Brakensiek, D.L.; Saxton, K.E. Estimation of soil water properties. Trans. ASAE **1982**, *25*, 1316–1320.
23. Hollis, J.M.; Jones, R.J.A.; Palmer, R.C. The effects of organic matter and particle size on the water retention properties of some soils in the West Midlands of England. Geoderma **1977**, *17*, 225–231.
24. Tran-Vinh-An; Nguba, H. Contribution a l'etude de l'eau utile de qelques sols dur Zaire. Sols Africana **1971**, *16*, 91–103.
25. Hamblin, A.P.; Davies, D.B. Influence of organic matter on the physical properties of some east Anglian soils of high silt content. J. Soil Sci. **1977**, *28*, 11–22.
26. Salter, P.J.; Haworth, F. The available water capacity of a sandy loam soil, II. the effects of farm yard manure and different primary cultivations. J. Soil Sci. **1961**, *12*, 335–342.
27. Schertz, D.L.; Moldenhauer, W.C.; Livingston, S.J.; Weesies, G.A.; Hintz, E.A. Effect of past soil erosion on crop productivity in Indiana. J. Soil Water Conserv. **1989**, *44*, 604–608.
28. Schjonning, P.; Christensen, B.T.; Carstensen, B. Physical and chemical properties of a sandy loam receiving animal manure, mineral fertilizer or no fertilizer for 90 years. Eur. J. Soil Sci. **1994**, *45*, 257–268.
29. Rawls, W.J.; Pachepsky, Y.A.; Ritchie, J.C.; Sobecki, T.M.; Bloodworth, H. Effect of soil organic carbon on soil water retention. Geoderma **2003**, *116*, 61–76.
30. Olness, A.; Archer, D. Effect of organic carbon on available water in soil. Soil Sci. **2005**, *170*, 90–101.
31. Mitchell, C.C.; Weserman, R.L.; Brown, J.R.; Peck, T.R. Overview of long-term agronomic research. Agron. J. **1991**, *83*, 24–29.
32. Moebius-Clune, B.N.; van Es, H.M.; Idowu, O.J.; Schindelbeck, R.R.; Moebius-Clune, D.J.; Wolfe, D.W.; Abawi, G.S.; Thies, J.E.; Gugino, B.K.; Lucey, R. Long-term effects of harvesting maize stover and tillage on soil quality. Soil Sci. Soc. Amer. J. **2008**, *72*, 960–969.
33. Bhogal, A.; Nicholson, F.A.; Chambers, B.J. Organic carbon additions: Effects on soil bio-physical and physicochemical properties. Eur. J. Soil Sci. **2009**, *60*, 276–286.
34. Paul, E.A.; Paustian, K.; Elliott, E.T.; Cole, C.V. *Soil Organic Matter in Temperate Agroecosystems: Long-Term Experiments in North America*; CRC Press: Boca Raton, 1997.
35. Mattson, K.G.; Swank, W.T.; Waide, J.B. Decomposition of woody debris in a regenerating, clear-cut forest in the southern Appalachian. Can. J. For. Res. **1987**, *17*, 712–721.
36. Rosenberg, N.J.; Izaurralde, R.C.; Malone, E.L. *Carbon Sequestration in Soils: Science, Monitoring and Beyond: Conference Proceedings: St. Michaels Workshop, Dec.* Battell Press: Columbus, OH; 1999.
37. Oulehle, F.; Evans, C.D.; Hofmeister, J; Krejci, R.; Tahovska, K.;Persson, T.; Cudlin, P.; Hruska, J. Major changes in forest carbon and nitrogen cycling caused by declining sulphur deposition. Global Change Biol. **2011**, *17*, 3115–3129.
38. Lawrence, G.B.; Shortle, W.C.; David, M.B.; Smith, K.T.; Warby, R.A.F.; Lapenis, A.J. Early indications of soil recovery from acidic deposition in U.S. red spruce forests. Soil Sci. Soc. Amer. J. **2012**, *76*, 1407–1417.
39. Houghton, J.T.; Ding, Y.; Griggs, D.C.; Noguer, M.; van der Linden, P.J.; Dai, X.; Maskell, K.; Johnson, C.A. *Climate Change 2001: The Scientific Basis;* Cambridge University Press: Cambridge, UK, 2001.
40. Rounsevelle, M.D.A.; Evans, S.P.; Bullock, P. Climatic change and agricultural soils: Impacts and adaptation. Clim. Change **1999**, *43*, 683–709.
41. Jenkinson, D.S.; Adams, D.E.; Wild, A. Model estimates of CO_2 emissions from soil in response to global warming. Nature **1991**, *351*, 304–306.

42. Cox, P.M.; Betts, R.A.; Jones, C.D.; Spall, S.A.; Totterdel, I.J. Acceleration of global warming due to carbon-cycle feedbacks in a coupled climate model. Nature **2000**, *408*, 184–187.
43. Carter, J.O.; Howden, S.M. *Global Change Impacts on the Terrestrial Carbon Cycle,* Working Paper Series 99/10, Report to the Australian Greenhouse Office, CSIRO Wildlife and Ecology: Canberra, Australia, **1999**.
44. Bellamy, P.H.; Loveland, P.J.; Bradley, R.I.; Lark, R.M; Kirk G.J.D. Carbon losses from all soils across England and Wales. Nature **2005**, *437*, 245–248.
45. Senthilkumar, S.; Basso, B.; Kravchenko, A.N.; Robertson, G.P. Contemporary evidence of soil carbon loss in the U.S. corn belt. Soil Sci. Soc. Amer. J. **2009**, *73*, 2078–2086.
46. Nafziger, E.D.; Dunker, R.E. Soil organic carbon trends over 100 years in the morrow plots. Agron. J. **2011**, *103*, 261–267.
47. Khan, S.A.; Mulvaney, R.L.; Ellsworth, T.R.; Boast, C.W. The myth of nitrogen fertilization for soil carbon sequestration. J. Environ. Qual. **2007**, *36*, 1821–1832.
48. Fantappié, M.; L'Abate, G.; Costantini, E.A.C. The influence of climate change on the soil organic carbon content in Italy from 1961 to 2008. Geomorphology **2011**, *135*, 343–352.
49. Smith, T.M.; Weishampel, J.F.; Shugart, H.H.; Bonan, G.B. The response of terrestrial c storage to climate change - modeling c-dynamics at varying temporal and spatial scales. Water Air Soil Pollut. **1992**, *64*, 307–326.
50. Schimel, D.S.; Braswell, B.H.; Holland, B.A.; McKeown, R.; Ojima, D.S.; Painter, T.H.; Parton, W.J.; Townsend, A.R. Climatic, edaphic, and biotic controls over the storage and turnover of carbon in soils. Global Biogeochem. Cycles **1994**, *8*, 279–293.
51. Torn, M.S.; Lapenis, A.G.; Timofeev, A.; Fischer, M.L.; Babikov, B.V.; Harden, J.W. Organic carbon and carbon isotopes in modern and 100-year-old-soil archives of the Russian steppe. Global Change Biol. **2002**, *8*, 941–953.
52. Johnston, A.E.; Poulton, P.R.; Coleman, K. Soil organic matter. Its importance in sustainable agriculture and carbon dioxide fluxes. Adv. Agronomy, **2009**, *101*, 1–57.
53. Conant, R.T.; Ryan, M.G.; Ågren, G.I.; Birge, H.E.; Davidson, E.A.; Eliasson, P.E.; Evans, S.E.; Frey, S.D.; Giardina, C.P.; Hopkins, F.M.; Hyvönen, R.; Kirschbaum, M.U.F.; Lavallee, J.M.; Leifeld, J.; Parton, W.J.; Megan Steinweg, J.; Wallenstein, M.D.; Martin-Wetterstedt, J.Å.; Bradford, M.A., Temperature and soil organic matter decomposition rates – synthesis of current knowledge and a way forward. Global Change Biol. **2011**, *17*, 3392–3404.
54. Hopkins, F.M.; Torn, M.S.; Trumbore, S.E. Warming accelerates decomposition of decades-old carbon in forest soils. Proc. Nat. Acad. Sci. **2012**, Early Edition, http://www.pnas.org/cgi/doi/10.1073/pnas.1120603109
55. Phillips, R.P.; Meier, I.C.; Bernhardt, E.S.; Grandy, A.S.; Wickings, K.; Finzi, A.C. Roots and fungi accelerate carbon and nitrogen cycling in forests exposed to elevated CO_2. Ecol. Lett. **2012**, *15*, 1042–1049.
56. Smith, P.; Fang, C.; Dawson, J.J.C.; Moncrieff, J.B.; Donald, L.S. *Impact of Global Warming on Soil Organic Carbon, Advances in Agronomy;* Academic Press: 2008; pp. 1–43.
57. Kirschbaum, M.U.F. The temperature dependence of organic matter decomposition: Seasonal temperature variations turn a sharp short-term temperature response into a more moderate annually averaged response. Global Change Biol. **2010**, *16*, 2117–2129.
58. Meehl, G.A.; Stocker, T.F.; Collins, W.D.; Friedlingstein, P.; Gaye, A.T.; Gregory, J.M.; Kitoh, A.; Knutti, R.; Murphy, J.M.; Noda, A.; Raper, S.C.B.; Watterson, I.G.; Weaver, A.J.; Zhao, Z.-C. Global climate projections. In *Climate Change: The Physical Science Basis. Contribution of Working Group I to the Fourth Assessment Report of the Intergovernmental Panel on Climate Change;* Solomon, S., Qin, D., Manning, M., Chen, Z., Marquis, M., Averyt, K.B., Tignor, M., Miller, H.L., Eds.; Cambridge University Press: Cambridge, United Kingdom, 2007; pp. 747–845.
59. Tebaldi, C.; Hayhoe, K.; Arblaster, J.M.; Meehl, G.A., Going to the extremes: An intercomparison of model-simulated historical and future changes in extreme events. Clim. Change **2006**, *79*, 185–221.

60. Chan, K.Y. Soil attributes and soil processes in response to the climate change. In *Soil Health and Climate Change*. Singh, B.P., Allen, D.E., Dalal, R.C. Eds.; Springer: Berlin, Soil Biol. 2011; *29,* 49–67.
61. McCarty, J.J.; Canziani, O.F.; Leary, N.A.; Dikken, D.J.; White, K.S. *IPCC Climate Change 2001: Impacts, Adaptation & Vulnerability*; Cambridge University Press, Cambridge UK. 2001.
62. Allen, D.E.; Singh, B.P.; Dalal, R.C.; Cowie, A.L.; Chan, K.Y. Soil health indicators under climate change: A review of current knowledge. In *Soil Health and Climate Change,* Singh, B.P., Allen, D.E., Dalal, R.C. Eds.; Springer: Berlin, Soil Biol. 2011; *29,* 25–55.
63. Pimentel, D.; Harvey, C.; Resosudarmo, P.; Sinclair, K.; Kurtz, D.; McNair, M.; Crist, S.; Shpritz, I.; Fitton, L.; Saffouri, R.; Blair, R. Environmental and economic costs of soil erosion and conservation benefits. Science **1995**, *267,* 1117–1123.
64. Huntington, T.G. Climate warming-induced intensification of the hydrologic cycle: A review of the published record and assessment of the potential impacts on agriculture. Advances in Agronomy **2010**, *109,* 1–53.

60. [illegible] soils and soil processes in response to the climate change. [illegible] Allen D.E., Dalal R.C. [illegible]

61. [illegible] Climate Change [illegible] vulnerability. Cambridge University Press, Cambridge [illegible]

62. [illegible] Dalal R.C. [illegible] Soil health indicators [illegible] climate change: a review [illegible] Climate Change [illegible] Springer [illegible]

63. [illegible] 1995 [illegible]

[illegible] agriculture [illegible]

34

Soil: Spatial Variability

Josef H. Görres
University of Vermont

Introduction

All soil properties, whether they be physical, biological, or chemical in nature, show considerable multi-scale variability, ranging from microbial scales (1–1000 μm) to regional (10–100 kilometers) and even continental scales (thousands of kilometers). Variability at all scales is high. The coefficients of variation (CVs—ratio of the standard deviation and mean) for individual properties may range from several percent to several hundred percent. Warrick[1] classified variations of physical soil properties with CVs less than 15%, between 15% and 50%, and greater than 50% as low, medium, and high, respectively. Bulk density and porosity were classed as low, whereas electrical conductivity, hydraulic conductivity, and solute concentrations were classified as high variability. For biological properties, similar ranges are reported in the literature. For example, CVs for phosphatase activity were reported as 39%.[2] For forest soils, the CV for nematode abundance was of the order of 100% as compared to moisture and organic matter with CVs of 10–40%.[3] Where earthworms are present, the CV of water percolation may vary from 10% to 100%.[4,5] To make matters more complex, some soil properties and their variability change seasonally[3] or indeed on a daily basis.

Soil scientists engaged in soil mapping, classification, and morphology may have to integrate over several scales to identify soil taxa. Soil taxonomy recognizes that transformations and translocations of soil materials connect soil horizons in depth. Evaluating the three-dimensional variation of soil properties in combination at the pedon-scale (meter-scale) may thus produce thousands of unique profiles, each representing a soil series, the finest taxonomic resolution in soil classification systems and the pedology analog of species in Linnaean taxonomy.

Both natural processes and human activity contribute to the spatial variability encountered in soils. Jenny's equation of soil formation is the most commonly used model of pedogenesis that attempts to understand differences among soils in terms of five spatially variable factors. These five factors are responsible for landscape to continental-scale variations in soil properties.[6] The mosaic of different soil series created by the superposition of Jenny's factors may be regarded as the fabric of soil variability which is further modified by land management.

Soil biological processes create very much smaller, microscale variations in properties other than those considered in soil classification. These variations can be tracked to particular origins.[7] For example, it has long been known that soil biota are more numerous in soil surrounding roots[8] and burrows of soil animals[9] than in the bulk soil. Microscale heterogeneity may cause the spatial separation of key nutrient cycling transformations that may drive many below-ground ecological processes.[10,11]

Any record of the spatial distribution of properties is a snapshot in time. Soil processes are strongly dependent on meteorological and climatic forces and thus respond to temporal variations in temperature, precipitation, wind speed, insolation, and relative humidity. Denitrification, the conversion of nitrate into a suite of nitrogen gases, for example, depends on the distribution of carbon at the microscale. However, denitrification is mainly an anaerobic process that is strongest under saturated conditions. Large rainfall events thus cause an increase in denitrification activity. Nitrification, on the other hand, is an aerobic process and thus occurs when soils are drying. As biochemical processes often occur at microsites, the diffusion of biochemical reaction products further links space and time.[11]

Soil variability has practical consequences for experimental designs, farming, and the implementation and design of best management practices. For example, precision farming employs variable rate technologies that strive to apply agrochemicals to soils based on georeferenced spatial variations of field fertility parameters. This may save farm resources and improve the environmental quality by reducing inputs of fertilizers and pesticides. Yet, environmental and economic impacts of variable rate agriculture may depend on many factors and it is not clear whether this technology always leads to a reduction in application rate on the field scale.[12] In environmental application, land managers and soil evaluators often face questions associated with landscape-scale variations, such as how site-specific soil evaluations need to be when determining soil suitability for land use projects.

What all applications that account for the spatial variability of soils have in common is that the scale of variation ought to be a focus. The scale germane to an application will have to be isolated to design an appropriate sampling strategy and determine the volume (sample support) of an individual soil sample.[13] Soil variability represents interesting challenges to soil scientists, making knowledge of spatial distribution of soil properties and processes paramount to current practice in both fundamental and applied pedology.

Facets of Spatial Variability in Soils

Soils, Soil Properties, Soil Processes, and Time

When discussing the spatial variability of soils, a distinction has to be made between the variability of soils, i.e., among taxonomic units, and soil properties, a measurable quantity. Pedologists classify terrestrial and subaqueous soils based on the vertical distribution of matter and its properties within a representative soil volume called a pedon.[14] Each soil series thus has a unique sequence of horizons with specific diagnostic properties that are intertwined with spatially and temporally variable processes within and outside of the soil. Soil survey maps are quintessentially two-dimensional as the nature of a soil series is given by the relationship and interpretation of properties in the vertical as well as the horizontal dimensions of its profile. Soil classification relies on the spatial, three-dimensional arrangement of measurable and relatively stable properties within a soil profile.[14]

However, stability and equilibrium are not one of the hallmarks of many soil properties. Spatial distributions change over time. Daily variations in soil moisture and temperature may engender rapid changes in food web dynamics and biochemical transformation rates responsible for much of the below-ground decomposition and nutrient cycling. Variations also occur at seasonal and inter-annual scales. Even longer term environmental variations relevant to climate and vegetation affect soil development in a way that may have tangible effects on soil loss.[15] Among these decade-long cycles, space weather, in particular the 22-year cycles of the solar magnetosphere and the concomitant variation in cosmic ray fluxes, may be responsible for droughts such as the Dustbowl episode in Southern U.S.A. in 1930.[16] While devastating for farmers, these drought episodes are responsible for a spatial redistribution of top soil materials. However, even at a single point in time, snapshots of soils present observers with sufficiently puzzling spatial variability.

The importance of the dependence of variability on time is best illustrated by how changes in soil properties may affect the variability of soil processes. It is well known that small differences in soil moisture content may translate into large differences in soil processes. The affected processes are physico-chemical and biological in nature. Moisture saturation at the center of aggregates in an otherwise dry soil may result in considerable denitrification rates.[17] Also, only small differences in moisture or the presence and location of macro-pores in soils may result in large differences in water fluxes through the soil.

Factors of Soil Formation: Intrinsic Sources of Spatial Variability

Early soil scientists were intrigued by the differences in soils that they encountered in their travels. Dukochaev (1846–1903), Hilgard (1833–1916), and Jenny (1899–1992) formulated theories in which independent factors could explain geographic variations in soils.[18] These factors are intrinsic or natural sources of variability. In an attempt to create a more quantitative foundation for soil science, Jenny's equation[6] expressed the magnitude of soil-forming processes in terms of five variables or factors: climate, parent material, topography, organisms, and time.

$$\text{Process} = F(\text{climate, topography, parent material, organisms, time}) \tag{34.1}$$

The five factors act on all soil-forming processes with each factor imbuing different qualities to the soil. Overlaying these qualities across the landscape differentiates among soil series, the spatial variability considered by soil taxonomy. At the continental scale, large-scale variations in climate, vegetation, and parent material give rise to characteristic patterns of soil orders, the coarsest soil taxonomic scale (Figure 34.1). For example, most spodosols are found in northern, temperate climes with boreal, coniferous forests, where precipitation is high and recent glaciation has left deposits of sands leading to distinct patterns of reduced and oxidized horizon environments (Figure 34.2a). In contrast, inceptisols have not developed very distinctive horizons due to a lack of time that the other factors have acted upon them. In Figure 34.2b, an agricultural inceptisol is shown with a horizon altered by plough action. This young soil (only thousands of years old) developed after the Laurentide ice sheet of the last glaciation melted and receded. In this soil, the combination of two parent materials, a fine silt mantle overlying glacial sand and gravel deposits, results in characteristic reduction-oxidation horizons that dominate the spatial variability in the depth of the soil. The fine silt material may retain water long enough above the gravel deposit to reduce iron oxides resulting in a horizon where the distinctive orange color of iron oxides is absent.

The five factors do not act independently, and Jenny's equation can thus not be understood as a linear algebraic equation. Toposequences are a good example of the interrelationship between soil-forming factors at the landscape scale. A special toposequence called a hydrosequence is shown in Figure 34.3. The landscape position along this hydrosequence determines not only elevation and, thus, the distance to ground water and the drainage class of the soil. It also determines vegetation, decomposition, and production rates of the ecosystems as the land surface slopes from an upland area into a wetland. Together, these factors are responsible for the variation of soils along the toposequence.

Action of *Homo sapiens*: Extrinsic Factors of Variability

Superimposed on the patchwork of intrinsic variation are extrinsic spatial patterns introduced by land management. At the landscape scale, the mosaic of agricultural fields (Figure 34.4) may impose an additional variability of soils. In addition, within each field, management such as tillage and irrigation may introduce further variability at the field scale. Notably, extrinsic patterns may persist for many years even when the practice causing the initial variation is abandoned. For example, ridge and furrow tillage in the nineteenth century may still have affected the crop yield pattern in 1911 at Rothamsted.[19]

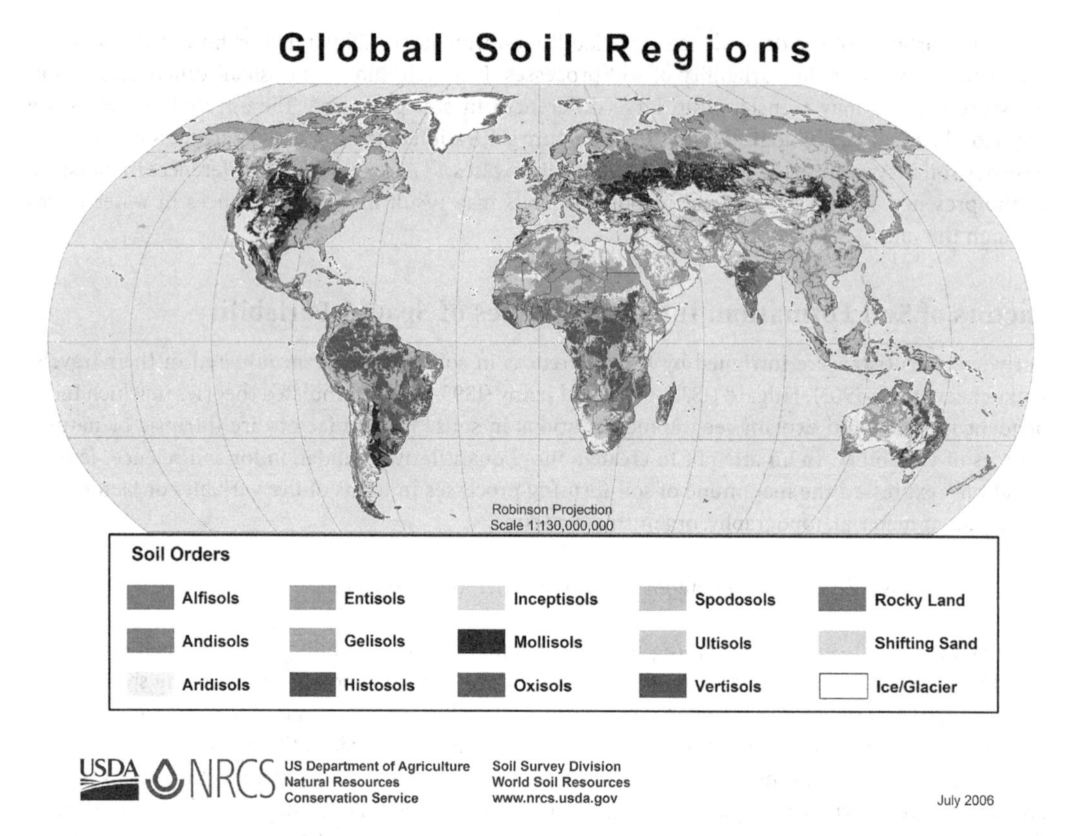

FIGURE 34.1 **(See color insert.)** The planetary distribution of soil orders across the land area. Distinct geographic patterns of soil orders emerge based on climate, topography, organisms, parent material, and time. **Source:** Reproduced with permission from Dr. Paul F. Reich, Geographer–World Soil Resources, USDA Natural Resources Conservation Service.

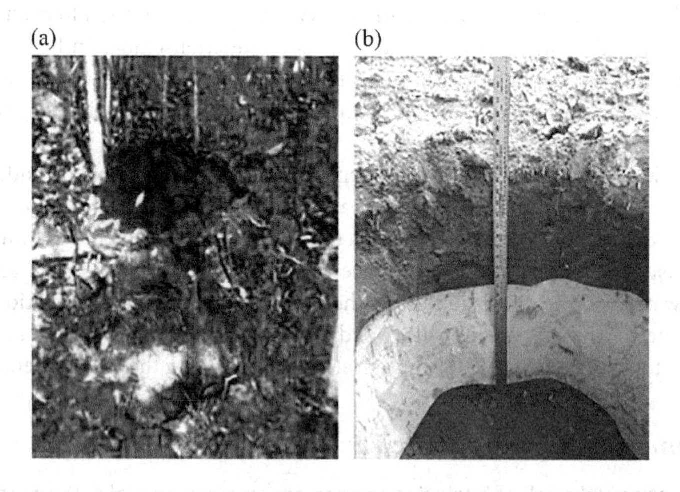

FIGURE 34.2 (**a**) Spodosol at Hubbard Brook, NH. Distinct horizontion with a "bleached" E horizon above a characteristic dark area B_{hs} horizon, indicating the translocation of iron and humus from the E and O horizons. Provided by Dr. Joel Tilley of the Agricultural and Environmental Testing Lab at the University of Vermont. (**b**) Inceptisol at the Peckham Farm, Kingston, RI. A silt cap deposited over coarse sand gravel deposits created conditions that favor reduction of iron oxides within the silt horizon resulting in a white, colorless layer and a streak of white iron oxides at the contact between the silt and the sand and gravel.

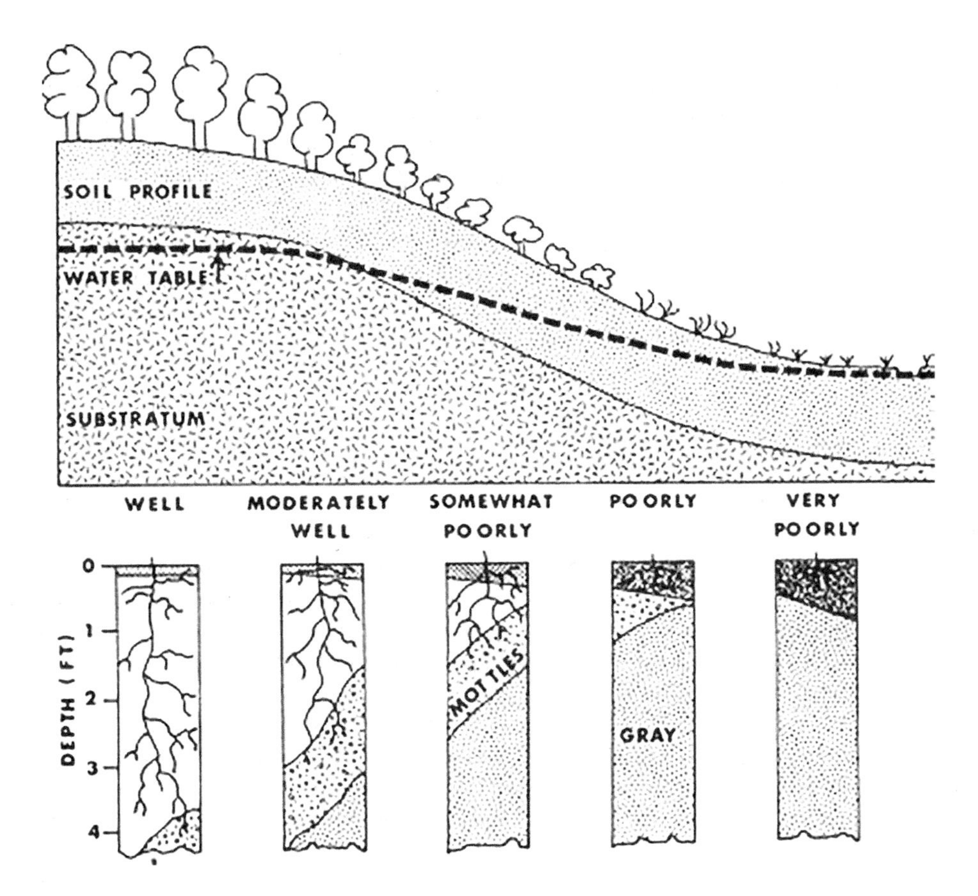

FIGURE 34.3 Soil profiles along a toposequence vary by elevation or more importantly with the distance to the water table.

Source: Rhode Island Agricultural Research Station 1988. Wright and Sautter. Soils of the RI Landscape, reproduced with permission of the Director, Rhode Island Agricultural Research Station.

FIGURE 34.4 Landscape-scale field-planting pattern, an extrinsic variation contributing to the spatial variability of soils.

Apart from the desired patterns of fertilizer application and planting, the use of machinery causes compaction which can also persist for many years.[20] Likewise, the hoof fall of large grazing animals may affect soil compaction in pastures with characteristic spatial variations that depend on grazing pattern, paddock geography, and animal stocking rate. Dung deposition patterns in the pasture add to the variability of pasture soils[21] and render the distribution of nutrient cycling processes uncertain.

These examples suggest that land management creates additional variability at the field scale. However, it can also be argued that land conversions may make soils more homogeneous. For example, the conversion from a heterogeneous land use mosaic to large fields that support monocultures such as corn or sunflowers would smooth variations at the landscape scale while increasing heterogeneity at the field scale.

Variation at the Microscale

The patterns of soil heterogeneity observed by soil surveys, i.e., those that are caused by soil-forming factors, essentially integrate soil properties over the pedon that has a volume of several meters cubed. Yet, the diversity of processes at much smaller scale is apparent even when observing differences at the soil surface. The pattern of vegetation may reveal small-scale variations in soil properties, a fact that is used in plant indicator systems.[22] Equally obvious at the soil surface are the patterns of excavation mounts of burrowing animals like ants and earthworms, which may coexist within centimeters of each other but have quite divergent microbial communities.[23]

Very small scale spatial variability is also regarded as the key to better understanding why soil faunal communities maintain high levels of diversity with much apparent functional redundancy.[24] Pore networks comprise a diversity of pore neck and chamber diameters that allow access only to those organisms that fit into the pores. The variability of pore diameter at the pm-scale is depicted in Figure 34.5, where the light-colored segmented areas represent the pore network of a sandy loam collected in Rhode Island. The physical pore structure interacts with biologically active "spheres," where soil organisms exert their influence on the soil and its microbial and faunal communities.[7] Drilosphere and rhizosphere are the soils surrounding earthworm burrows and plant roots, respectively. The enrichment of the drilosphere with detrital resources welded to the burrow walls by earthworms and modification of

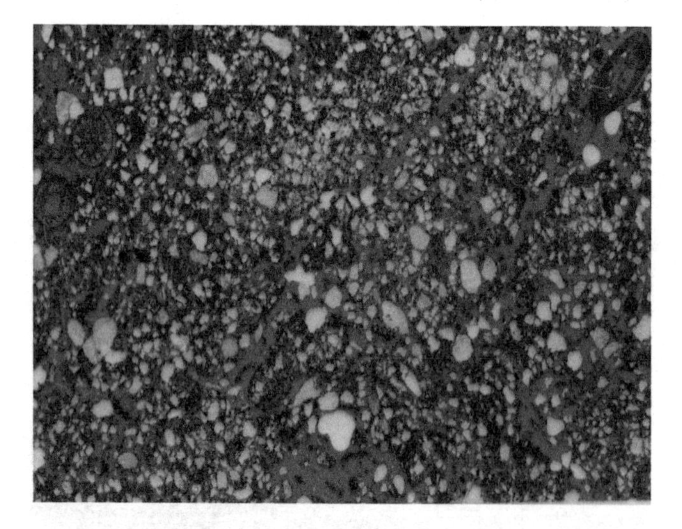

FIGURE 34.5 Micrograph of a thin section of sandy loam core showing the small-scale spatial variation in pore sizes. Pores are segmented in a light color. The network shows how the physical pore habitat varies even at the millimeter scale. The vertical extent of the image is 1 cm.

the soil pore structure[25] causes a large increase in biological activity.[9,26] Similarly, resource enrichment can cause increases in microbial populations in the rhizosphere.[8] Here, local photosynthate exudations prime soil biological and chemical activity.

Hattori and Hattori[27] introduced the concept of microhabitats of protozoa and microorganisms that are located within soil aggregates, which Beare et al.[7] called the "aggregatusphere." Now, with molecular techniques, it is possible to establish the spatial variability of microorganisms within single aggregates revealing great diversity at the millimeter scale.[28]

The spatial distribution of pore spaces (also called the "porosphere") and its partitioning into air and water-filled pores may select for specific soil organisms. Moisture controls the level of competition and predator–prey relationships that occur among soil organisms. Naturally, the relative fraction of air-filled and water-filled pores changes as soils dry and wet up, and thus, the mosaic of habitat of water and air-based soil organisms changes too. The concept of habitable pore space[29,30] encapsulates this variability at scales of pore diameters ranging from tens to hundreds of micrometers. At these scales, soil pores are microhabitats for protozoa[27] and nematodes.[25,31] However, there are different interpretations as to what happens to these habitable spaces when drying and wetting cycles rapidly alter their habitability for the aerial and the aquatic communities that inhabit soil pores.[32]

The idea that microsites can significantly affect bulk soil processes suggests that measurement may be scale dependent. Parkin[10] found that a soil core (100 cm^3-scale) had very high denitrification activity. A single, small woodchip was isolated as the source of the high denitrification rate measured for the core. Parkin's experiment demands that sampling support be considered in designs that are to estimate a soil property or process. Similarly, the moisture distribution within aggregates should be considered when examining denitrification in an otherwise dry soil.[17]

Concepts of Variability

Variability is often conceptualized as the superposition of intrinsic and extrinsic factors, some of which were discussed earlier. However, there are other ways to look at soil variability. One such approach distinguishes between stochastic variations and deterministic variations that when added together are responsible for the variation in soil properties.

Deterministic variations are those that can be predicted from soil processes, and they may be spatially dependent at various scales. Spatial dependence is introduced by the translocation of matter and energy from one location to another or by the extent of large parent material deposits or even by vegetation patterns. Stochastic variations are those that cannot be predicted and are much like white noise and thus not spatially dependent. For spatial interpretations, the deterministic variables may have to be removed so that random variations can be properly characterized.

Webster[33] argued that with sufficient knowledge about all soil processes, all variability is deterministic and that only incomplete knowledge and the sheer complexity of interacting soil processes may make variations in soil appear random. Notwithstanding philosophical argumentation about random and deterministic variations, apparent spatial dependence or independence may require different methods of estimating a central tendency and the variation about it.

One measure that is often used to describe variability is the *CV*, which measures the variation as standard deviation, σ, standardized by the mean, μ:

$$CV = \frac{\sigma}{\mu}$$

The *CV*, however, requires that sample points are independent and may be falsely applied when there is any degree of spatial dependence. This casts some doubt on *CVs* estimated at different scales as soil properties may have different levels of spatial dependence at different scales.

Sampling and Quantifying Variability

Quantification of the variability of soil properties is more complex than it may appear. Even when designing the sampling method, important decisions need to be made. Apart from the classical equation for the number of samples that need to be taken to determine the mean of a sample distribution at a given confidence level and accuracy, the physical size of the sample support (the physical volume or mass of a sample specimen) becomes important as it will determine the smallest variations that can be resolved.

Sometimes, the nature of the assay will determine the sample support. For enzyme activity assays, only a couple of grams of soil are necessary. To get reliable enzyme activity information, several soil samples need to be homogenized and several replicates may need to be analyzed. The purpose of homogenization and analytical replication is to smooth small-scale variations and increase the certainty of estimation at a given scale. For measurements of bulk density, cores are taken that are typically 5–10 cm in diameter and 10 cm long. Variations that are contained within that volume are averaged out, or smoothed. Large-scale examples of spatial averaging come from remote sensing where luminosity is averaged over the area of pixels that are tens to thousands of meters in length. Selection of the satellite and sensor allows matching the scale of resolution with the research or management need. Spatial averaging is an essential tool not only for improving the reliability of data, but also focusing on the scale of the relevant variation.

Given a sufficient number of samples that were randomly collected, first-order statistics can describe soil variability in terms of statistical moments of the sample distribution: mean, variance, skewness, and kurtosis. This assumes that measured soil properties are independent of each other at the sample locations and for the scale represented by the sample support.

Many soil properties are spatially dependent. First-order, or classical, estimates of mean and variance can become biased when spatial relationships, so-called second-order properties, are not taken into account. A second-order measure commonly used in soil science is the semivariogram, although other methods such as power spectra derived from Fourier harmonic decomposition have also been used.[19] A semivariogram measures one half of the covariance between soil property values of sample pair classes that are classed by pair-separation distances called lags. The further samples are apart, the greater their semivariance and the less related they are in space unless there are repeating patterns. Theoretically, the semi variance is limited to a maximum value that is reached when the lag exceeds the range of spatial dependence. To parameterize the semivariogram, an analytical semivariance function is fitted that best describes the variation. For spatially dependent samples, the semivariance functions increase as lag increases until the variance reaches a constant value called the "sill" (Figure 34.5). The lag value at which the sill is reached is called "range" and gives the maximum distance up to which the measured soil property is spatially correlated. Thus, for truly random variations, the sill variance is reached at lag 0. The semivariance parameter at lag 0 is also called "nugget variance." It serves as a good aggregate estimator of error due to sampling uncertainty, unresolved variations, and analytical errors.

However, estimates of second-order properties should be carried out with some care because just as there are rules for sampling spatially independent sample sets, there are rules for measuring the variability of spatially dependent samples. The Nyquist sampling theorem states that any variation potentially resolved by the sampling support needs to be sampled with a frequency of greater than twice the wavelength or characteristic length of the variation. If the rule is not obeyed, the signal cannot be adequately reconstructed or worse the signal is misrepresented by an *alias* signal. For soil science, this means that undersampling a deterministic variation can result in a biased estimate of the mean and variance and potentially a misrepresentation of a pattern. An example for this is the soil moisture distribution across a plowed field (Figure 34.6). The solid line in Figure 34.7 shows the original, exhaustive 230-point sample set for soil moisture. The sample interval was 10 cm. The stippled line is based on 32 data points randomly selected from the original. This represents an average sampling interval of just under the wavelength, associated with a ridge and furrow pattern. The range of the original is 50 cm, but 90 cm for the under-sampled variation. In addition to the overestimation of the range distance, the sill variance is overestimated by 30%. This error is not captured by the nugget variance, which usually

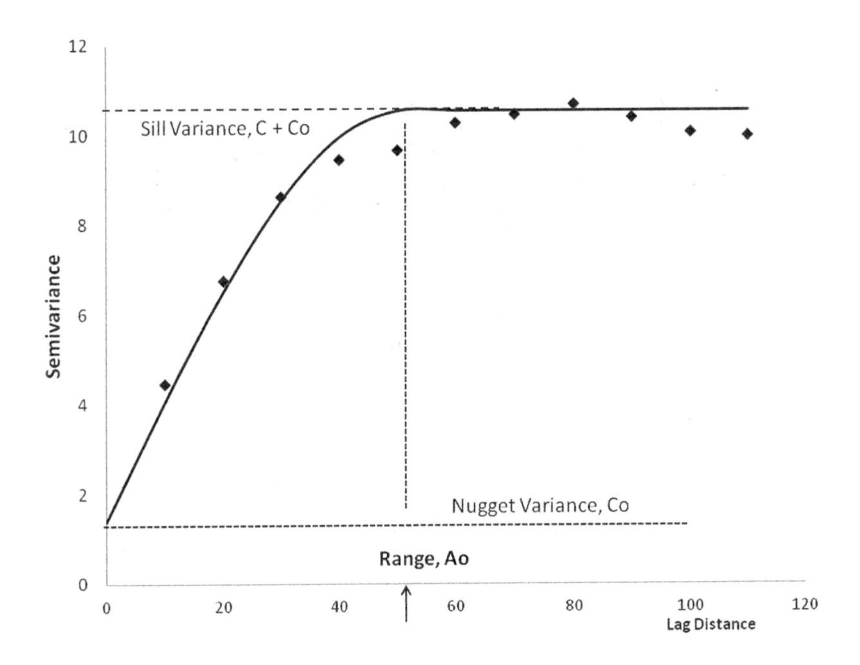

FIGURE 34.6 Diagram of a semivariance function. There are three spatial parameters most often estimated with semivariance analyses. The distance, A_o, where the data is no longer spatially dependent is called range. The maximum semivariance, $C+C_o$, is called sill and comprises the variance due to spatial variations, C, plus the nugget variance, C_o, which is an aggregate measure of uncertainty due to analytical and sampling errors.

is a good estimator of spatial estimation errors, but is zero for the under-sampled moisture variation in this case.

For complete sample sets where sample points abut each other, the Nyquist theorem does not need to be invoked as the sample support would smooth out or integrate over any small unresolved variations. This kind of smoothing can be accomplished by taking several samples from each location, followed by combining and homogenizing the samples. Prior to analysis satellite technology may also provide unabridged sample sets, where small-scale variations in luminosity are integrated over the pixel support. Apart from several specialized measurements (e.g., light detection and ranging [LIDAR]), environmental satellite imagery is of low resolution and suitable only for large-scale investigations. Much progress needs to be made before remotely sensed data can replace *in situ* sampling for the analysis of some properties.

While it is clearly important to obtain as many samples as possible to satisfy needs for accuracy and improved inferential power, the cost of sampling and analysis often competes with the requirements of the central limit theorem and the Nyquist sampling theorem. Cost may prohibit high-frequency sampling across an area of interest. As a result a bias toward long wavelength, variations may develop, exposing the estimates to aliasing of patterns. To include and improve the estimates at shorter scales, random sampling schemes have been devised that include nests of closely spaced variations.[26] In their study of physical and biological properties, 25×25 m plots were subdivided into twenty-five 5×5 m plots. A random core sample was taken from each plot. However, at five locations nine samples were collected in 3×3 sample nests. The nested samples were 10 cm apart. In this way, variations at the meter and centimeter scale could be resolved.[34]

Kriging

To estimate regional means of spatially correlated soil properties, the semivariance values are used in a spatial interpolation method called kriging. Kriging gives unbiased means and minimizes the error

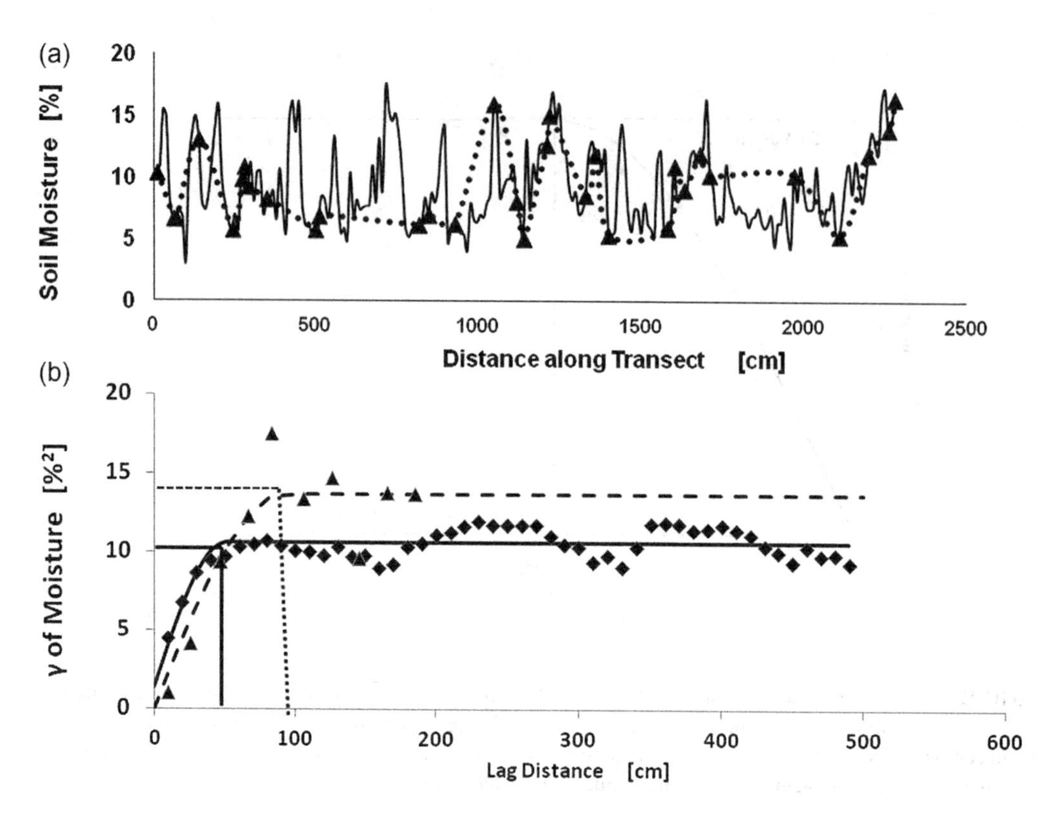

FIGURE 34.7 (a) Exhaustive trace of soil moisture variation across a plowed field (solid line) with a randomly selected subset of data points (stippled line and triangular markers) at an average spacing just less of the wavelength of the ridge-furrow pattern (Gorres, unpublished data). This is a violation of the Nyquist sampling theorem. The sampled variation mimics or aliases a variation with much longer spatial characteristics. (b) The semivariance function of the more exhaustively sampled moisture profile (solid line and diamond markers) has a spatially correlated range of 50 cm while the semivariance of the aliased variation (triangular markers and broken line) is 90 cm long overestimating both the correlated range and the sill variance. Note that the nugget variance of the aliased signal is zero, showing that when signals are undersampled, the nugget variance is not a good indicator of the estimation error made.

of estimation of regional parameters. The method was developed as a way to estimate the yield of ore deposits. In geological investigations, depositional patterns are large scale and aliasing may not be as much of a problem as in soils. Thus, the true error of the estimation of regional means with kriging depends strongly on the quality of the data collected and the care given to considering the spatial scale of patterns, even though kriging will mathematically minimize estimation error.

Some Applications of Spatial Estimation Methods

Spatial variations in soil properties are particularly important when the agricultural or environmental management require targeted measures. For example, nutrients need to be managed differently on a sandy soil than a clay soil. Sometimes environmental risks can be better assessed when a spatially explicit distribution of soil properties are available. There is a wide range of applications that use the spatial variability of soils in order to optimize estimates of properties or assess environmental risk.

One of the foremost commercial applications of the analysis of spatial patterns of soil properties is in variable rate agriculture, also known as precision agriculture. Improved productivity has been attributed to applying fertilizers at rates tailored to the previous year's yields or soil fertility status measured and spatially referenced using GPS. In field areas where fertility is low, more fertilizer is applied than in areas where it is high. Not only does this approach improve productivity and economic return, but it can also reduce nutrient loading to ground and surface water.

Losses of nutrients from fields are not only a risk to water quality but they also represent valuable farm resources. Spatial variations of soil properties are thus considered in farm nutrient management plans. Soil management tools such as the N leaching index and P index are used to estimate the risk of nutrient losses. In the U.S.A., many states assign the nitrate leaching index based on the hydrologic group of the soil, a landscape scale measure of infiltration potential, as well as the regional rainfall patterns.

At much larger scales, the spatial variability of soil alkalinity has been incorporated into approaches to regulating acid air pollution. The critical load concept is central to this. The critical load of acid deposition is reached when observable changes in the ecosystem occur, i.e., when the acid challenge can no longer be buffered by the ecosystem. The critical load concept relates the atmospheric deposition of SOx and NOx to the rate of base cation release from soil minerals during weathering as well as other factors. Weathering is a soil-forming process and thus climate, parent material, organisms, and topography impose a spatial pattern, which has to be accounted for when mapping critical loads of acid deposition for a region. The typical resolution of critical load maps is one to several kilometers. Once critical loads have been determined, exceedance of the critical load for a map unit can be calculated and emissions can be regulated.

Increasingly, variations in soil properties are being used in combination with soil process models such as the Leaching Estimation and Chemistry Model (LEACHM) or Nitrogen Loss and Environmental Assessment Package (NLEAP) in pollution risk assessment. For some applications, it is sufficient to treat the distribution of soil properties as independent. In this case, the means and measures of uncertainty of a particular soil process can be simulated by randomly selecting soil properties from their respective cumulative distribution functions as inputs to a soil process model. The repetition of these simulations gives a distribution of values for the soil process from which statistical moments can be derived. For example, LEACHN, a submodel of LEACHM that estimates nitrogen and carbon dynamics, was parameterized using soil properties randomly selected from cumulative distribution functions of measured soil organic matter, bulk density, texture, and field capacity to simulate the mean of NO_3-N in water extracted at a hypothetical production well.[35] This approach also allowed an estimate of the probability that the water pumped at the well exceeded the EPA drinking water standard of 10 mg NO_3-N/L under different land use scenarios.

Watershed-scale model estimates of runoff and erosion have been improved using spatially explicit representations of the landscape. These models use remotely sensed data or spatially referenced soil survey data to establish a spatial context rather than assuming random distributions of soil property values across the landscape. For example, the incorporation of soil information into a watershed model improved the estimate of storm hydrographs by accounting for delays in runoff contributions to stream flow.[36]

Conclusions

Soils exhibit high levels of variability. CVs vary from less than 10% to greater than 100%. The magnitude of the CV varies with the scale of measurement. Classical statistics can parameterize spatial distributions of spatial variations in terms of first-order statistical moments when the variations are spatially independent. However, more often than not, variations are spatially dependent making it necessary to evaluate a spatial correlation between points. Any analysis of spatial heterogeneity that does not consider the properties of spatial patterns in the design of measurement and monitoring schemes will likely produce erroneous information.

References

1. Warrick, A.W. Spatial variability. In *Environmental Soil Physics*; Hillel, D., Ed.; Academic Press: New York, 1998; 655–676.

2. Amador, J.A.; Glucksman, A.; Lyons, J.; Görres, J.H. Spatial distribution of phosphatase activity within a riparian forest. Soil Sci. **1997**, *162* (11), 808–825.
3. Görres, J.H.; DiChiaro, M.; Lyons, J.; Amador, J.A. Spatial and temporal patterns of soil biological activity in a forest and an old field. Soil Biol. Biochem. **1998**, *30* (2), 219–230.
4. Subler, S.; Baranski, C.A.; Edwards, C.M. Earthworm additions increased short term nitrogen availability and leaching in two grain-crop agro ecosystems. Soil Biol. Biochem. **1997**, *29* (3), 413–421.
5. Dominguez, J.; Bohlen, P.J.; Parmelee, R.W. Earthworms increase nitrogen leaching to greater soil depths in row crop agroecosystems. Ecosystems **2004**, *7* (6), 672–685.
6. Jenny, H. *Factors of Soil Formation: A System of Quantitative Pedology*; McGraw Hill Book Company: New York, NY, USA, 1941; 281 pp.
7. Beare, M.H.; Coleman, D.C.; Crossley, D.A.; Hendrix, P.F.; Odum, E.P. A hierarchical approach to evaluating the significance of soil biodiversity to biochemical cycling. Plant Soil **1995**, *170* (1), 5–22.
8. Rouatt, J.W.; Katznelson, H.; Payne, T.M.B. Statistical evaluation of the rhizosphere effect. Soil Sci. Soc. Am. J. **1960**, *24* (4), 271–273.
9. Savin, M.C.; Görres, J.H.; Amador, J.A. Dynamics of soil microbial and micro faunal communities as influenced by macropores, litter incorporation, and the anecic earthworms *Lumbricus terrestris* (L.). Soil Sci. Soc. Am. J. **2004**, *68* (1), 116–124.
10. Parkin, T.B. Soil microsites as a source of denitrification variability. Soil Sci. Soc. Am. J. **1987**, *51* (5), 1194–1199.
11. Drury, C.F.; Voroney, R.P.; Beauchamp, E.G. Availability of NH_4^+-N to microorganisms and the soil internal N-cycle. Soil Biol. Biochem. **1991**, *23* (2), 165–169.
12. Batte, M.T. Factors influencing the profitability of precision farming systems. Soil Water Conserv. **2000**, *55* (1), 12–18.
13. Parkin, T.B.; Starr, J.L.; Meisinger, J.J. Influence of sample size on measurement of soil denitrification. Soil Sci. Soc. Am. J. **1987**, *51* (6), 1492–1501.
14. Soil Survey Staff. *Soil Taxonomy: A Basic Guide for Soil Classification for Making and Interpreting Soil Surveys, 2nd Ed.*; USDA-NRCS: Washington, DC, 1999.
15. Nearing, M.A.; Pruski, F.F.; O'Neal, M.R. Expected impacts of climate change on soil erosion rates. Soil Water Conserv. **2005**, *59* (1), 43–50.
16. Kallenrode, M.-B. Droughts in the Western U.S. In *Space Physics: An Introduction to Plasmas and Particles in the Heliosphere Magnetospheres*; Springer: Berlin, Germany, 2004; 397–398.
17. Sexstone, A.J.; Revsbech, N.P.; Parkin, T.B.; Tiedje, J.M. Direct measurement of oxygen profiles and denitrification rates in soil aggregates. Soil Sci. Soc. Am. J. **1985**, *49* (3), 645–651.
18. Fanning, D.S.; Fanning, M.C.B. The factors of soil formation—Overview. In *Soil Morphology, Genesis and Classification*; John Wiley and Sons: New York, USA, 1989; 287–290.
19. McBratney, A.B.; Webster, R. Detection of a ridge and furrow pattern by spectral analysis of crop yield. Int. Stat. Rev. **1981**, *49* (1), 45–52.
20. Raghavan, G.S.V.; McKyes, E.; Chassé, M.; Merineau, F. Development of compaction patterns due to machinery operation in an orchard soil. Canad. J. Plant Sci. **1976**, *56* (3), 505–509.
21. Afzal, M.; Adams, W.A. Heterogeneity of soil mineral nitrogen in pasture grazed by cattle. Soil Sci. Soc. Am. J. **1991**, *56* (4), 1160–1166.
22. Ellenberg, H.; Weber, H.E.; Duell, R.; Wirth, V.; Werner, W.; Paulissen, D. Indicator values of plants in Central Europe. Scripta Geobotanica **1992**, *18*, 160–166.
23. Amador, J.A.; Görres, J.H. Microbiological characterization of the structures built by earthworms and ants in an agricultural field. Soil Biol. Biochem. **2007**, *39* (8), 2070–2077.
24. Ettema, C.H.; Wardle, D.A. Spatial soil ecology. Trends Ecol. Evol. **2002**, *17* (4), 177–183.
25. Görres, J.H.; Amador, J.A. Partitioning of habitable pore space in earthworm burrows. J. Nematol. **2010**, *42* (1) 68–72.

26. Görres, J. H.; Savin, M.; Amador, J.A. Dynamics of carbon and nitrogen mineralization, microbial biomass, and nematode abundance within and outside the burrow walls of anecic earthworms (*Lumbricus terrestris*). Soil Sci. **1997**, *162* (9), 666–671.
27. Hattori, T.; Hattori, R. The physical environment in soil microbiology: An attempt to extend principles of microbiology to soil microorganisms. CRC Crit. Rev. Microbiol. **1976**, *4* (4), 423–461.
28. Mummey, D.; Holben, W.; Six, J.; Stahl, P. Spatial stratification of soil bacterial populations in aggregates of diverse soils. Microb. Ecol. **2006**, *51* (3), 404–411.
29. Elliott, E.T.; Anderson, R.V.; Coleman, D.C.; Cole, C. V. Habitable pore space and microbial trophic interactions. Oikos **1980**, *35* (3), 327–335.
30. Hassink, J; Bowmann, L.A.; Zwart, K.B. Relationship between habitable pore space, soil biota and mineralization rates in grassland soils. Soil Biol. Biochem. **1993**, *25* (1), 47–55.
31. Neher, D.A.; Weicht, T.R.; Savin, M.C.; Görres, J.H.; Amador, J.A. Grazing in a porous environment. 2. Nematode community structure. Plant Soil **1999**, *212* (1), 85–99.
32. Görres, J.H.; Savin, M.C.; Neher, D.A.; Weicht, T.R.; Görres, J.H. Grazing in a porous environment: 1. The effect of soil pore structure on C and N mineralization. Plant Soil **1999**, *212* (1), 75–83.
33. Webster, R. Is soil variation random? Geoderma **2000**, *97* (3–4), 149–163.
34. Amador, J.A.; Wang, Y.; Savin, M.C.; Görres, J.H. Small-scale variability of physical and biological soil properties in Kingston, Rhode Island. Geoderma **2000**, *98* (1), 83–94.
35. Görres, J.H.; Gold, A. Incorporating spatial variability into GIS to estimate nitrate leaching at the aquifer scale. J. Environ. Qual. **1996**, *25* (3), 491–498.
36. Zhu, A.X.; Mackay, D.S. Effects of spatial detail of soil information on watershed modeling. J. Hydrol. **2001**, *248* (1), 54–77.

35

Soil: Taxonomy

Mark H. Stolt
University of Rhode Island

Introduction

One could argue that of all the natural resources—soil, water, and air are the only essential resources. Soil, like any natural resource, needs to be classified in order to best utilize and conserve the resource. Fanning and Fanning[1] suggested fairly simple reasons for classification: to provide a means to organize our knowledge; and to provide a system to retrieve the information for any number of purposes. Soils have been classified for almost as long as humans have lived in communities[2] One of the primary reasons was to assess the land for agriculture; those with the best soils for agriculture were taxed the greatest amount because they owned the best land. Since such classification approaches were not designed around the natural physical, chemical, and morphological properties of the soils (taxonomic properties), but based on their productivity or other use assessments, these soil classification approaches are termed interpretive classification systems.[1]

Agriculture is also the root of most soil taxonomic systems. For example, in Soil Taxonomy (the current system used in the United States and many other areas of the world), in deciding which properties to use as criteria for classifying the soil, the property with the most effect on agriculture is the one that is chosen.[3] Although throughout history agriculture has driven the design of a soil classification system, there are many other uses of soil taxonomy. This is clear in the title of the U.S. soil classification system; although typically referred to as just "Soil Taxonomy," the complete title is "Soil Taxonomy: A Basic System for Making and Interpreting a Soil Survey." In every U.S. soil survey there is a list of use and management interpretations for each soil taxa. These include many nonagronomic interpretations such as use of the soil for construction materials, recreational purposes, supporting roads and buildings, waste disposal treatment, constructing earthen dams, silviculture, etc. Thus, the ultimate purpose of soil taxonomy is to provide a means to assess the potential of the soil resource to serve any number of purposes through soil surveys. In this entry, we will focus on taxonomic soil classification systems.

Soil Taxonomy over the Last Century

Soil science and taxonomy, as we know them today, are little more than a century old. Prior to the late 1800s, soils were considered a surface mantle of loose and weathered rock.[4] V.V. Dokuchaiev, a Russian geologist working mostly between 1870 and the turn of the 20th century, articulated our current concept of soil: a natural body that has formed over time through a series of physical, chemical, and biological processes acting upon a given soil parent material. These processes are governed by the landscape setting, climate, and soil biology. Jenny[5] called these the five state factors of soil formation: climate, organisms, relief, parent materials, and time. Thus, soils with similar parent materials and biology that have formed under similar climates and landscape settings, will have similar properties, and thus taxonomic classification.

Of the taxonomic soil classification systems prior to the turn of the 20th century, Dokuchaiev's system was the most influential. His works were translated to German by Glinka[6] and later translated from German to English by C.F. Marbut.[1,7–9] Marbut considered many of the ideas and concepts of soils at that time relative to mapping soils and developed a classification system for the United States that was published in its final form in 1935.[10] Followers of Dokuchaiev, especially Sibertsev.[11] further developed Dokuchaiev's system that later became the foundation of the soil classification system that was used in the United States between 1938 and 1965.[1] The 1938 Yearbook system[12] was based primarily on Dokuchaiev's concepts of soil orders and were named Zonal, Intrazonal, and Azonal soils in the 1938 Yearbook.[13] Schaetzl and Anderson[14] called the 1938 Yearbook the first serious attempt of classifying soils in the United States. Zonal soils essentially matched the concept of "normal soils" described by Dokuchaiev and more fully articulated by Marbut[15]: soils that have developed in better drained conditions over enough time to be considered to have reached maturity under current climatic conditions. Azonal soils were the young soils developing in areas such as shallow to bedrock, floodplains, or aridic sands. Intrazonal soils were controlled by wetness or chemical composition such as salts. Examples and summaries of the 1938 Yearbook system can be found in Fanning and Fanning[1] and Schaetzl and Anderson.[14]

Although the 1938 Yearbook was used for many years in the United States, serious flaws were widely recognized in the system. There were six categories in the hierarchy, but the suborder (second) and family (fourth) categories went essentially unused.[16] Likewise, the series-level category failed in the hierarchical scheme; since it would not always match with a higher-level taxa. Some soils went unrecognized, such as some of those formed from volcanic ash or shrink-swell clays.[1] Another issue was that classification was based on unaltered (virgin) soils, thus a soil that had been cultivated or eroded was classified into a taxa that represented what it would have been like in a virgin form.[14] Because of these issues a more quantitative and less subjective system was needed.

In the early 1950s, Guy Smith began to work toward a new classification system.[17] Over the next decade, six versions (labeled approximations) of the system were circulated among soil scientists mostly in the United States.[18] The 7th approximation[19] was circulated worldwide at the World Soil Congress in 1960. At first there was little acceptance of the new system, but Simonson[20] noted regarding his paper published in Science "evidence of growing interest was indicated by requests for a few thousand reprints of my 1962 paper." By 1965, the system was adopted for USDA-SCS soil surveys and a decade later published as Soil Taxonomy[3]

A list of guidelines for the development of an effective soil classification system was given within the 7th Approximation[19] and later revised for inclusion in Soil Taxonomy[3] In general, the guidelines suggested that the system be flexible enough that soils that were not currently identified could easily be added; the system could be used worldwide; the properties used to classify the soil will be measured (not necessarily by instruments); the focal point of the properties measured are soil genesis related; if a choice needs to be made between which of two properties to use in the classification criteria, the one with the most affect on plant growth be used; and properties that are easily changed by anthropogenic activities such as plowing, be only used at the lowest of levels when classifying the soil.

The Soil That We Classify

Our concept of soil depends on our background and experiences[21] For example, someone with a potted plant may say the plant is supported by soil; and many would agree. A soil scientist would argue that this concept of soil is analogous to saying a cup of water taken from the lake shore is a lake. No one would agree with that! Knox[22] reviewed the various bodies of soil materials that could be considered soil. These ranged from individual particles, like a sand grain, to landscape-level bodies that could be mapped out at a 1:24000 scale. That pot of soil materials described above would be considered a "hand specimen" by Knox.[22] In this entry, we will use the USDA-NRCS[23] definition of soil: a collection of natural bodies; composed of mineral and organic materials; that supports growth of plants outdoors or show evidence of soil forming (pedogeneic) processes. Within this context, our collection natural bodies are the series of horizons that vary with depth and over an area of the landscape on the order of 1 to 10 m^2. Such a three-dimensional area of soil is called a pedon[19,21,24] Most soils are described using a 2-dimensional view of a pedon, called a soil profile (Figure 35.1). In most current soil taxonomic systems, a pedon is the smallest soil unit that is classified.

Soil Taxonomy into the 21st Century

Soil Taxonomy has six hierarchical categories: order, suborder, great group, subgroup, family, and series (Table 35.1). The first five categories are set up as a key so that every soil can be classified, the last taxa in the key is the default. In most cases, classification to the order, great group, or subgroup level in mineral soils is predicated on the presence or absence of diagnostic horizons. Those diagnostic horizons forming at the soil surface are called epipedons. Subsurface diagnostic horizons are typically in the position

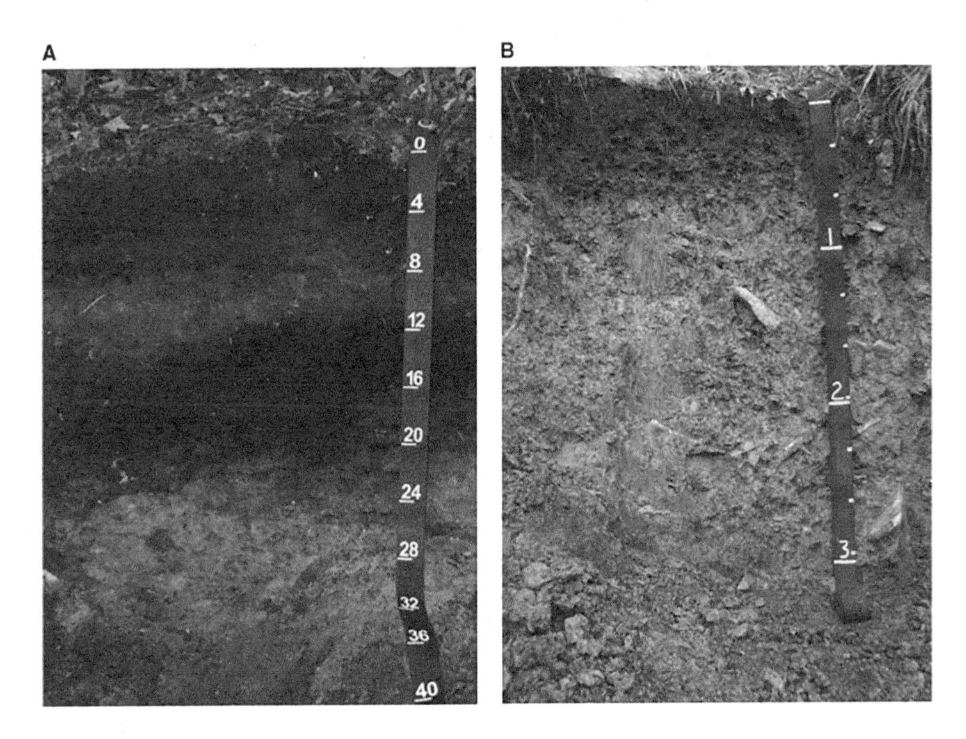

FIGURE 35.1 (See color insert.) (A) A Typic Epiaquod in Rhode Island. The tape is in inches. The spodic diagnostic subsurface horizon varies in thickness and depth. Near the tape the spodic horizon occurs from 10 to 20 inches. The soil is saturated to the surface for parts of year, thus the aquic suborder. The water is held, or perched, above a restrictive layer (epi) starting at 36″. There are no other diagnostic features (Typic). **(B)** An Umbric Endoaquult in Pennsylvania. The soil has an umbric epipedon, an argillic diagnostic subsurface horizon, and a base saturation <35%. The soil is saturated throughout the profile and to the soil surface (Endoaqu) for extended periods of the year. The small marks on the tape are 10 cm increments. The large marks are in feet.

(Continued)

FIGURE 35.1 (CONTINUED) **(See color insert.)** **(C)** Fluvaquentic Eutrudept on a Pennsylvania floodplain. The tape is in cm. Fluvaquentic indicates the soil is saturated during the year for a considerable amount of time above 60 cm and there is buried soil organic carbon (note the buried A horizon between 80 and 90 cm). The soil has a cambic diagnostic subsurface horizon (20 to 80 cm) with a base saturation >60% (Eutr). The climate is humid, thus a Udept. **(D)** Calcic Haplustalf in Texas. The small increments on the tape are 10 cm apart. The larger marks are one-foot apart. The soil has calcic and argillic diagnostic subsurface horizons. The base saturation is greater than 35% and the moisture regime is semiarid (ustic).

TABLE 35.1 The Architecture of Soil Taxonomy

There are six categories from the highest (least detail) to lowest (most detail) level: order, suborder, great group, subgroup, family, and series. The names for classes in the order, suborder, great group and subgroup categories are all coined terms using Greek, Latin, French, or similar roots. The soil orders are listed in the order they key out, a limited number of examples are provided for the other categories.

Order: Names of the twelve orders consist of three or four syllables and every name ends in the suffix "sol."

Order name	Formative element	Derivation
Entisol	ent	Nonsense syllable, think of recent
Vertisol	ert	L. *Verto*, turn, think of invert
Inceptisol	ept	L. *inceptum*, think of inception
Aridisol	id	L. *aridus*, dry, think of arid
Mollisol	oll	L. *mollis*, soft, think of mollify
Spodosol	od	Gk. *Spodos*, wood ash, think of Podzol*
Alfisol	alf	Nonsense syllable, think of Pedalfer**
Ultisol	ult	L. *ultimus*, last, think of ultimate

(*Continued*)

TABLE 35.1 (*Continued*) The Architecture of Soil Taxonomy

Order name	Formative element	Derivation
Oxisol	ox	F. *oxide*, think of oxide
Andisol	and	Gk. *Andes*, think andesite
Histosol	ist	Gk. *histos*, tissue, think of Histology
Gelisol	el	L. *gelare*, to freeze, think of gel

*Podzol-classic term used for soils with spodic horizons; **Pedalfer-one of the two taxa in Marbut's (1935) highest category.
Suborder: The name of every suborder is a two-syllable term consisting of a prefix syllable with some specific connotation plus the formative element from the order to which the suborder belongs. The examples are for Ultisols, Inceptisols, and Vertisols, respectively.

Formative element	Connotation	Example
Aqu	Aquic moisture regime	Aquult
Ud	Udic moisture regime	Udept
Ust	Ustic moisture regime	Ustert

Great group: The name of each great group is constructed by adding a second prefix element with a specific connotation to the two-syllable term which is the name of the suborder to which the great group belongs.

Formative element	Connotation	Example
Umbra	An umbric epipedon	Umbraquult
Eutr	High base status	Eutrudept
Calci	A calcic horizon	Calciustert

Subgroup: Names of subgroups are binomial. An adjective (sometimes two) is used to modify the name of the great group to give the name of each subgroup within that great group.

Adjective	Connotation	Example
Plinthic	Presence of plinthite	Plinthic Umbraquult
Arenic	Thick sandy surface	Arenic Eutrudept
Lithic	Shallow to rock	Lithic Calciustert

Family: A sequence of descriptive adjectives is used in family names. The common examples are used to describe particle-size classes, mineralogy classes, activity classes, and soil temperature classes.

An example is Coarse-loamy, mixed, active, mesic, Typic Dystrudept. Coarse-loamy (particle-size class), mixed (mineralogy class), active (activity class), mesic (soil temperature class), Aquic Dystrudept (subgroup).
Series: is a family that falls within a given range of characteristics. Series names are typically derived from towns or localities that are near where the soils are found (i.e., Newport). The range in characteristics are related to morphological properties such as thickness of the B horizons, soil colors, rock fragment contents, soil textures, etc. The range in characteristics are not found in Soil Taxonomy but are accessed online: http://soils.usda.gov/technical/classification/osd/index.html

of the B horizons in the profile. The standard for the epipedon is the mollic because soils with mollic epipedons typically are excellent for agriculture. Most of the soils in the Great Plains in the United States and the Steppe regions in Russia have mollic epipedons. These are the soils that Dokuchaiev studied in Russia and formulated his early concepts of soils. Mollic epipedons are thick (typically 25 cm or more), dark in color, rich in organic matter, and rich in basic cations (base saturation >50%). Umbric epipedons are similar to the mollic except more acidic, having a lower base status. Histic epipedons are typically 20–40 cm thick and dominated by organic soil materials. There are also folistic (unsaturated histic epipedons), mellanic (associated with volcanic ash), plaggen (heavily manured for centuries), and anthropic (P rich from long-term human occupation) epipedons. In general, the ochric epipedon is the default, meaning soils that do not meet the criteria for the other six epipedons have an ochric epipedon (very new soils may have no epipedon).

In Soil Taxonomy, there are 18 diagnostic subsurface horizons. Only the more common are briefly discussed here so that the reader can understand the concept and follow the examples in Table 35.1. Most diagnostic subsurface horizons are related to the accumulation of soil constituents in the B horizon position of the soil profile. The accumulations may be of silicate clay (argillic), sesquioxides bound to organic matter (spodic), calcium carbonate (calcic), gypsum (gypsic), and salts more soluble than gypsum (salic). Certain diagnostic subsurface horizons are dense or cemented by the constituent that has accumulated such as petrocalcic, petrogypsic, or a duripan (cemented by silica that has reprecipitated in the B horizon). Younger soils may not have had enough time to form any of the diagnostic horizons previously mentioned but have B horizons recognized by a simple change in color. These "color" B horizons typically meet the criteria for a cambic diagnostic horizon.

In keying out soils to the order level, Gelisols are the first possibility. The important criterion is the presence of permafrost (generally within a meter of the soil surface). Histosols key out next, these soil are dominated by organic soil materials (at least 40 cm) in the upper 80 cm. Spodosols follow, having spodic diagnostic horizons. Andisols have andic soil properties (high exchange capacity and water holding capacity). Oxisols are extremely weathered and dominated by Fe and Al oxides (oxic horizon). Vertisols are dominated by shrink-swell clays that results in a self-mixing soil. Aridsols may have a number of diagnostic horizons such as argillic, calcic, cambic as long as they are in aridic (dry) moisture regimes. Ultisols have argillic horizons and low base saturation (<35%). Mollisols have mollic epipedons and a base saturation throughout the profile greater than 50%. These soils may have calcic or argillic horizons. Alfisols almost always have argillic horizons, but unlike Ultisols their base saturation is >35%. Inceptisols are weakly developed soils typically with cambic horizons. Entisols are the least developed, characterized by a lack of diagnostic horizons or epipedons other than ochric, and key out last.

Because soil moisture (or lack of) has a dramatic effect on soil morphology, suborders typically bring the wetness attributes or soil moisture states related to climate (soil moisture regimes) into the classification (see examples in Table 35.1 and Figure 35.1). These moisture regimes include aquic (wet), udic (humid), ustic (semiarid), and xeric (dry parts of the year and moist other parts).

There were only a few major changes to Soil Taxonomy over the years. In 1975, there were 10 orders; that changed to 12 with the addition of Gelisols (soils with permafrost) and Andisols (soils formed in volcanic ash type materials). The definition of soil was changed in the 2nd edition.[25] Previously by definition, a soil was required to be able to support rooted plants.[3] In the new edition, soils, as long as they showed evidence of pedogenesis, were considered soils even if they did not support rooted plants. This departure from the agricultural roots of Soil Taxonomy was designed to accommodate soils in the arctic that were permanently covered with snow or ice[26,27] or soils that were permanently under water (subaqueous soils).[28]

In 1983, field versions of Soil Taxonomy were first published. The field versions essentially just included what was needed to key the soil out to the family level, much of the supplemental materials of the classification system was left out. These versions were called the *Keys to Soil Taxonomy.* As we learned more and more about the soils, changes in the taxonomy were introduced in revised versions of *Keys to Soil Taxonomy.* For example, taxa were included in the 11th edition of *Keys to Soil Taxonomy*[23] to accommodate subaqueous soils.[28] Although by design Soil Taxonomy was meant to be flexible enough so that new soils could be added when deemed necessary,[3] some argue that over time it might be too big and cumbersome to be useful any longer. For example, the first *Keys to Soil Taxonomy*[29] was a 4″ × 9″ paperback that easily fit in your back pocket for use in the field (<250 pages). By 1996, the 7th edition of *Keys to Soil Taxonomy* in the same 4″ × 9″ format was well over 600 pages.

One area of soils that continues to be difficult to classify in Soil Taxonomy are the anthropogenic soils (i.e., soils developed in materials moved by man).[30] In an attempt to resolve this issue, an International Committee on Anthropogenic Soils (ICOMANTH) was formed with the charges defining appropriate classes in Soil Taxonomy for soils that have their major properties derived from human activities. The first activities of this committee were reported in the 1st circular letter dated 1995 (http://clic.cses.vt.edu/icomanth/03-AS_Circulars.pdf). Efforts of the committee resulted in new horizon designations for soils with restrictive properties because of the presence of a continuous man-made material such as cement (master horizon designation M) and horizons containing human artifacts such as glass, brick, or cement (subordinate horizon designation u). These changes appeared in the 10th edition of *Keys to Soil Taxonomy*.[31] Suggestions for classifying such soils have primarily focused on subgroups such as Spolic, Urbic, or Garbic[1,32] or a separate order with Soil Taxonomy— Anthrosols.[33] Efforts toward classifying anthropogenic soils continue through ICOMANTH such that at the present time there are seven circular letters describing proposals, changes, and related activities of the committee (http://clic.cses.vt.edu/icomanth/circlet.htm).

Other National and International Classification Systems

There are numerous other taxonomic soil classification systems. Buol et al.[34] reviewed a number of these including the Russian, French, Belgian, British, Canadian, Australian, Brazilian, and Chinese. Many of these systems take elements directly or indirectly from Soil Taxonomy.[25] For example, at the order level the Australian system uses "Organosols" instead of Histosols.[35] Seven of the diagnostic horizons or features in the Chinese system are taken directly from Soil Taxonomy.[36] The Chinese and Australian systems have soil orders for anthropogenic soils.[35,36]

The World Reference Base[37] is the only true international soil classification system. This system, known as the WRB, was endorsed and adopted by the International Union of Soil Sciences (IUSS) as the classification system for soil correlation and international communication. The intention of the WRB was not to provide an alternative for any national soil classification system, but to serve as a soil classification translator for communication at an international level.[37] During the development of the WRB, a focused effort was put forth to include as much nomenclature from Soil Taxonomy,[25] and other major national soil classification systems, as possible.

The WRB is a two-tiered system. At the highest category are 32 Reference Soil Groups. A number of these are essentially the same as the soil orders in Soil Taxonomy; examples of these include Vertisols, Histosols, Podzols (Spodosols), Cryosols (Gelisols), and Chernozems (Mollisols). Like Soil Taxonomy,[25] diagnostic horizons are used as criteria in many of the taxa. Although their definitions are not identical, a number of the WRB diagnostic horizons are conceptually the same as those in Soil Taxonomy. Some examples in include mollic, ochric, histic, gypsic, calcic, and cambic.

The second tier of the WRB is a series of qualifiers that are added as a prefix or suffix to the reference soil groups to identify in greater detail properties of the particular soil. Each reference soil group is assigned a list of qualifiers that may be used. In the WRB[37] an example is given for a Cryosol soil reference group: "Histic Turbic Cryosol (Reductaquic, Dystric)." Histic and Turbic are the prefix qualifiers indicating that the Cryosol has a histic epipedon and evidence of cryoturbation (mixing by freeze-thaw), respectively. Reductaquic and Dystric are the suffixes indicating the soil is both saturated and reduced, and has a base saturation less than 50%, respectively. Unlike Soil Taxonomy,[25] climatic parameters are not applied in the classification of soils in the WRB and the system is not designed for mapping soils at resource-level scales.

Summary and Conclusions

Soil is one of the basic and essential natural resources. Thus, some form of soil classification has been around since humans formed communities. The early systems were primarily based on a use assessment of the soils for a single purpose such as agriculture productivity (i.e., low yields, moderate yields, or high yields). These systems are often referred to as interpretive soil classifications. Taxonomic classification of soils is based on the natural physical, chemical, and morphological properties of the soil and can provide many use and management interpretations. Application of taxonomic classification for soil resources is through the use of soil surveys. The best example of a taxonomic system for mapping and assessment of the soil resource is Soil Taxonomy. This is a hierarchical system that is set up like a key so that every soil can be classified. There are six categories in the system with order the highest (least amount of detail) and series the lowest (greatest amount of detail). Although Soil Taxonomy is used worldwide, many countries have their own soil classification system. There is also a truly international system called the World Reference Base (WRB). The WRB is designed as an international translator for systems such as Soil Taxonomy and was not meant to provide classifications for resource soil mapping.

Current issues with Soil Taxonomy appear to be related to the lack of significant taxa for anthropogenic soils, its field applications, and the rapidly expanding number of taxa. Issues with field applications center around the need for considerable laboratory data to correctly classify a soil. A possible resolution is to develop guidelines based on reasonable assumptions that can assist the user in applying Soil Taxonomy in the field without any laboratory data. As we learn more and more about soils, the number of taxa in Soil Taxonomy continues to grow. Currently, proposals are being considered to add taxa to accommodate the classification of some anthropogenic soils. Some users feel that continued additions leaves the classification system too cumbersome to key through to find the correct taxa. Whether this is a problem or not is debatable, but correcting such a problem may require reformatting the hierarchical system. An approach to resolve this issue that is currently being considered is new classification system using some of the format of the WRB and some of Soil Taxonomy. As the world's population continues to grow resource managers will turn more and more to soils information to best utilize this essential resource. As such, developing an understanding of soil taxonomy will continue to be a need for natural resource users and managers.

References

1. Fanning, D.S.; Fanning, M.C.B. *Soil Morphology, Genesis, and Classification;* John Wiley and Sons: New York, NY, 1989.
2. Simonson, R.W. Soil classification in the United States. Science **1962**,*137,* 1027–1034.
3. Soil Survey Staff. *Soil Taxonomy: A Basic System of Soil Classification for Making and Interpreting Soil Surveys, USDA-NRCS Agricultural Handbook 436*; U.S. Government Printing Office: Washington, DC, 1975.
4. Simonson, R.W. Historical aspects of soil survey and soil classification. Part 1. 1899–1910. Soil Survey Horizons **1986a**, *27* (1), 3–11.
5. Jenny, H. *Factors of Soil Formation.* McGraw Hill Book Co. Inc.: New York, NY, 1941.
6. Glinka, K.D. *Die Typen der Bodenbildung: Ihre Kalssification und Geographische Berbreitung.* Gebruder Borntrager: Berlin, 1914.
7. Marbut, C.F. *The Great Soil Groups of the World and Their Development. Translated from Glinka, 1914.* Edwards Brothers: Ann Arbor, MI, 1927.
8. Smith, G.D. Historical development of soil taxonomy-background. In *Pedogenesis and Soil Taxonomy.* Wilding, L.P., Smeck, N.E.; Hall, G.F.; Eds.; Elsevier, Amsterdam, 1983, 23–49 pp.
9. Simonson, R.W. Historical aspects of soil survey and soil classification. Part 2. 1911–1920. Soil Surv. Horizons **1986b**, *27* (2), 3–9.
10. Marbut, C.F. Soils of the United States. In *Atlas of American Agriculture;* Baker, O.E., Eds.; USDA, U.S. Government Printing Office: Washington, D.C., 1935; 12–15 pp.
11. Sibertsev, N.M. *Russian Soil Investigations. Israel Programs for Science Translations Jerusalem. 1966. Translated from Russian by N. Kaner*; U.S. Department of Commerce: Springfield, VA, 1901.
12. Baldwin, M., Kellogg, C.E.; Thorp, J. Soil classification. In *Yearbook of Agriculture: Soils and Men;* USDA, U.S. Government Printing Office: Washington, DC, 1938; 979–1001 pp.
13. Simonson, R.W. Historical aspects of soil survey and soil classification. Part 4. 1931–1940. Soil Survey Horizons **1986c**, *27* (4), 3–10.
14. Schaetzl, R.; Anderson, S. *Soils: Genesis and Geomorphology*; Cambridge University Press: New York, NY, 2005.
15. Marbut, C.F. A scheme of soil classification. In *Proceeding of the 1st International Congress of Soil Science;* Washington, DC., 1928; 1–31 pp.

16. Simonson, R.W. Historical aspects of soil survey and soil classification. Part 5. 1941–1950. Soil Survey Horizons **1987a**, *28* (1), 1–8.
17. Cline, M. *Soil Classification in the United States. Agronomy Mimeo 79-12;* Cornell University: Ithaca, NY, 1979.
18. Simonson, R.W. Historical aspects of soil survey and soil classification. Part 6. 1951–1960. Soil Survey Horizons **1987b**, *28* (2), 39–46.
19. Soil Survey Staff. *Soil Classification, a Comprehensive System—7th Approximation;* USDA. U.S. Government Printing Office: Washington, D.C, 1960.
20. Simonson, R.W. Historical aspects of soil survey and soil classification. Part 7. 1961–1970. Soil Survey Horizons **1987c**, *28* (3), 77–84.
21. Simonson, R.W. Concept of soil. Adv. Agronomy. **1968**, *20*, 1–47.
22. Knox, E.G. Soil individuals and soil classification. Soil Sci. Soc. Am. Proc. **1965**, *29*, 79–84.
23. Soil Survey Staff. *Keys to Soil Taxonomy*, 11th Ed.; United States Department of Agriculture Natural Resources Conservation Service. Government Printing Office: Washington, D.C, 2010.
24. Simonson, R.W.; Gardner, D.R. The concepts and functions of the pedon. Transactions of the 7th International Congress of Soil Science. 1960; 129–131.
25. Soil Survey Staff. *Soil Taxonomy: A Basic System of Soil Classification for Making and Interpreting Soil Surveys*, 2nd Edn.; USDA-NRCS Agricultural Handbook 436, U.S. Government Printing Office: Washington, DC, 1999.
26. Bockheim, J.G. Soil development rates in the Transantarctic Mountains. Geoderma **1990**, *47*, 59–77.
27. Bockheim, J.G. Properties and classification of cold desert soils from Antarctica. Soil Science Soc. Am. J. **1997**, *64*, 224–231.
28. Stolt, M.H.; Rabenhorst, M.C. Subaqueous soils. In *Handbook of Soil Science*, 2nd Ed.; Huang, P.M.; Li, Y.; Sumner, M.E., Eds.; CRC Press: Boca Raton, FL, 2011; 36.1–36.14 pp.
29. Soil Survey Staff. *Keys to Soil Taxonomy;* United States Department of Agriculture Natural Resources Conservation Service, Government Printing Office: Washington, D.C, 1983.
30. Kimble, J.M., Aherns, R.; Bryant, R. *Classification, Correlation, and Management of Anthropogenic Soils*; USDA-NRCS, National Soil Survey Center: Lincoln, NE, 1999.
31. Soil Survey Staff. *Keys to Soil Taxonomy*, 10th Ed.; United States Department of Agriculture Natural Resources Conservation Service, Government Printing Office: Washington, D.C, 2006.
32. Pouyat, R.V.; Ellland, W.R. The investigation and classification of humanly-modified soils in the Baltimore ecosystem study. In *Classification, Correlation, and Management of Anthropogenic Soils.* Kimble, J.M., Aherns, R., Bryant, R., Eds.; USDA-NRCS, National Soil Survey Center: Lincoln, NE, 1999; 141–154 pp.
33. Bryant, R.B.; Galbraith, J.M. Incorporating anthropogenic processes in soil classification. In *Soil Classification: A Global Desk Reference;* Eswaren, H.; Rice, T.; Aherns, R.; Stewart, B.A., Eds.; CRC Press: Boca Raton, FL, 2003; 57–65 pp.
34. Buol, S.W.; Southard, R.J.; Graham, R.C.; McDaniel, P.A. *Soil Genesis and Classification*, 5th Ed.; Iowa State Press: Ames, Iowa, 2003.
35. Isbell, R.F. *The Australian Soil Classification System. Australian Soil and Land Survey Handbook;* CSIRO Publishing: Collingwood, Australia, 1998.
36. Gong, Z.; Zhang, G.; Luo, G. The Anthrosols in Chinese soil taxonomy. In *Classification, Correlation, and Management of Anthropogenic Soils;* Kimble, J.M., Aherns, R., Bryant, R., Eds.; USDA-NRCS, National Soil Survey Center: Lincoln, NE, 2003; 40–51 pp.
37. Food and Agriculture Organization of the United Nations. *World Reference Base for Soil Resources*; FAO: Rome, 2006.

36

Soil: Microbial Ecology

Sara Wigginton and
Jose A. Amador
University of Rhode Island

Introduction

Microbial ecology—the interactions among soil microorganisms (bacteria, archaea, fungi, and viruses; Figure 36.1), among microorganisms and soil fauna, and between microorganisms and their environment—is at the center of many of the ecosystem services provided by soil (Table 36.1). These include (i) the formation of soil organic matter, which results in C sequestration and impacts soil physical and chemical properties essential for plant growth; (ii) biological nitrogen fixation, which turns biologically inert N_2 gas in the atmosphere into plant-available N; (iii) biogeochemical cycling of elements, which makes nutrients such as N, P, and S available to plants; (iv) production and consumption of greenhouse gases (CO_2, CH_4, N_2O), which partly controls the composition of the atmosphere and impacts climate change; and (v) control of plant pathogens, which affects crop production globally. These interactions take place in the context of a foodweb (Figure 36.2), where microorganisms interact with other soil organisms as prey, predator, competitor, or collaborator. The physical and chemical properties of soil—a variably porous, opaque medium made up of mineral particles and organic matter that interact with water, heat, nutrients, and contaminants—help define the niches available for organisms to live, and constrain the size and activities of soil organisms, and thus the transfer of C, nutrients, and energy in the soil ecosystem.

Here, we discuss basic microbial metabolism and describe the main groups of soil microorganisms based on their morphology, and carbon and energy needs. We also describe modern nucleic acid-based methods of quantifying and describing soil microbial communities. We then discuss the ecological interactions among soil organisms within the soil foodweb, and how these affect microbial community dynamics and the flow of C, energy, and nutrients.

Microbial Metabolism

An understanding of microbial metabolism is essential to understand how the soil environment affects where microorganisms live and the activities they can carry out. Microorganisms are made up of thousands of organic molecules with different cellular functions. Carbon (C) makes up the backbone of these compounds, and one of the main metabolic divisions among organisms is how they obtain C for biomass synthesis and energy. A vast majority of soil microorganisms are *heterotrophic*: they get carbon from other organisms and derive energy from processing organic compounds to drive the synthesis

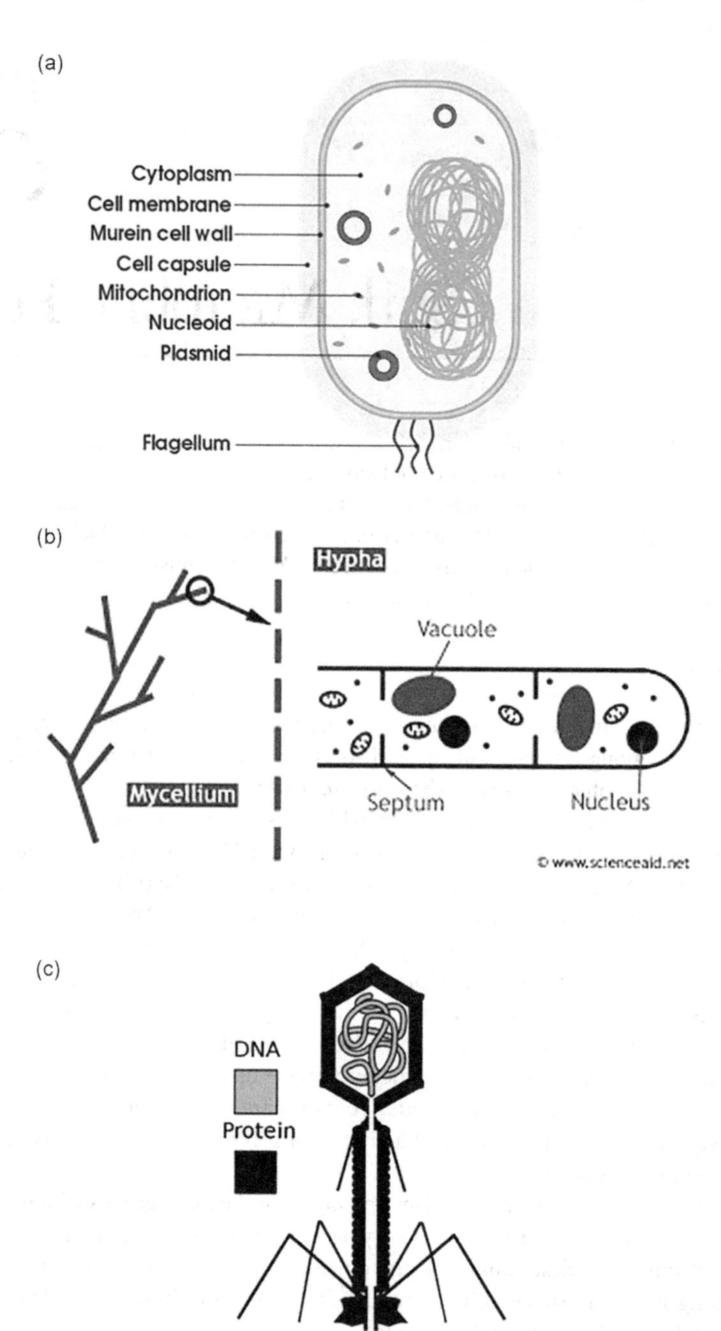

FIGURE 36.1 **(See color insert.)** Morphology of bacteria (**a**), fungal hyphae (**b**), and bacteriophage (**c**). (Picture credit: (a) Image by Domdomegg, distributed under a CC-BY 4.0 license. (b) Image from ScienceAid.net. (c) Image by Mike Jones, distributed under a CC BY-SA 2.5 license.)

of biomass. In contrast, some soil microorganisms are *autotrophic*: they get carbon from CO_2, as plants do. Autotrophic microorganisms derive energy from processing of inorganic compounds, used to drive the synthesis of biomass. The difference between heterotrophic and autotrophic metabolism has consequences for interactions in soil. For example, heterotrophic and autotrophic organisms do not compete for C sources.

TABLE 36.1 Examples of Important Biogeochemical Processes Carried Out by Soil Microorganisms

Process	Group(s) Involved	Representative Equation
Natural polymer decomposition	Bacteria, fungi	$(C_6H_{12}O_6)_n \rightarrow C_6H_{12}O_6$ Cellulose glucose
Nutrient mineralization	Bacteria, archaea, fungi	$R\text{-}PO_4^{3-} \rightarrow R\text{-}H + PO_4^{3-}$
Nutrient immobilization	Bacteria, fungi	$R\text{-}H + PO_4^{3-} \rightarrow R\text{-}PO_4^{3-}$
Nitrogen fixation	Bacteria	$N_2 \rightarrow 2NH_3$
Nitrification	Bacteria, archaea	$NH_3 \rightarrow [N_2O]NO_2^- \rightarrow NO_3^-$
Denitrification	Bacteria	$2NO_3^- \rightarrow N_2O \rightarrow N_2$
Methane oxidation	Bacteria	$CH_4 \rightarrow CO_2$
Methanogenesis	Archaea	$CO_2 + H_2 \rightarrow CH_4$ $CH_3COOH \rightarrow CH_4 + CO_2$
Metal reduction and oxidation	Bacteria, fungi, archaea	$FeOOH \leftrightarrow Fe^{2+}$
Degradation of synthetic compounds	Bacteria, fungi, archaea	$C_6H_5Cl \rightarrow CO_2, Cl^-$ Chlorobenzene

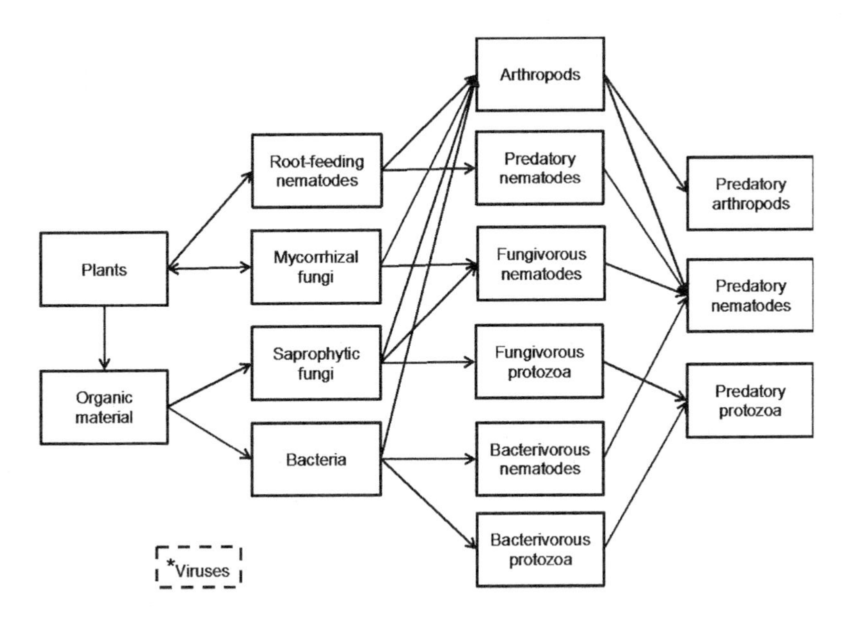

FIGURE 36.2 A simplified soil foodweb showing interactions among soil organisms. Archaea are likely affected by the same interactions as bacteria. Viruses can infect plants, bacteria, fungi, and protozoa.

The synthesis and breakdown of organic molecules—proteins, membrane components, and DNA—that make up microbial cells requires energy. Microorganisms generate energy by transferring electrons from molecules that have a high concentration of electrons (*electron donors*) to those that have a low concentration of electrons (*electron acceptors*). As electrons are transferred from one molecule to another via the *electron transport chain* across the cell membrane, a pH gradient is established between the inside and outside of the cell. This gradient is coupled with the synthesis of ATP (adenosine triphosphate), a high-energy compound that is a source of energy for metabolic processes. We classify microorganisms metabolically based on the *terminal electron acceptor* (TEA) they use. The TEA for *aerobic* microorganisms is oxygen (O_2), which yields the most energy, and is the only TEA strict aerobic microorganisms can use. In contrast, *anaerobic* microorganisms use compounds other than O_2, including nitrate (NO_3^-), manganese (Mn), iron (Fe), sulfate (SO_4^{2-}), and CO_2, which yield progressively less energy than oxygen. Many soil microorganisms—referred to as *facultative anaerobes* or *facultative aerobes*—can switch between O_2 and other TEAs. Differences in the ability to use different TEAs have consequences for the type of habitat that can be occupied by soil microorganisms. They also affect the ability of soil organisms to extract energy from substrates, and thus the size of their populations and communities.

Groups of Soil Microorganisms

A teaspoon of soil contains between 100 million and 1 billion *bacteria* and thousands of bacterial species, making them the most numerous organisms in soil. They are *prokaryotic* (= without a nucleus), unicellular, and range in diameter from 0.5 to 1.0 µm (Figure 36.1). Bacteria can be autotrophic or heterotrophic, with the latter using a wide range of natural and synthetic compound for C and energy. They are the most metabolically versatile members of the soil biota: they can use the whole range of TEAs, which allows them to live everywhere within the soil. Bacteria are aquatic and live in soil pore water or on the surface of soil particles that make up soil pores. They reproduce by fission, creating two cells from a single cell. Many bacteria can survive adverse environmental conditions, such as desiccation and temperature extremes, by reducing their metabolic requirements or by producing spores that are resistant to environmental stresses. They are the main drivers of biogeochemical cycles in soil (Table 36.1) and are responsible for symbiotic biological nitrogen fixation. A small number of soil bacteria are plant and animal pathogens, and a few are human pathogens, such as *Bacillus anthracis*, which causes anthrax. The size and composition of bacterial communities is generally constrained by the availability of organic C and electron acceptors, by predation by bacterivorous nematodes and protozoa, and by infection by *bacteriophage*—viruses that infect bacteria (Figure 36.2).

Archaea are prokaryotic, unicellular, aquatic microorganisms that resemble bacteria in size and morphology, which led to their misclassification until very recently. They differ from bacteria in the structure of the molecules that make up their cell membranes, which allows them to live in a broad range of extreme conditions. Like bacteria, archaea can be autotrophic or heterotrophic, and can use a range of TEAs. Relatively little is known about the ecology of archaea. Many species are extremophiles that thrive in extremes of temperature, pressure, pH, or salinity. In soil, a number of archaeal species are autotrophs that oxidize ammonium to nitrite for energy, like their bacterial counterpart. Although they carry out the same process, their niches are differentiated by a preference for environments with low ammonium concentrations by ammonia-oxidizing Archaea (Jia and Conrad, 2009). The size and composition of archaeal communities are likely controlled by the same factors that control bacterial communities.

Soil *fungi* are *eukaryotic* (= with a nucleus) and can be unicellular or multicellular, with dozens of species of fungi found in a gram of soil. Multicellular fungi form strands (*hyphae*), which together form filaments (*mycelia*) that are visible to the naked eye (Figure 36.1). Hyphae are between 2 and 10 µm in diameter, and unicellular species (yeasts) are ~4 µm in diameter. Filamentous fungi live in larger soil pores or displace soil particles as they grow through the soil, whereas yeasts live in smaller, water-filled pores. Most fungi are *saprophytic*, feeding on organic detritus in soil. They are aerobes, although some unicellular species are facultative anaerobes. Fungi reproduce by budding, or through sexual or asexual spore production. In forest soils, fungal biomass is generally larger than that for bacteria, whereas in agricultural soils, it tends to be smaller than bacterial biomass, likely due to physical disruption of mycelia and lower C inputs in the latter. In contrast, filamentous fungi are less affected by lack of moisture than bacteria or archaea. Fungi play various roles in the soil ecosystem (Frąc et al., 2018), including (i) decomposition of detritus—especially polymers like cellulose and lignin—and consequent biogeochemical cycling of C, nutrients, and trace elements (Table 36.1); (ii) improving the ability of vascular plants to acquire water, phosphorus ,and other nutrients through mycorrhizal associations; and (iii) as pathogens of plants, including economically important crops, as well as insects and other animals. The availability and diversity of organic C substrates are important controls on the size and composition of soil fungal communities, as are predatory pressures from fungivorous nematodes, mites, and other soil fauna (Figure 36.2).

Viruses range in size from 20 to 400 nm, with 10^7–10^9 viral particles present in 1 g of soil (Emerson, 2019). They consist of a molecule of nucleic acid (DNA or RNA) covered in a protein coating (Figure 36.1). They do not carry out any of the metabolic processes that other microorganisms do, and cannot reproduce on their own. Instead, they use the reproductive machinery of their hosts to reproduce, lysing the host cell to release the viral particles. Viruses can live under oxic and anoxic conditions, and are susceptible to environmental stresses, including temperature, pH, and low moisture. They infect mammals, plants, insects, and bacteria, affecting their survival and thus population dynamics. Viruses that infect bacteria and archaea (*bacteriophage*) are important in controlling the size of their communities and in interspecies gene transfer, which in turn affects biogeochemical cycles (Armon, 2011).

Measuring the Size and Structure of Soil Microbial Communities

Until recently, our knowledge of the size and structure of microbial populations and communities was limited to those microorganisms that could be grown in the laboratory. These culture-dependent methods underestimated the size of microbial communities by a factor of ~100 and misrepresented their composition (Torsvik and Øvreås, 2002). Culture-dependent methods have largely been replaced by nucleic acid-based analyses, which vary in complexity, expense, and the kinds of information obtained. Arguably, the most powerful method is *metagenomic analysis*, which allows for the extraction and analysis of all the genetic material in mixed environmental samples, such as soil. The ability to sequence—determine the order of base pairs in a genetic sample—and investigate the entire genetic fingerprint of a soil sample has ushered in a new era of soil microbial ecology that has resulted in new insights into the inner workings of the soil ecosystem.

Metagenomic analyses rely on two recent advances in molecular biology: (i) the advent of the *polymerase chain reaction* (*PCR*), which allows for quick, inexpensive production of multiple copies of a nucleic acid sequence, and (ii) inexpensive, rapid methods for determining the sequence of base pairs in these copies, referred to as *high-throughput sequencing* (Figure 36.3). The ability to make copies of gene segments using PCR is the basis for *quantitative PCR* (*qPCR*), a technique used to enumerate soil microorganisms that bypasses the need for culturing them.

The two main metagenomic approaches are (i) *target gene sequencing* and (ii) *whole metagenome sequencing*. Target gene sequencing involves extracting all the DNA or RNA from a sample and targeting a small portion of the genetic code for sequencing. Much of the work that has been done in soil metagenomics has targeted the 16S ribosomal (16S rRNA) gene of bacteria and archaea, and the 18S rRNA gene of fungi and other eukaryotes. This sequence is often used to identify the taxonomic composition of these communities because it is ubiquitous and has changed at a much slower rate than the rest of the microbial genome. Sequencing these genes is thus an invaluable tool for identifying unculturable organisms at the genus and even species level, and is likely the most effective and efficient means to assess the diversity and phenology of soil microbial communities.

Genes coding for proteins that carry out specific functions—such as enzymes, can also be targeted in metagenomic sequencing. For example, studies of biological nitrogen fixation may involve sequencing the nitrogenase iron marker gene *nifH* in a soil sample to explore this metabolic function in soil, which provides information about the functional potential and diversity. If instead of DNA the RNA is targeted for sequencing, we can identify and sequence *nifH* genes that are being actively transcribed, which yields information about gene expression and activity. Studying gene expression, known as *metatranscriptomics*, allows further insight to ecosystem function and the links between function and taxonomic identity.

Whole metagenome "shotgun" sequencing involves subdividing all the DNA or RNA in a sample into small strands that are sequenced and then reassembled using computational methods. This method generates long sections of DNA or RNA (sometimes entire genomes) from all the organisms

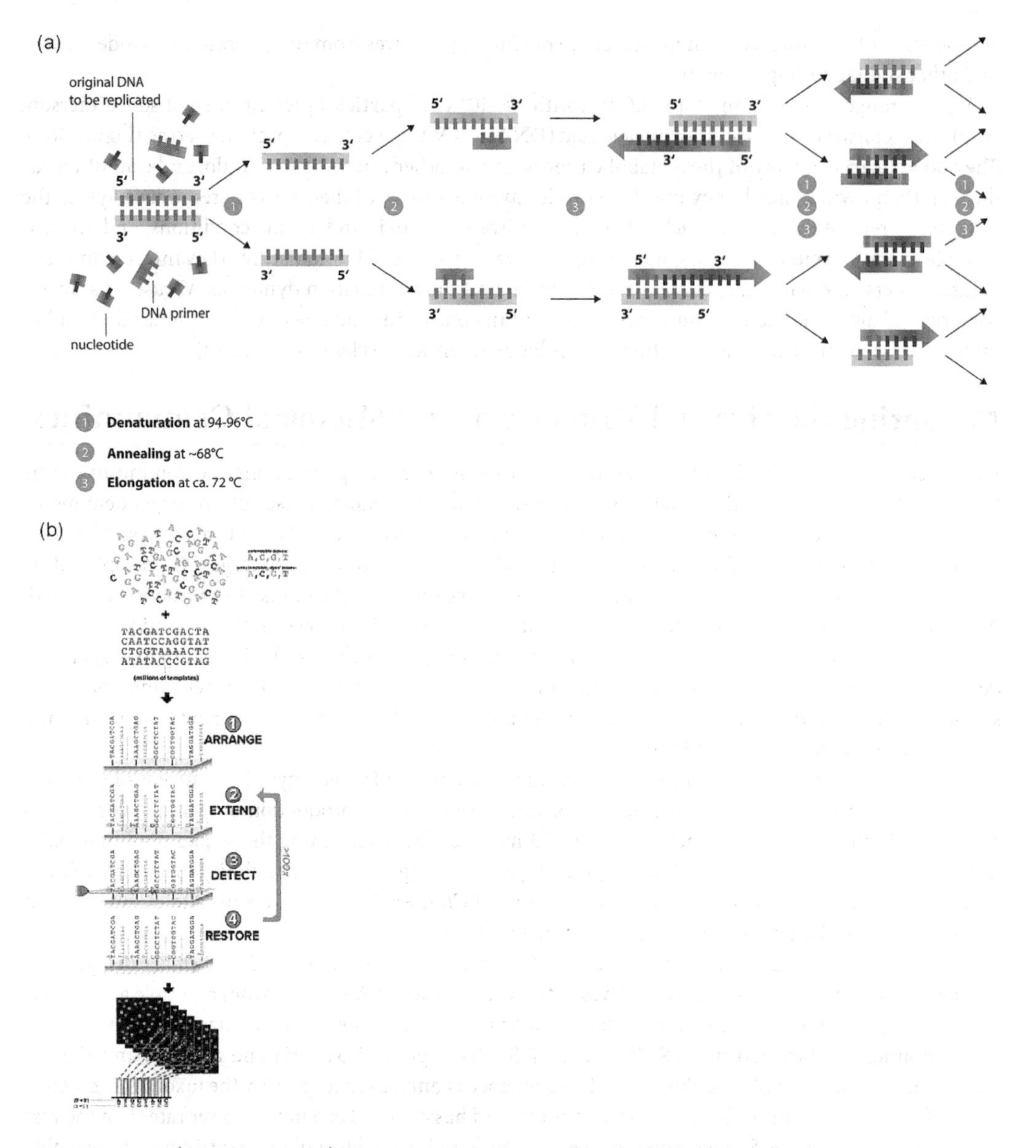

FIGURE 36.3 (See color insert.) (**a**): Steps involved in making multiple copies of DNA or RNA sequences using the PCR. Primers have a sequence that is complementary to the target sequence in either RNA or DNA. (**b**): High-throughput sequencing process, which is used to sequence PCR products quickly and inexpensively. (Image credit: (a): Image by Enzoklop, distributed under a CC BY-SA 3.0 license. (b): This image was reproduced from Figure 36.1 in Price et al (2018). It is licensed under Creative Commons Attribution 4.0 International License, http://creativecommons.org/licenses/by/4.0/.)

in a soil sample. Although this approach is more expensive and computationally complex than target gene sequencing, exploring large segments of an organism's genetic code allows researchers to assess the breadths of functional genes present in a community, as well as which functional genes co-occur in a species' genome. Whole-genome sequencing has also been shown to better predict overall microbial diversity than 16S sequencing (Ranjan et al., 2016).

Both target gene and whole-genome sequencing produce large amounts of data that cannot be analyzed using simple computational methods. The need to quickly analyze these datasets has led to the development of *bioinformatics*—a field based on the application of computational methods and computer software for analysis of large biological datasets. These include 16S datasets containing thousands of sequences, and computational comparisons of these sequences with reference databases, such as the SILVA database, a comprehensive and open-source database containing millions of 16S sequences (Quast et al., 2012). Bioinformatics allows researches to take the raw base pair genetic data generated by sequencing and translate it into information about the structure and function of microbial communities.

The last step in the analysis of genetic data is interpretation. A soil microbiologist may be interested in the evolutionary history of an organism (*phylogeny*) or the organism's grouping and classification (*taxonomy*). Using bioinformatics software, researchers can calculate community diversity metrics including *alpha diversity*, which answers questions such as "How many different microbial species are present in a sample?" and "Are some species dominant within a community?" They can also assess *beta diversity*, which allows comparison of species composition among samples, answering the question "How similar/different is the microbial composition among samples?" The ultimate goal of these analyses is often to connect microbial community structure to ecosystem function. While great strides have been made in this area, making this connection in soil communities remains a challenge (Bowen et al., 2014).

Metagenomics has improved our understanding of the structure and function of soil microbial communities, revealing an ecosystem that is more diverse and complex than previously appreciated, and produced global databases of soil organisms. Among the interesting complexities that have been revealed are (i) the paradigm of extraordinarily high diversity and dissimilarity between two soil ecosystems, and (ii) the ubiquitous nature of a very few taxa found in nearly all of earth's soils. Pairing environmental data with genetic information have also given insight into the habitat preferences of ubiquitous soil microorganisms (Delgado-Baquerizo et al., 2018).

Ecological Interactions and Microorganisms in the Soil Foodweb

Soil microorganisms are involved in a variety of ecological interactions with other microorganisms as well as with plants and soil fauna (Table 36.2). Biological nitrogen fixation—one of the most important natural processes for turning inert atmospheric N_2 into N available to crops—results from *symbiosis* between bacteria in the genera *Rhizobium* and *Bradyrhizobium* and legumes, which results in a species-specific infection of the roots. The bacterium resides in nodules found in the roots and oxidizes N_2 to NH_4, which can then be used by the plant. In turn, the plant provides organic C and protection for the bacterium.

Most plants form a *mutualistic* relationship with filamentous fungi known as mycorrhizae. The fungus infects the roots and transports nutrients and water from the surrounding soil to the plant. The plant, in turn, provides organic C for the fungus. Unlike the symbiotic relationship of N-fixing bacteria, mycorrhizal associations are much less stringent in terms of the particular plant and fungal species involved. Over 80% of plant species studied have been shown to have a mycorrhizal association (Wang and Qiu, 2006).

There may be *competition* among heterotrophic microorganisms for carbon substrates which they need for energy and biomass. Most microorganisms are capable of utilizing only a limited number of substrates, such that they should be direct competition among them. Ecological theory predicts that

TABLE 36.2 Example Ecological Interactions among Soil Microorganisms, Plants, and Soil Fauna

Type of Interaction	Example	Organisms Involved	Resources Involved
Symbiosis	N_2 fixation	Bacteria and plants	NH_3 and organic C
Mutualism	Mycorrhizae	Fungi and plants	Nutrients, water, and organic carbon
Competition	Carbon limitations	Microorganisms	Organic C substrates
Predation	Grazing	Protozoa, nematodes, and bacteria	Bacterial C and N

direct competition will eventually result in dominance by a single species. However, the large diversity of microorganisms that carry out the same function in soil, such as starch-degrading bacteria or cellulose-degrading fungi, runs counter to this prediction. This apparent paradox is resolved when we realize that direct competition among soil microorganisms is lessened by the fact that they are separated in space. The variably porous nature of soil allows for microorganisms that carry out the same function to live in different soil pores. This results in greater functional redundancy and ecosystem stability, and underscores the importance of physical soil properties like texture in shaping soil ecosystems.

As we have indicated earlier, fungi and bacteria are subject to *predation* by microbivorous nematodes and protozoa. Predation affects the size and age distribution of the prey population. In addition, predators excrete part of the N and P found in microbial prey in mineral forms, increasing the availability of nutrients to plants and other microorganisms.

The soil foodweb illustrates the interactions of microorganisms with other microorganisms and with plants and soil fauna (Figure 36.2). Unlike a food chain, where a single interaction takes place among organisms, there is considerable redundancy in the ecological interactions in a foodweb, which results in greater stability in ecosystem function. These interactions result in the cycling of carbon, nutrients, and energy that is essential for the functioning of natural and engineered soil ecosystems (Scheu, 2002). For example, the carbon fixed by plants, and the nitrogen fixed by biological nitrogen fixation flows through the foodweb as a result of interactions among soil organisms that result in decomposition and release of nutrients, allowing for growth and energy of microflora and fauna. Carbon processing by microorganisms results in the formation of stable soil organic matter, which physically stabilizes the soil and improves habitat for plants and microorganisms.

References

Armon, R. 2011. Soil bacteria and bacteriophages. In: Witzany, G. (ed.) *Biocommunication in Soil Microorganisms* (*Soil Biology*), vol. 23. Springer, Berlin, Heidelberg, pp. 67–112.

Bowen, J.L., A.R. Babbin, P.J. Kearns, and B.B. Ward. 2014. Connecting the dots: Linking nitrogen cycle gene expression to nitrogen fluxes in marine sediment mesocosms. *Front. Microbiol.* 5: 1–10. doi: 10.3389/fmicb.2014.00429.

Delgado-Baquerizo, M., A.M. Oliverio, T.E. Brewer, A. Benavent-González, D.J. Eldridge, et al. 2018. Bacteria found in soil. *Science* 325: 320–325. doi: 10.1126/science.aap9516.

Emerson, J.B. 2019. Soil viruses: A new hope. *mSystems* 4 (3): e00120-19. doi: 10.1128/mSystems.00120-19.

Frąc, M., S.E. Hannula, M. Bełka, and M. Jędryczka. 2018. Fungal biodiversity and their role in soil health. *Front. Microbiol.* 9: 707.

Jia, Z., and R. Conrad. 2009. Bacteria rather than Archaea dominate microbial ammonia oxidation in an agricultural soil. *Environ. Microbiol.* 11: 1658–1671. doi: 10.1111/j.1462-2920.2009.01891.x.

Price, K. S., Svenson, A., King, E., Ready, K., and Lazarin, G. A. 2018. Inherited cancer in the age of next-generation sequencing. *Biol. Res. Nurs.* 20 (2): 192–204. doi: 10.1177/1099800417750746.

Quast, C., E. Pruesse, P. Yilmaz, J. Gerken, T. Schweer, et al. 2012. The SILVA ribosomal RNA gene database project: Improved data processing and web-based tools. *Nucleic Acids Res.* 41 (D1): D590–D596. doi: 10.1093/nar/gks1219.

Ranjan, R., A. Rani, A. Metwally, H.S. McGee, and D.L. Perkins. 2016. Analysis of the microbiome: Advantages of whole genome shotgun versus 16S amplicon sequencing. *Biochem. Biophys. Res. Commun.* 469 (4): 967–977. doi: 10.1016/j.bbrc.2015.12.083.

Scheu, S. 2002. The soil food web: Structure and perspectives. *Eur. J. Soil Biol.* 38: 11–20.

Torsvik, V. and L. Øvreås. 2002. Microbial diversity and function in soil: From genes to ecosystems. *Curr. Opin. Microbiol.* 5: 240–245.

Wang, B. and Y.L. Qiu. 2006. Phylogenetic distribution and evolution of mycorrhizas in land plants. *Mycorrhiza* 16: 299–363. doi: 10.1007/s00572-005-0033-6.

37

Soil Invertebrates: Responses to Forest Types in Changbai Mountains

Chen Ma and
Xiuqin Yin
Northeast Normal University

Introduction

Soil invertebrates play crucial roles in nutrient cycling and energy flows of belowground ecosystems and participate in soil ecosystem services.[1,2] On one hand, soil invertebrates can affect many forests' environmental parameters via crushing litter,[3] excreting feces,[4] and microbial modification.[5] On the other hand, due to the narrow geographic scope of their activities, as well as their weak migrational abilities, soil invertebrates are able to sensitively respond to environmental variation.[6,7] Currently, the majority of studies have focused on the essential role that soil invertebrates play in organic matter decomposition, nutrient mineralization, and plant productivity in the forest ecosystem.[8]

In the mountainous area of Northeast Asia, forests are the main terrestrial land cover type, but various natural and human disturbances have altered almost all forest stands.[9] The primeval forests have been replaced mostly by secondary forests in northeastern China, and characteristics of the forest communities have been considerably changed.[10] When the primeval forests are replaced by the secondary forests, it will change litter quality and soil properties, and this will significantly affect nutrient cycling energy flows at belowground ecosystem level.[11] Consequently, a better understanding of the relationship between forest types and belowground system is crucially needed to develop biodiversity guidelines for managed forests.

However, despite the fact that the evidence of a close relationship between soil invertebrates and forest vegetation is substantial, the relationships between soil invertebrates and forest ecosystem remain a matter of debate,[12] for instance, the effects of mixed forests on soil invertebrate communities and soil processes, and the relation between forest richness and soil invertebrates diversity.[13] Consequently, the responses of soil invertebrates to forest types cannot be ignored. To better understand the responses of soil invertebrates to different forest types, we selected five forest types in the Changbai Mountains of China. Soil invertebrates were collected in primeval coniferous broad-leaved mixed forests (PMFs), secondary coniferous broad-leaved mixed forests (SMFs), secondary broad-leaved forests (SBF), secondary *Quercus mongolica* forests (SQFs), and secondary shrub forests (SSFs).

Distribution Patterns of Soil Invertebrates

In the different forest types of Changbai Mountains, the soil invertebrates belonging to 72 taxa, 24 orders (suborders), 8 classes, and 3 phyla were collected from the five forest types. Fifty-two taxa (53,404 ind. m^2) of soil invertebrates were collected in the PMFs; 49 taxa (66,883 ind. m^{-2}) in the SMFs; 43 taxa (44,035 ind. m^{-2}) in the SBFs; 48 taxa (57,711 ind. m^{-2}) in the SQFs; and 52 taxa (36,792 ind. m^{-2}) in the SSFs. SMF had the maximum abundance (66,883 ind. m^{-2}), and the minimum abundance was found in the SSF (36,792 ind. m^{-2}). The dominant taxa were found to be Oribatida (33.69%) and Isotomidae (25.91%), and the common taxa included Actinedida (9.70%), Gamasida (6.77%), Pseudachorutidae (5.93%), Hypogastruridae (5.13%), Entomobryidae (3.16%), Sminthuridae (1.49%), and Diptera larva (1.01%). The other 63 taxa were considered to be rare and accounted for 7.21% of the total number of individuals.

Fifty-two taxa of the soil invertebrates were collected in the PMFs and SSFs, the highest among the five forest types. The previous studies revealed that the main factors that could potentially affect soil invertebrates' diversity included the quality, quantity, and species of plant litter.[14] Comparing with other forests, PMF had the greatest richness of plant species and the most complex community structure of vegetation in the Changbai Mountains.[15] These provided a favorable condition for soil invertebrates in the PMF. Due to the fact that the SFF is in the primary stage of a secondary succession, a variety of pioneer species have emerged. Some previous studies have demonstrated that new species of plants could establish a microenvironment and change the characteristics of the soil humus, which would consequently supply increasing amounts of food for soil invertebrates.[16,17] We also observed that 66,883 ind. m^{-2} of soil invertebrates were found in the (SMFs), the highest abundance among the five forest types. The intermediate disturbance hypothesis indicates that the intermediate disturbance can promote species diversity in ecological communities and permit more numerous species' invasion and colonization.[18] Comparing with other forest types in the Changbai Mountains, a relatively intermediate disturbance was found in the SMFs , and it may promote the richness of soil invertebrates. Additionally, the lowest number of soil invertebrates' taxa (43 taxa) was observed in the SBF. It is known that increased forest stand densities can result in higher amounts of water and nutrients in the lower levels,[19] which restrict the growing of the understory plants.[20] The SBF is mainly composed of *Acer mono* and *Betula platyphylla*, and a large number of saplings are in there; thus, it had a highest shrub coverage and a single tree species composition. This resulted in a singleness of the litter species and a deterioration of the soil environment, which consequently decreased species richness of soil invertebrates.

Taxonomic Compositions of Soil Invertebrates

Taxonomic compositions of soil invertebrates from different forest types were summarized in Figure 37.1a. Oribatida and Isotomidae were the dominant taxa in all of forest types in this study. However, the percentages were not identical in each forest type. In particular, the Oribatida was absolutely predominant in the SQF, accounting for 53.68% of the total. In different forest types, the sequences of some of soil invertebrates' taxa varied. For example, Actinedida was the dominant taxon in the SMF and SSF, whereas it was the common taxon in the other types of forest. In addition, the Pseudachorutidae was found to be the dominant taxon type only in the SBF.

Venn diagrams of unique and shared taxa demonstrated that all five forest types had 24 taxa in common (Figure 37.1b), which contributed to 46.15%–55.81% of the full set of taxa in each forest type. Ptiliidae, Cicadellidae, Hahniidae, and Drosophilidae were unique taxa in the PMF; Neanuridae and Lygaeidae were unique taxa in the SMF; Bibionidae was the only unique taxa in the SBF; Curculionidae and Theridiosomatidae were unique taxa in the SQF; Coccinellidae and Cucujidae were unique taxa in the SSF. Previous studies have determined that the taxonomic compositions of soil invertebrates are dependent on the environmental factors and biological interactions[21,22]. Due to the fact that all of the sites are located in the Changbai Mountains, they have the same flora and fauna. Consequently, a

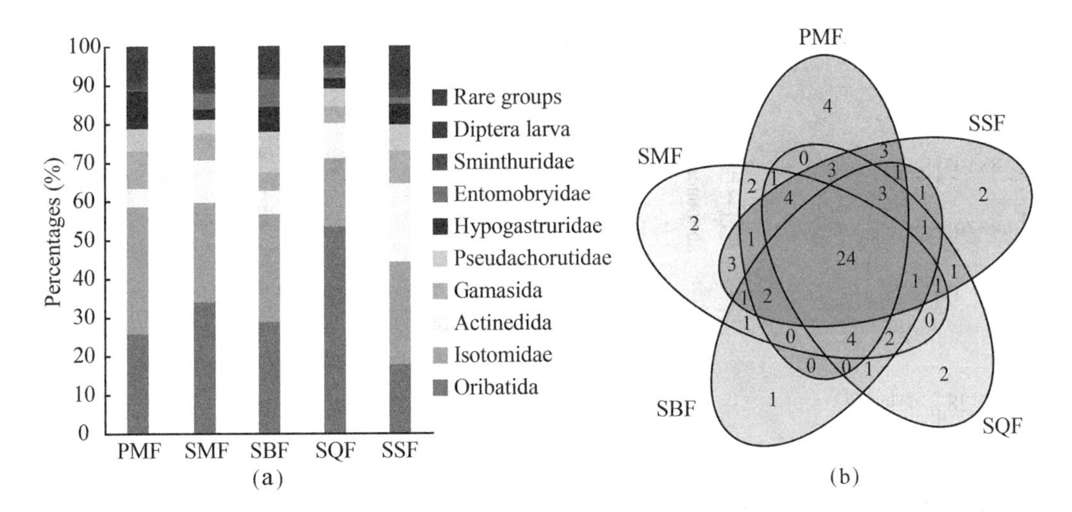

FIGURE 37.1 **(See color insert.)** Characteristics of soil invertebrates' communities. (**a**) The relative abundance was defined as the percentage of soil invertebrates' community in different forest types of Changbai Mountains. (**b**) Venn diagram of the number of shared and unique soil invertebrates' taxa in different forest types of Changbai Mountains. The numbers in the circles indicate either the unique number of taxa in a forest or the number of shared taxa in the overlapping regions; PMFs, primeval coniferous broad-leaved mixed forests; SMFs, secondary coniferous broad-leaved mixed forests; SBFs, secondary broad-leaved forests; SQF, secondary *Q. mongolica* forests; SSFs, secondary shrub forests.

similarity of soil invertebrates' taxonomic compositions was found in this study. Additionally, large numbers of Ptiliidae were collected in the PMF, which were considered to be a rare taxon on the Chinese Mainland[23]. It is known that soil temperature and moisture could significantly affect feeding activity of soil invertebrates.[24] The previous study has indicated that the Ptiliidae live only in moist environments, such as algae, rotten wood, fungus, and excrement.[25] PMF had wetter environment and thicker litter layer; thus, good conditions were created for the Ptiliidae. In addition, due to the rarity of Ptiliidae in the Chinese mainland, its species taxonomy and ecological distribution required further investigation.

Seasonal Variations of Soil Invertebrates

During different seasons, distribution patterns of soil invertebrates' communities were found to be different in the five forest types (Figure 37.2). In spring, the abundance of soil invertebrates in SQF (62,032 ind. m^{-2}) was significantly higher than SBF (33,555 ind. m^{-2}) and SMF (33,246 ind. m^{-2}) ($p < 0.05$), and significant differences were not found among the five forest types ($p > 0.05$) regarding the richness levels. During summer, there were no significant differences observed among the five forest types ($p > 0.05$) regarding the abundance and richness levels. During autumn, the abundance of soil invertebrates in SSF (43,443 ind. m^{-2}) was only significantly lower than those in SMF ($p < 0.05$); for the rest of forests, no differences were found, and no significant differences were found among the five forest types regarding the richness levels of soil invertebrates ($p > 0.05$). The previous research study has reported that phytophagous and omnivorous arthropods are more active and had a higher abundance in autumn.[26] In agreement with previous studies of soil invertebrates, the seasonal variations of soil invertebrates' communities were found to be mainly related to the variation of temperature and moisture.[27] This investigation was conducted in the eastern part of northeastern China, which was influenced by stable weather systems during the autumn; thus, the temperature and moisture levels are neither high nor low. Therefore, fine temperature and moisture were created for soil invertebrates.

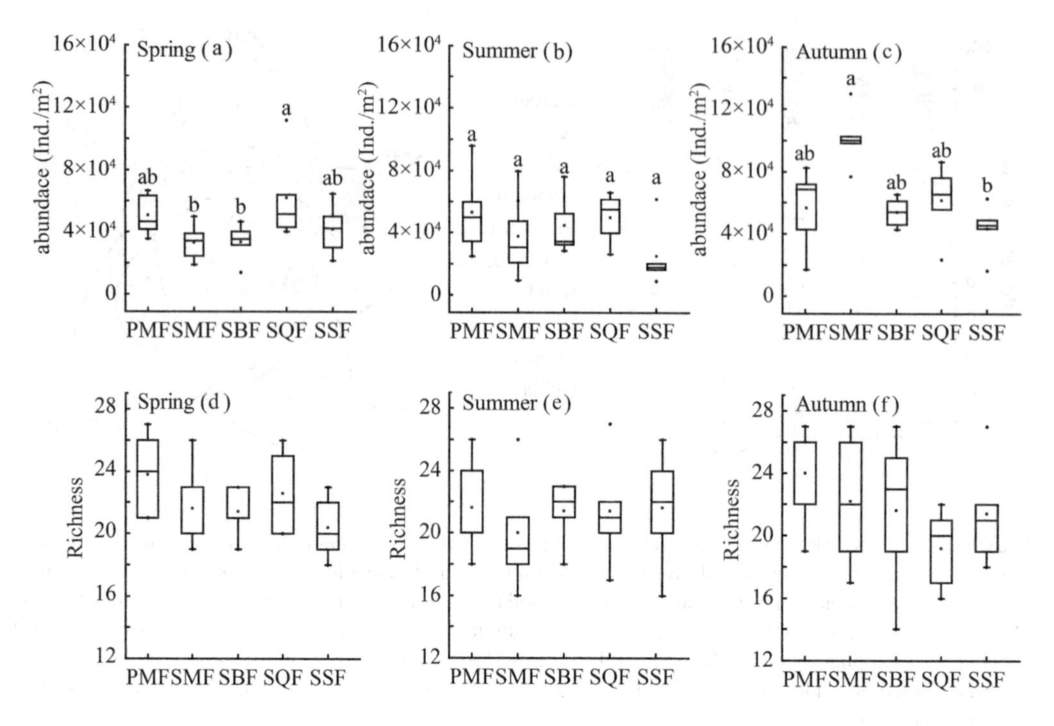

FIGURE 37.2 Box-whisker plots illustrating medians (line in the box), 25th and 75th percentiles (box), 10th and 90th percentiles (outer lines), and counts outside the latter percentiles (dots) of abundance (a–c) and richness (d–f) of soil invertebrates among different forest types of Changbai Mountains in each season. The same letter(s) indicate no significantly different between each forest type at $p<0.05$ by one-way ANOVA. PMFs, primeval coniferous broad-leaved mixed forests; SMFs, secondary coniferous broad-leaved mixed forests; SBFs, secondary broad-leaved forests; SQFs, secondary *Q. mongolica* forests; SSFs, secondary shrub forests.

Conclusions

In summary, different forest types affected the distribution patterns of soil invertebrates. The taxonomic composition of soil invertebrates slightly varied among the different forest types in the Changbai Mountains. Seasonal variations were exhibited in the abundance and richness levels, with the abundance levels in autumn slightly higher in summer. The findings of this study have implications for the relationship between different forest types and soil invertebrates in the temperate forest ecosystem, and also can provide some help to develop biodiversity guidelines for managed forests.

References

1. Lavelle P, Barot S, Blouin M, Decaëns T, Jimenez JJ, Jouquet P. 2007. Chapter 5, Earthworms as key actors in self-organized soil systems. In: Cuddington KJEB, Wilson WG, Hastings A (eds.) Ecosystem Engineers: From Plants to Protists. Theoretical Ecology Series. Cambridge, MA: Academic Press, vol. 4, pp. 77–106.
2. Yin X, Song B, Dong W, Xin W, Wang Y. 2010. A review on the eco-geography of soil fauna in China. *J Geogr Sci*. 20(3): 333–346.
3. Joly FX, Coq S, Coulis M, Nahmani J, Hättenschwiler S. 2018. Litter conversion into detritivore faeces reshuffles the quality control over C and N dynamics during decomposition. *Funct Ecol*. 32(11): 2605–2614.

4. Coulis M, Hättenschwiler S, Coq S, David JF. 2016. Leaf litter consumption by macroarthropods and burial of their faeces enhance decomposition in a mediterranean ecosystem. *Ecosystems.* 19: 1104–1115.
5. Joly FX, Coulis M, Gérard A, Fromin N, Hättenschwiler S. 2015. Litter-type specific microbial responses to the transformation of leaf litter into millipede feces. *Soil Biol Biochem.* 86: 17–23.
6. van Straalen NM. 1998. Community structure of soil arthropods as a bioindicator of soil health. In: Pankhurst CE, Doube BM, Gupta VVSR (eds.) *Biological Indicators of Soil Health.* Oxford and New York, CAB International, pp. 235–264.
7. Coyle DR, Nagendra UJ, Taylor MK, Campbell JH, Cunard CE, Joslin AH, Mundepi A, Phillips CA, Callaham MA. 2017. Soil fauna responses to natural disturbances, invasive species, and global climate change: Current state of the science and a call to action. *Soil Biol Biochem.* 110: 116–133.
8. Elie F, Vincenot L, Berthe T, Quibel E, Zeller B, Saint-André L, Normand M, Chauvat M, Aubert M. 2018. Soil fauna as bioindicators of organic matter export in temperate forests. *For Ecol Manage.* 429: 549–557.
9. Kolbek J, Šrůtek M, Elgene OB. 2003. *Forest Vegetation of Northeast Asia.* Dordrecht: Springer.
10. Xu WD, He XY, Chen W, Liu C. 2004. Characteristics and succession rules of vegetation types in Changbai Mountain. *Chin J Ecol.* 23(5): 162–174.
11. Celentano D, Rousseau GX, Engel VL, Zelarayán M, Oliveira EC, Araujo ACM, de Moura EG. 2017. Degradation of riparian forest affects soil properties and ecosystem services provision in Eastern Amazon of Brazil. *Land Degrad Dev.* 28: 482–493.
12. Sławska M, Bruckner A, Sławski M. 2017. Edaphic Collembola assemblages of European temperate primeval forests gradually change along a forest-type gradient. *Eur J Soil Biol.* 80: 92–101.
13. Korboulewsky N, Perez G, Chauvat M. 2016. How tree diversity affects soil fauna diversity: A review. *Soil Biol Biochem.* 94: 94–106.
14. Sauvadet M, Chauvat M, Brunet N, Bertrand I. 2017. Can changes in litter quality drive soil fauna structure and functions? *Soil Biol Biochem.* 107: 94–103.
15. Wang YQ, Wu ZF, Feng J. 2015. *Geographical and Ecological Security of the Changbai Mountains.* Changchun: Northeast Normal University Press.
16. Hansen RA. 1999. Red oak litter promotes a microarthropod functional group that accelerates its decomposition. *Plant Soil.* 209(1): 37–45.
17. Thoms C, Gattinger A, Jacob M, Thomas FM, Gleixner G. 2010. Direct and indirect effects of tree diversity drive soil microbial diversity in temperate deciduous forest. *Soil Biol Biochem.* 42: 1558–1565.
18. Connell JH. 1979. Intermediate-disturbance hypothesis. *Science.* 204(4399): 1345.
19. Mcdowell NG, Adams HD, Bailey JD, Kolb TE. 2007. The role of stand density on growth efficiency, leaf area index, and resin flow in southwestern ponderosa pine forests. *Can J For Res.* 37(2): 343–355.
20. Paluch JG, Gruba P. 2010. Relationships between local stand density and local species composition and nutrient content in the topsoil of pure and mixed stands of silver fir (Abies alba Mill.). *Eur J For Res.* 129(4): 509–520.
21. Copley J. 2000. Ecology goes underground. *Nature.* 406: 452–454.
22. Yin X, Ma C, He H, Wang Z, Li X, Fu G, Liu J, Zheng Y. 2018. Distribution and diversity patterns of soil fauna in different salinization habitats of Songnen Grasslands, China. *Appl Soil Ecol.* 123: 375–383.
23. Ma ZX, Chen DL, Zhang J. 2011. A new record on Acrotrichis sp. in the Main Land of China. *J Northwest Normal Univ.* 47: 95–96.
24. Konstantinb G, Tryggve P, Andreid P. 2008. Effects of soil temperature and moisture on the feeding activity of soil animals as determined by the bait-lamina test. *Appl Soil Ecol.* 39(1): 84–90.

25. Hall E. 2005. Ptiliidae. In: Beutel RG (ed.) *Handbook of Zoology, Volume 4. Arthropoda: Insects: Part 38. Coleoptera, Beetles. Volume 1: Morphology and Systematics.* New York: Walter de Gruyter & Co, pp: 251–261.
26. Tack AJM, Dicke M. 2013. Plant pathogens structure arthropod communities across multiple spatial and temporal scales. *Funct Ecol.* 27(3): 633–645.
27. Peña-Peña K, Irmler U. 2016. Moisture seasonality, soil fauna, litter quality and land use as drivers of decomposition in Cerrado soils in SE-Mato Grosso, Brazil. *Appl Soil Ecol.* 107: 124–133.

38

Spatial–Temporal Distribution of Soil Macrofauna Communities: Changbai Mountain

Xiuqin Yin
Northeast Normal University

Yeqiao Wang
University of Rhode Island

Introduction

Soil macrofauna play crucial roles in nutrient cycling and energy flows of belowground ecosystems and participate in soil ecosystem services.[1] On the one hand, soil arthropods can increase nutrients mineralization via physical fragmentation, excreting feces, and microbial modification. On the other hand, soil arthropods are able to sensitively respond to environmental changes due to the relatively weaker ability of their activities.[2] A wide range of animals have been shown to follow gradients of biodiversity in latitude and altitude,[3] while small organisms such as protists have shown no such patterns.[4,5] Because of body size and their interaction with soil protozoa and microorganism, soil macrofauna represent a useful model to study patterns of species distribution of the "missing link" between soil protozoa and "large animals."

Elevation has the same effect on climate and vegetation and soil as latitude, which inevitably affect the structure and dynamics of soil fauna communities.[6,7] However, detailed studies on the effect of elevation on soil macrofauna communities have been scarce. The change of soil macrofauna compositions and diversities had been investigated in the mountain ecosystems of a few regions. Previous researches had indicated that elevation was one of decisive factors affecting species richness and the structure and function of soil fauna community.[8,9] The diversity of soil fauna communities decreased[10] or varied as a unimodal type with the increasing elevation.[11] However, due to the differences in location, altitudinal range, vegetation types, and sampling gradient interval of mountain systems, there was a lack of common agreement on the distribution pattern of soil macrofauna with increasing elevation. More investigations deemed necessary to study the characteristics of soil macrofauna communities along elevation variation.

The Changbai Mountain is one of the few well-conserved natural ecosystems in the temperate zone. The vertical zonality of vegetation in this mountain is known to be a mirror of the vegetation latitudinal or horizontal zonation from temperate to frigid zones on the Eurasian continent. Major studies on the soil macrofauna communities in the Changbai Mountain have mostly been performed since the 1980s in one of the vegetation zones rather than in the entire vertical zone system.[12] Therefore, the objective of this study was to investigate the characteristics of the soil macrofauna communities in the vertical zone system of the Changbai Mountain and reveal how soil macrofauna communities performed in altitudinal zonality by comparing the diversity and spatial–temporal distribution of soil macrofauna communities along the elevation.

Difference of Soil Macrofauna Communities in the Vertical Zone of the Mountain System

A total of 614 individuals belonging to 36 taxonomic groups were collected from coniferous and broadleaved mixed forests, 469 individuals in 30 taxonomic groups were collected from coniferous forests, 273 individuals in 18 taxonomic groups were collected from subalpine dwarf birch forests, and 60 individuals in 15 taxonomic groups were collected from alpine tundra. The dominant taxonomic groups in the coniferous and broadleaved mixed forests include Geophilidae, Formicidae, Lithobiidae, and Juliformia, which together accounted for 60.1% of the total individuals. The dominant taxonomic groups in the coniferous forests include Formicidae, Juliformia, and Lithobiidae, which together accounted for 53.3% of the total individuals. The community is dominated by Elateridae larve, Staphylinidae, and Lithobiidae (together for 48.36%) in the subalpine dwarf birch forests and dominated by Diptera larve and Tomoceridae larve (together for 43.34%) in the alpine tundra. This study concluded that Lumbricidae, Juliformia, Elateridae larve, Diptera larve, Carabidae larve, and Agelenidae were presented in all of the studied vegetation zones except the alpine tundra. These taxonomic groups could be considered as inadaptable inhabitants to the alpine tundra in higher elevation area. Moreover, Brachycera larvae were more abundant in the alpine tundra than others and that could be the result of selective adaptation of this taxonomic group to the habitat. Although diversity of soil macrofauna community varied insignificantly among the seasons in each vegetation zone, soil macrofauna reached their maximums of the abundance (22), richness (9 taxonomic groups), and diversity index (1.8854) in the summer in the alpine tundra zone against mostly in the spring or autumn in the other three zones. The seasonal activities of soil macrofauna in the mountain were limited by the effect of environmental factors such as climate and vegetation, as well as living habit of soil macrofauna.

Vertical Zonality of Soil Macrofauna Communities in the Mountain System

In each season, the abundance and richness of the soil macrofauna also decreased with the ascending elevation (Table 38.1). The diversity indexes were higher in coniferous and broadleaved mixed forests (2.3028~2.4497) and coniferous forests (1.9186~2.6010) of the low elevations than those in the subalpine dwarf birch forests (2.2226~2.3448) and alpine tundra (1.5881~1.8854) of higher elevations (Table 38.1). The abundance, richness, and diversity of the soil macrofauna varied significantly ($P < 0.05$) among the four vegetation zones during each season; however, the variations of these indexes were insignificant among three seasons in each zone ($P > 0.05$). The probability distribution curve of the richness and diversity of soil macrofauna communities were higher in coniferous and broadleaved mixed forests and coniferous forests than those in subalpine dwarf birch forests and alpine tundra (Figures 38.1 and 38.2), which indicated that the diversity of soil macrofauna decreased with the increasing elevation.

At the same time, we found that the dominant taxonomic groups of the soil macrofauna observed differed from those of the mountain with the vertical vegetation zones system in the subtropical,

TABLE 38.1 Quantitative Characteristics of Soil Macrofauna Community Diversity in the Four Vegetation Zones of the Changbai Mountain

	Spring			Summer			Autumn			Total		
Vegetation Zones	A	R	D	A	R	D	A	R	D	A	R	D
A	225	34	2.3028	248	20	2.4374	141	20	2.4497	614	36	2.5824
B	218	21	1.9186	128	20	2.6010	123	18	2.3852	469	30	2.5589
C	76	15	2.3448	90	12	2.2767	107	15	2.2226	273	18	2.4168
D	18	7	1.5811	22	9	1.8854	20	7	1.6230	60	15	2.3083

A, the abundance; R, the richness of taxonomic group; D, the Shannon–Wiener diversity index.

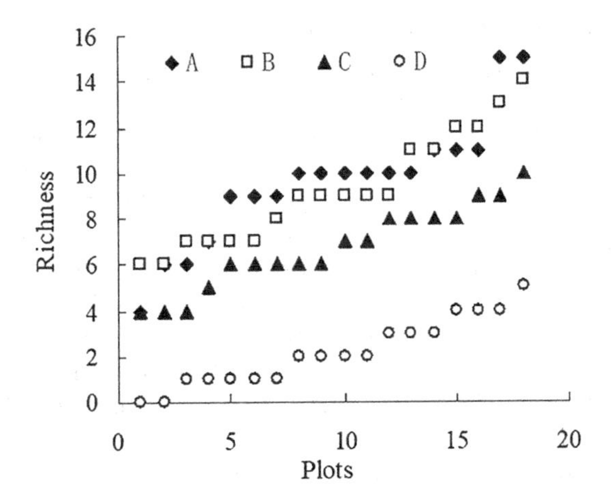

FIGURE 38.1 Probability distribution curve of the richness of soil macrofauna in the four vegetation zones on the northern slope of the Changbai Mountains. A, Coniferous and broadleaved mixed forest zone. B, Coniferous forest zone. C, Subalpine dwarf birch (Betula ermanii) forest zone. D: Alpine tundra zone.

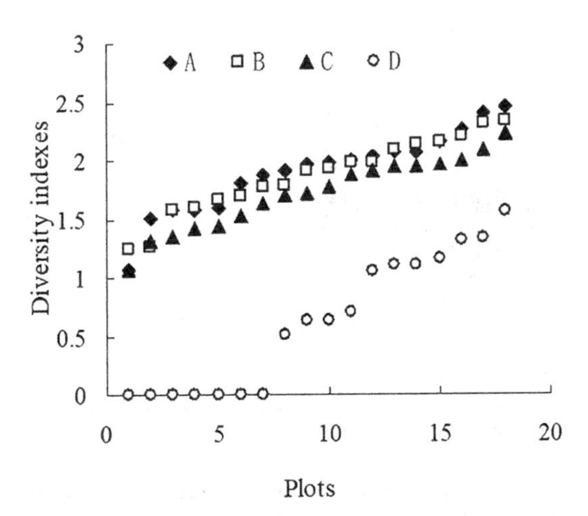

FIGURE 38.2 Probability distribution curve of the diversity indexes of soil macrofauna communities in the four vegetation zones on the north slope of the Changbai Mountains. A, Coniferous and broadleaved mixed forest zone. B, Coniferous forest zone. C, Subalpine dwarf birch (Betula ermanii) forest zone. D, Alpine tundra zone.

e.g., dominated by Hymenoptera, Diptera, and Coleoptera in the Wuyi Mountains;[13] dominated by Coleoptera, Diptera, Diplopoda, Malacostraca and Oligochaeta in the Hengduanshan Mountains in the warm temperate zone;[14] and dominated by Enchytraeidae in the cold temperate regions[15] of China. These just gave an example of the influence of altitudinal and latitudinal on the dominant taxonomic groups composition of soil macrofauna in a monsoon climate region. This study confirmed that the diversity of the soil macrofauna communities varied significantly among vegetation zones along elevation. However, a study in the Gaoligong Mountains demonstrated a greater diversity in the mid-elevation area comparing to the lower and higher elevation areas.[11] The results offered a clue that the distribution of soil macrofauna communities had the vertical zonality trend in the mountain ecological system which reflected the pattern of latitudinal zonality. Further studies along elevation gradients should be investigated in the mountain systems of different latitudes.

Major Factors of Controlling Soil Macrofauna in the Vertical Zone of the Mountain System

Redundancy analysis (RDA) was conducted to examine the major factors affecting sampled plot habitats and the seven major taxonomic groups of the soil macrofauna. The Monte Carlo significance test revealed that 61.3% and 80.4% of the variance of species data and species–environment relation were explained in the first axis and 11.2% and 14.7% in the second axis, respectively. The mean annual precipitation (AP), annual radiation quantity (AR), and altitude (AL) were positively correlated with Axisl, and mean annual temperature (AT) was negatively correlated with Axisl. The soil organic matter content (OM) and soil moisture content (SM) were positively correlated with Axis2, and pH was negatively correlated with Axis2 (Figure 38.3). The AP, AL, AR, and AT were the major factors to determine the environmental conditions of the plot habitats. The distributions of Elateridae larve and Staphylinidae were influenced by the SM and pH; however, Geophilidae, Lithobiidae, Agelenidae, Juliformia, and Formicidae were influenced by the AT, AL, AR, and AP (Figure 38.3).

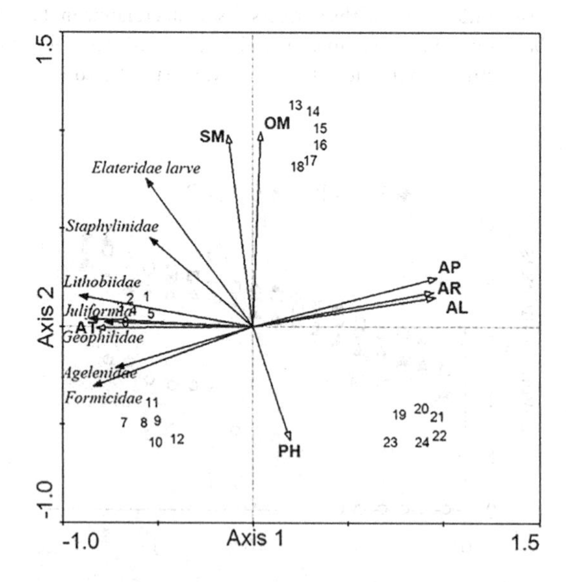

FIGURE 38.3 RDA two-dimensional ordination diagram of the two axes, showing the relationship among the sample habitats, soil macrofauna, and environmental factors. The samples ($n=24$) were represented by Arabic numerals, and the environmental factors were represented as follows: AL, altitude; AR, annual radiation quantity; AT, mean annual temperature; AP, mean annual precipitation; SM, soil moisture content; OM, organic matter content; PH, the value of soil pH.

TABLE 38.2 Results of Correlation Analysis between Environmental Factors and Soil Macrofauna Diversity

	AL	AR	AT	AP	SM	OM	PH
A	−0.889**	−0.889**	0.889**	−0.889**	0.238	0.420*	−0.316
R	−0.863**	−0.863**	0.863**	−0.863**	0.191	0.348	−0.319
D	−0.786**	−0.786**	0.786**	−0.786**	0.258	0.443*	−0.371

AL, altitude; AR, annual radiation quantity; AT, mean annual temperature; AP, mean annual precipitation; SM, soil moisture content; OM, soil organic matter content; PH, the value of soil pH; A, the abundance of individuals; R, the richness of taxonomic group; D, the Shannon–Wiener diversity index.
*$P < 0.05$.
**$P < 0.01$.

It had been proposed that there was a close relationship between soil property and soil macrofauna diversity.[16,17] However, the abundance, richness, and diversity of the soil macrofauna were correlated with the investigated three climate factors and altitude, and uncorrelated with the investigated three soil factors ($r<0.443$, Table 38.2). The RDA results also suggested that terrain and climate factors had stronger effect on the activities of major taxonomic groups (Figure 38.3). Therefore, we suggest that in a vertical system of the mountain environment, the change of climate factors along elevation leads to changes of vegetation type and ecological environment, which have stronger influence on soil macrofauna community in comparison with soil micro-environment.

Conclusions

In the vertical system of the Changbai Mountains, the abundance, richness, and Shannon–Wiener index of soil macrofauna communities decreased with the increasing elevation. The composition and diversity of the soil macrofauna had significant difference among vertical vegetation zones and insignificant among the seasons. The diversity and distribution of soil macrofauna communities were observably influenced by mean annual precipitation, altitude, and annual radiation quantity and mean annual temperature.

References

1. Yin, X. Q., Song, B., Dong, W. H., Xin, W. D., Wang, Y. 2010. A review on the eco-geography of soil fauna. China. *J Geogr Sci.* 20: 333–346.
2. Liiri, M., Häsä, M., Haimi, J., Setälä, H. 2012. History of land-use intensity can modify the relationship between functional complexity of the soil fauna and soil ecosystem services: A microcosm study. *Appl Soil Ecol.* 55: 53–61.
3. Dominguez, A., Bedano, J. C., Becker, A. R., Arolfo, R. V. 2014. Organic farming fosters agroecosystem functioning in Argentinian temperate soils: Evidence from litter decomposition and soil fauna. *Appl Soil Ecol.* 83: 170–176.
4. Finlay, B. J. 2002. Global dispersal of free-living microbial eukaryote species. *Science.* 296: 1061–1063.
5. Fontaneto, D., Ficetola, G. F., Ambrosini, R., Ricci, C. 2006. Patterns of diversity in microscopic animals: are they comparable to those in protists or in larger animals? *Glob Ecol Biogeogr.* 15: 153–162.
6. Laossi, K. R., Barot, B., Carvalho, D., Desjardins, T., Lavelle, P., Martins, M., Mitja, D., Rendeiro, A. C., Roussin, G., Sarazin, M., Velasquez, E., Grimald, I. M. 2008. Effects of plant diversity on plant biomass production and soil macro fauna in Amazonian pastures. *Pedobiologia.* 51: 397–407.
7. Wardle, D. A., Bardgett, R. D., Klironomos, J. N., Setala, H., Vander Putten, W. H., Wall, D. H. 2004. Ecological link ages between above ground and belowground biota. *Science.* 304: 1629–1633.

8. Lieberman, D., Lieberman, M., Peralta, R. 1996. Tropical forest structure and composition on a largescale altitudinal gradient in Costa Rica. *J Ecol.* 84: 137–152.
9. Hou, W. L., Fan, H. 2002. Ecological series of soil animal in Darlidai Mountain. *Chin Geogr Sci.* 12: 378–382.
10. Yan, Y., Li, K., Fang, Y. 2010. Tropical forest structure and composition on a large scale altitudinal gradient in Costa Rica Journal of Ecology. *Chin J Ecol.* 29: 1754–1767.
11. Xiao, N. W., Liu, X. H., Ge, F., Ouyang, Z. Y. 2009. Research on soil faunal community composition and structure in the Gaoligong Mountains National Nature Reserve. *Acta Ecol Sin.* 29: 3576–3584.
12. Yin, X. Q., Jiang, Y. F., Tao, Y., An, J. C., Xin, W. D. 2011. Eco-geographical distribution of soil fauna in Pinus koraiensis mixed broadleaved forest of the Changbai Mountains. *Sci Geogr Sin.* 31: 935–940.
13. Wang, S. J., Ruan, H. H., Wang, J. S., Xu, Z. S., Wu, Y. Y. 2010. Composition structure of soil fauna community under the typical vegetations in the Wuyi Mountains, China. *Acta Ecol Sin.* 30: 5174–5184.
14. Wu, P. F., Liu, S. R., Liu, X. L. 2012. Composition and spatio-temporal changes of soil macroinvertebrates in the biodiversity hotspot of northern Hengdunshan mountains, China. *Plant Soil.* 357: 321–338.
15. Zhang, X. P., Huang, L. R., Jiang, L. Q. 2008. Characteristics of macro-soil fauna in forest ecosystem of northern Da Xing'anling Mountains. *Geogr Res.* 27: 509–518.
16. Schlaghamerský, J., Devetter, M., Háněl, L., Tajovský, K., Starý, J., Tuf, I.H., Pižl, V. 2014. Soil fauna across Central European sandstone ravines with temperature inversion: From cool and shady to dry and hot places. *Appl Soil Ecol.* 83: 30–38.
17. Rantalainen, M. L., Haimi, J., Fritze, H., Pennanen, T., Setälä, H. 2008. Soil decomposer community as a model system in studying the effects of habitat fragmentation and habitat corridors. *Soil Biol Biochem.* 40: 853–863.

IV

Landscape Change and Ecological Security

39

Ecological Factors Influencing the Landscape Epidemiology of Tickborne Disease

Heather L. Kopsco and Thomas N. Mather
University of Rhode Island

Introduction

Perhaps the question most commonly asked by the public when it comes to ticks is: "how bad will they be this year?" People seem to associate tick abundance with the weather conditions of the preceding season, which usually means winter in the northern latitudes. If it was a harsh winter, people are hopeful, assuming that ticks were killed. If mild, then perhaps fretful, as they reason that more ticks likely survived. If wet, tick survival is imagined to be greater. Yet, if too wet, some perceive that to be detrimental for ticks. If snow cover was high or of longer duration, some attribute that as a harsh winter with direct cause and effect negative impacts on ticks, while others consider that it may have shielded the ticks from cold, killing conditions. There are considerable and diverse *public perceptions* about the environmental drivers of tick abundance, tick encounter rates, and incidence of tickborne disease.

The science of accurately predicting tick risk is still largely in its infancy or, at least, is mostly lacking in actionable reliability when it comes to public health policy. We contend that the question being asked—"how bad will ticks be this year?"—needs some unpacking and re-framing before a useful answer can be offered. There are many types of ticks, each posing their own level of disease risk and each possessing their own environmental drivers. Unpacking the question "how bad..." to focus on a particular tick species and its abundance in a particular geographic location is probably the appropriate first step in re-framing the question. It also is probably good to remember that the drivers of tick abundance, for any type of tick and in any geographic region, are likely to include not only *survival* but also *reproduction* and in some cases *movement* or *migration*. Thus, presence, absence, and availability of key reproductive hosts for the various life stages of different tick species likely is as important a consideration as tick survival is. Taking this a bit further, tick reproduction may even be influenced by the suitability of the blood meal that these key reproductive hosts provide, as the host immune response to components of tick saliva can potentially reduce blood feeding success in some cases (Wikel 1982; Wilson 1994). Tick migration, especially the intrusion of ticks into new regions and new habitats, is greatly influenced by the dynamic interplay of tick biology, host availability, as well as environmental permissivity. While these general ecological rules likely remain relatively stable, the outcome of tick activity is thrown into question by the advent of changing climate variables across landscapes like rainfall, temperature, and humidity.

Tick Biology

Ticks are obligate ectoparasites that feed on the blood of a wide variety of host species, spanning birds, reptiles, and mammals. Collectively, ticks can be found on every continent and are responsible for vectoring numerous human and non-human animal pathogens including bacteria, viruses, protozoal species, and even triggering allergies and toxin-induced paralysis. In the United States, ticks are of increasing medical and veterinary importance as tick-vectored disease cases climb to surpass hundreds of thousands of reports annually (Rosenberg et al. 2018). As arthropods, tick physiology and behavior are directly affected by the external environment and might, therefore, reasonably be expected to be predictable by examining environmental variables. Here, we provide a brief overview of the components of tick biology that are highly relevant to conditions that vary across landscapes.

Hard Ticks versus Soft Ticks

Two major divisions of ticks (Acari: Parasitiformes: Ixodidae) exist, differentiated by basic morphology: "soft ticks" (Family: Argasidae) and "hard ticks" (Family: Ixodidae). A third family has been named (Nuttalliellidae), but it contains a single species and is only found in bird and mammal nests in regions of South Africa (Keirans et al. 1976). Soft ticks live worldwide in dry environments and contain 57 species within four genera (Horak et al. 2002). Their outer layer lacks the hard shield and densely chitinous outer coating of the hard tick, and it instead appears more like a raisin. The mouthparts are located on the ventral side of their body surface, and soft ticks can blood-feed multiple times in each stage; when engorged, they are capable of increasing 10–15 times their unfed size. Commonly found examples of this tick include the spinose ear tick (*Otobius megnini*), a pest of domestic pets and cattle, and a relapsing fever tick (*Ornithodoros hermsi*) that can transmit relapsing fever *Borrelia* to humans and other animals.

Hard ticks are organized into six genera, including the genus *Ixodes*, which comprises 241 species and is the largest known tick genus (Horak et al. 2002). Within *Ixodes* are also four closely related species of major medical concern to humans and domestic animals: the blacklegged (deer) tick (*Ixodes scapularis*) and western blacklegged tick (*Ixodes pacificus*) in North America, the sheep tick or castor bean tick (*Ixodes ricinus*) in Europe, and the taiga tick (*Ixodes persulcatus*) in Europe and western Asia.

Our focus here is mainly on ixodid ticks due to their particularly important role as vectors of human and animal disease.

Morphology

Hard ticks are composed of three major anatomical segments: (i) the mouthparts (gnathosoma), (ii) the body (idiosoma), and (iii) the legs (Figure 39.1). The blood-feeding mouthparts are attached to the body by the basis capituli. They contain a jagged needle-like structure called the hypostome sheathed between protective palps and saw-like chelicerae, which are used to cut into the tissue layers of a host to provide blood access to the hypostome. The top layer of the idiosoma (the cuticle) is composed of chitin and

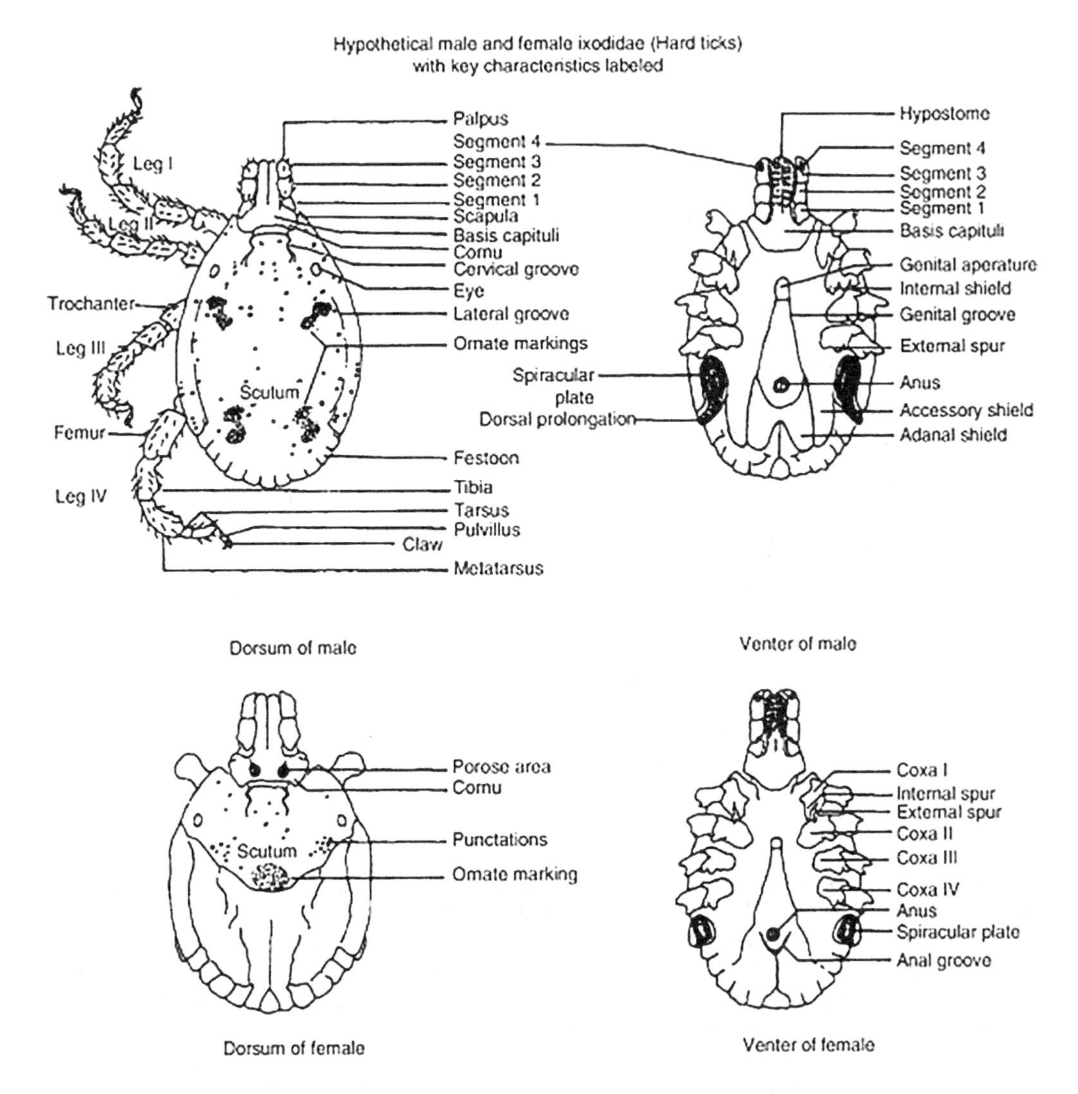

FIGURE 39.1 Hypothetical male and female ixodid ticks with key characteristics labeled. (Coons and Rothschild 2008, used with permission from Springer.)

covered in sclerotized plates to prevent dessication. The legs are attached to the ventral portion of the anterior podosoma by joints called coxae and can be folded in against the body for additional protection against desiccation. The two most anterior legs contain the Haller's organ which allows ticks to sense their environment and locate potential hosts and mates using odor detection (e.g. detection of carbon dioxide), thermosensation, and mechanosensation.

Life Cycle

All hard ticks undergo four distinct life stages: egg, six-legged larva, eight-legged nymph, and eight-legged adult (either male or female) (Figure 39.2). With the exception of the egg phase, each tick life stage feeds once before molting, but an unfed or partially unfed tick can be forcibly removed from a host and yet go on to feed again. It is presumed that interrupted feeding is common only among males as they can mate multiple times and continue to take intermittent blood meals while on a host (Bremer et al. 2005). However, studies documenting tick movement among hosts in close quarters (Little et al. 2007

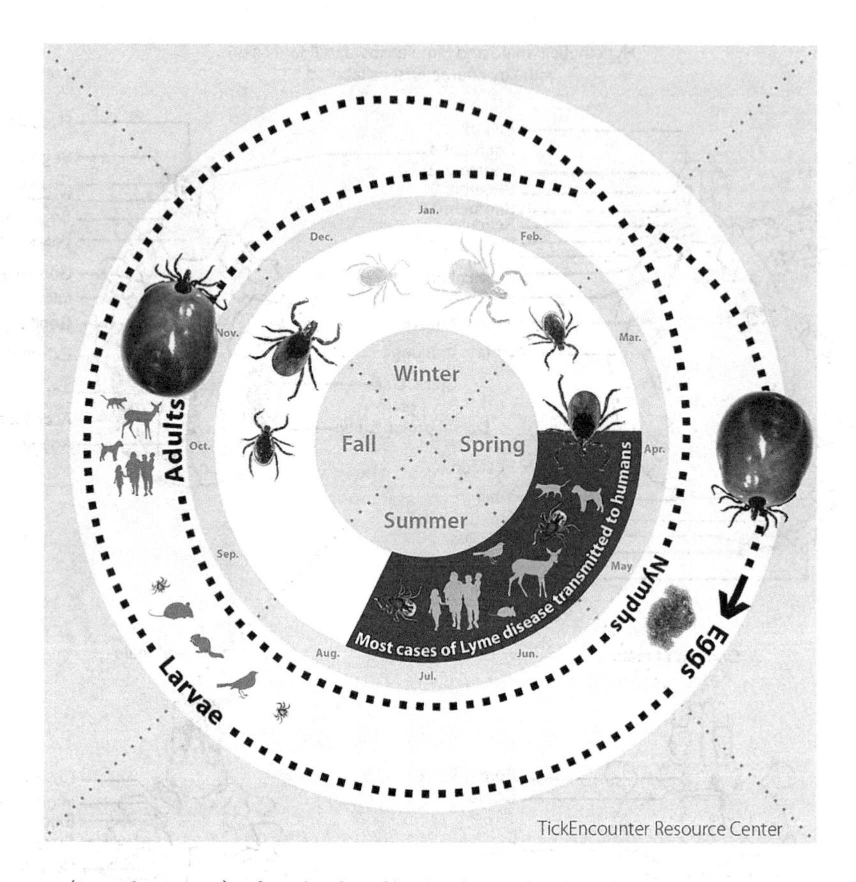

FIGURE 39.2 (See color insert.) Life cycle of *Ixodes scapularis* (a.k.a. blacklegged or deer tick) in the northeast/mid-Atlantic/upper mid-western United States as an example of a typical three-host ixodid tick life stage. Larval deer ticks are active in August and September, but these ticks are pathogen free. Ticks become infected with pathogens when larvae (or nymphs) take a blood meal from infectious animal hosts. Engorged larvae molt over winter and emerge in May as poppy-seed-sized nymphal deer ticks. Most cases of Lyme disease are transmitted from May through July, when nymphal-stage ticks are active. Adult-stage deer ticks become active in October and remain active throughout the winter whenever the ground is not frozen. Blood-engorged females survive the winter in the forest leaf litter and begin laying their 1,500 or more eggs around Memorial Day (late May). These eggs hatch in July, and the life cycle starts again when larvae become active in August.

and molecular evidence of multi-host meals in both larval- and nymph-stage ticks (Moran Cadenas et al. 2007; Allan et al. 2010; Collini et al. 2016) suggest that interrupted feeding may be a more common phenomenon than thought and has the potential to pose a risk of accelerated intrastadial pathogen transmission to incidental hosts.

Adult tick mating behavior is largely described by whether they are suborder Metastriata or Prostriata. In the Metastriata group, most male ticks (e.g. genera *Amblyomma*, *Dermacentor*, and *Rhipicephalus*) need to feed prior to mating in order to mature their sperm. *Ixodes* is the only genus within Prostriata, and males of most species do not require any blood meal to mate. Instead, they attach to a host to wait for females or take small sips of blood to briefly rehydrate. As a result of their general lack of feeding, *Ixodes* males have much shorter longevity than metastriate males and are less risky as transmitters of disease agents.

Most ixodid ticks have an obligate three-host life cycle (Figure 39.2). Both larvae and nymphs seek hosts within forest leaf litter and feed on small animals including rodents, birds, and reptiles for several days before detaching and molting. Larger mammals such as white-tailed deer are parasitized by all stages but are the primary host for adult-stage ticks, and thereby serve as the reproductive host for many tick species. Other ixodid tick species exhibit two- and one-host life cycle strategies. Larvae and nymphs of two-host ticks feed upon a single host for these first two stages and drop off before molting into adults. Adults then find a different host for a final blood meal and mating. Examples of two-host ticks include those in the *Hyalomma* genus that often parasitize smaller mammals and birds as larvae and nymphs and feed on large ungulates as adults. The main hosts for all one-host ticks, where a tick completes all stages on the same host, are large ungulates including camels (e.g. *Hyalomma dromedarii*), moose (e.g. *Dermacentor albipictus*), and cattle (e.g. *Rhipicephalus microplus*). Because of the intense blood loss of a one-host tick infestation, animals can be easily weakened during a time of poor nutrient availability and sometimes die from either anemia or secondary infections (Jones et al. 2019). Moose in North America experience epizootic outbreaks of winter ticks during their fall breeding season that can lead to more than a 50% reduction in calves in springtime (Jones et al. 2019). While humans and domestic animals can encounter each of these types of ticks, they are considered to be incidental hosts in all cases and are not part of the natural life cycle.

Ecological Factors Affecting Tick Population Dynamics

In all biological systems there are ecological rules that regulate the number of plants and animals in any given space, and so it is with ticks. Ticks feed on the blood of their hosts, and therefore, host availability and abundance can impact tick populations as well as risk for tickborne disease. Similarly, tick *reproduction*, *survival*, and *movement* are critical factors that help determine the abundance and spatial distribution of ticks and tick encounter rates. A few examples are offered below to help the reader think about and begin to appreciate the ecological rules of ticks.

Tick Reproduction

It is the female tick that contributes most to the growth of any tick population. Among hard ticks, fully blood engorged female *Amblyomma* and *Hyalomma* species can produce 10,000 eggs or more while species of *Dermacentor* and *Ixodes* typically lay 2,000–5,000 eggs (Oliver 1989). In either case, these levels of fecundity for ticks can lead to explosive increases in tick density or distribution. Fortunately, in most cases, ticks generally also experience high mortality rates and examples of survival factors will be discussed below.

Hosts of different species and life-stages of ticks vary widely, with smaller animals often serving as preferred hosts for immature tick stages. Some types of ticks are rather host restricted, while others have quite broad host ranges. Adult tick stages of most of the epidemiologically important species of ticks distributed across north-temperate regions of the globe feed on medium-sized and large mammals. Ungulates, including various species of deer are the preferred hosts of many species of adult ixodid ticks. In North America, adult *I. scapularis, I. pacificus, Amblyomma americanum, and A. maculatum* feed principally on white-tailed or other species of deer, and these hosts are permissive to feeding large numbers of ticks at any one time. Deer populations across North America were decimated to near extinction due to hunting pressure, clear-cutting of forests, and other loss of suitable habitat during the 18th and 19th centuries but had recovered and even expanded by the 1970s and 1980s. It is the re-establishment and continued growth and spread of deer populations into suburban and semi-urban landscapes that is largely implicated for the current-day epidemic of tickborne diseases (Spielman et al. 1985).

Tick Survival

Various macroclimatic conditions may influence and even limit tick distributions, and it is essential to understand the factors that naturally regulate tick survival in order to accurately predict tickborne disease risk and design effective tick management programs. Temperature and humidity extremes are both commonly cited as factors limiting tick survival, but for many species, especially those in the genus *Ixodes*, it is the desiccating conditions of lower humidity that influence survival most. Due to their small size and high surface-to-volume ratio, the nymph stage, in particular, is most susceptible to desiccation when questing for their next blood meal (Stafford 1994; Rodgers et al. 2007). Nymphal *I. scapularis* can desiccate in less than 48 hours if unable to imbibe enough water moisture from partially saturated air (Needham and Teel 1991). Extreme variability of nymphal *I. scapularis* abundance in similar habitats suggests that additional environmental factors also may regulate tick population survival and abundance (Mather et al. 1996). However, field studies on the relationship between environmental moisture, tick populations, and Lyme disease incidence have produced conflicting results. While investigators have documented associations between climatic variables such as drought and precipitation events and tick populations (Jones and Kitron 2000), other studies have found only modest or no relationship between precipitation and tick numbers (Ostfeld et al. 2006a,b; Schulze et al. 2009). One plausible explanation for these confounding results is that the weather variables measured did not consistently reflect the environmental conditions experienced by the ticks in their adapted landscape. In laboratory studies, Rodgers et al. (2007) reported significant nymphal *I. scapularis* mortality after the ticks were continuously exposed to atmospheric moisture below an 82% relative humidity (RH) threshold for longer than 8 hours, even when RH was subsequently increased to favorable levels above 95% RH. Berger et al. (2014a) conclusively demonstrated with direct field observations collected over a 14-year period that an increased number of tick-adverse moisture events (TAMEs), defined as sub-82% threshold humidity episodes >8 hours in duration, correlated with naturally lowered nymphal blacklegged tick abundance, thereby greatly reducing risk of tick encounter. It should be noted, however, that while other species of ticks have adapted different tolerances for desiccating conditions, the concept of a humidity threshold likely applies as a critical determinant of tick survival.

Selective pressures of tick survival mediated by temperature and humidity also can affect geographical trends in tick host-seeking behavior and disease rates. For example, host-seeking nymphs in southern U.S. populations of *I. scapularis* remain below the leaf litter surface, while northern nymphs seek hosts on leaves and twigs at the top of the litter surface. Tick mortality due to desiccation stress is greater under southern than under northern conditions, and this selective pressure appears to have resulted in altered host-seeking behavior. A person walking through the woods in the north would directly encounter host-seeking nymphal ticks on leaves and twigs at the litter surface, while the same activity in the south would be less likely to result in human encounter with immature ticks because the person would remain mostly on top of the leaf litter and would not encounter the questing ticks below the surface (Arsnoe et al. 2015). This behavioral difference appears to result in decreased tick contact with humans in the south and fewer cases of Lyme disease.

Tick Movement

Local abundance of ticks also may be influenced by their movement to or from a defined area as well as the ecological factors specific to that area. Lacking wings, tick movement occurs either by their crawling or by passive transport on hosts. Their small size typically limits crawling to just short distances (one to up to tens of meters). Most of longer-range movement of ticks involves their attachment to hosts. Tick density may increase regionally or even locally where feeding ticks detach from their infested hosts. Hosts that either regularly migrate through a specific region or perhaps concentrate diurnally (Mather and Spielman 1986) or seasonally have the potential to affect the distribution of ticks in the landscape and the risk of tick encounters by people and pets (Ogden et al. 2008; Palomar et al. 2012). For example, immature blacklegged ticks *(Ixodes scapularis)* have been removed from more than 60 bird species in the northeastern U.S. (Anderson et al. 1986), and higher concentrations of host-seeking nymphal- and adult-stage *I. scapularis* are often found in river watersheds and other bird migratory corridors with adequate forest understory and large forest patch area (Reed et al. 2003; Parker et al. 2017). Adult-stage ticks frequently attach to and engorge on medium- and large-sized mammals, such as deer, and these hosts can both distribute engorged females over greater distances and also increase tick abundance depending on their movements and specifically their bedding behaviors. Engorged ticks frequently detach while hosts are resting, and each engorged female tick is capable of laying 1,500 or more eggs, depending on the quantity and quality of the blood meal.

Ecological Dynamics of Tickborne Disease Are Associated with Landscape Features

While mosquitoes are notoriously considered to be the world's deadliest animals, ticks hold the title for transmitting the greatest variety of arthropod-borne disease germs (Sonenshine and Roe 2014). For a tickborne disease to be transmitted, three components, called the *nidus of infection*, are required to co-occur: (i) ticks that can become infected with a pathogen and are capable of transmitting an infectious dose to a host, (ii) host animals that can carry the pathogen and pass it to a feeding tick, and (iii) hosts that can become infected with the pathogen when an infected tick feeds (Reisen 2010).

Landscape features like vegetation, landscape composition, and land-use patterns are strongly linked to the nidus of tickborne infection because various habitat types support different reproductive and reservoir hosts and vector tick populations, as well as differentially influence the interactions among these nidi components (Asghar et al 2016; MacDonald et al. 2017; Ferrell and Brinkerhoff 2018). While tick species are considered to be spatially autocorrelated at large scales (e.g. county, state, regional), fine-scale distribution (e.g. municipality, town) is not necessarily autocorrelated and can be much more difficult to predict due to the variable combination of landscape factors (Pardanani and Mather 2004). Satellite vegetation indices like the Normalized Difference Vegetation Index (NDVI), the Enhanced Vegetation Index (EVI), and Normalized Difference Water Index (NDWI) have been used to assess vegetation structure and density in relation to tick density. Studies have reported that these indices correlate well with suitable tick habitat and are effective predictors of tick populations and cases of Lyme disease (Barrios González 2013; Ozdenerol 2015).

Landscape heterogeneity, configuration, and connectivity play a large role in the dynamics of tickborne disease transmission and maintenance. For example, forest cover is considered to be one of the largest drivers of *I. scapularis* density (Ferrell and Brinkerhoff 2018), and the specific vegetation structure impacts the abundance of ticks found on the main Lyme bacteria reservoir, the white-footed mouse (*Peromyscus leucopus*) (Adler et al. 1992). When white-tailed deer are present, larval *I. scapularis* were found predominantly in woody vegetation, while nymphal *I. scapularis* were associated with grassy and herbaceous vegetation (Adler et al. 1992). However, after deer are removed from an environment, these habitat structure associations disappear.

Natural and anthropogenic alteration of contiguous landscapes into fragments is largely associated with outbreaks and continued maintenance of tickborne disease cycles. The dilution hypothesis predicts that increases in the number of patches of forest versus open grassland-type habitat can increase both the abundance of tick vectors as well as the availability of highly competent hosts for tickborne diseases, leading to enhanced pathogen transmission (Ostfeld and Keesing 2000). In a similar way, patch ratios of forest versus agricultural moorland were found to impact the transmission of louping ill virus (LIV) from *Ixodes ricinus* ticks to moorland-based red grouse (Jones et al. 2013). In this case, as forest/moorland patch heterogeneity increased, increasing the amount of forest edge through fragmentation, there was greater abundance of ticks and greater rate of transmission of LIV to grouse because ticks were able to travel further out from the forest due to deer movement (Jones et al. 2013). However, other studies have found forest fragmentation to be negatively correlated with tick density and pathogen prevalence at large scales (Perez et al. 2016; Ferrell and Brinkerhoff 2018). The lack of consistency surrounding investigations into the effect of fragmentation on tick and disease density suggest that these variables may be scale and even system dependent.

Other landscape features like riparian zones and rivers can also impact the dispersal and occurrence of ticks and tickborne disease. Depending on the size and permanence of the river, it can serve as a barrier to host movement and tick dispersal. The Connecticut and Hudson rivers are considered to have slowed the westward expansion of *I. scapularis* in Connecticut and New York, respectively, evidenced by differences in Lyme disease rates on the eastern and western shores (Reisen 2010). However, riparian zones also provide suitable habitat for tick hosts including migrating birds, white-tailed deer, and rodents and thus are correlated with a high probability of established populations of *I. scapularis* (Guerra et al. 2002).

Socio-ecological processes are rarely discussed in connection with ticks and disease but likely influence tickborne disease epidemiology. For example, the "attractiveness" of natural spaces to humans may help determine transmission rates of disease agents from natural cycles to incidental hosts (i.e. humans or domestic animals). Cases of tickborne encephalitis (TBE) in Sweden were mapped and modeled in an effort to isolate factors involved with disease distribution. Areas representing the highest density of cases were not only associated with the relevant wildlife hosts and disease reservoirs, but also strongly correlated with abundance of vacation homes and roads providing easy access to forested regions (Zemies et al. 2014).

The Effects of a Changing Climate on the Landscape Ecology of Tickborne Disease

We have illustrated how the cycle of tickborne disease operates on numerous ecological principles, so it is no surprise that the impact of shifting climate norms has the potential to alter tick activity and distribution based on increasing temperatures and varying precipitation trends. For example, localized and regional drought conditions produce different effects depending on the tick vector and its biology. Ixodid ticks may experience higher mortality during drought or even seasonal dry weather events because of high humidity requirements and intolerance for desiccation. Both nymphal *I. scapularis* host-seeking activity as well as survival are significantly impacted by dry and extreme temperature conditions. It is well documented that *I. scapularis* nymphs cannot tolerate relative humidity below roughly 90% in laboratory conditions (Stafford 1994; Rodgers et al. 2007; Ostfeld and Brunner 2015; Ginsberg et al. 2017). However, during adverse climate conditions like extensive summer droughts and heat waves, nymphs demonstrate an acute behavioral response and descend into or remain in humid microclimates within leaf litter to protect themselves from desiccation and are less likely to encounter a human or pet host (Lindsay et al. 1999; Vail and Smith 1998). Berger et al. (2014a) found that the total number of TAMEs in June was negatively correlated with the overall nymphal tick density for that season. Additionally, a high number of TAMEs was correlated with large early season to late season tick ratios, which could suggest tick mortality, but could also be evidence of ticks having found small mammal hosts and therefore were not available to be collected from the environment.

Burtis et al. (2016) examined the effect of consecutive days of high temperature (>25°C) and no precipitation on nymphal *I. scapularis* questing activity and then compared these to the inter-annual variation of human Lyme disease cases in the following years to determine whether the reduction in activity was short-term behavioral strategy or evidence of mortality (i.e. reduced tick population in the following year). They found that the number of hot, dry summer days both negatively impacted the density of questing nymphs and were inversely correlated with the number of new Lyme disease cases reported in long-term endemic areas (Burtis et al. 2016). However, the number of hot, dry days did not reduce the number of ticks found on small mammals, nor did it appear to have any negative impacts on the next season's tick population.

In newly endemic regions (e.g. the Midwest), this weather measurement had no effect on inter-annual variation in human Lyme disease (Burtis et al. 2016). There was also no effect on cases of Lyme disease or the density of nymphal ticks in the following year in either long-term or recently endemic regions when examining hot, dry days and larval ticks (Burtis et al. 2016). The linear model showed that hot, dry days were negatively associated with incidence of human Lyme disease in the same year in historically endemic zones but not in recently expanded zones (Burtis et al. 2016).

Evidence from these studies demonstrates that a reduction in nymphal *I. scapularis* tick abundance during unfavorable (high temperature, low RH) weather conditions appears to be mostly due to a behavioral survival strategy, but it is reasonable to conclude that mortality will occur should the relative humidity (RH) of leaf litter be reduced below 82% quickly or for long enough that nymph ability to seek out small mammal hosts is reduced (Berger et al. 2014b). In the southeastern United States, *I. scapularis* is present, but human infections are lower than those in north-eastern and upper mid-western regions of the country. Nymphal *I. scapularis* exhibits differences in questing behavior and host preference in these regions, with the northern variety questing higher and opting to feed on small rodents. The southern genotype remains below the leaf litter and often feeds on reptiles that are poorly reservoir competent (Arsnoe et al. 2015).

Naturally occurring as well as accidental wildfires in the western United States are becoming an increasing concern as climate change continues to alter temperatures and precipitation patterns because they result in fire events that burn much greater spatial areas and more vegetation than do prescribed fires (Neary et al. 1999). Some studies have demonstrated that natural fires reduce abundance of certain tick species, including lone star ticks (Davidson et al. 1994; Cully 1999). Some researchers have found that wildfires remove ticks because they directly kill them or burn necessary ground cover and obligate hosts flee the area (Stafford et al. 1998); others have found that no difference occurs in tick questing between burned and unburned plots (Padgett et al. 2009). The effects of a natural wildfire on local tick and wildlife abundance can also produce a paradoxical effect, similar to what controlled investigations have found (Mather et al. 1993). In the first year after a large California wildfire, the number of questing adult and nymphal (*Ixodes pacificus*) ticks was significantly higher in the burned plots than in the control plots, but these numbers fell drastically in the last two years of the study (MacDonald et al. 2018). There were also far fewer small animal hosts in the burned plots than in the unburned plots in that first year post fire, suggesting that the ticks that were found questing had survived the fire in the soil and were now host seeking because their typical hosts had either fled or perished in the fire (MacDonald et al. 2018). Reasonably then, the first year after a fire could be particularly risky for any humans or pets hiking through a burnt area because of the increased chance of encountering ticks.

A poleward spread of some tick vectors occurs as temperatures at high latitudes and elevations become more favorable for their survival and reproduction. The northward expansion of the European tick vector *Ixodes ricinus* is strongly associated with milder winters, earlier springs, and warmer autumnal seasons, as is the expansion of the North American Lyme vector (*Ixodes scapularis*) into southern Canada (Beck et al. 2000; Wu et al. 2013). However, at a more fine scale, elevation is not correlated with increased risk (Ferrell and Brinkerhoff 2018). These movements put new human populations at risk for a variety of diseases including Lyme borreliosis, babesiosis, and tickborne encephalitis (TBE). However, temperatures that rise too quickly may result in poleward contraction to limited regions because vectors cannot tolerate the conditions and increasing temperatures may disrupt delicate disease transmission cycles (Jones et al. 2013; Ogden and Lindsay 2016). An example of pathogen-vector cycle interruption as a result of increased temperatures has been observed in Europe with TBE in *I. ricinus* and *I. persulcatus* (Randolph and Rogers 2000). In order for these ticks to become infected with TBE, an infected nymph must be feeding on the same rodent host at the same time as an uninfected larva because rodents do not maintain systemic infection. Transmission cycles are dependent on hot summers to mature eggs and rapidly-cooled autumns to force eggs into diapause so that newly hatched larvae and nymphs both emerge concurrently in the spring. Randolph and Rogers (2000) built a model comparing current TBE distribution and relevant climate factors, comparing 30 years of climate data (mean min. and max. temperatures, rainfall, and saturation vapor pressure variables) to cases of TBE infection. Their model predicted with 86% accuracy the distribution of TBE and demonstrated that hot, dry summers were increasingly driving the occurrence of the disease into high-altitude regions while reducing incidence at the lower altitudes (Randolph and Rogers 2000). Due to the predicted lack of rapid autumnal cooling necessary to delay larval emergence until the spring in more northern regions of Europe (Scandinavia), rates of human cases of TBE are predicted to reduce in those areas (central and eastern Europe).

Large-scale climate data are becoming easily accessible to help forecast and monitor vector-borne disease prevalence and distribution as climate conditions change. Burtis et al. (2016) utilized the Centers for Disease Control and Prevention's (CDC) Wide-ranging Online Data for Epidemiological Research (WONDER) database to examine the effect of spatiotemporal climate effects on tick populations at a broad resolution. This database provides free-to-the-public NOAA weather station data for the entire U.S. that can be broken down by multiple spatial and temporal boundaries (https://wonder.cdc.gov/). While using ground-truthing methods to produce these types of robust models may be costly and time consuming, the economic impact in preventing the effects of devastating outbreaks with increasing certainty may outweigh the initial costs.

Conclusions

A growing knowledge of the basic ecological rules that govern complex cycles of tickborne disease is beginning to provide opportunities for targeted tick management strategies. However, due to the complex interconnectedness of tick abundance, host ecology, weather, climate, and landscape variables, we suggest that evidence-based integrated tick management approaches are the most sensible means of area-wide tick control and human disease mitigation. The most effective strategies currently are typically grouped into personal protection (i.e. repellents, protective dress, tick checks), habitat-targeted tick reduction (i.e. insecticide treatments, landscape management, host population manipulation), and host-targeted acaricide use (i.e. treated rodent nesting material, treated bait boxes and feeders). Targeted tick reduction practices increasingly are being focused on reducing tick reproduction and decreasing tick survival.

Targeting Tick Reproduction and Survival

As we are gaining a better understanding of the evolving landscape ecology of tick reproductive hosts, like white-tailed deer in the case of blacklegged and Lone Star ticks, this knowledge can largely help explain the expanding epidemic of tickborne disease. It also may hold a key for reversing the "more ticks in more places" phenomenon at the root of the current tick problem.

Even though deer may not be competent reservoirs for many of the tickborne infections plaguing people and pets, as tick reproductive hosts they add to the force of infection by supplying ticks, which are the critical agents of tickborne pathogen spillover from reservoir hosts. Deer in more places provide the opportunity for tick reproduction to occur in more places; more tick reproduction equals greater spillover potential. In a scenario where deer are less dense, ticks and the risk for disease are largely restricted to core foci at the intersection where deer aggregate and the habitat is permissive for tick survival. But as deer herds increase in number, individual animals carrying engorged female ticks begin moving out of these core foci in search of food. This serves to expand the size of core foci until they begin to coalesce with others nearby, eventually filling in even marginally permissive landscapes with ticks and increasing the likelihood of more uniformly distributed dense tick infestations and human tick encounters (Telford 2017). For ticks that rely on deer for reproduction, deer reduction, perhaps aggressive deer reduction, should be expected to eventually shrink the core foci so that they once again are more dispersed than coalesced. But reproductive host removal strategies, such as various deer reduction campaigns, are not likely to deliver short-term relief from encountering ticks and should be considered a longer-term, albeit more sustained solution.

Reduced tick survival can be achieved directly through landscape alterations that remove permissive environments or by attacking ticks using habitat-targeted chemical or bio-control applications. Landscape management techniques, including, vegetation clearing, mowing, burning, and leaf litter removal, have been proven to, at least temporarily, reduce the abundance of host-seeking ticks (Wilson 1986; Schulze et al. 1995; Mather et al. 1993; MacDonald et al. 2018; Fischhoff et al. 2019). Traditional pesticide sprays or granule applications utilizing synthetic pyrethroids (e.g. bifenthrin, deltamethrin, cyfluthrin) or organophosphate/carbamate (e.g. carbaryl) active ingredients applied principally to tick-permissive habitats are highly effective means of area-wide tick control for numerous species, demonstrating near 100% nymph and larval-stage knockdown (Schulze et al. 1991, 2001). However, concern surrounding the toxicity of these chemicals to non-target insects and vertebrates has resulted in the exploration and development of minimal risk products; however, these have demonstrated varying degrees of efficacy. Included in a list of those with demonstrated lethal efficacy on ticks are a derivative of a bark oil of the Alaskan yellow cedar (*Cupressus nootkatensis*; nookatone) and entomopathogenic fungi like *Metarhizium anisopliae* (Met-52) (Benjamin et al. 2002).

Various host-targeted strategies for tick control have been developed as alternative methods to area-wide sprays but have been shown to work best as part of an integrated treatment strategy. Two strategies targeting rodents, tick control tubes containing permethrin-treated cotton nest-building material (Mather et al. 1987) and bait boxes that attract rodents with food causing them to self-apply tick-killing fipronil (Dolan et al. 2004), have been successful in reducing the proportion of *I. scapularis* nymphs infected with *Borrelia burgdorferi* and the agent that causes human anaplasmosis (*Anaplasma phagocytophilum*) but not in reducing the overall number of questing ticks (Stafford 1992). Another example of host-targeted tick reduction is the USDA-ARS "4-Poster," a deer feeding station designed to self-apply acaricide directly to deer for preventing successful feeding of adult ticks. Wearing effective tick repellent, especially wearing permethrin-treated clothing (Miller et al. 2011), and administering tick killing protection for pets are examples of targeted personal protection. Newer host-targeted strategies, including host genetic manipulations, new delivery platforms, and wildlife-targeted vaccines are also in development.

References

Adler, G.H.; Telford, S.R.; Wilson, M.L.; Spielman, A. Vegetation structure influences the burden of immature *Ixodes dammini* on its main host, *Peromyscus leucopus*. *Parasitology* 1992, 105: 105–110.

Allan, B.F.; Goessling, L.S.; Storch, G.A.; Thach, R.E. Blood meal analysis to identify reservoir hosts for *Amblyomma americanum* ticks. *Emerging Infectious Diseases* 2010, 16(3): 433–440.

Anderson, J.F.; Johnson, R.C.; Magnarelli, L.A.; Hyde, F.W. Involvement of birds in the epidemiology of the Lyme disease agent *Borrelia burgdorferi*. *Infection and Immunity* 1986, 51(2): 394–396.

Arsnoe, I.M.; Hickling, G.J.; Ginsberg, H.S.; McElreath, R.; Tsao, J.I. Different populations of blacklegged tick nymphs exhibit differences in questing behavior that have implications for human Lyme disease risk. *PLoS ONE* 2015, 10(5): e0127450.

Asghar, N; Petersson, M; Johansson, M; Dinnetz, P. Local landscape effects on population dynamics of *Ixodes ricinus*. *Geospatial Health* 2016, 11: 283.

Barrios González, J.M. *Spatio-Temporal Modelling of the Epidemiology of Nephropathia Epidemica and Lyme Borreliosis*. KU Leuven, Leuven. 2013.

Beck, L.R.; Lobitz, B.M.; Wood, B.K. Remote sensing and human health: New sensors and new opportunities. *Emerging Infectious Diseases* 2000, 6(3): 217–227.

Benjamin, M.A.; Zhioua, E.; Ostfeld, R.S. Laboratory and field evaluation of the entomopathogenic fungus *Metarhizium anisopliae* (Deuteromycetes) for controlling questing adult *Ixodes scapularis*(Acari: Ixodidae). *Journal of Medical Entomology* 2002, 39: 723–728.

Berger, K.A.; Ginsberg, H.S.; Dugas, K.D.; Hamel, L.H.; Mather, T.N. Adverse moisture events predict seasonal abundance of Lyme disease vector ticks (*Ixodes scapularis*). *Parasites and Vectors* 2014a, 7: 181.

Berger, K.A.; Ginsberg, H.S.; Gonzalez, L.; Mather, T.N. Relative humidity and activity patterns of *Ixodes scapularis* (Acari: Ixodidae). *Journal of Medical Entomology* 2014b, 51(4): 769–776.

Bremer, W.G.; Schaefer, J.J.; Wagner, E.R.; Ewing, S.A.; Rikihisa, Y.; Needham, G.R.; Jittapalapong, S.; Moore, D.L.; Stitch, R.W. Transstadial and intrastadial experimental transmission of *Ehrlichia canis* by male *Rhipicephalus sanguineus*. *Veterinary Parasitology* 2005, 131: 95–105.

Burtis, J.C.; Sullivan, P.; Levi, T.; Oggenfuss, K.; Fahey, T.J.; Ostfeld, R.S. The impact of temperature and precipitation on blacklegged tick activity and Lyme disease incidence in endemic and emerging regions. *Parasites and Vectors* 2016, 9: 606.

Collini, M.; Albonico, F.; Rosa, R.; Tagliapietra, V.; Arnold, D.; Conterno, L.; Rossi, C.; Mortarino, M.; Rizzoli, A.; Hauffe, H.C. Identification of *Ixodes ricinus* blood meals using an automated protocol with high resolution melting analysis (HRMA) reveals the importance of domestic dogs as larval tick hosts in Italian alpine forests. *Parasites and Vectors* 2016, 9: 638.

Coons, L.B.; Rothschild, M. Ticks (Acari: Ixodida). In Capinera, J.L. (ed.) *Encyclopedia of Entomology*. Springer, Dordrecht. 2008.

Cully, J.F, Jr. Lone star tick abundance, fire, and bison grazing in tallgrass prairie. *Journal of Range Management* 1999, 52: 139–144.

Davidson, W.R.; Siefken, D.A.; Creekmore, L.H. Influence of annual and biennial prescribed burning during March on the abundance of *Amblyomma americanum* (Acari: Ixodidae) in central Georgia. *Journal of Medical Entomology* 1994, 31: 72–81

Dolan, M.C.; Maupin, G.O.; Schneider, B.S.; Denatale, C.; Hamon, N.; Cole, C.; Zeidner, N.S.; Stafford, K.C. Control of immature *Ixodes scapularis* (Acari: Ixodidae) on rodent reservoirs of *Borrelia burgdorferi* in a residential community of southeastern Connecticut. *Journal of Medical Entomology* 2004, 41: 1043–1054.

Ferrell, A.M.; Brinkerhoff, R.J. 2018. Using landscape analysis to test hypotheses about drivers of tick abundance and infection prevalence with *Borrelia burgdorferi*. *International Journal of Environmental Research and Public Health* 2018, 15(737): 1–14.

Fischhoff, I.R.; Keesing, F.; Pendleton, J.; DePietro, D.; Teator, M.; Duerr, S.T.K.; Mowry, S.; Pfister, A.; LaDeau, S.L.; Ostfeld, R.S. Assessing effectiveness of recommended residential yard management measures against ticks. *Journal of Medical Entomology* 2019. 56(5): 1420–1427.

Ginsberg, H.S.; Albert, M.; Acevedo, L.; Dyer, M.C.; Arsnoe, I.M.; Tsao, J.I.; Mather, T.N.; LeBrun, R.A. Environmental factors affecting survival of immature *Ixodes scapularis* and implications for geographical distribution of Lyme Disease: The climate/behavior hypothesis. *PLoS ONE* 2017, 12(1): e0168723.

Guerra, M.; Walker, E.; Jones, C.; Paskewitz, S.; Cortinas, M.R.; Stancil, A.; Beck, L.; Bobo, M.; Kitron, U. Predicting the risk of Lyme disease: Habitat suitability for *Ixodes scapularis* in the North Central United States. *Emerging Infectious Diseases* 2002, 8(3): 289–297.

Horak, I.G.; Camicas, J.L.; Keirans, J.E. The Argasidae, Ixodidae, and Nuttallielidae (Acari: Ixodidae): A world list of valid tick names. *Experimental and Applied Acarology* 2002, 28: 27–54.

Jones, B.; Grace, D.; Kock, R.; Alonso, S.; Rushton, J.; Said, M.Y. Zoonosis emergence linked to agricultural intensification and environmental change. *Proceedings of the National Academy of Sciences* 2013, 110: 8399–404.

Jones, C.J.; Kitron, U.D. Populations of *Ixodes scapularis* (Acari: Ixodidae) are modulated by drought at a Lyme disease focus in Illinois. *Journal of Medical Entomology* 2000, 37: 408–415.

Jones, H.; Pekins, P.; Kantar, L.; Sidor, I.; Ellingwood, D.; Lichtenwalner, A.; O'Neal, M. Mortality assessment of moose (*Alces alces*) calves during successive years of winter tick (*Dermacentor albipictus*) epizootics in New Hampshire and Maine (USA). *Canadian Journal of Zoology* 2019, 97: 22–30.

Keirans, J.E.; Clifford, C.M.; Hoogstraal, H.; Easton, E.R. Discovery of *Nuttalliella namaqua* Bedford (Acarina: Ixodoidea: Nuttallielidae) in Tanzania and re-description of female based on scanning electron microscopy. *Annals of the Entomological Society of America* 1976, 69: 926–932.

Lindsay, L.R.; Mathison, S.W.; Barker, I.K.; McEwen, S.A.; Gillespie, T.J.; Surgeoner, G.A. Microclimate and habitat in relation to *Ixodes scapularis* (Acari: Ixodidae) populations on Long Point, Ontario, Canada. *Journal of Medical Entomology* 1999, 36: 255–262.

Little, S.E.; Hostetler, J.; Kocana, K.M. Movement of *Rhipicephalus sanguineus* adults between co-housed dogs during active feeding. *Veterinary Parasitology* 2007, 150: 139–145.

MacDonald, A.J.; Hyon, D.W.; Brewington, J.B., III; O'Connor, K.E.; Swei, A.; Briggs, C.J. Lyme disease risk in southern California: Abiotic and environmental drivers of *Ixodes pacificus* (Acari: Ixodidae) density and infection prevalence with *Borrelia burgdorferi*. *Parasites and Vectors* 2017, 10: 7.

MacDonald, A.J.; Hyon, D.W.; McDaniels, A.; O'Connor, K.E.; Swei, A.; Briggs, C.J. Risk of vector tick exposure initially increases, then declines through time in response to wildfire in California. *Ecosphere* 2018, 9(5): e02227.

Mather, T.N.; Spielman, A. Diurnal detachment of immature deer ticks (*Ixodes dammini*) from nocturnal hosts. *American Journal of Tropical Medicine and Hygiene* 1986, 35: 182–186.

Mather, T.N.; Ribeiro, J.M.C.; Spielman, A. Lyme disease and babesiosis: Acaricide focused on potentially infected ticks. *American Journal of Tropical Medicine and Hygiene* 1987, 36: 609–14.

Mather, T.N.; Duffy, D.C.; Campbell, S.R. An unexpected result from burning vegetation to reduce Lyme disease transmission risks. *Journal of Medical Entomology* 1993, 30: 642–645.

Mather, T.N.; Nicholson, M.C.; Donnelly, E.F.; Matyas, B.T. Entomologic index for human risk of Lyme disease. *American Journal of Epidemiology* 1996, 144: 1066–1069.

Miller, N.J.; Rainone, E.E.; Dyer, M.C.; González, M.L.; Mather, T.N. Tick bite protection with permethrin-treated summer weight clothing. *Journal of Medical Entomology* 2011, 48: 327–333.

Moran Cadenas, F.M.; Rais, O.; Humair, P.F.; Douet, V.; Moret, J.; Gern, L. Identification of host bloodmeal source and *Borrelia burgdorferi* sensu lato in field-collected *Ixodes ricinus* ticks in Chaumont (Switzerland). *Journal of Medical Entomology* 2007, 44(6): 1109–1117.

Neary, D.G.; Klopatek, C.C.; DeBano L.F.; Folliott, P.F. Fire effects on belowground sustainability: A review and synthesis. *Forest Ecology and Management* 1999, 122: 51–71.

Needham, G.R.; Teel, P.D. Off-host physiology ecology of Ixodid ticks. *Annual Review of Entomology* 1991, 36: 659–681.

Ogden, N.H.; Lindsay, L.R. Effects of climate and climate change on vectors and vector-borne diseases: Ticks are different. *Trends in Parasitology* 2016, 32(8): 646–656.

Ogden, N.H.; Lindsay, L.R.; Hanincová, K.; Barker, I.K.; Bigras-Poulin, M.; Charron, D.F.; Heagy, A.; Francis, C.M.; O'Callaghan, C.J.; Schwartz, I.; Thompson, R.A. Role of migratory birds in introduction and range expansion of *Ixodes scapularis* ticks and of *Borrelia burgdorferi* and *Anaplasma phagocytophilum* in Canada. *Applied Environmental Microbiology* 2008, 74(6): 1780.

Oliver, J.H., Jr. Biology and systematics of ticks (Acai: Ixodida). *Annual Review of Ecology, Evolution, and Systematics* 1989, 20: 397–430.

Ostfeld, R.S.; Keesing, F. Biodiversity series: The function of biodiversity in the ecology of vector-borne zoonotic diseases. *Canadian Journal of Zoology* 2000, 78: 2061–2078.

Ostfeld, R.S.; Brunner, J.L. Climate change and *Ixodes* tick-borne diseases of humans. *Philosophical Transactions of the Royal Society of Biology* 2015, 370: 20140051.

Ostfeld, R.S.; Canham, C.D.; Oggenfus, K.; Winchcombe, R.J.; Keesing, F. Climate, deer, rodents, and acorns as determinants of variation in Lyme disease risk. *PLoS Biology* 2006a, 4: e145.

Ostfeld, R.S.; Price, A.; Hornbostel, V.L.; Benjamin, M.A. Keesing, F. Controlling ticks and tick-borne Zoonoses with biological and chemical agents. *BioScience* 2006b, 56(5): 383–394.

Ozdenerol, E. GIS and remote sensing use in the exploration of Lyme disease. *International Journal of Environmental Research and Public Health* 2015, 12(12): 15182–15203.

Padgett, K.A.; Casher, L.E.; Stephens, S.L.; Lane, R.S. Effect of prescribed fire for tick control in California chaparral. *Journal of Medical Entomology* 2009, 46: 1138–1145

Palomar, A.M.; Santibáñez, P.; Mazuelas, D.; Roncero, L.; Santibáñez, S.; Portillo, A.; Oteo, J.A. Role of birds in dispersal of etiologic agents of Tick-borne Zoonoses. *Emerging Infectious Diseases* 2012, 18(7): 1188–1191.

Pardanani, N.; Mather, T. Lack of spatial autocorrelation in fine-scale distributions of *Ixodes scapularis* (Acari: Ixodidae). *Journal of Medical Entomology* 2004, 41(5): 861–864.

Parker, C.M.; Miller, J.R.; Allan, B.F. Avian and habitat characteristics influence tick infestation among birds in Illinois. *Journal of Medical Entomology* 2017, 54(3): 550–558.

Perez, G.; Bastian, S.; Agoulon, A.; Bouju, A.; Durand, A.; Faille, F.; Lebert, I.; Rantier, Y.; Plantard, O.; Butet, A. Effect of landscape features on the relationship between *Ixodes ricinus* ticks and their small mammal hosts. *Parasites and Vectors* 2016, 9(20): 1–18.

Randolph, S.E.; Rogers, D.J. Fragile transmission cycles of tick-borne encephalitis virus may be disrupted by predicted climate change. *Proceedings of the Royal Society of Biology* 2000, 267(1454): 1741–1744.

Reed, K.D.; Meece, J.K.; Henkel, J.S.; Shukla, S.K. Birds, migration and emerging Zoonoses: West Nile virus, Lyme disease, influenza A and enteropathogens. *Clinical Medicine and Research* 2003, 1(1): 5–12.

Reisen, W.K. Landscape epidemiology of vector-borne diseases. *Annual Reviews in Entomology* 2010, 55: 461–483.

Rodgers, S.E.; Zolnik, C.P.; Mather, T.N. Duration of exposure to suboptimal atmospheric moisture affects nymphal blacklegged tick survival. *Journal of Medical Entomology* 2007, 44: 372–375.

Rosenberg, R.; Lindsey, N.P.; Fischer, M.; Gregory, C.J.; Hinckley, A.F.; Mead, P.S.; Paz-Bailey, G.; Waterman, S.H.; Drexler, N.A.; Kersh, G.J.; Hooks, H.; Partridge, S.K.; Visser, S.N.; Beard, C.B.; Petersen, L.R. Vital Signs: Trends in reported vectorborne disease cases – United States and territories, 2004–2016. *MMWR Morbidity Mortality Weekly Report* 2018, 67: 496–501.

Schulze, T.L.; Taylor, G.C.; Jordan, R.A.; Bosler, E.M.; Shisler, J.K. Effectiveness of selected granular acaricide formulations in suppressing populations of *Ixodes dammini* (Acari, Ixodidae)—short-term control of nymphs and larvae. *Journal of Medical Entomology* 1991, 28: 624–629.

Schulze, T.L.; Jordan, R.A., Hung, R.W. Suppression of subadult *Ixodes scapularis* (Acari: Ixodidae) following removal of leaf litter. *Journal of Medical Entomology* 1995, 32: 730–733.

Schulze, T.L.; Jordan, R.A.; Hung, R.W.; Taylor, R.C.; Markowski, D.; Chomsky, M.S. Efficacy of granular deltamethrin against *Ixodes scapularis* and *Amblyomma americanum* (Acari: Ixodidae) nymphs. *Journal of Medical Entomology* 2001, 38: 344–346.

Schulze, T.L.; Jordan, R.A.; Schulze, C.J.; Hung, R.W. Precipitation and temperature as predictors of local abundance of *Ixodes scapularis* (Acari: Ixodidae) nymphs. *Journal of Medical Entomology* 2009, 46: 1025–1029.

Sonenshine, D.; Roe, R.M. Ticks, people, and animals. *Biology of Ticks*. D.E. Sonenshine and R.M. Roe eds. Vol. 1. 2nd Edition. Oxford University Press, Oxford. 2014, 5–15.

Spielman, A; Wilson, M.L.; Levine, J.F.; Piesman, J. Ecology of *Ixodes dammini*-borne human babesiosis and Lyme disease. *Annual Review of Entomology* 1985, 30: 439–460.

Stafford, K.C. Third-year evaluation of host-targeted permethrin for the control of *Ixodes dammini* (Acari: Ixodidae) in southeastern Connecticut. *Journal of Medical Entomology* 1992, 29: 717–720.

Stafford, K.C. Survival of immature *Ixodes scapularis* (Acari, Ixodidae) at different relative humidities. *Journal of Medical Entomology* 1994, 31: 310–314.

Stafford, K.C.; Ward, J.S.; Magnarelli, L.A. Impact of controlled burns on the abundance of *Ixodes scapularis* (Acari: Ixodidae). *Journal of Medical Entomology* 1998, 35: 510–513.

Telford, S.R. III. Deer reduction is a cornerstone of integrated deer tick management. *Journal of Integrated Pest Management* 2017, 8(1): 25; 1–5.

Vail, S.G.; Smith, G. Air temperature and relative humidity effects on behavioral activity of blacklegged tick (Acari: Ixodidae) nymphs in New Jersey. *Journal of Medical Entomology* 1998, 35: 1025–1028.

Wikel, S.K. Immune responses to arthropods and their products. *Annual Review of Entomology* 1982, 27: 21–48.

Wilson, M.L. Reduced abundance of adult *Ixodes dammini* (Acari: Ixodidae) following destruction of vegetation. *Journal of Economic Entomology* 1986, 79: 693–696.

Wilson, M.L. Population ecology of tick vectors: Interaction, measurement, and analysis. In *Ecological Dynamics of Tick-Borne Zoonoses*. D.E. Sonenshine and T.N. Mather, eds. Oxford University Press, Oxford. 1994, 20–44.

Wu, X.; Duvvuri, V.R.; Louc, Y.; Ogden, N.H.; Pelcat, Y.; Wu, J. Developing a temperature-driven map of the basic reproductive number of the emerging tick vector of Lyme disease, *Ixodes scapularis,* in Canada. *Journal of Theoretical Biology* 2013, 319: 50–61.

Zemies, C.B.; Olsson, G.E.; Hjerqvist, M.; Vanwambeke, S.O. Shaping zoonosis risk: Landscape ecology vs. landscape attractiveness for people, the case of tick-borne encephalitis in Sweden. *Parasites and Vectors* 2014, 7: 370–380.

40

Yeqiao Wang
University of Rhode Island

Zhengfang Wu
Northeast Normal University

Hong S. He
Northeast Normal University
University of Missouri

Shengzhong Wang, Hongyan Zhang, Jiawei Xu, and Shusheng Wang
Northeast Normal University

Ecological Security: Changbai Mountains, China

Introduction

Ecological security is an essential cornerstone for the sustainability of human societies and nature ecosystems. Ecological security depends on the balance between human demands and actions in consumption and alteration of resource base and the sustainability, vulnerability, and resilience of environmental systems that provide ecosystem services. Since the very beginning, human societies have interacted and coevolved with other forms of life, i.e., plants, animals, and microorganisms, and learned that their well-being depended on the sustainability of the resource systems. Many ecological security issues, concerns, and actions have been reflected under the framework of "Agenda 21"—the Rio Declaration on Environment and Development, adopted by world leaders in the 1992 U.N. Conference on Environment and Development (World Commission on Environment and Development, 1987; National Research Council, 1999). The concept and implementation have been identified as issues of environmental security (Allenby, 2000; Floyd and Matthew, 2013).

Pirages (2005) suggested that ecological security rests on preserving four interrelated dynamic equilibriums between human and nature. Those include (i) human demands on resources and the ability of nature to provide services, (ii) human population and pathogenic microorganisms, (iii) human populations and those of other plant and animal species, and (iv) the size and growth rates of various human populations. Any type of significant breakdown of the equilibriums will threaten ecological security. This is a particular concern for regions with sensitive and fragile ecosystems and ecotones, with trans-region and cross-boundary movements such as transportation of pollutants through water and air systems and migration of human populations and other species, and with potential intra- and inter-state conflicts in demographic, environmental, political, and resource issues (Wang et al., 2012).

Studies about ecological and environmental security issues have been reported in recent years in China (Grumbine, 2014; Wang et al., 2015a). Case studies include, e.g., the land ecological security assessments in Tibetan plateau (Zhao et al., 2006); in Xiaolangdi Reservoir Region (Su et al., 2010); in Dali (Ye et al., 2011); in Guangzhou, Shanghai, and Dongguan (Gong et al., 2000; Su et al., 2011; Lin et al., 2016); and in Dianchi Lake (Wang et al., 2015b). Zhang and Xu (2017) evaluated ecological and environmental security in terrestrial ecosystems of China. The study revealed spatial diversities and causes of ecological and environmental problems in different regions and identified hotspot areas at the danger level.

Inventory and monitoring of ecosystem conditions become critically important under the facts of intensified human-induced land-use and land-cover change, fragmentation of habitats and effects of climate changes. This is particularly true for the regions that contain areas with significance of biological diversity and in the trans-boundary sensitive international regions. With characteristics described above, natural resources and ecological security of the Changbai Mountain region in Northeast Asia deserve attention (Wang, et al., 2012; Lu et al., 2014).

In a large spatial context, the Changbai Mountains is a mountain range that extends along the border between northeast China, North Korea, and Russia. The range consists of paralleled sections of mountains. It extends toward southwest connecting the Qian-Shan Mountains in the Liaodong Peninsula. Toward the north and the northwest is the Northeast China Plain, a major agricultural production base of China. Toward northeast, it connects the Sikhote-Alin Mountains in the Russian Far East. Geologically the region is on the border of the Pacific competent zone. The Himalayan tectonic movement since the Miocene Epoch had resulted in volcanic eruptions and hence the formation of a typical volcanic geomorphology composed of volcanic cones, inclined plateaus, and lava table lands (Zhau et al., 1991; Wang et al., 2003). The range is mostly between 800 and 1,500 meters in elevation and has the typical forest and agricultural ecosystems in Northeast Asia. The highest section and the most representative of this ecoregion is the Changbai Mountain Nature Reserve (CMNR) (Figure 40.1).

The CMNR is situated between 41°41′ and 42°26′N latitudes and 127°42′ and 128°17′E longitudes in China side on the border. The CMNR occupies 196,465 hectares of lands with the largest protected temperate forests that support a significant species gene base and biodiversity in Northeast Asia. The summit cups a crater lake, i.e., Lake of Heaven, with a spectacular view of surrounding landscape (Figure 40.2). The CMNR was established in 1961 and admitted into the UNESCO's Man and Biosphere Program in 1979. The ecological conditions of the Changbai Mountain have long been the subject of focus of the scientific community of the world (Stone, 2006).

The most unique feature of the CMNR is its vertical distribution of forest and ecosystem zones due to elevation change. The climate and terrain conditions support four distinctive vertically distributed vegetation zones within the CMNR. The needle- and broad-leaf mixed forest zone is distributed between 600 and 1,600 meters. The dominant tree species include Korean pine (*Pinus koraiensis*) and temperate hardwoods such as aspen (*Poplus davidiana*), birch (*Betula platyphylla*), basswood (*Tilia amurensis*), oak (*Quercuc mongolica*), maple (*Acer mono*), and elm (*Ulmus propinqua*), among others. The larch (*Larix olgensis*) and Changbai pine (*Pinus sylvestris var. sylvestriformis*) form the "bright" coniferous forests on the upper elevation of this vegetation zone. The evergreen "dark" coniferous forest zone is distributed between 1,600 and 1,800 meters with dominant species of spruce (*Pieca jezoensis, Pieca koreana*) and fir (*Abies nephrolepis*). Between 1,800 and 2,100 meters is distributed the zone of subalpine dwarf birch (*Betula ermanii*) forest with other species such as *Larix olgensis*. The Alpine tundra zone is distributed between 2,100 and 2,400 meters with representative species such as short Rhododendron shrubs (*Rhododendron chrysanthum Pall*) and *Vaccinium uliginosum* L. (Figure 40.3). This unique and distinctive vertical zonal pattern of vegetation and the ecosystems showcase a condensed configuration and composition of temperate and boreal forests found across Northeast Asia.

Natural Resources and Ecological Security Concerns

This region has a long and rich history of cultural diversity co-existing between multiple ethnic groups in China. The Changbai Mountain was regarded as the birthplace of ancestors of the Qing dynasty (1622–1912) and therefore restricted as the forbidden area of sacred place during the time. Because of such an imperial ban, the forests in the Changbai Mountain were protected from being logged, and it

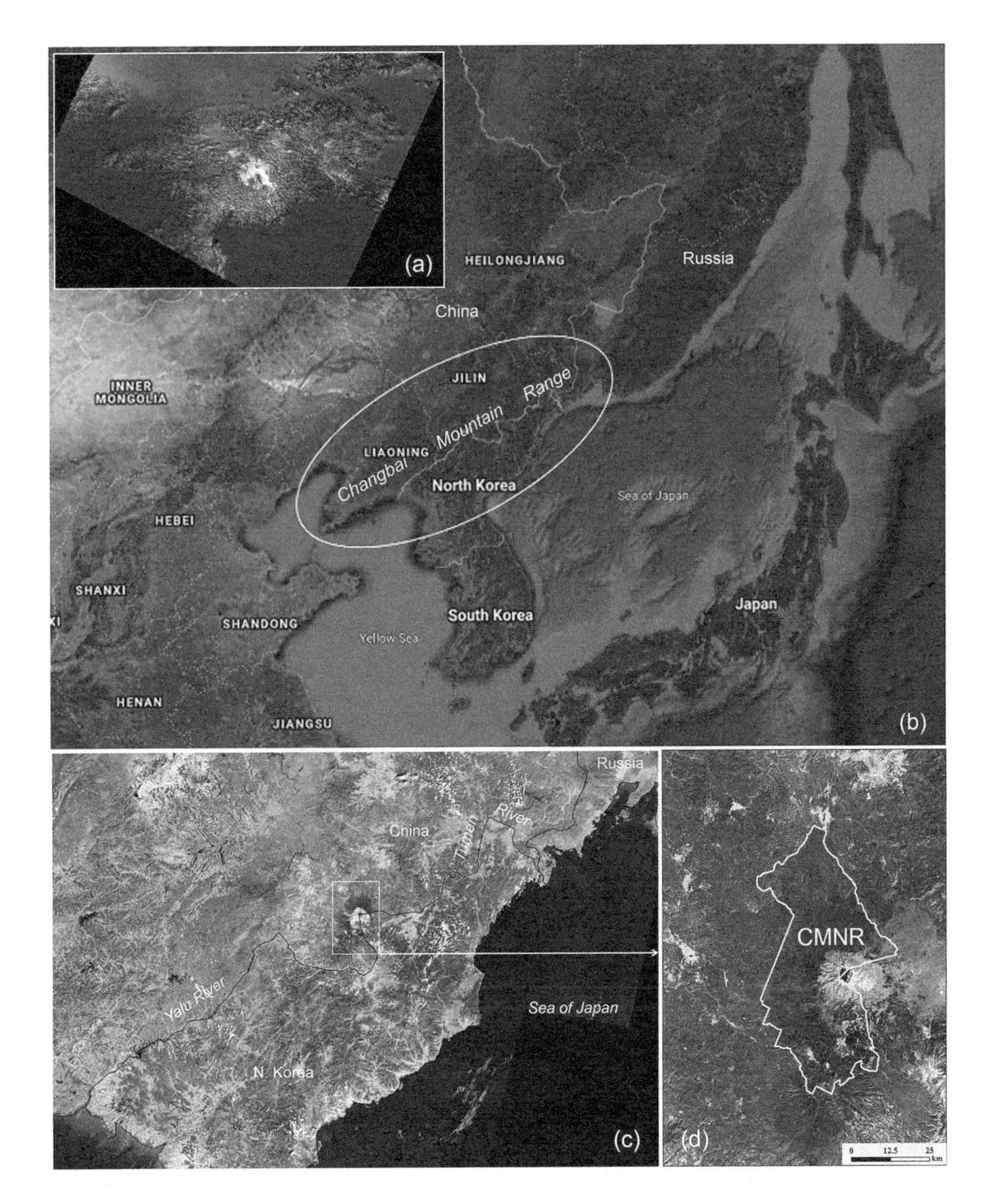

FIGURE 40.1 The Changbai Mountain range consists of paralleled sections of mountains (**a**). It extends toward the west connecting the Qian-Shan Mountains in the Liaodong Peninsula in China and toward the south connecting the Korean Peninsula (**b**). In the north is the Northeast China Plain, and in the east connects the Sikhote-Alin Mountains in the Russian Far East (b) and (**c**). The Sentinel-2 Multispectral Instrument (MSI) imagery acquired September 25, 2017 (Bands 4, 3, 2 in RGB display) illustrates the landscape characterization within and surrounding the CMNR (**d**).

helped preserve the biodiversity of representative native flora and fauna that inhabited the mountain. Therefore the mountainous ecosystems were mostly intact at the time when the CMNR was established, and the regional development in adjacent areas was relatively late in comparison to the other regions in Northeast China.

FIGURE 40.2 (See color insert.) The summit of the Changbai Mountain cups a crater lake (**a**) and has a spectacular volcanic and subalpine tundra landscape (**b**). (Photos: Yeqiao Wang.)

The Changbai Mountain is the origin of the Yalu, Tuman and Songhua rivers. The Yalu and Tumen rivers serve the border between China and North Korea. The water resources and water environments are among focal and sensitive international issues of the region.

Social and natural systems have a profound complexity of interconnections. Anthropogenic impacts are always among key issues for regional ecological security concerns. The Changbai Mountain region has experienced dramatic transition and demographic change in the past several decades. Those changes can be reflected in immigration and human population increase; in socioeconomic development and improved living status; in change of land-use practices; and in changes of government policies regarding socioeconomic development, environmental regulations, and resource management. Effects from human activities are particularly important for such an international conjunction region with cross-border movements of people and commodities and with discrepancies in wealth and with cultural and linguistic diversity.

Trans-boundary pollution through air and water systems is one of the many ecological security issues that this region is facing due to rapid economic development, urbanization, and consumption of energy (Schreurs and Pirages, 1998). Trans-boundary pollution causes degradation of air, water, and soil; depletion of resources; and loss of biodiversity. Issues related to water resources, water quality, land use, and management within the watersheds of border rivers need to be monitored and studied.

FIGURE 40.3 **(See color insert.)** The climate and terrain conditions support distinctive vertically distributed vegetation zones within the CMNR. Field photos illustrate the Alpine tundra zone between 2,100 and 2,400 meters with representative species such as short Rhododendron shrubs (**a**), the zone of subalpine dwarf birch (*Betula ermanii*) forest between 1,800 and 2,100 meters (**b**), the evergreen coniferous forest zone between 1,600 and 1,800 meters with dominant species of spruce (*Pieca jezoensis, Pieca koreana*) and fir (*Abies nephrolepis*) (**c**), and the needle- and broad-leaf mixed forest zone between 600 and 1,600 meters (**d**). (Photos: Yeqiao Wang.)

Mountain ranges often create ecotones and provide habitats and boundaries because of terrain induced variations of climatic conditions. Ecotonal areas are most sensitive to changes in environmental conditions and are usually where changes first occur (IPCC, 2001; Neilson, 1993; Rusek, 1993; Risser, 1993, 1995). Integration of Earth observation data with field-based multidisciplinary studies can help overcome a number of challenges and offer a starting point on the path toward ecological security at the ecotone regions.

Geologically the region is situated at the edge of the East Asian continent along the Pacific competent zone. The so-called Tianchi volcanic zone had one of the largest eruptions in the last 2,000 years and has been active since Cenozoic time (Liu, 1997; Zielinski et al., 1994). A study concluded that the volcanic eruption is intracontinental rather than active continental margin or intraplate (Wang et al., 2003). The volcano in the Changbai Mountain has been quiet since minor eruptions in 1597, 1688, and 1702. It is still a high-risk volcano under intensive monitoring (Stone, 2006) by the Chinese scientists and international volcano observation networks.

As volcanic eruption is another major regional concern in ecological security, satellite interferometric synthetic aperture radar (InSAR) can be employed for monitoring ground surface deformation associated with volcanic and other geological hazards (Lu et al., 2012). It should be pointed out that some volcanic eruptions are preceded by periods of little or no deformation (e.g., Lu et al., 2010). This suggests that short-term forecasting based on InSAR-mapped deformation alone might be difficult in some cases. InSAR can effectively track surface deformation in the long run but might not provide a robust indication of an ensuing eruption in the months or weeks beforehand. Experience shows that short-term precursors such as localized deformation, seismicity, and changes in volcanic gas emission commonly are observed when a shallow magma reservoir nears rupture or as magma intrudes surrounding rock (e.g., Sparks, 2003). Such precursors can be detected by *in-situ* sensors, and they typically provide days to months of warning. Therefore, effective monitoring and hazard mitigation for volcanoes requires the integration and analysis of multiple remotes sensing, geophysical, and geochemical datasets in near-real time.

Hazardous events at Changbai Mountain volcano include landslides from steep volcano flanks and floods, which need not be triggered by eruptions as well as eruption-induced events such as fallout of volcanic ash and lava flows. A proximal hazard zone of about 20 km in diameter centering at the existing Changbai Mountain volcanic cone could be affected within minutes of the onset of an eruption or large landslide. Distal hazard zones could be inundated by lahars (generated either by melting of snow and ice during eruptions or by large landslides) flowing down the slopes of Changbai Mountain volcano and valleys. Fallout of tephra from eruption clouds can affect areas hundreds of kilometers downwind. The proximal hazard zone extends to a ballistic range of about 10-km radius. Due to a large crater lake existing on top of the Changbai Mountain, any magma interaction will produce very explosive outputs. Areas prone to inundation by future lahars can be calculated from numerical modeling of an existing DEM (Iverson et al., 1998). Distal hazard zones based on 3 scenarios of lahar volumes of 10 million cubic meters, 100 million cubic meters, and 1 billion cubic meters were mapped for the Changbai Mountain volcano with indication of nearby population centers (Figure 40.4; Lu et al., 2012).

Landscape patterns are one of the key measures of ecological security. Landscape patterns that are composed of strategic portions and positions with significance in safeguarding and controlling certain ecological processes are considered ecologically secure (Yu, 1996). Landscape patterns and the dynamics can dramatically influence a host of biological, biophysical, and biochemical resources. Landscape patterns and change reflect status of ecosystem connectivity and habitat quality and are often a central component to assessing changes in other resources such as water quality, flora and fauna, terrestrial vertebrates, and vegetation communities. Understanding the magnitude and pattern of land-use and land-cover change helps establish a landscape context for the protected areas. This is particularly important for the internationally sensitive border regions. Remote sensing has been proven the most effective approach in the monitoring of landscape characterization and dynamics through land use

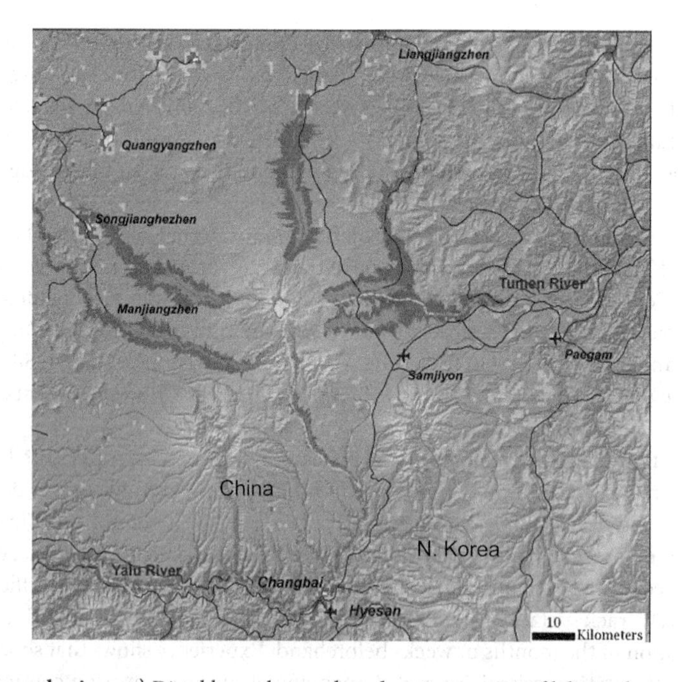

FIGURE 40.4 **(See color insert.)** Distal hazard zones based on 3 scenarios of lahar volumes of 10 million cubic meters, 100 million cubic meters, and 1 billion cubic meters were mapped for the Changbai Mountain volcano with indication of nearby population centers of larger than 100 per km^2 and 1,000 per km^2. (Adapted from Lu et al., 2012.)

and land-use mapping and the change analysis (Zhang, et al., 2014; Zong et al., 2018; Guo et al., 2018; Wu et al., 2018).

Conclusion Remarks

Ecological security is essential for the sustainability of any human and nature systems. Ecological security depends on the balance between human demands and actions in consumption and alteration of resource base and the sustainability, vulnerability, and resilience of environmental systems that provide ecosystem services. This is particularly true for the regions with significance in biological diversity and trans-boundary sensitive international regions with different political regimes and social administration systems.

One of the central challenges in understanding the equilibriums is to establish a data framework that can support multidisciplinary studies and quantitative modeling and facilitate inventory and monitoring of the critical components contributing to the ecological security issues, such as demographic change, invasive species and biodiversity, deforestation and land degradation, urbanization and intensified land-use change, environmental pollution, food production, freshwater supply, health and disease, rural development, and climate change.

Integration of remote sensing, geospatial modeling, and field-based approaches helps reveal the structure of natural resources, the past and current status, and their change in association with human-induced alterations for informed management of the resource base to address ecological security concerns.

References

Allenby, B.R., 2000. Environmental security: Concept and implementation. *International Political Science Review*, 21: 5–21.

Bing, L., Z. Endi, Z. Zhenhua, L. Yu, 2008. Preliminary monitoring of Amur tiger population in Jilin Hunchun national nature reserve. *Acta Theriologica Sinica*, 28: 333–341.

Floyd, R., R. Matthew, 2013. *Environmental Security: Approaches and Issues*. Routledge, Abingdon.

Grumbine, R.E., 2014. Assessing environmental security in China. *Frontiers in Ecology and the Environment*, 12: 403–411.

Gong, J.Z., Y.S. Liu, B.C. Xia, G.W. Zhao, 2000. Urban ecological security assessment and forecasting, based on a cellular automata model: A case study of Guangzhou, China. *Ecological Modelling*, 220: 3612–3620.

Guo, X., H. Zhang, Y. Wang, H. He, Z. Wu, Y. Jin, Z. Zhang, J. Zhao, 2018. Comparison of the spatio-temporal dynamics of vegetation between the Changbai Mountains of eastern Eurasia and the Appalachian Mountains of eastern North America. *Journal of Mountain Science*, 15(1). doi:10.1007/s11629-017-4672-9.

Intergovernmental Panel on Climate Change (IPCC), 2001. In: Working Group II: Impacts, Adaptation and Vulnerability; Chapter 19-Vulnerability to Climate Change and Reasons for Concern: A Synthesis. Smith, J.B., Schellnhuber, H.J., Mirza, M.M.Q. (Eds.). Section 3-Impacts on Unique and Threatened Systems; Subsection 3-Biological Systems; Unit 3-Ecotones (19.3.3.3), pp. 932–933.

Iverson, R., Schilling, S., Vallance, J., 1998. Objective delineation of lahar-inundation hazard zones. *GSA Bulletin*, 110(8): 972–984.

Lin, Q., J. Mao, J. Wu, W. Li, J. Yang, 2016. Ecological security pattern analysis based on InVEST and Least-Cost Path Model: A case study of Dongguan Water Village. *Sustainability*, 8: 172.

Liu, Q.J., 1997. Structure and dynamics of the subalpine coniferous forest on Changbai Mountain, China. *Plant Ecology*, 132: 97–105.

Lu, Z., D. Dzurisin, J. Biggs, C. Wicks Jr., S. McNutt, 2010, Ground surface deformation patterns, magma supply, and magma storage at Okmok volcano, Alaska, inferred from InSAR analysis: 1. inter-eruptive deformation, 1997–2008. *Journal of Geophysical Research*, 115: B00B03, doi:10.1029/2009JB006969.

Lu, Z., D. Dzurisin, H.S. Jung, 2012. Monitoring natural hazards in protected lands using interferometric synthetic aperture radar (Ch. 19). In *Remote Sensing of Protected Lands*, Y. Wang (Ed.). CRC Press, Boca Raton, FL, pp. 439–471.

Lu, X., J. Zhang, X. Li, 2014. Geographical information system-based assessment of ecological security in Changbai Mountain region. *Journal of Mountain Science*, 11: 86–97.

National Research Council, 1999. *Our Common Journey: A Transition toward Sustainability*. National Academy Press, Washington, D.C.

Neilson, R.P., 1993. Transient ecotone response to climate change: Some conceptual and modelling approaches. *Ecological Applications*, 3: 385–395.

Pirages, D., 2005. From limits to growth to ecological security. In *From Resource Security to Ecological Security: Exploring New Limits to Growth*, D. Pirages and K. Cousins (Eds.). The MIT Press, Cambridge, MA, pp. 1–20.

Risser, P.G., 1993. Ecotones at local to regional scales from around the world. *Ecological Applications*, 3: 367–368.

Risser, P.G., 1995. The status of the science examining ecotones. *BioScience*, 45: 318–325.

Rusek, J., 1993. Air-pollution-mediated changes in alpine ecosystems and ecotones. *Ecological Applications*, 3: 409–416.

Schreurs M.A., D. Pirages, 1998. Ecological security and the future if inter-state relations in Northeast Asia. In *Ecological Security in Northeast Asia*, M.A. Schreurs and D. Pirages (Eds.). Yonset University Press, Seoul, pp. 11–22.

Sparks, S., 2003. Forecasting volcanic eruptions. *Earth and Planetary Science Letters*. Frontiers in Earth Science Series, 210: 1–15.

Stone, R., 2006. A threatened nature reserve breaks down Asian borders. *Science*, 313: 1379–1380.

Su, S., X. Chen, S. DeGloria, J. Wu, 2010. Integrative fuzzy set pair model for land ecological security assessment: A case study of Xiaolangdi Reservoir Region, China. *Stochastic Environmental Research and Risk Assessment*, 24: 639–647.

Su, S., D. Li, X. Yu, Z. Zhang, Q. Zhang, R. Ziao, J. Zhi, J. Wu, 2011. Assessing land ecological security in Shanghai (China) based on catastrophe theory. *Stochastic Environmental Research and Risk Assessment*, 25: 737–746.

Wang, Y., C. Li, H. Wei, X. Shan, 2003. Late Pliocene–recent tectonic setting for the Tianchi volcanic zone, Changbai Mountains, northeast China. *Journal of Asian Earth Sciences*, 21: 1159–1170.

Wang, Y., Z. Wu, H. Zhang, J. Zhang, J. Xu, X. Yuan, Z. Lu, Y. Zhou, J. Feng, 2012. Remote sensing assessment of natural resources and ecological security of the Changbai Mountain Region in Northeast Asia (Ch. 11). In *Remote Sensing of Protected Lands*, Y. Wang (Ed.), CRC Press, Boca Raton, FL, pp. 203–232.

Wang, S., W. Meng, X. Jin, B. Zheng, L. Zhang, H. Xi, 2015a. Ecological security problems of the major key lakes in China. *Environmental Earth Sciences*, 74: 3825–3837.

Wang, Z., J. Zhou, H. Loaiciga, H. Guo, S. Hong, 2015b. A DPSIR model for ecological security assessment through indicator screening: A case study at Dianchi Lake in China. *PLoS ONE*, 10: e0131732.

World Commission on Environment and Development, 1987. *Our Common Future*. Oxford University Press, New York.

Wu, M., H. He, S. Zong, X. Tan, H. Du, D. Zhao, K. Liu, Y. Liang, 2018. Topographic controls on vegetation changes in Alpine Tundra of the Changbai Mountains. *Forests*, 9: 756.

Ye, H., Y. Ma, L. Dong, 2011. Land ecological security assessment for Bai autonomous prefecture of Dali based using PSR model–with data in 2009 as case. *Energy Procedia*, 5: 2172–2177.

Yu, K., 1996. Security patterns and surface model in landscape ecological planning. *Landscape Ecology*, 36: 1–17.

Zhang, H., E. Xu, 2017. An evaluation of the ecological and environmental security on China's terrestrial ecosystems. *Scientific Reports*, 7: 811. doi:10.1038/s41598-017-00899-x.

Zhang, J., F. Liu, G. Cui, 2014. The efficacy of landscape-level conservation in Changbai Mountain Biosphere Reserve, China. *PLoS ONE*, 9(4): e95081.

Zhao, Y., X. Zou, H. Cheng, H. Jia, Y. Wu, G. Wang, C. Zhang, S. Gao, 2006. Assessing the ecological security of the Tibetan plateau: Methodology and a case study for Lhaze County. *Journal of Environmental Management*, 80: 120–131.

Zhau, S.D., G. Zhau, 1991. Management of Changbai Mountain Biosphere Reserve: The present conditions, problems, and perspectives. *Mountain Research and Development*, 11: 168–169.

Zielinski, G., P. Mayewski, L. Meeker, S. Whitlow, M.S. Twickler, M. Morrison, D.A. Meese, A.J. Gow, R.B. Alley, 1994. Record of volcanism since 7000B.C. from the GISP Greenland ice core and implications for the volcano-climate system. *Science*, 264: 948–952.

Zong, S., H. He, K. Liu, H. Du, Z. Wu, Y. Zhao, H. Jin, 2018. Typhoon diverged forest succession from natural trajectory in the treeline ecotone of the Changbai Mountains, Northeast China. *Forest Ecology and Management*, 407: 75–83.

41

Ecological Security: Land Use Pattern and Simulation Modeling

Xun Shi
Dartmouth College

Qingsheng Yang
Guangdong University of Business Studies

Introduction

The term "ecological security" appeared in the late 1980s.[1,2] There is no consensus yet on its definition,[3] but that has not prevented the research under this topic from growing into a fairly active area in the past two decades.[4–7] While the environmental, ecological, and resource problems addressed by the research are not new, its perspective and emphasis may be somewhat different from those of conventional environmental and ecological studies. The diverse work in this area can be summarized as in the following text.

Expanding the Growth-Limit Perspective

Represented by the book *The Limits to Growth,*[8] a classical conceptualization attributes contemporary environmental problems to the depletion of *resources,* including energy, food, and capability of dispersing waste, as a result of fast-growing population and industrialization. The growth-limit notion triggered fierce debate between the resource-pessimists and techno-optimists.[5] A few decades later, the crises caused purely by resource scarcity have not materialized, but humankind is not facing a less challenging environmental condition. While empirical observations indicate that for the foreseeable future resource scarcity might be a relatively minor source of human suffering, infectious disease, conflict among peoples, starvation, and various kinds of environmental disasters have become major causes of premature human deaths and disabilities, and meanwhile, the extinction of many flora and fauna species has become a serious problem. All these have raised the level of concern over future ecological security.[5] This general concern is presented by Pirages as four interrelated dynamic equilibriums:

1. Between human populations living at higher consumption levels and the ability of nature to provide resources and services
2. Between human populations and pathogenic microorganisms

3. Between human populations and those of other plant and animal species
4. Among human populations

Pirages considers that whenever any of these equilibriums is disrupted either by changes in human behavior or in nature, insecurity will increase.[5]

Raising to the Security Level

A highlight of the conception of ecological security is that it raises concern about the environmental/ecological issues to a qualitatively new level, the level of national security.[2] Conventionally, national security mainly refers to the protection of societies and states from predatory neighbors. However, ecosystemic challenges such as plagues, pestilence, pollution, blizzards, floods, and droughts, often aided and abetted by intemperate human behavior, over time, have been responsible for killing and injuring much larger numbers of human beings.[4] Thus, Westing defines a broader human security as being composed of two intertwined components: political security and environmental/ecological security, where political security includes military, economic, and social/ humanitarian subcomponents, and ecological security has protection-and utilization-oriented subcomponents.[9] The protection requirement refers to safeguarding the quality of the human environment, and the utilization requirement means providing a sustaining basis for any exploitation (harvesting or use) of a renewable natural resource.

Geographic Scale

Ecological security is temporally dynamic and has spatial structured[10] and is thus closely associated with the geographic scale. Typically, ecological security has been studied at the global, national, and regional/city levels. Globalization is making the boundaries formed by physical, political, and cultural barriers much more porous and less important, which seems to have brought about ecological risk and insecurity.[4] For example, the dramatically increased number of people and quantities of goods moving rapidly from place to place is facilitating the unintended and often destructive spread of plants and pests into new environments and ecosystems, which is disturbing the long-established equilibrium between people and pathogens, and is causing the rapid spread of new and resurgent diseases.[4] At the national and regional/city scales, over the background of resource constraints and climate change, national security, infrastructure "protection," and economic competitiveness are being overlaid with concerns around energy security, constraints on water resources, the growth of diseases, increased flood risks, and multiple aspects of demographic shifts. National and regional/city governments are seeking the ability to ensure *secure* access to the resources needed for the country's, region's, and city's replications.[6]

Urban Ecological Security and Land Use Pattern

Urban ecological security (UES) refers to the contemporary conditions within which cities must actively seek to reproduce their economic, social, and material fabric.[6] The term reflects the emergence of the situation that ecological security is increasingly becoming an issue particularly at an urban scale, which is the consequence of four interrelated sets of pressures:[6] First, cities are responsible for disproportionate levels of consumption of resources and greenhouse gas emissions. For example, despite accounting for around half of the world's population, cities are responsible for around 75% of energy consumption and 80% of greenhouse gas emissions. Second, consequently, cities are likely to be chief amongst the *victims* of resource constraints and climate change. For example, many coastal and river-side cities may suffer flooding and the health consequences of the urban heat island effect. Third, cities give the potential contexts for the response to issues of resource constraint and climate change. For example, cities may be ideal places to demonstrate and experiment with decentralized energy and water technologies, and new urban mobility and transportation systems based on, for example, hydrogen and biodiesel. Finally, in an era of intensified economic globalization and competition between places, urban coalitions should be able to anticipate, shape, and respond strategically to national priorities.

Urban land use and land cover changes (LUCC) can be both causes and results of ecological and environmental problems. Thus, strategic urban/regional plans, which are targeting and are to be reflected by LUCC, ought to support sustainable development and promote ecological security by addressing existing as well as newly introduced environmental problems.[10] Methods to assess urban planning and the resulting growth scenarios, as well as to forecast the resulting urban ecological security, are a critical requirement for the sustainable development of an area.[10]

Simulation and Modeling of Land Use Pattern for Assessing Urban Ecological Security

Cellular automata (CA), integrated with geographic information systems (GIS), are an effective technical approach to simulating complex urban land use.[11–17] For UES studies, the CA-GIS simulation can serve two purposes: 1) predicting an UES situation in the future based on the simulation of land use dynamics, and 2) experimenting with planning interferences that may affect UES by creating different "what-if" scenarios.

CA consist of a collection of discrete cells that represent spatial units, each in one of a finite number of states. The state of each cell evolves through a number of discrete time steps controlled by a set of transition rules. These rules define how a cell will evolve based on its own state and the states of its neighboring cells.[17] CA models can generate complex global patterns with simple local rules. While the rules are only applied at the neighborhood level, they may be able to represent the impact of factors at different (i.e., local, regional, and global) spatial scales.[13,14,18,19] The transition rules are key inputs in a CA model.

Gong et al. developed a CA model for predicting the ecological security of Guangzhou, a big city located in one of the most economically active regions of China.[10] The model couples CA with ecological security assessment to identify insecure cells within the region and simulate their future development. The *digital space* of the model is composed of square grids, each with a size of 100 × 100 m. The status of each cell was designated as 1 for *not secure* or 0 for *secure*. The cell values were determined according to a previously conducted ecological security assessment for Guangzhou for 1990–2005. During the modeling, the grids evolved as the simulation map generated from one iteration sequentially being used as the input for the next iteration. The transition rules that determine if the status of a cell would change were represented by an integrated vector. These rules calculated the transition probability based on the cell's current status, the neighborhood effect, and the urban planning factor. Validation using historical data showed that the accuracy of the model could reach 72.1%. Using the situation of 2005 as the starting state, Gong et al. ran simulations with the model to predict ecological security for 2020. The predicted ecological security situation of Guangzhou shows that while certain measures in the governmental development planning may gentle the decreasing trend of UES of the city, such a trend will continue into 2020. A landscape pattern analysis suggested a more scattered and homogenous distribution of land use in Guangzhou's urban landscape, as well as considerable variation among different districts of the city. The modeled UES highlights the need to make ecological protection an integral part of urban planning.

The CA-GIS-based modeling can also be used to conduct experiments about effects of different planning measures on UES. This idea can be implemented by incorporating the planning measures into the transition rules of CA. For example, if a goal of the planning is to restrict excessive expansion of big cities, and meanwhile maintain the total magnitude of regional urban development, one can translate this policy into a rule that suppresses the transition probability of a non-urban cell if the number of urban cells within its neighborhood is above a threshold. The maps in Figure 41.1 illustrate such a simulation.

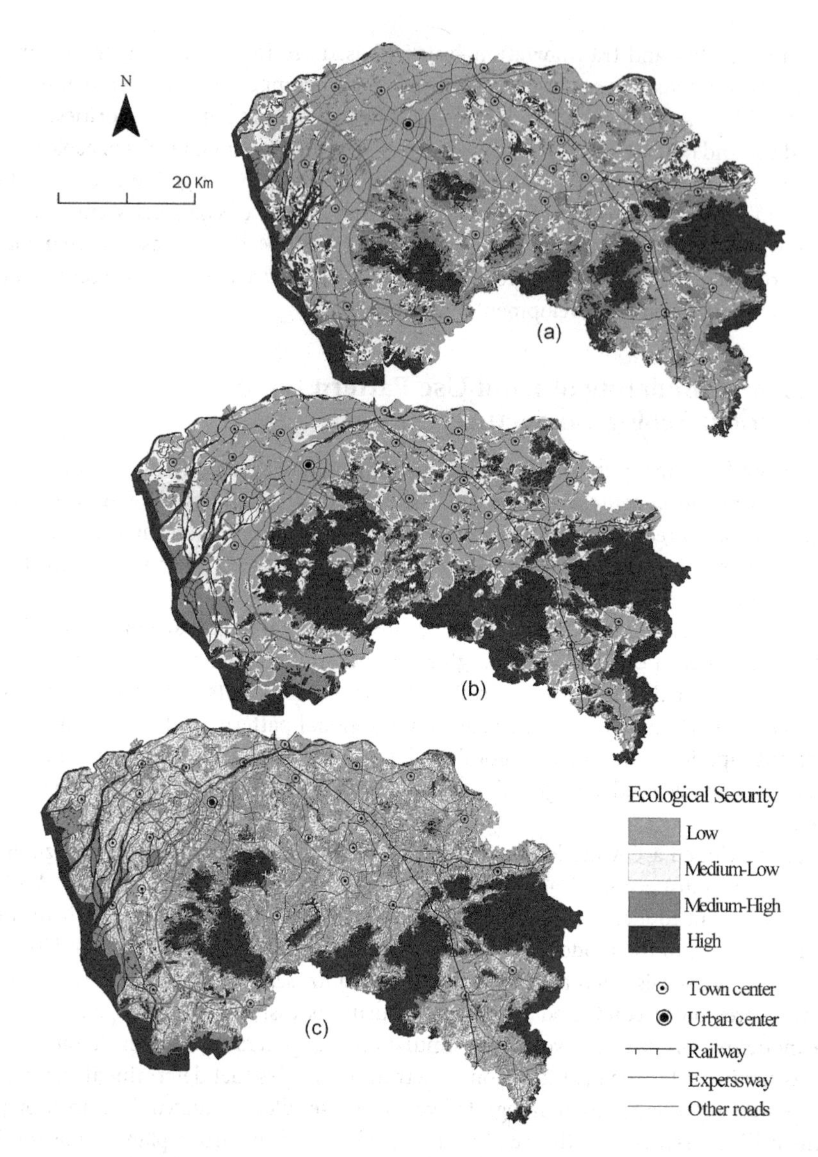

FIGURE 41.1 GIS-CA-based simulation of urban ecological security under certain planning measures for Guangzhou City, China: (**a**) The result of ecological security assessment based on the actual land use situation in 2005; (**b**) The simulated ecological security for 2005 solely based on the historical land use trend (i.e., without any restrictions from planning); (**c**) The simulated ecological security for 2005 based on the historical trend and controlled by a local development threshold.

Summary

The concept of ecology security raises concern about environmental/ecological problems to the level of national security, and recognizes this security issue at different geographic scales, from global to regional and city. At the city level, land use dynamics can be both cause and consequence of a severe ecological security situation. CA integrated with GIS are an effective approach to: 1) predicting future land use dynamics of an area based on the historical trend; and 2) modeling the effect or consequence of certain planning measures or policies on urban land use. Therefore, CA-GIS models should be useful in decision support for planning and policy-making regarding urban ecological security.

References

1. Myers, N. The environmental dimension to security issues. Environmentalist **1986**, *6* (4), 251–257.
2. Timoshenko, A.S. Ecological security: The international aspect. Pace Environ. Law Rev. **1989**, *7* (1), 151–160.
3. Zou, C.; Shen, W. Advances in ecological security. Rural Eco-Environ. **2003**, *19* (1), 56–59.
4. Pirages, D.C.; DeGeest, T.M. From international to global relations. In *Ecological Security: An Evolutionary Perspective on Globalization*; Rowman & Littlefield Publishers: Lanham, USA, 2004, 1–28.
5. Pirages, D. From limits to growth to ecological security. In *From Resource Scarcity to Ecological Security: Exploring New Limits to Growth*; Pirages, D., Cousins, K., Eds.; The MIT press: Cambridge, USA, 2005, 1–19.
6. Hodson, M.; Marvin, S. Urban ecological security: A new urban paradigm? Int. J. Urban Reg. Res. **2009**, *33* (1), 193–215.
7. Hodson, M.; Marvin, S. Urbanism in the anthropocene: Ecological urbanism or premium ecological enclaves? City **2010**, *14* (3), 298–313.
8. Meadows, D.H.; Meadows, D.L.; Randers, J.; Behrens, W. *The Limits to Growth*; University books: New York, USA, 1972.
9. Westing, A.H. The environmental component of comprehensive security. Bull. Peace Proposals **1989**, *20* (2), 129–134.
10. Gong, J.; Liu, Y.; Xia, B.; Zhao, G. Urban ecological security assessment and forecasting, based on a cellular automata model: A case study of Guangzhou, China. Ecol. Model. **2009**, *220* (24), 3612–3620.
11. Batty, M.; Xie, Y. From cells to cities. Environ. Plann. B Plann. Des. **1994**, *21* (7), 531–548.
12. Clarke, K.C.; Hoppen, S.; Gaydos, L. A self-modifying cellular automaton model of historical urbanization in the San Francisco Bay area. Environ. Plann. B Plann. Des. **1997**, *24* (2), 247–261.
13. Wu, F.; Webster, C.J. Simulation of land development through the integration of cellular automata and multicriteria evaluation. Environ. Plann. B Plann. Des. **1998**, *25* (1), 103–126.
14. Li, X.; Yeh, A.G. Neural-network-based cellular automata for simulating multiple land use changes using GIS. Int. J. Geogr. Inf. Sci. **2002**, *16* (4), 323–343.
15. Li, X.; Yeh, A.G. Data mining of cellular automata's transition rules. Int. J. Geogr. Inf. Sci. **2004**, *18* (8), 723–744.
16. Li, X.; Yeh, A.G. Modelling sustainable urban development by the integration of constrained cellular automata and GIS. Int. J. Geogr. Inf. Sci. **2000**, *14* (2), 131–152.
17. Yang, Q.; Li, X; Shi, X. Cellular automata for simulating land use changes based on support vector machines. Com-put. Geosci. **2008**, *34* (6), 592–602.
18. Li, X.; Shi, X.; He, J.; Liu, X. Coupling simulation and optimization to solve planning problems in a fast developing area. Ann. Assoc. Am. Geogr. **2011**, *101* (5), 1032–1048.
19. Yu, P. *Multiple Criteria Decision Making: Concepts, Techniques and Extensions*; Plenum Press: New York, USA, 1985.

References

42

Effects of Volcanic Eruptions on Forest in Changbai Mountain Nature Reserve

Jiawei Xu
Northeast Normal University

Yeqiao Wang
University of Rhode Island

Introduction

The Changbai Mountain Nature Reserve (CMNR) is located on border area between China and North Korea. The CMNR is famous for having a unique set of vertically distributed ecosystems that surround the volcanic summit. Geologically the region is situated at the edge of the East Asian Continent along the Pacific competent zone. The so-called Tianchi volcanic zone has been active since Cenozoic time (Liu, 1987; Liu et al., 1992a,b) and had one of the largest eruptions in the last 2,000 years (Machida et al., 1990; Liu et al., 1998). A study concluded that the volcanic eruption is intracontinental rather than active continental margin or intraplate (Wang et al., 2003; Lei and Zhao, 2005). The volcano in the Changbai Mountain has been quiet since minor eruptions in 1597, 1688, and 1702. It is still a high-risk volcano under intensive monitoring (Stone, 2006) by the Chinese scientists and international volcano observation networks.

CMNR hosts a unique vertically distributed zonal vegetation pattern due to the climate and terrain conditions (Wang et al., 2011). The vegetation distribution in the CMNR reflects the trajectories of after-math regeneration and succession post significant volcanic eruptions in the most recent several 100 years. Volcanic eruptions not only destroyed the vegetation, but also altered the soil contents, landform, and hydrology, which in further affected vegetation recovery and succession. Therefore the timeline of the most recent significant volcanic eruption that devastated the forest ecosystems is the key to understand the history of the vegetation recovery and succession.

Studies reported that recent intensive eruption at the Changbai Mountain was about 800 years ago (Liu et al., 1999) and between 600 and 1000 A.D. (Gill et al., 1992). Results from examining the SO_4^{2-} concentration of Greenland ice (GISP2) suggested that the eruption time was about 1026 A.D. (Zielinski et al., 1994). Results from examining carbonized wood at about 800 m elevation in the northern slope and below 1,000 m in the western slope suggested that the eruption time was 1050 ± 70 A.D., 1120 ± 70 A.D., and 1410 ± 80 A.D., respectively (Zhao, 1981). The results from ^{14}C suggested that the eruption time was 1215 A.D. (Liu et al., 1997). Other studies suggested that time should be about 1,200 years ago (Liu, 1999), 1199–1200 A.D. (Cai et al., 2000), and 910–1435 A.D. (Guo et al., 2001), respectively. The common understanding is that the eruption that occurred at about 1215 A.D. had the devastated destruction of the vegetation of the CMNR. Other minor eruptions such as those occurred in 1625, 1373, 1401, 1573, 1668, 1702, and 1903 affected the eastern slope only but had no significant disturbance on the vegetation in the current CMNR areas (Liu and Wang, 1992; Zhao, 1980; Kayama, 1943). With the above timeline, we analyzed the traces and effects of past volcanic eruptions on forests for an understanding of spatial-temporal patterns of the vegetation in CMNR. We conducted the study by a combination of intensive field investigation augmented by remote sensing observations.

Methods

Remote Sensing Data and the Analysis

In order to reveal spatial distribution and temporal change of forest types, we employed time series Landsat data, high-spatial-resolution IKONOS data, and topography information from Shuttle Radar Topography Mission (SRTM) data. Landsat and Sentinel-2 satellite imageries revealed vegetation patterns and land-cover changes surrounding the protected CMNR in the past 40 years (Figure 42.1).

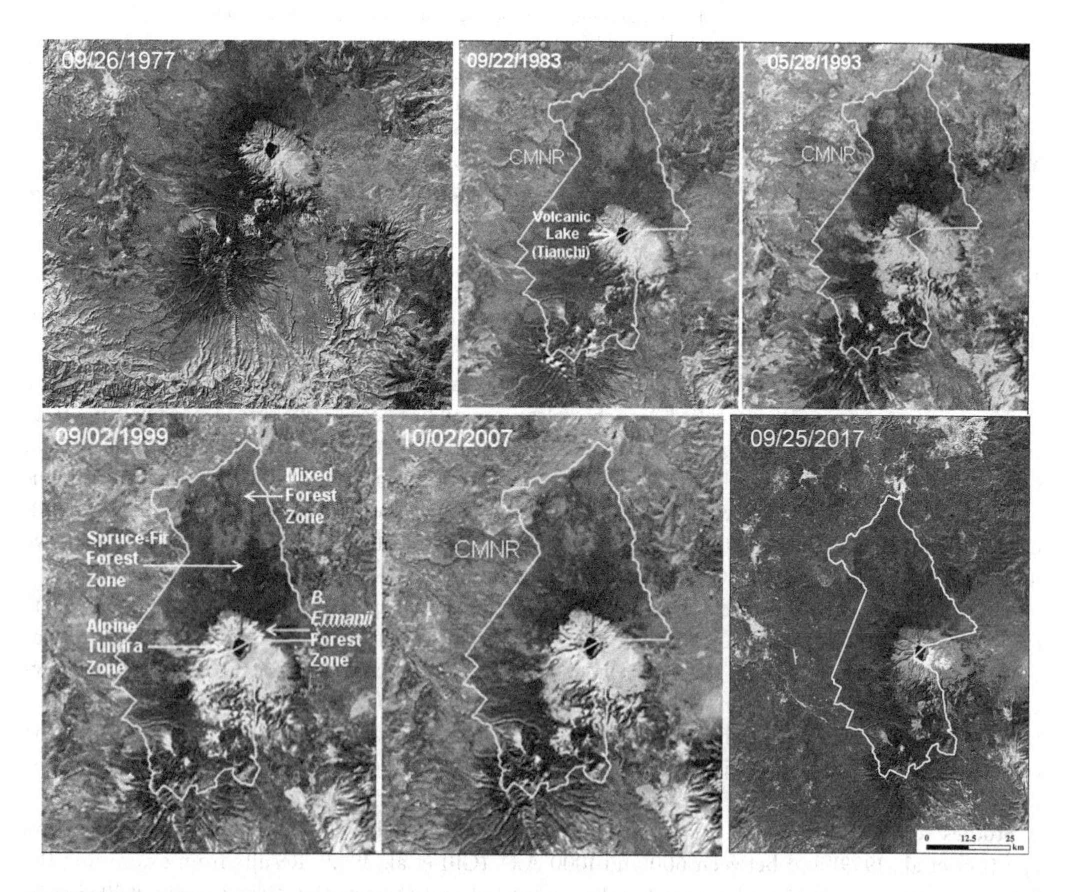

FIGURE 42.1 **(See color insert.)** Landsat and Sentinel-2 satellite imagery data revealed vegetation patterns and change from 1977 to 2017 within and adjacent to the protected CMNR.

Fine-spatial-resolution IKONOS data illustrated detailed spatial distributions of forest types. SRTM and the derivatives of elevation and slope/aspect provide guidance and control for extraction of information about vertical distribution of forest types and their changes (Figure 42.2).

One of the primary objectives of this study was to differentiate forest categories based on species composition and structural characteristics in different vertical zones using remote sensing data. Integration of Landsat data and SRTM-derived variables has been reported for habitat classification and change detection (Sesnie et al., 2008). We conducted unsupervised and stratified classification on Landsat and IKONOS data, respectively. We first obtained generalized vegetation and land cover types from classification of Landsat data. The categories include broad-leaf forest; needle- and broad-leaf mixed forest; dark coniferous forest; bright coniferous forest; subalpine birch forest; tundra, barren, and volcanic rocks; water and volcanic lake; agricultural land; and urban and developed land. We then employed the stratified classification (Wang et al., 2007) using SRTM-derived elevation and slope/aspect as the control to extract forest types within each of the vertical zones. We paid special attention to the dominant and representative species in different vertical zones and in different slopes and aspects. GPS-guided *in situ* observation and documentation provided guidance for the classification process. We referenced GIS boundary of CMNR to separate land cover types between protected land and adjacent areas to reveal land cover changes. For taking advantage of high-spatial-resolution IKONOS data, we conducted resolution merge using 1-m panchromatic and 4-m multispectral data. The resolution merged dataset helped identify effectively representative forest species, in particular the transition areas between vertical vegetation zones (Figure 42.3). We added additional subcategories to identify detailed vegetation distributions for classification of the IKONOS data.

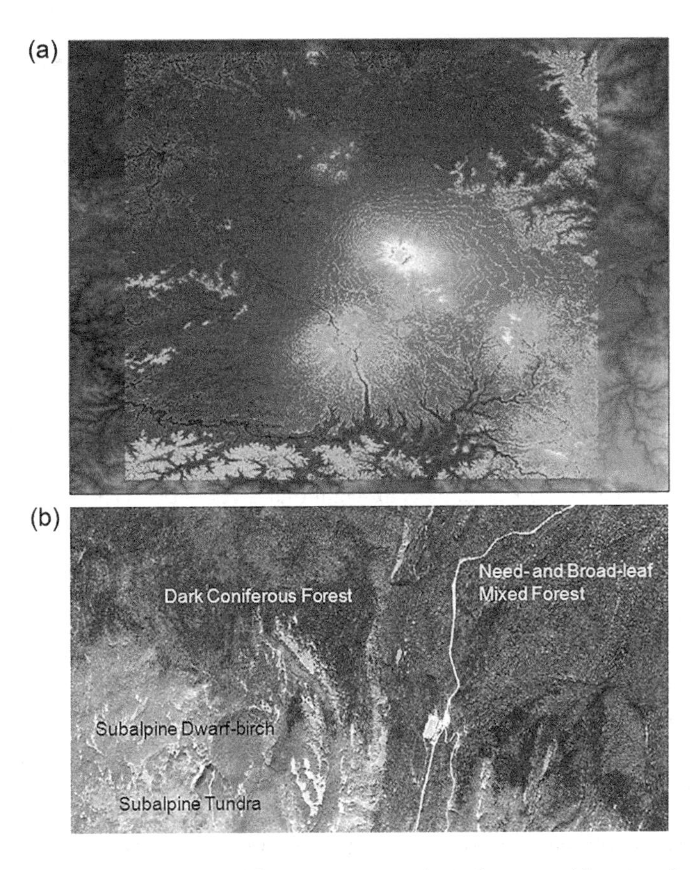

FIGURE 42.2 **(See color insert.)** SRTM and derivative topographic information (elevation, slope, aspects) provide zonal guidance and reference (**a**) to differentiate vegetation distribution from high-resolution IKONOS data (**b**).

FIGURE 42.3 **(See color insert.)** Image classification of IKONOS data to identify representative species in transition areas between vertical vegetation zones. Field survey confirmed the colonization of pioneer species of subalpine *B. ermanii* and *L. olgensis* into the tundra zone in high altitudes of the northern slope.

Survey of Forest Stands and Carbonized Wood Buried in Volcanic Ash

We conducted *in situ* survey by gradient pattern in different vertical zones. We set three 20 × 20 m plots in each of the needle- and broad-leaf mixed forests from altitude 700–1,100 m; the coniferous forest zone from altitude 1,100 to 1,700 m; the subalpine *Betula ermanii* forest zone from altitude 1,700 to 2,000 m; and the Alpine tundra zone from altitude 2,000 to 2,600 m. We documented the basic status such as altitude, slope degree, species and density for tree and shrub layers, regeneration layer, and herb layer. In addition, we set 8 × 3 20 × 20 m plots for eight major vegetation types, including, Korean pine and broad-leaf mixed forest, Korean pine and spruce-fir forest, typical spruce-fir forest, *B. ermanii* and spruce-fir forest, oak forest, larch forest, subalpine larch, and low mountain larch for documenting basic status listed above. We also set 4 × 3 20 × 20 m plots in the dark coniferous forest between 1,300 and 1,400 m in different slope aspects for the documentation of species and density for tree, shrub, regeneration, and herb layers. We then integrated *in situ* and remote sensing observations for revealing the trace of volcanic eruptions and the impacts on vegetation.

Carbonized wood buried in volcanic ash is the evidence of pre-eruption forest structure, species composition, spatial distributions and temporal variations, and the impacted areas (Figure 42.4). In order to understand the pre-eruption vegetation structure and distribution, we sampled and documented carbonized wood in species and abundance at different altitudes and slope/aspects of the volcanic summit of the CMNR. We surveyed the areas at the altitudes of 974, 1,426, 1,731 and 2,236 m in the northern slope; 951, 1,493, 1,760 and 2,093 m in the western slope; and 936, 1,472, 1,753 and 2,311 m in the southern slope.

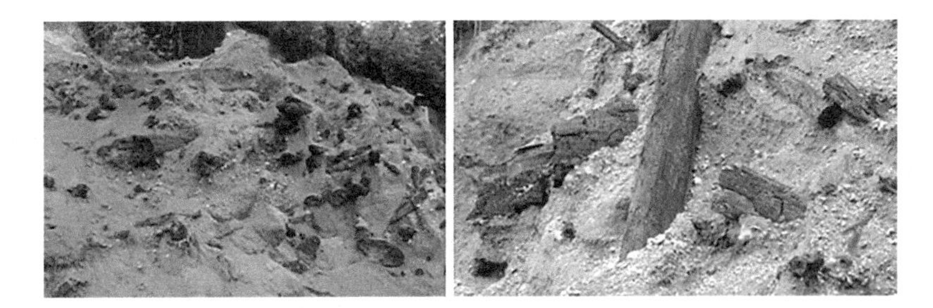

FIGURE 42.4 (See color insert.) Samples of carbonized woods in species and abundance at different altitudes and slope/aspects of the area surrounding the volcanic summit were referenced to reconstruct pre-eruption vegetation structure and distribution.

Results

Remote Sensing Vegetation Mapping

The classification of Landsat images and the derivatives of vegetation maps reveal spatial distribution and temporal variations. In particular, when augmented by SRTM derived elevation model data, the maps illustrate the spatial distribution and pattern of different forest zones in different altitudes and slope/aspects. The observed land-cover changes during 1972–1983, 1983–1993, 1993–1999, and 1999–2007 from Landsat data mostly resulted from deforestation and agricultural and urban land development. These included significant changes immediately adjacent to the border of the CMNR, which indicate the result of enforced land protection and management of the reserve. There has been no change of tree line and tundra line in the CMNR since the early Landsat data acquisition in 1972. A recent study reached the same conclusion with observed increase of absolute normalized difference vegetation index (NDVI) values (Zhang et al., 2008). The study acknowledged that the increase of absolute NDVI might be attributed to multiple factors, such as climate warming, CO^2 fertilization effect, and biases from remote sensing data.

Spatial distribution of vertical vegetation zones, forest types and landscape configuration and composition provide information about the status of vegetation and the impacts from past volcanic eruptions. High-spatial-resolution vegetation mapping from IKONOS data reveal forest structure for specific species. In particular, high-spatial-resolution information confirmed, in an aerial context, the colonization of subalpine *B. ermanii* into the tundra zone in the high altitudes of the northern slope (Figure 42.3).

Spatial Variation of Vegetation Distribution

The volcanic eruptions had significant impact on the alpine vegetation than that at lower altitudes. Almost all alpine and subalpine vegetation was destroyed by the last devastated eruption that caused descent of forest timberline between vertical zones. Some species disappeared due to the impacts. For example, although samples of dwarf Siberian pine (*Pinus pumila*) were observed in abundance in carbonized wood, there is still no sign of this species in the high-altitude area of the CMNR today. Remote sensing images and *in situ* documentation suggested that although vegetation distribution appears in relatively completed vertical zones, its uniqueness is quite obvious. Landsat images show that the vertical zonal pattern of vegetation is more complete in the northern slope than the southern and eastern slopes. The *Larix olgensis* and *Pinus koraiensis* forests are widely distributed and often seen in pure stands. The spruce (*Pieca jezoensis*) and fir (*Abies nephrolepis*) forests exist in pure stands in the northern slope. The *B. ermanii* forest zone is extensively established in the northern slope but not so on the eastern slope. The difference in spatial distribution results from volcanic impacts.

The study on the carbonized wood samples suggested that the current distribution of forest types is in the same pattern as that of pre-eruptions. The *Pinus koraiensis* is the constructive species with mixed broad-leaf tree species. The tree line between needle- and broad-leaf mixed forests and dark coniferous forests is consistent, which suggested that there was no significant climate change in the past several 100 years. The Landsat data from 1972 to 2007 also illustrates that the tree line has been consistent in the recent years under other impacting factors such as human disturbances and scenarios of regional warming under global climate change. However the tree line and tundra line were higher than the current ones before volcanic eruptions, especially in the eastern slope. The altitude of current Alpine timberline is about 2,100 and 2,000 m in the northern and eastern slopes, respectively. The current upper tundra line is between 2,500–2,600 m and 2,400 m in the northern and eastern slopes, respectively. The highest altitude of residual wood on eastern slope is 2,160 m. There are scattered forest patches above 2,100 m. In the northern slope, *B. ermanii* forests can reach up to 2,200 m altitude. The distributions suggest that the tree line could be higher than 2,200 m in the northern slope and 2,100 and 2,300 m in the western and southern slopes, respectively. The increasing trend of tree-line altitude is testified by the observations that the forest ages are younger in the higher altitude. As simulated by hydrothermal index (Xu, 1985), the potential timberlines are higher than the forest distributions today. The lowered timberline and tundra line reflect the impacts of volcanic eruptions on vegetation. It also indicates that the vegetation recovery is still in progress.

Extended Subalpine Dwarf-Birch and Larch Forests

B. ermanii and *L. olgensis* are considered as pioneer species. The *B. ermanii*- and *L. olgensis*-dominated forest has the characteristics of being in different stages of post-eruption succession. Remote sensing images reveal the spatial distribution and locations of the *B. ermanii* forests. The evidence observed in field surveys show that *B. ermanii* trees are older in age than coniferous species in the mixed forest zone. The remains of *B. ermanii*, however, are rarely found in carbonized wood samples in the volcanic ash. Studies indicate that the pre-eruption spatial distribution of *B. ermanii* was smaller in area than the current extension. *Betula ermanii* forest should not be regarded as the product of volcanic eruptions as it exists in the other area of the northeast China and Russia. *Betula ermanii* is the species that can adapt to harsh habitats such as poor soil, extreme drought, and cold and windy conditions than common coniferous forest. *Betula ermanii* forest in the northern slope is largely mixed with *L. olgensis* which has obvious pioneer characteristics. Field survey results indicate that the mixed *L. olgensis* forest is older in age, and they are gradually exiting the communities. The *L. olgensis* forests in lower altitudes have been invaded by coniferous forests as the succession progresses. The *Pieca jezoensis* and *B. ermanii* forests will take over after time. Therefore the extensive distribution of the *B. ermanii* forest zone reflects the impacts of volcanic eruptions and post-eruption succession.

The larch species, an important constructing community plant in boreal and high-altitude forest, is distributed narrowly around the world. The larch species has a wide extension and is mainly distributed from 500 to 1,950 m altitude (Yu et al., 2005). The areas of larch forest are extensively distributed from needle- and broad-leaf forest zone to *B. ermanii* forest zone, forming the interzonal vegetation. The wide distribution of *L. olgensis* suggested the uniqueness of vegetation changes affected by volcanic eruptions. *L. olgensis* is a species that prefers light; tends to be resistant to cold and humidity; and coexists with *Pinus koraiensis*, *Pieca jezoensis*, and *Abies nephrolepis*, among other species. The number of *Pieca jezoensis* and *Abies nephrolepis* trees increase toward high altitudes, and the number of *Pinus koraiensis* trees increase toward lower altitudes. *L. olgensis* does not show a clear zonal distribution in the CMNR. The subalpine larch is the secondary forest vegetation formed by destroyed subalpine coniferous forest, which is mainly distributed at the altitude between 1,100 and 1,700 m of the eastern and northeastern slopes. The most recent volcanic eruptions at 1597, 1668, and 1702 had no destructive impacts on the vegetation of the entire area. Those eruptions destroyed part of forest and produced large bared land on the eastern slope, which is observable from Landsat images. Hence, *L. olgensis* can result from recent volcanic eruptions.

Conclusions

The integrated information from remote sensing and *in situ* survey concluded that the current vegetation distribution and structure in the CMNR reflect the effects of historical volcanic eruptions and succession process, in particular at the high-altitude areas. Historical volcanic eruptions not only imposed direct influences on vegetation, but also affected indirectly and the altered natural conditions such as strata, terrain, and hydrology. Landsat remote sensing data and the derivatives of vegetation maps reveal the pattern of vertical zonal vegetation distribution. The zonal pattern is more completed in the northern slope but not quite obvious in the eastern slope. However plenty of carbonized wood samples within volcanic ash are observable in places in the eastern slope, suggesting volcanic impacts.

Spectral features from Landsat data are capable of discriminating forest types and secondary successional stages (Foody and Hill, 1996). However vegetation classification can be limited when ecologically important differences in forest structure and composition are spectrally similar (Castro et al., 2001; Helmer et al., 2002; Pedroni, 2003). Stratified classification employed SRTM-derived topography information to map vegetation and land-cover types associated with vertical zonal distributions effectively.

With the integrated information, we conclude the following: (i) The current vertical zonal vegetation distribution is consistent with that prior to the volcanic eruptions. The current vegetation base bands are following the same distribution followed about 800 years ago. (ii) The rareness of *B. ermanii* in carbonized wood suggests that pre-eruption distribution of *B. ermanii* was smaller in area than the current extension. (iii) Volcanic eruptions destroyed high-altitude vegetation, lowered the tree line, and affected species composition. The species such as dwarf Siberian pine (*Pinus pumila*) was wiped out, and the vegetation recovery is still in progress. (iv) The differences in silvan vegetation composition between CMNR and the similar settings of cold temperate and mid-temperate zones suggest the impacts of past volcanic eruptions on the current vegetation structures. The wide distribution of *L. olgensis* testifies the uniqueness of vegetation change affected by volcanic eruptions. All the above demonstrate that the distinctive quality of vegetation in the CMNR is determined by the influence of volcanic eruptions on the ecosystems. Traces of volcanic eruptions exist in forest ecosystems in the CMNR after 800 years of the last devastating volcanic eruption.

References

Cai, Z., D. Jin, and L. Ni, 2000. The historical record discovery of 1199–1200 AD large eruption of Changbaishan Tianchi volcano and its significance. *Acta Petrologica Sinica*, 16(2): 191–193.

Castro, K.L., G.A. Sánchez-Azofeifa, and B. Rivard, 2001. Monitoring secondary tropical forest using space-born data: Implications for Central America, *International Journal of Remote Sensing*, 24: 1853–1894.

Foody, G.M. and R.A. Hill, 1996. Classification of tropical forest classes from Landsat TM data, *International Journal of Remote Sensing*, 17: 2353–2367.

Gill, J., C. Dunlap, and M. McCurry, 1992. Large- volume, mid- latitude, Cl- rich volcanic eruption during 600~1000 AD, Baitoushan, China, Climate, Volcanism and Global Change. *American Geophysical Union Chapman Conference*, March 23–27, Hito, HI, pp. 1–10.

Guo, Z., J. Liu, and S. Sui, et al., 2001. Total estimated of effusive volcanic gas of Baitoushan volcano and its significance of 1199–1200 AD, *Science in China Series D: Earth Sciences*, 31(8): 668–675.

Helmer, E.H., O. Ramos, T.D.M. López, M. Quiñones, and W. Diaz, 2002. Mapping the forest type and land cover of Puerto Rico, a component of the Caribbean biodiversity hotspot, *Caribbean Journal of Science*, 38: 165–183.

Kayama, N., 1943. Forest trees in foothills of Baitoushan before volcanic eruptions, *Botanical Magazine (Tokyo)*, 57(679): 258–273.

Lei. J. and D. Zhao, 2005. P-wave tomography and origin of the Changbai intraplate volcano in Northeast Asia, *Tectonophysics*, 397: 281– 295.

Liu, J.Q., 1987. Study on geochronology of the Cenozoic volcanic rocks in northeastern China, *Acta Petrologica*, 4: 21–31 (in Chinese with English abstract).

Liu, J. 1999. *Volcanes of China*, Science Press, Beijing, pp. 1–219.

Liu, Q. and Z. Wang, 1992. Recent volcanic eruption and vegetation history of alpine and subalpine of Changbai Mountain, *Forest Ecosystem Research*, 6: 57–62.

Liu, R.X., W.J. Chen, J.Z. Sun, and D.M. Li, 1992a. K-Ar chronology and tectonic settings of Cenozoic volcanic rocks in China, In: Liu, R.X. (Ed.), *Chronology and Geochemistry of Cenozoic Volcanic Rocks in China*, Seismological Press, Beijing, pp. 1–43 (in Chinese).

Liu, R.X., J.T. Li, H.Q. Wei, D.M. Xu, and Q.F. Yang, 1992b. Changbaishan Tianchi Volcano - A recent dormant volcano for potential dangerous eruptions, *Acta Geophysica*, 35: 661–664 (in Chinese with English abstract).

Liu, R., S. Qiu, L. Cai, et al., 1997. Age research of the most recent large eruption of Changbaishan Tianchi volcano and its significance, *Science in China Series D: Earth Sciences*, 27(5): 437–441.

Liu, R.X., H.Q. Wei, and J.T. Li, 1998. *Recent Eruptions of the Changbaishan Tianchi Volcano*, Scientific Press, Beijing, pp. 1–159 (in Chinese).

Liu, R., Q. Fan, and H. Wei, 1999. The research of active volcanoes in China, *Geological Review*, 45 (Sup.): 3–15.

Machida, H., H. Morwaki, and D.C. Zhao, 1990. The recent major eruption of Changbai Volcano and its environmental effects, Geographical reports of Tokyo Metropolitan University, 25: 1–20.

Pedroni, L., 2003. Improved classification of Landsat Thematic Mapper data using modified prior probabilities in large and complex landscapes, *International Journal of Remote Sensing*, 24: 91–113.

Sesnie, S.E., P.E. Gessler, B. Finegan, and S. Thess; 2008. Integrating Landsat TM and SRTM-DEM derived variables with decision trees for habitat classification and change detection in complex neotropical environments, *Remote Sensing of Environment*, 112: 2145–2159.

Stone, R. 2006. A threatened nature reserve breaks down Asian borders, *Science*, 313: 1379–1380.

Wang, Y., C. Li, H. Wei, and X. Shan, 2003. Late Pliocene–recent tectonic setting for the Tianchi volcanic zone, Changbai Mountains, northeast China, *Journal of Asian Earth Sciences*, 21: 1159–1170.

Wang, Y., M. Traber, B. Milestead, and S. Stevens, 2007. Terrestrial and submerged aquatic vegetation mapping in Fire Island National seashore using high spatial resolution remote sensing data, *Marine Geodesy*, 30(1): 77–95.

Wang, Y., Z. Wu, X. Yuan, H. Zhang, J. Zhang, J. Xu, Z. Lu, and J. Feng, 2011. Remote sensing assessment of natural resources and ecological security of the Changbai Mountain Region in Northeast Asia. In: Wang, Y. (Ed.), *Remote Sensing of Protected Lands*, CRC Press, Boca Rato, FL, pp. 203–232.

Xu, W., 1985. Heat index of Kira and its application of vegetation in China, *Chinese Journal of Ecology*, 3: 35–39.

Yu, D., S. Wang, and L. Tang, 2005. Relationship between tree-ring chronology of Larix olgensis in Changbai Mountains and the climae change, *Chinese Journal of Applied Ecology*, 16(1): 14–20.

Zhang, Y., M. Xu, J. Adams, and X. Wang, 2008. Can landsat imagery detect tree line dynamics? *International Journal of Remote Sensing*, 30(5): 1327–1340.

Zhao, D., 1980. Vertical distribution of the vegetation belts in Changbaishan Mountain, Forest ecosystem research station of Changbai Mountain of Chinese academy of sciences. *Forest Ecosystem Research*, China Forestry Publishing House, Beijing, 1: 65–70.

Zhao, D., 1981. Preliminary investigation on relation between volcano eruption of Changbai Mountain and the evolution of its vegetation[A], Forest ecosystem research station of Changbai Mountain of Chinese academy of sciences, *Forest Ecosystem Research*, 2: 81–87.

Zielinski, G., P. Mayewski, L. Meeker, et al., 1994. Record of volcanism since 7000B.C. from the GISP Greenland ice core and implications for the volcano-climate system, *Science*, 264: 948–952.

43

LANDIS PRO Forest Landscape Model

Hong S. He
University of Missouri

Wen J. Wang
Chinese Academy of Sciences

Introduction

Forest landscape models (FLMs) are spatially interactive, computer simulation models that simulate forest stand dynamics and forest landscape processes (FLPs) at landscape scales (He et al. 2017). The fundamental difference between FLMs and other terrestrial biosphere models (TBMs) (e.g., stand dynamic, biogeographical, biogeochemical, and land surface models, and dynamic global vegetation model (DGVM) (Fisher et al. 2014; Shifley et al. 2017) is that FLMs simulate FLPs as spatially contiguous and interacting processes at a fine resolution (He 2008). FLPs are spatially contiguous processes that include seed dispersal and natural and anthropogenic disturbances such as wild and prescribed fire, windstorm, insect outbreak, disease spread, timber harvesting, silvicultural activities, and urban expansion (He et al. 2011). FLPs may spread over a group of spatially contiguous cells (pixels) as a function of vegetation, terrain, fuel, and wind on these cells, but because of the need to simulate FLPs and site-scale forest dynamics, FLMs typically operate at relatively fine spatial resolutions (e.g., 30–300 m cell size) and moderate temporal resolution (e.g., yearly).

FLMs vary greatly on how they simulate cell-level vegetation dynamics, because direct plugging in a stand dynamic model (gap model) or an ecosystem process model into a FLM at each site would render the FLM computationally impractical due to the Big O Notation (He 2008). Here, we discuss an FLM, LANDIS PRO, and its design of simulating cell-level dynamics. we also present a few examples showing LANDIS PRO simulation results.

LANDIS PRO Overview

LANDIS PRO is a raster-based FLM that evolved over 20 years from the LANDIS model family (Mladenoff and He 1999; He and Mladenoff 1999a; Mladenoff 2004; He et al. 2005; Scheller et al. 2007; Yang et al. 2011; Wang et al. 2013) (Figure 43.1). Within each raster cell, the model records number of trees for each species age class. Tree size (diameter at breast height or DBH) for each species age class is derived from user-defined empirical age–DBH relationships for each land type (or ecoregion

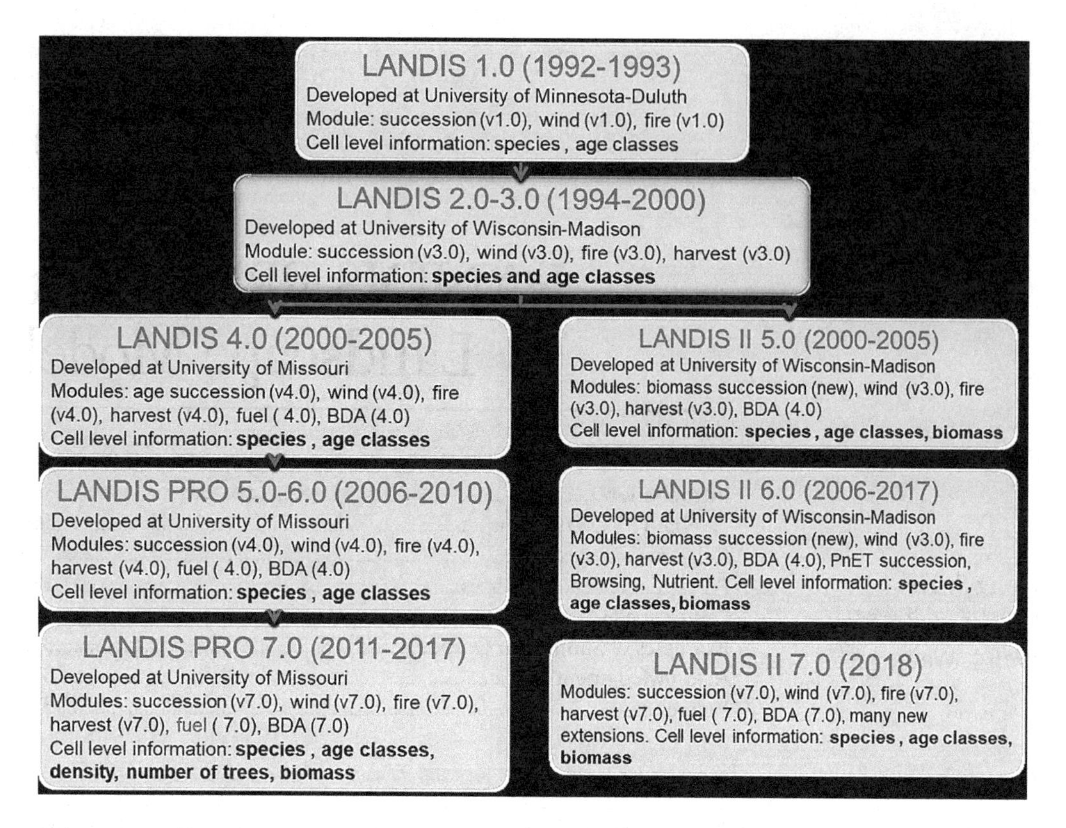

FIGURE 43.1 LANDIS forest landscape model development (1992–2019).

at regional scales), which accounts for the effects of varying resource availabilities and species establishment (Figure 43.2). With both tree density and size information, LANDIS PRO derives key stand parameters including density, basal area, stocking, and biomass by species and by size class for each raster cell. Such information at each cell forms wall-to wall representations of forest attributes across the simulation area. LANDIS PRO simulation results allow for straightforward comparisons with forest inventory data, allowing seamlessly use of such data to construct the initial forest conditions, calibrate model parameters, and validate simulation results (Figure 43.2).

Cell-Level Succession

LANDIS PRO simulates forest composition and structure changes incorporating species-, stand-, and landscape-scale processes (Figure 43.2). Species-scale processes include tree growth, establishment, resprouting, and mortality and are simulated via species' vital attributes using theories of forest stand dynamics (Oliver and Larson 1996). For example, tree growth is simulated as DBH increment (species growth curves) in addition to age increment. DBH increment is simulated using empirical lognormal relationships of DBH to age or calibrated locally using forest inventory data by land type.

Stand-scale processes contain density- and size-related resource competition that regulates seedling establishment and self-thinning. Each cell has the physical space to support growth of limited number of trees or maximum growing capacity that is determined by each land type. The competition intensity is quantified by growing space occupied (GSO), which is estimated by summing the total growing space required to support all trees within each raster cell. The growing space is derived from the stand density index (Reineke 1933) using the tree density and size information. Stand development patterns

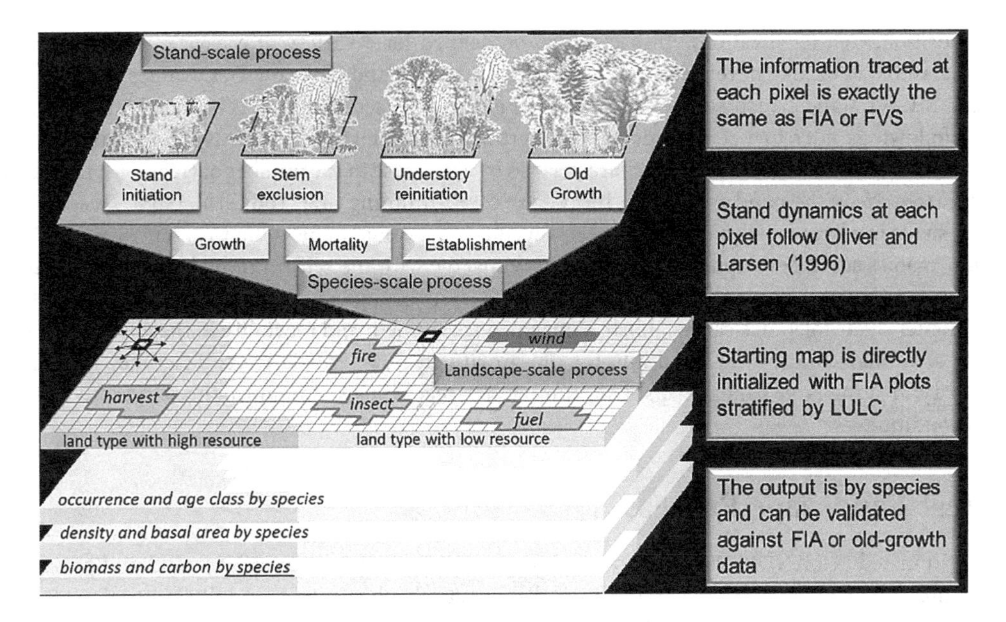

FIGURE 43.2 **(See color insert.)** The conceptual design of LANDIS PRO that simulates forest change incorporating species-, stand-, and landscape-scale processes. (Modified from Wang et al. 2014a.)

are governed by GSO and simulated to follow stand initiation, stem exclusion, understory re-initiation, and old-growth stages (Oliver and Larson 1996) (Figure 43.3). Different GSO thresholds are defined to regulate seedling establishment in the stand initiation stage, where seedlings can only become established before stands reach fully occupied: (i) open grown, (ii) partially occupied, (iii) crown closure, (iv) fully occupied (maximum growing space occupied, MGSO) (Figure 43.3). These GSO thresholds can be adjusted for each land type (or ecoregion) and can be modeled outside LANDIS PRO platform to incorporate the effects of varying MGSO resulting from environmental change (climate change,

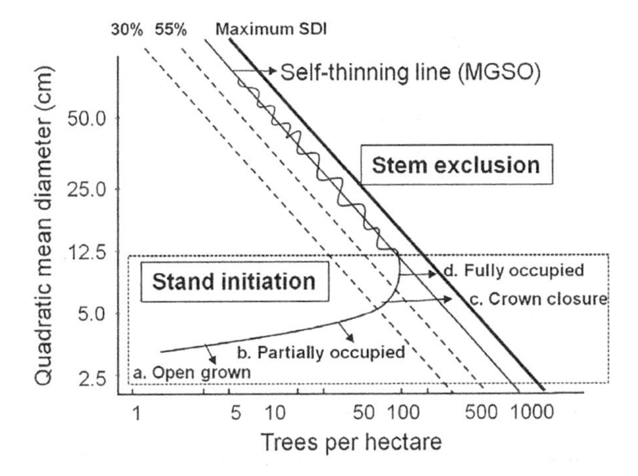

FIGURE 43.3 The Reineke density diagrams (Reineke 1933) and Yoda's 3/2 self-thinning rule (Yoda et. al 1963) are combined to display the stand development patterns governed by GSO. The maximum size density lines represents 100% growing space, and other size-density combinations are expressed as a percentage of the maximum resource availability at different stand development stages: (1) stand initiation stage, (2) stem exclusion stage, (3) understory reinitiation stage, and (4) old-growth stage (Oliver and Larson 1996). (From Wang et al. 2014a.)

N deposition, CO_2 fertilization). They can also be defined for a variety of ecosystems. For example, a woodland system may never reach the crown closure stage and have low GSO values. Once stands exceed MGSO, stands reach stem exclusion stage; meanwhile, self-thinning is initiated and continues to the following understory re-initiation and old growth stages (Figure 43.3). LANDIS PRO implements self-thinning, where tree mortality is characterized by a decrease in the number of trees with increasing average tree size in the stand and follows theYoda's 3/2 self-thinning rule (Yoda et al. 1963). Trees that are small, shade intolerant, or approaching their longevity can be outcompeted first via self-thinning. Such design enables stand development trajectories to converge with the self-thinning line and move along the line from lower right to upper left (Figure 43.3). During the understory re-initiation stage, seedlings with higher shade tolerance can establish. Continued tree growth and mortality in the absence of exogenous disturbance moves the stand into the old-growth stage, where old trees die as they reach their longevity, creating large canopy gaps that promote tree regeneration and move the stand into uneven-aged condition.

Forest Landscape Processes

FLPs simulated in LANDIS PRO include seed dispersal (exotic species invasion), wild and prescribed fire, wind (hurricane and tornado), insects and diseases, timber harvesting, and silvicultural treatments as independent modules (Figure 43.2). Seed dispersal is simulated using a dispersal kernel determined by species-specific maximum dispersal distances, where the probability of seed dispersal to every cell is calculated using a negative exponential decay function (He and Mladenoff 1999b). The total number of potential germination seeds (NPGS) for each species reaching a given raster cell is accumulated from all available mature trees within the dispersal kernel. The NPGS can be derived from Burns and Honkala (1990) and calibrated to ensure the predicted species density and basal area are consistent with forest inventory data. Thus, LANDIS PRO accounts for dispersal limitation and seed availability that can constrain species distributions under rapidly changing environment.

LANDIS PRO disturbance modules, which simulate FLPs as plugin extensions, interact with cell-level species and age class information using frequency, intensity, and size information either generated from historic data or modeled under future climatic conditions (He et al. 2005; Wang et al. 2013).

Landscape Heterogeneity

A heterogeneous landscape is stratified into land types in LANDIS PRO (also called ecoregions for broad-scale studies), which capture the coarse-level (coarse grain) spatial heterogeneity in resource availabilities (MGSO) and species assemblages (species establishment probability). Seed dispersal and disturbances result in the intermediate-level (fine grain, within or between land types) heterogeneity in forest composition and structure. Finally, site-scale processes (e.g. competition, establishment) result in the fine-level (within raster cell) heterogeneity in forest composition and structure. LANDIS PRO is a 64-bit computer model that runs on Windows 10, whose software package and a detailed user's guide can be downloaded from http://landis.missouri.edu.htm.

Examples of Simulation Output

The study area was over 10,000,000 ha and comprised of 410,033,200 cells with a resolution of 90 m (0.81 ha). Boundaries corresponded to FIA Survey Unit 5 in Arkansas, USA. This area encompasses the Ozark and Boston Mountains in Arkansas. The area is characterized as mixed hardwood-pine forest dominated by oak (Quercus spp.), hickory (Carya spp.), and shortleaf pine (Pinus echinata Mill.). The dominant hardwood species include white oak (Q. alba L.).

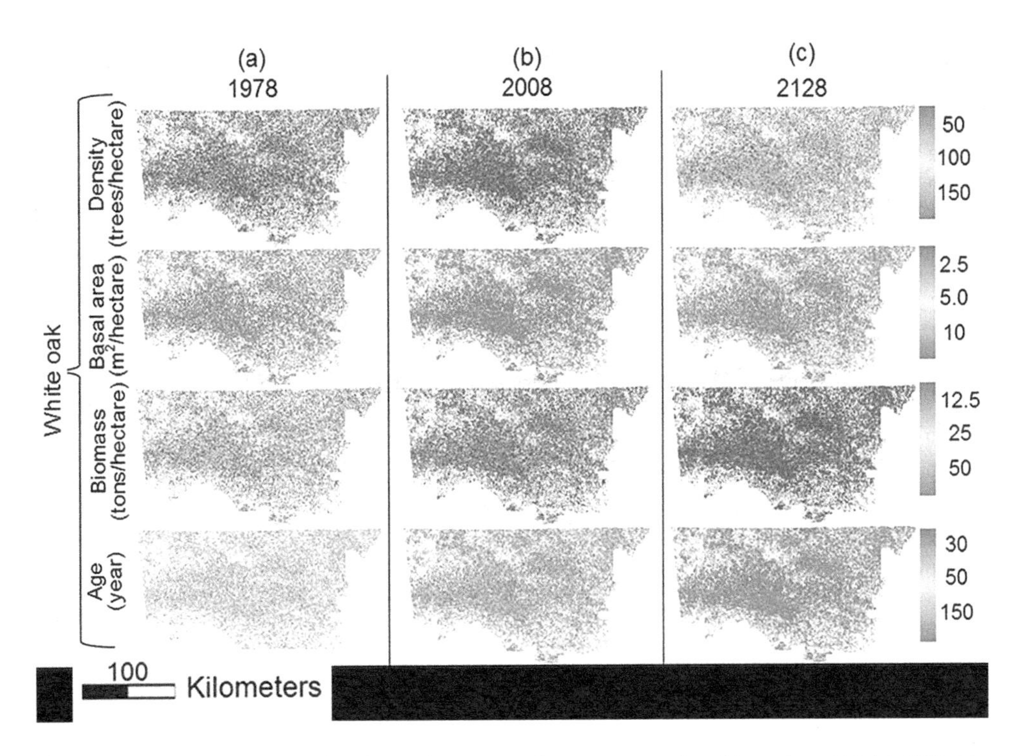

FIGURE 43.4 Simulated density, basal area, biomass, and age structure for white oak at 1978 (**a**), 2008 (**b**), and 2128 (**c**) in Northern Arkansas (Wang et al. 2014b).

The initial landscape can be derived from historic (1978) forest inventory data (point) and land cover (raster) data using the Landscape Builder software (Dijak 2013). The example shows that key forest stand attributes of white oak density, basal area, biomass, and age can be derived spatially (Figure 43.4). The 1978 map is derived from using 70% of forest inventory data and verified using the remaining 30% of forest inventory data (Figure 43.4a). Once the model passed the verification, it projects in the case, for 150 years at 5-year time steps, including examples of year 2008 (Figure 43.4b) and year 2128 (Figure 43.4c).

Example shown here use 2008 forest inventory data to check against 2008 model prediction, before making future predictions. Simulation results show that, under no disturbance and management, with forest aging to old-growth, white oak density decreases, while basal area and biomass increase over time. Heterogeneity in all forest attributes exists for white oak across the study area.

Simulated results can be summarized into simple graph for all species and species groups (Figure 43.5). These results have gone through the verification and validation processes at both landtype level and the entire landscape level (Wang et al. 2014b). Compared to previous forest landscape modeling efforts, this kind of vigorous model verification and validation has enhanced model credibility and reliability.

Acknowledgments

We thank numerous researchers and programmers in developing LANDIS PRO. This research was funded by the U.S. Forest Service Northern Research Station and Southern Research Station, the University of Missouri Agricultural Experimental Research Program, Chinese Academy of Sciences, and Northeastern Normal University.

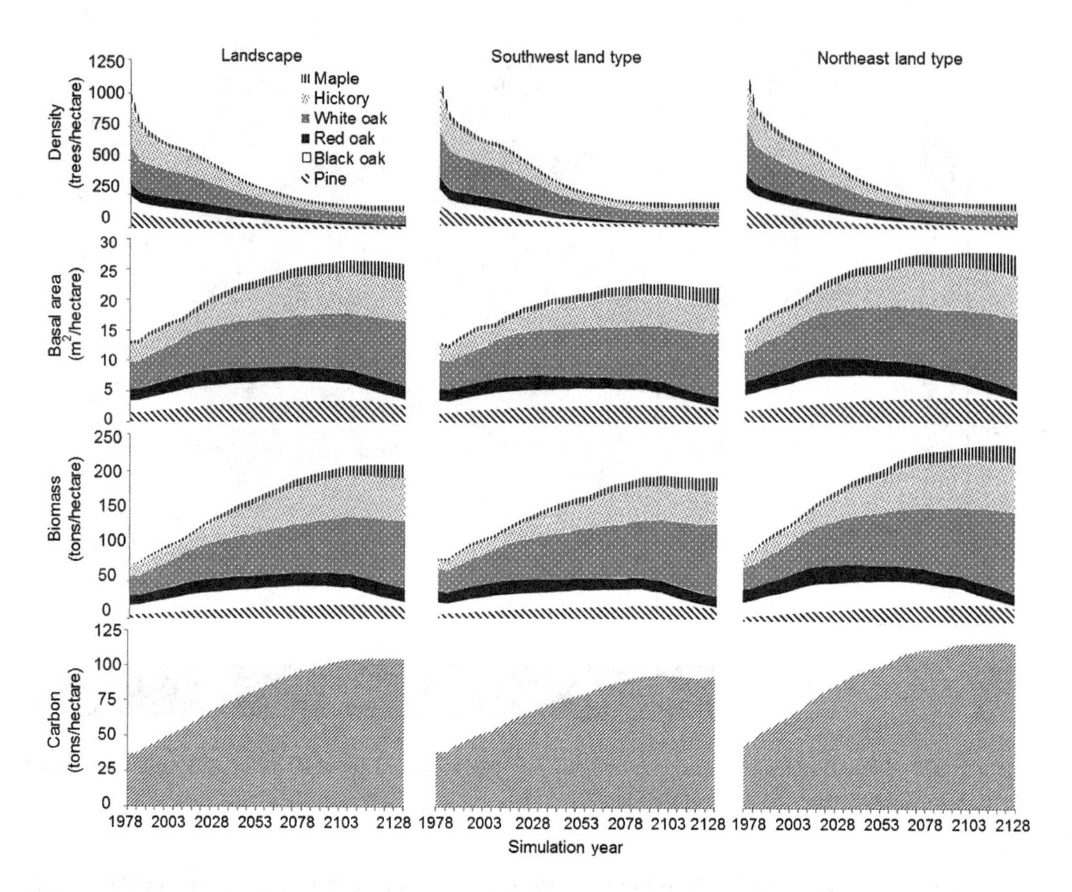

FIGURE 43.5 Simulated density, basal area, and biomass by species group and carbon of total species at landscape- and land-type scales over 150 simulation years in Northern Arkansas (Wang et al. 2014b).

References

Burns, R. M., Honkala, B. H. (technical coordinators). 1990. *Silvics of North America: 1. Conifers; 2. Hardwoods*. Agriculture Handbook 654. USDA, Forest Service, Washington, D.C.

Dijak, W. 2013. Landscape builder: Software for the creation of initial landscapes for LANDIS from FIA data. *Computational Ecology and Software* 3 (2): 17–25.

Fisher, J. B., Huntzinger, D. N., Schwalm, C. R., Sitch, S. 2014. Modeling the terrestrial biosphere. *Annual Review of Environment and Resources* 39: 91–123.

He, H. S., Mladenoff, D. J. 1999a. Spatially explicit and stochastic simulation of forest landscape fire disturbance and succession. *Ecology* 80 (1): 81–99.

He, H. S., Mladenoff, D. J. 1999b. The effects of seed dispersal in the simulation of long-term forest landscape change. *Ecosystems* 2 (4): 308–319.

He, H. S., Li, W., Sturtevant, B. R., Yang, J., Shang, Z. B., Gustafson, E. J., Mladenoff, D. J. 2005. LANDIS, a spatially explicit model of forest landscape disturbance, management, and succession—LANDIS 4.0 User's guide. USDA Forest Service North Central Research Station General Technical Report. NC-263.

He, H. S. 2008. Forest landscape models, definition, characterization, and classification. *Forest Ecology and Management* 254: 484–498.

He, H. S., Yang, J., Shifley, S. R., Thompson, F. R. 2011. Challenges of forest landscape modeling - Simulating large landscapes and validating results. *Landscape and Urban Planning* 100 (4), 400–402.

He, H. S., Gustafson, E. J., Lischke, H. 2017. Modeling forest landscapes in a changing climate: theory and application. *Landscape Ecology* 32: 1299–1305. Springer: Dordrecht. DOI:10.1007/s10980-017-0529-4.

Mladenoff, D. J., He, H. S. 1999. Design and behavior of LANDIS, an object-oriented model of forest landscape disturbance and succession. In Mladenoff D. J., Baker W. L. (editors), *Advances in Spatial Modeling of Forest Landscape Change: Approaches and Applications.* Cambridge University Press, Cambridge.

Mladenoff, D. J. 2004. LANDIS and forest landscape models. *Ecological Modelling* 180: 7–19.

Oliver, C. D., Larson B. C. 1996. *Forest stand dynamics. Update edition.* John Wiley & Son: New York.

Reineke, L. H. 1933. Perfecting a stand density index for even-aged forests. *Journal of Agricultural Research* 46: 627–638.

Scheller, R. M., Domingo, J. B., Sturtevant, B. R., Williams, J. S., Rudy, A., Gustafson, E. J., Mladenoff D. J. 2007. Design, development, and application of LANDIS-II, a spatial landscape simulation model with flexible temporal and spatial resolution. *Ecological Modelling* 201: 409–419.

Shifley, S. R., He, H. S., Lischke, H., Wang, W. J., Jin, W. C., Gustafson, E. J., Thompson, J. R., Thompson III, F. R., Dijak, W. D., Yang, J. 2017. The past and future of modeling forest dynamics: from growth and yield curves to forest landscape models. *Landscape Ecology* 32 (7).

Yang, J., He, H. S., Shifley, S. R., Thompson, F. R., Zhang, Y. 2011. An innovative computer design for modeling forest landscape change in very large spatial extents with fine resolutions. *Ecological Modelling* 222 (15): 2623–2630.

Yoda, K., Kira T., Ogama H., Hozumi K. 1963. Self-thinning in overcrowded pure stands under cultivate and natural conditions. *Journal of Biology* 14: 107–129.

Wang, W. J., He, H. S., Spetich, M. A., Shifley, S. R., Thompson III, F. R., Larsen, D. R., Fraser, J. S., Yang J. 2013. A large-scale forest landscape model incorporating multi-scale processes and utilizing forest inventory data. *Ecosphere* 4 (9): 106. DOI:10.1890/ES13-00040.1.

Wang W. J., He, H. S., Fraser, J. S., Thompson III, F. R., Shifley, S. R., Spetich, M. A. 2014a. LANDIS PRO: A landscape model that predicts forest composition and structure changes at regional scales. *Ecography* 37 (3): 225–229. DOI: 10.1111/j.1600-0587.2013.00495.x.

Wang, W. J., He, H. S., Spetich, M. A., Shifley, S. R., Thompson, F. R., Dijak, W. D., Wang, Q. 2014b. A framework for evaluating forest landscape model predictions using empirical data and knowledge. *Environmental Modelling & Software* 62: 230–239. DOI: 10.1016/j.envsoft.2014.09.003.

44

Landscape Pattern and Change by Stone Wall Feature Identification

Rebecca Trueman
and Yeqiao Wang
University of Rhode Island

Introduction

In present day New England region of northeastern United States, stone walls are the most noticeable relics existing as evidence of the historical agricultural civilization that once flourished between the 18th and 19th centuries. New England stone walls are a result of the integrated histories of nature and humans. The geologic setting, climate, and human land use worked together to create an ideal environment for soil disturbance and the unburying of till stones. For New Englanders, the solution to fields covered in stone year after year was to create stone walls, also known as "linear landfills" [1]. When these stone walls were originally formed on the landscape, they were placed along edges of fields and in conjunction with existing wood fencing. Later on, due to a reduction in forest resources and shifting cultural norms, stone walls served as property boundaries and were secured for animal confinement. "Good fences make good neighbors" became a common sentiment [2].

Stone walls have previously been considered as a factor in studies pertaining to historical land use and change [3], as well as landscape characterization [4]. With advancements in remote sensing technologies, long linear features such as stone walls can be identified remotely and quantified across large areas. In this chapter, we present a case study that integrated the available record of land use and land cover (LULC), parcel boundaries, protected open space, and roads with high-resolution remote sensing imagery to assess the persistence of human land use, based on the temporal distribution of stone walls and recent characterizations of the natural and anthropogenic landscapes.

Methods

Study Site

The study was confined to the town of New Shoreham (Block Island), Rhode Island, located 14.5 km south of the Rhode Island mainland. Block Island is part of the end moraine deposited in the Late Wisconsin by the Laurentide ice sheet, which retreated ~18,000 years ago [5]. The Manisseans, a Niantic tribe of the Native Americans, inhabited the area prior to a group of 16 European settlers from the Massachusetts Bay Colony in 1661 [6]. The land was quickly cleared for settlement, agriculture, and husbandry, leading to the exposure of till stones and subsequent building of stone walls. In 1886, it was estimated that over 482 km (300 mi) of stone wall were contained on Block Island [7]. As agriculture slowed in the 20th century, the culture transformed, and preservation and tourism became and continues to be prioritized on Block Island. In 1991, Block Island was named by the Nature Conservancy as one of 12 "The Last Great Places" in the Western Hemisphere, increasing the spotlight on this Island [8]. As of 2015, records and calculations indicate that 44.8% of the island's land area is conserved, 36.4% through deeded protection and 9.8% through regulation.

Data and Analysis

Geospatial data for stone wall feature identification include digital true color orthophotography, digital aerial photography, and historical topographic Map. Other data include LULC maps of 1988, 1997, 2003/2004, and 2011; historical topographic map of 1990; and maps of protected open space and parcels (Table 44.1).

The analysis for this study included several steps: the identification of stone walls; classification of LULC; incorporation of existing vector geographic information system (GIS) data, e.g., parcel

TABLE 44.1 Summary of the Sources of Geospatial Data Used in This Study

Dataset	Year	Format	Resolution	Coding	Source
			Data for Stone Wall Identification		
Digital true color orthophotography	2003/2004	Raster	0.6 m (1.97 ft)	N/A	Rhode Island Geographic Information System (RIGIS)
Digital aerial photography	2008	Raster	0.10 m (0.33 ft)	N/A	RIGIS
Digital true color orthophotography	2011	Raster	0.15 m (0.50 ft)	N/A	RIGIS
Historical topographic map	1900	Raster	1.28 m (4.20 ft)	N/A	Town of New Shoreham, RI
			Data for Temporal LULC Classification		
LULC	1988	Polygon	2,023.43 m² (0.5 acre)	Modified Anderson Level 3	RIGIS
LULC	1995	Polygon	2,023.43 m² (0.5 acre)	Modified Anderson Level 3	RIGIS
LULC	2003/2004	Polygon	2,023.43 m² (0.5 acre)	Modified Anderson Level 3	RIGIS
LULC	2011	Polygon	2,023.43 m² (0.5 acre)	Modified Anderson Level 3	RIGIS
			Ancillary Data		
Protected open space	2013	Polygon	N/A	N/A	Town of New Shoreham, RI
Parcels	2013	Polygon	N/A	N/A	Town of New Shoreham, RI
Roads	2014	Line	N/A	N/A	RIGIS

boundaries, roads, and conservation land; and then assessment of stone walls in relation to those data of the natural and human-defined landscape.

We created the 2011 stone wall distribution, considered as the current distribution of stone walls, by GIS delineation features identified through user visualization and pattern recognition using digital true color orthophotography. Stone wall features are clearly visible along open fields, urban areas, and under canopy (Figure 44.1). Other data, e.g., 2008 aerial imagery and Google Earth data, were also incorporated into the analysis to identify and confirm stone walls for the current distribution, such as those less visible under canopy.

We were given access to a historical topographic map of Block Island circa 1900 from the Town of New Shoreham (Figure 44.2). This topographic map depicted stone walls which we used to create the historical distribution of stone wall dataset. We were then able to complete an assessment comparing the data as follows: matching between 2011 and 1900, removed after 1900, and built after 1900.

Using the existing LULC dataset, we determined temporal land cover change from 1988 to 2011. For simplification purposes, the four LULC datasets of 1988, 1995, 2003–2004, and 2011 were recoded to seven general land cover categories, including urban, agriculture, brushland, forest, water, wetland, and barren. We converted the LULC datasets to a 45 m spatial resolution to be consistent with the 2,023.43 m^2 minimum mapping unit (MMU) of the original data. Distribution of land among seven classes for each of the four datasets was quantified.

To integrate the current stone wall distribution with the temporal LULC dataset, we calculated the length of stone wall within each of the seven classes for the LULC dataset. To further assess the placement of stone walls on the landscape, we integrated the 2013 parcel boundaries of Block Island. Coincidence of parcel boundaries and 2011 stone walls was quantified by determining which parcel boundaries are adjacent, i.e., within 3 m, to a stone wall. Additionally, we determined the amount of

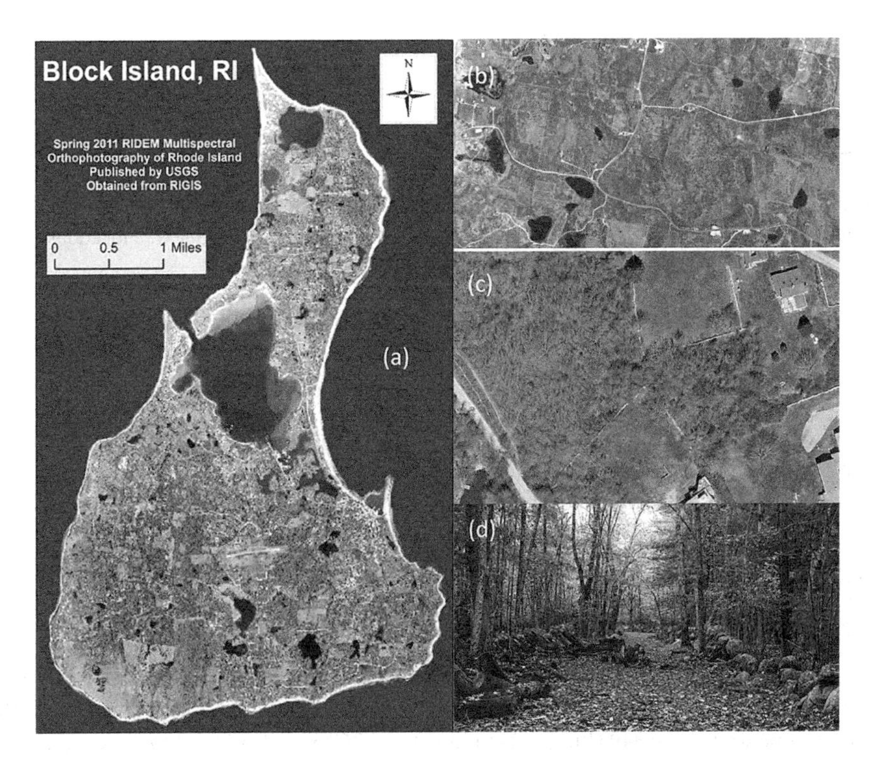

FIGURE 44.1 **(See color insert.)** True color orthophotography data of the study area were collected in the spring of 2011 at a pixel resolution of 0.1524 m (**a**). The historical aerial photo of 1951 (**b**), recent high-spatial-resolution orthophotography of 2008 (**c**), and a field photo illustrate stone wall features on the landscape (**d**).

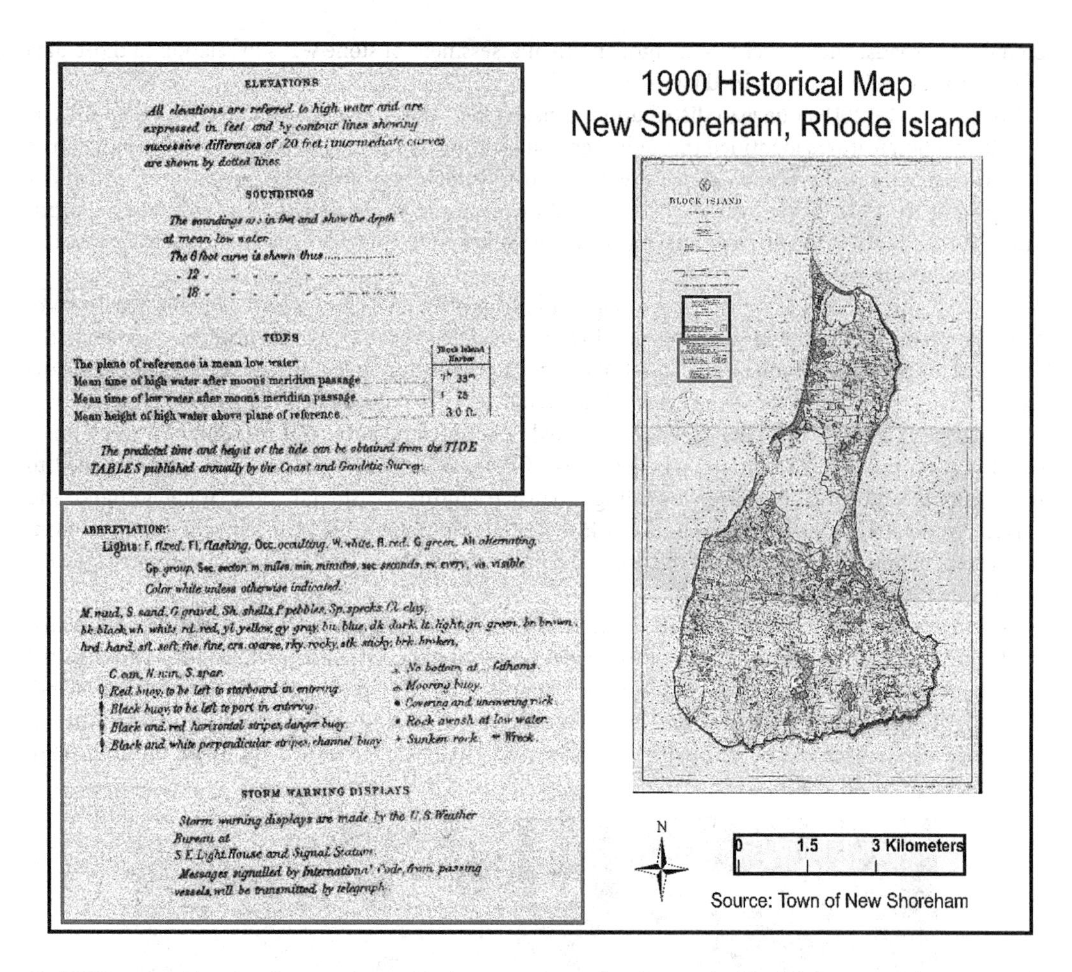

FIGURE 44.2 The 1900 historical topographic map, New Shoreham, Rhode Island, was used to create the historical stone wall distribution. Dashed lines symbolized in the map represent stone wall locations as of 1900.

stone walls which border a parcel boundary in each of the five distributions, i.e., current, historical, matching, removed, and built. Data of 2013 protected open space and 2014 roads were also incorporated during this study.

For field verification of the 2011 stone wall dataset, stone walls on Block Island were surveyed using a Trimble GeoXT running Terrasync Pro GPS with an accuracy of <1 m differential correction. Results of the stone wall ground survey were used to determine the accuracy of the 2011 stone wall distribution based on the stone walls identified in the field which were not initially identified in the data.

Results

Temporal Change in Distribution of Stone Walls

A temporal dataset of stone walls on Block Island was created based on the visual interpretation of the data, which includes a stone wall distribution for 1900 (Figure 44.3) and 2011 (Figure 44.4). The result revealed that stone walls were 349.1 km (14.2 km km^{-2}) and 260.6 km (10.6 km km^{-2}), respectively, for the years 1900 and 2011.

By determining the spatial distribution of stone walls as of 1900 and 2011, we were able to quantify how stone walls changed in different time periods. Analysis of the stone wall dataset determined that

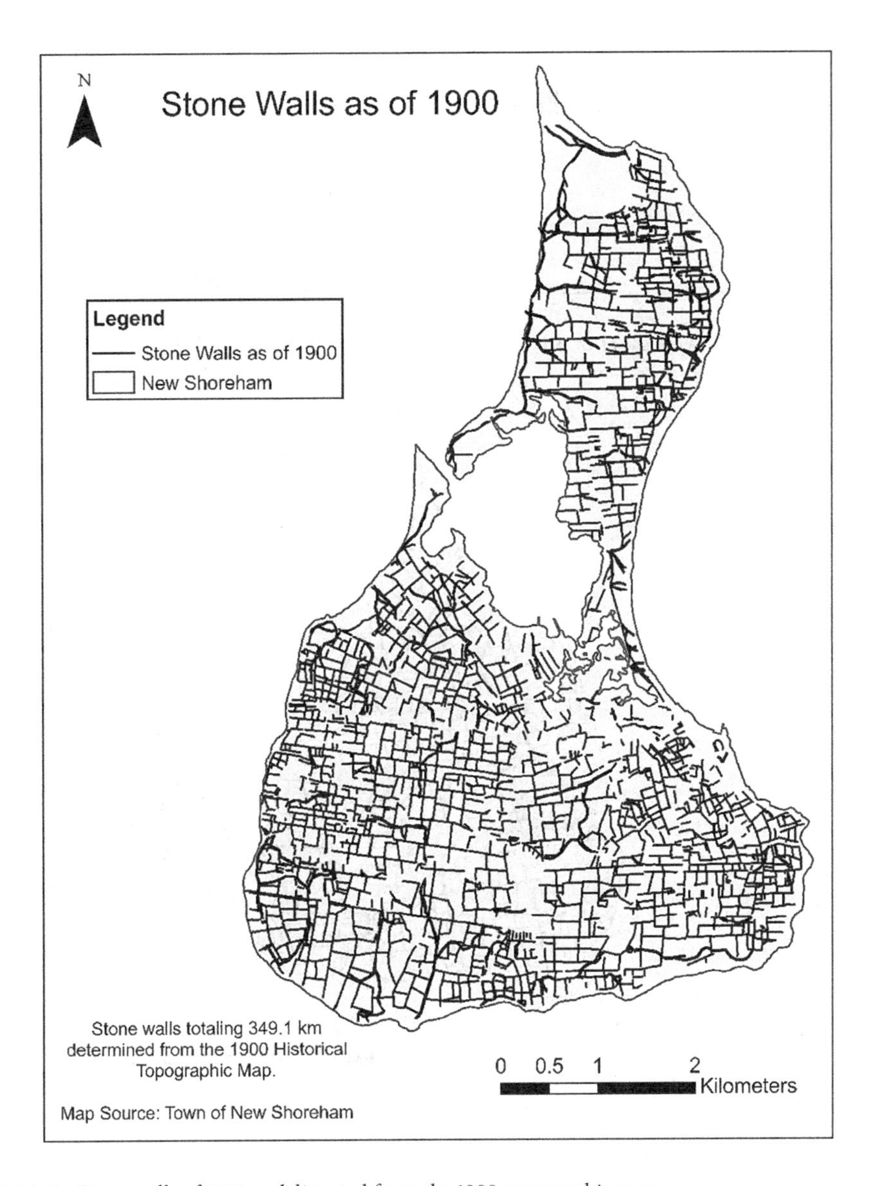

FIGURE 44.3 Stone walls of 1900 as delineated from the 1900 topographic map.

195.8 km of stone walls matched between 1900 and 2011, 65.3 km of stone walls were removed from 1900 to 2011, and 153.3 km of stone walls were built between 1900 and 2011, respectively. It was noted that 27.1 km of the 65.3 km (43%) removed stone walls were parallel and within 10 m of current roads on Block Island.

Stone Walls and Temporal Land Cover

LULC change assessment revealed that, from 1988 to 2011, the total change was approximately 31%, which was the same as LULC change for the Rhode Island mainland during the same time period. On Block Island, the most abundant LULC type was consistently urban which increased from 25% to 37% from 1988 to 2011. Throughout the temporal range, urban, forest and water increased, while agriculture, bushland, wetland, and barren lands decreased.

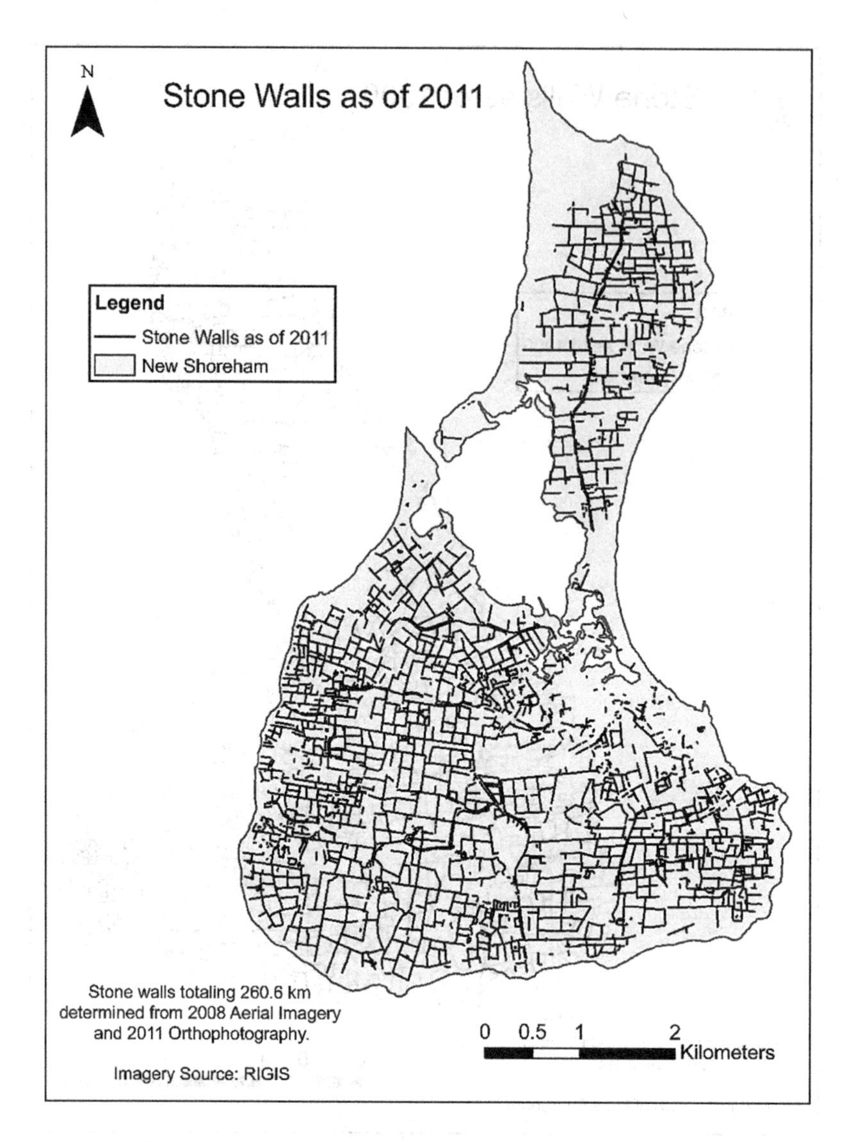

FIGURE 44.4 Stone walls of 2011 as delineated from the 2011 true color orthophotography with field verifications.

We then quantified the distribution of 2011 stone walls within each LULC class in the temporal dataset. Stone walls were found in each of the seven LULC categories. Stone walls were more abundant in the LULC classes of urban and agriculture; less abundant in water, wetland, and barren lands; and proportional to brushland and forest. Additionally, the percent of stone walls increased in urban and forest; decreased in brushland, wetland, and barren lands; and stayed consistent within 1% in agriculture and water from 1988 to 2011.

Stone Walls and the Anthropogenic Landscape

As of 2013, the ownership of the land in Block Island was distributed into 2,208 individual parcels, with an average parcel area of 10,750 m^2. Using the 2011 stone wall data, it was determined that 1,788 parcels (81%) were in part bordered by a stone wall. Additionally, 208 km of stone walls (80%) from the 2011 data border a parcel boundary. Assessment of the parcel boundaries with the other stone wall data determined that 234 km of stone walls (67%) from the 1900 stone wall data, 158 km (81%) of stone walls from the matching stone wall data, 79 km (52%) of stone walls from the removed stone wall data, and 54 km (83%) of stone walls from the built stone wall data border parcel boundaries.

As of 2013, there were 12 organizations that owned almost 35% of the land on Block Island as protected open space, as identified on the 2013 protected open space map. Using the 2011 stone wall data, it was determined that this 35% protected open space contained 37% of the stone walls.

Ground Verification

The field verification calculated an accuracy of 86.6% for the 2011 stone wall data. Of the 26.88 km of stone walls identified in the field, 23.29 km were also identified within the 2011 stone wall dataset, while the other 3.59 km of stone walls were not but were then added to the final 2011 data of stone walls. Other than those missing, there were no incorrectly mapped stone walls in the 2011 dataset identified within the surveyed area. Additionally, based on access during the ground survey, 12.12 km (45.09%) of the stone walls surveyed in the field are within 7.5 m of roads from the 2014 road map.

Conclusions

Temporal Distribution of Stone Walls

The successful creation of a temporal dataset for Block Island in 1900 and 2011 allowed for the subsequent assessment of stone wall distribution in relation to natural and anthropogenic LULC types. Results of this study support connections between the presence of stone walls and characterization of the temporal landscape. However, there are also inaccuracies and limitations to consider when forming conclusions.

The topographic map created in 1900 for Block Island was symbolized using dashed lines for stone walls and solid lines for roads as indicated from the map key. However, metadata does not exist to know the procedure the cartographer used in the map creation. In this study, we created the 1900 stone wall distribution by delineation of the dashed lines from this topographic map. Dashed lines to indicate a stone wall were not found along roads, i.e., solid lines, and therefore, the assumption in this study is that stone walls were not located along roads as of 1900. Based on our review of historical literature, we know that the absence of stone walls along roads would not be realistic. And in fact if stone walls were located along roads, the total amount of stone walls would increase to a representation closer to the 1886 estimate of 482.80 km (300 mi), compared to our identified total of 349.1 km [7]. These adjustments to the 1900 stone wall distribution would get carried forward to the results of the matching, removed, and built stone wall distributions. For example, the 27.1 km (43%) of stone walls within the dataset of built stone walls located parallel to roads might be moved to the matching or removed distributions. For this study, we did not have the data to calculate the accuracy of the 1900 stone wall distribution, and therefore, the accuracies of the data for stone walls matching, removed, and built between 1900 and 2011 also remain unknown. While specific quantities of stone walls may be in question, it is clear that the distribution of stone walls is not stagnant. The trend from 1900 to 2011 shows a decrease in stone wall abundance over time, yet there is evidence that stone walls continue to be built.

The accuracy of 2011 stone wall distribution was essential to this study and justified the laborious process of digitizing stone walls with the use of the 2011 orthophotography and other available data. Based on field-verified stone walls, the 2011 stone wall distribution has an accuracy of 86.6%. The inaccuracies within the 2011 stone wall data were likely confound to stone walls under forest canopy, which were difficult to identify in both the imagery and in the field. To minimize inaccuracies, the 2008 imagery, which possesses 10.16 cm (4 in.) spatial resolution compared to the 198.12 (6 in.) resolution of the 2011 imagery, was assessed to confirm and identify stone walls. While both the 2011 and 2008 imagery has been collected in the spring, the 2008 leaf cover was less dense and greatly assisted in stone wall identification under canopy. However, by incorporating the 2008 imagery into this study, it was assumed that stone walls were unchanged from 2008 to 2011. It became evident that stone walls were both moved and built between 2008 and 2011.

Stone Walls and Land Cover Change

While the temporal stone wall distribution spans from 1900 to 2011, based on data availability, the LULC change study only spans from 1988 to 2011. The distribution of 2011 stone walls within the seven LULC classes from 1988 to 2011 generally supports what would be expected based on facts surrounding the initial creation of stone walls, land use suitability, and current land use practices. In the late 18th and 19th centuries, stone walls were initially distributed around field borders and property boundaries, lands that would be classified as agriculture and urban. This study identified that there continues to be an emphasis on the maintenance and creation of stone walls among these land use types. It was also determined that a lower proportion of stone walls was located in less suitable land, such as water, wetlands, barren lands, and sandy areas along the coast. Stone wall distribution from 1988 to 2011 remained unchanged in brushland and forest land cover areas.

Stone Walls and the Anthropogenic Landscape

Parcel boundaries as of 2013 were assessed in relation to 2011 stone walls by quantifying the percentage of current parcels surrounded by a stone wall and the length of stone walls which were surrounding parcel boundaries. With 81% of the parcels on Block Island at least in part bordered by a stone wall, it is expected that these features would influence the majority of land owners. This influence could range from landscape maintenance to property value, and from aesthetic perspective to ecological characteristics. Perhaps a more telling connection is that 80% of stone walls from the 2011 data border a parcel boundary. The potential accuracies in the matching, removed, and built stone wall distributions would get carried through to the analysis of these data in relation to parcel boundaries. Through correction and further assessment, these local-scale relationships as well as larger-scale connections between temporal distributions of parcels and stone walls could become clearer.

We determined that 37% of stone walls as of 2011 were contained within protected open space, emphasizing the role of conservation, more specifically the influence of humans in the protection of the landscape and therefore the minimization of land use change. Over time, stone wall conservation would be a result of human value placed on both stone walls and lands which contain stone walls. Based on this study, there is a clear appreciation for both stone walls and lands. Additionally, the state of Rhode Island places specific value on stone wall conservation enacted through the RI General Law § 45-2-39.1 and RI General Law § 11-41-32, which give penalty to theft of a stone wall, and RI General Law § 44-3-43, which gives tax exemption to owners of certain types of historic stone walls. Conservation of stone walls benefits those who seek to further understand the relationships between stone walls and the environment, as well as the influence of stone walls on land use change and function.

References

1. Thorson, Robert. *Stone by Stone: The Magnificent History in New England's Stone Walls.* Bloomsbury Publishing: New York (2009).
2. Frost, Robert. Mending Wall. *North of Boston.* Henry Holt & Co.: New York (1915).
3. Cronon, William. *Changes in the Land: Indians, Colonists, and the Ecology of New England.* Macmillan: New York (2011).
4. Wessels, Thomas. *Reading the Forested Landscape. A Natural History of New England.* The Countryman Press: Woodstock, VT (1997).

5. Boothroyd, Jon C. and Les Sirkin. *The Quaternary Geology of Block Island and Adjacent Regions.* Rhode Island Natural History Survey: Kingston, RI (2002): 13–27.
6. Rosenzweig, L., R. Duhaime, A. Mandeville, and P. August. (eds). Ecological geography of Block Island. In: P. Paton, et al. (eds.) *The Ecology of Block Island.* Rhode Island Natural History Survey: Kingston, RI (2002): 3–11.
7. Livermore, Samuel Truesdale. *A History of Block Island.* The Case, Lockwood & Brainard Co: Hartford, CT (1877).
8. Paton, P. W., L. L. Gould, P. V. August, and A. O. (eds). Frost. Introduction. In: P. Paton, et al. (eds.) *The Ecology of Block Island.* Rhode Island Natural History Survey: Kingston, RI (2002): 7–8.

45

Remote Sensing of Urban Dynamics

Wenting Cao and
Yuyu Zhou
Iowa State University

Introduction

Urban areas are the places dominated by the built environment (i.e., human-made surfaces) such as roads, driveways, parking lots, and rooftops. More than 50% of global population lives in urban areas, and this proportion will increase to 68% by 2050 [1]. Urbanization promotes various economic and social developments, while it challenges the sustainable goals of development in various fields such as public health, urban heat island, air quality degradation, and biodiversity loss. Therefore, it is of great significance to map urban dynamics for urban planning and environmental management. With the development of remote sensing techniques, satellite observations become a major data source for mapping urban dynamics. This chapter aims to present the current progress of remote sensing applications in mapping urban dynamics.

Global Urban Dynamics

Because cities are well lit and the rural areas are dark during night, satellite observations of nighttime lights (NTLs) are an effective data source for mapping urban dynamics at the global level. The Defense Meteorological Satellite Program's Operational Linescan System (DMSP/OLS) stable NTL data are the most widely used NTL data source for such purpose. DMSP/OLS provides a historical record of the years 1992–2013 with visible band Digital Numbers (DNs) ranging from 0 to 63 and a spatial resolution of 30 arc-seconds (~1 km at the equator and 0.8 km at 40°N). Various methods were developed to address the challenges and issues in NTL-based urban mapping. From the spatial dimension, vegetation-based spatial adjustment [2], detection frequency thresholds [3], and threshold-based approaches [4] were used for mitigating the saturation effects in urban core and blooming effects on the urban–rural boundary. From the temporal prospective, Elvidge et al. [5] built the most widely used framework of intercalibration for the annual NTL product, and several subsequent studies [6–8] were carried out to improve this framework. Urban extent is usually identified as pixels with DN value higher than a pre-defined threshold; however, the thresholds vary in different regions due to the different development levels. Therefore, Zhou et al. [9,10] developed a cluster-based method to estimate the optimal threshold for each urban cluster. Recently, Zhou et al. [11] improved the cluster-based method to generate temporally and spatially consistent global urban dynamics, finding that the percentage of urban land increased from around 0.2% (1992) to 0.5% (2013) at the global level.

Multi-spectral remote sensing observations covering the visible, near-infrared, and shortwave-infrared ranges (e.g., Moderate Resolution Imaging Spectroradiometer (MODIS) and Landsat) can provide more details of urban-built environments compared with NTL data. Several maps of the global urban land have been produced using MODIS [12–14] and Landsat [15–17] and supervised classification algorithms. Schneider et al. [13,14] mapped the global urban extent of circa 2001–2002 at a 500 m spatial resolution from MODIS, and Gong et al. [16] and Chen et al. [17] produced the Landsat-based 30 m global land cover maps for 2000 and 2010. However, these maps are only available for specific years (e.g., 2000 and 2010) and were produced based on different criteria, which brings difficulties in generating a consistent global urban land expansion over a long historical period. Recently, Liu et al. [15] developed multi-temporal global urban land maps from Landsat images for the 1990–2010 period with a 5-year interval using a method of normalized urban areas composite index. They found that the global urban land had increased from 450.97 ± 1.18 thousand km^2 in 1990 to 747.05 ± 1.50 thousand km^2 in 2010.

National and Regional Applications

There are extensive applications of remote sensing in urban dynamics mapping at regional and national scales. These studies provided an exhaustive insight into the magnitude, location, geometry, and spatial pattern of urban dynamics, which are of great importance to the regional environmental management and urban planning.

Landsat time series imageries are the most widely used data source for its long record and moderate temporal (16 days) and spatial (30-m) resolutions. Although numerous studies have investigated urban dynamics using Landsat imageries, most of them were conducted at relatively coarse temporal intervals (e.g., 5 years, 10 years, or decades) using traditional multi-temporal classification approaches (e.g., supervised classification or visual interpretation). Such studies may not be able to capture urban change paces well, especially in rapidly developing regions. Recently, efforts were made in several studies to capture detailed urban dynamics at finer temporal intervals using time series images. Representative examples include Li et al. [18–20] in Beijing and USA, Zhang et al. [21,22] in the Pearl River Delta, and Sexton et al. [23] and Song et al. [24] in the Washington, D.C.–Baltimore metropolitan region. In particular, Li et al. [19,20] developed a change detection framework for mapping the annual urban extent (1985–2015) in the conterminous US using all Landsat time series data.

NTL observations were also widely used in national and regional urban dynamics mapping. Utilizing the multi-temporal DMSP/OLS NTL data and additional regional socioeconomic information, most of national studies were performed in large countries such as the United States [25,26], China [8,27,28], and India [29]. Moreover, NASA and NOAA launched the Suomi National Polar Partnership (SNPP) satellite carrying the Visible Infrared Imaging Radiometer Suite (VIIRS) instrument in 2011. The improvements of VIIRS Day/Night Band (DNB) over DMSP/OLS include a higher spatial resolution, in-flight calibration, better imaging capability of low light, and reduced spatial blooming and pixel saturation [30]. The VIIRS/DNB data are available for a global coverage at the monthly level since April 2012, which provides insights into more detailed urban dynamics. VIIRS/DNB data recently have been used to examine the urban dynamics in regional studies. For example, Shi et al. [31] found a better performance of VIIRS over OLS in mapping urban areas in China.

Local Applications

Although global and regional urban dynamics around the world have been revealed successfully using satellite remote sensing, local research is still needed to account for the inner-urban dynamics. The advent of high-spatial resolution (<10 m) satellite images, aerial photographs, social media, and LiDAR data makes it possible to reduce the mixed pixel problem for more accurate information of urban land [32].

High-spatial resolution data acquired from both space-borne and airborne sensors, such as SPOT, WorldView, and Quickbird, contain information such as texture, shape, and context. Such type of data stimulate new applications of urban dynamics mapping. Recently, high-spatial resolution data have been used to identify urban functional zones [33], road extraction [34], and detection of slums [35] using methods of object-oriented image analysis and image texture measures. Moreover, social media data contain rich geospatial information of human activities collected publicly from taxi, smart card check-in, social media, and mobile phone data. Such social media data have been used in deriving the information of urban intra-functions [36]. In addition, LiDAR data have been increasingly used to extract urban building heights for a better understanding of morphology and density of urban buildings [37]. Therefore, the integration of high-resolution images, social media data, and LiDAR data could play an important role in mapping urban dynamics at the local scale.

Conclusions

This chapter presents representative applications of remote sensing observations in urban dynamics mapping at the global, regional, and local scales. Remote sensing plays an important role in mapping the distribution and dynamics of the urban land around the world. A majority of efforts have been made for mapping the urban land dynamics over large areas using coarse-spatial-resolution and medium-spatial-resolution images such as NTL, MODIS, and Landsat time series data. These studies show that urban land has increased significantly during the past decades, especially in China and India, the two largest developing countries. With the availability of high-resolution images, social media data, and LiDAR data, which have demonstrated the ability to detect the inner-urban structure and forms, the inner-urban dynamics mapping becomes possible.

Despite much progress made in previous studies, efforts are still needed in several research areas for improving the scientific understanding of urban dynamics in the rapidly urbanizing world. First, high-frequency temporal analysis (e.g., monthly or seasonal) can enable remote sensing observations for monitoring fast urban changes and disturbance events. In most of previous studies using time series analysis, only one observation or composite in a year was used to map the urban changes, which might not capture the changes within the year. Moreover, the temporal frequency of observations can be improved by fusing different remote sensing observations such as Sentinel and Landsat. Second, urban form and structure, the physical characteristics of urban areas including the shape, size, density, and configuration of urban settlements [38] are essential for sustainable development. Previous studies mainly focused on the two-dimensional urban extent, while three-dimensional urban forms, such as the height or volume of buildings, remain largely unexplored. With the ever-increasing amount of high-resolution remote sensing data, capable computational techniques, and more advanced algorithms, the assessment of the three-dimensional urban form dynamics will become possible. Such scientific information of three-dimensional urban form dynamics can provide fundamental information for intelligent urban management.

References

1. Department of Economic and Social Affairs, Popular Division, World urbanization prospects: The 2018 revision. 2018: United Nations.
2. Zhang, Q., C. Schaaf, and K.C. Seto, The Vegetation Adjusted NTL Urban Index: A new approach to reduce saturation and increase variation in nighttime luminosity. *Remote Sensing of Environment*, 2013, **129**: pp. 32–41.
3. Small, C., F. Pozzi, and C. Elvidge, Spatial analysis of global urban extent from DMSP-OLS night lights. *Remote Sensing of Environment*, 2005, **96**(3–4): pp. 277–291.
4. Elvidge, C.D., et al., Radiance calibration of DMSP-OLS low-light imaging data of human settlements. *Remote Sensing of Environment*, 1999, **68**(1): pp. 77–88.

5. Elvidge, C., et al., A fifteen year record of global natural gas flaring derived from satellite data. *Energies*, 2009, **2**(3): pp. 595–622.
6. Li, X. and Y. Zhou, A stepwise calibration of global DMSP/OLS stable nighttime light data (1992–2013). *Remote Sensing*, 2017, **9**(6): p. 637.
7. Liu, Y., et al., Correlations between urbanization and vegetation degradation across the World's metropolises using DMSP/OLS nighttime light data. *Remote Sensing*, 2015, **7**(2): pp. 2067–2088.
8. Liu, Z., et al., Extracting the dynamics of urban expansion in China using DMSP-OLS nighttime light data from 1992 to 2008. *Landscape and Urban Planning*, 2012, **106**(1): pp. 62–72.
9. Zhou, Y., et al., A cluster-based method to map urban area from DMSP/OLS nightlights. *Remote Sensing of Environment*, 2014, **147**: pp. 173–185.
10. Zhou, Y., et al., A global map of urban extent from nightlights. *Environmental Research Letters*, 2015, **10**(5): pp. 1–11.
11. Zhou, Y., et al., A global record of annual urban dynamics (1992–2013) from nighttime lights. *Remote Sensing of Environment*, 2018, **219**: pp. 206–220.
12. Schneider, A., et al., Mapping urban areas by fusing multiple sources of coarse resolution remotely sensed data. *Photogrammetric Engineering and Remote Sensing*, 2003, **69**(12): pp. 1377–1386.
13. Schneider, A., M.A. Friedl, and D. Potere, A new map of global urban extent from MODIS satellite data. *Environmental Research Letters*, 2009, **4**(4): pp. 44003–44011.
14. Schneider, A., M.A. Friedl, and D. Potere, Mapping global urban areas using MODIS 500-m data: New methods and datasets based on 'urban ecoregions'. *Remote Sensing of Environment*, 2010, **114**(8): pp. 1733–1746.
15. Liu, X., et al., High-resolution multi-temporal mapping of global urban land using Landsat images based on the Google Earth Engine Platform. *Remote Sensing of Environment*, 2018, **209**: pp. 227–239.
16. Gong, P., et al., Finer resolution observation and monitoring of global land cover: First mapping results with Landsat TM and ETM+ data. *International Journal of Remote Sensing*, 2012, **34**(7): pp. 2607–2654.
17. Chen, J., et al., Global land cover mapping at 30m resolution: A POK-based operational approach. *ISPRS Journal of Photogrammetry and Remote Sensing*, 2015, **103**: pp. 7–27.
18. Li, X., P. Gong, and L. Liang, A 30-year (1984–2013) record of annual urban dynamics of Beijing City derived from Landsat data. *Remote Sensing of Environment*, 2015, **166**: pp. 78–90.
19. Li, X., et al., Mapping annual urban dynamics (1985–2015) using time series of Landsat data. *Remote Sensing of Environment*, 2018, **216**: pp. 674–683.
20. Li, X., et al., A national dataset of 30-m annual urban extent dynamics (1985-2015) in the conterminous United States. *Earth System Science Data*, July 2019. doi: 10.5194/essd-2019-107.
21. Zhang, L. and Q. Weng, Annual dynamics of impervious surface in the Pearl River Delta, China, from 1988 to 2013, using time series Landsat imagery. *ISPRS Journal of Photogrammetry and Remote Sensing*, 2016, **113**: pp. 86–96.
22. Zhang, L., Q. Weng, and Z. Shao, An evaluation of monthly impervious surface dynamics by fusing Landsat and MODIS time series in the Pearl River Delta, China, from 2000 to 2015. *Remote Sensing of Environment*, 2017, **201**: pp. 99–114.
23. Sexton, J.O., et al., Urban growth of the Washington, DC–Baltimore, MD metropolitan region from 1984 to 2010 by annual, Landsat-based estimates of impervious cover. *Remote Sensing of Environment*, 2013, **129**: pp. 42–53.
24. Song, X.-P., et al., Characterizing the magnitude, timing and duration of urban growth from time series of Landsat-based estimates of impervious cover. *Remote Sensing of Environment*, 2016, **175**: pp. 1–13.
25. Imhoff, M., et al., A technique for using composite DMSP/OLS "City Lights" satellite data to map urban area. *Remote Sensing of Environment*, 1997, **61**(3): pp. 361–370.

26. Elvidge, C.D., et al., National trends in satellite-observed lighting. *Global Urban Monitoring and Assessment through Earth Observation*, 2014, **23**: pp. 97–118.
27. Ma, T., et al., Quantitative estimation of urbanization dynamics using time series of DMSP/OLS nighttime light data: A comparative case study from China's cities. *Remote Sensing of Environment*, 2012, **124**: pp. 99–107.
28. Ma, T., et al., Night-time light derived estimation of spatio-temporal characteristics of urbanization dynamics using DMSP/OLS satellite data. *Remote Sensing of Environment*, 2015, **158**: pp. 453–464.
29. Pandey, B., P.K. Joshi, and K.C. Seto, Monitoring urbanization dynamics in India using DMSP/OLS night time lights and SPOT-VGT data. *International Journal of Applied Earth Observation and Geoinformation*, 2013, **23**: pp. 49–61.
30. Elvidge, C.D., et al., Why VIIRS data are superior to DMSP for mapping nighttime lights. *Proceedings of the Asia-Pacific Advanced Network*, 2013, **35**: pp. 62–69.
31. Shi, K., et al., Evaluation of NPP-VIIRS night-time light composite data for extracting built-up urban areas. *Remote Sensing Letters*, 2014, **5**(4): pp. 358–366.
32. Hsieh, P.-F., L.C. Lee, and N.-Y. Chen, Effect of spatial resolution on classification errors of pure and mixed pixels in remote sensing. *IEEE Transactions on Geoscience and Remote Sensing*, 2001, **39**(12): pp. 2657–2663.
33. Zhang, X. and S. Du, A Linear Dirichlet Mixture Model for decomposing scenes: Application to analyzing urban functional zonings. *Remote Sensing of Environment*, 2015, **169**: pp. 37–49.
34. Valero, S., et al., Advanced directional mathematical morphology for the detection of the road network in very high resolution remote sensing images. *Pattern Recognition Letters*, 2010, **31**(10): pp. 1120–1127.
35. Kit, O., M. Lüdeke, and D. Reckien, Texture-based identification of urban slums in Hyderabad, India using remote sensing data. *Applied Geography*, 2012, **32**(2): pp. 660–667.
36. Liu, Y., et al., Social sensing: A new approach to understanding our socioeconomic environments. *Annals of the Association of American Geographers*, 2015, **105**(3): pp. 512–530.
37. Yu, B., et al., Automated derivation of urban building density information using airborne LiDAR data and object-based method. *Landscape and Urban Planning*, 2010, **98**(3–4): pp. 210–219.
38. Williams, K., Urban form and infrastructure: a morphological review. Technical Report 2014.

46

Simulation of Post-Fire Vegetation Recovery

Yeqiao Wang
University of Rhode Island

Y. Zhou
Iowa State University

J. Yang
University of Kentucky

H. He
University of Missouri

Introduction

Wildfires have been in growing numbers and environmental impacts in the United States and around the world, due to climate change, fuel buildup, and other causes (Westerling et al. 2006). Wildland fire is an important natural disturbance, which greatly affects ecosystem productivity, biodiversity, forest landscape dynamics, and climate systems with uncertainties (Turner 2010, Robinson 1989). Although as a natural process, wildfire can be beneficial to ecosystem functions, direct fire impacts and secondary effects such as erosion, landslides, introduction of invasive species, and changes in water quality are often disastrous. Pattern, severity, and timing of wildfires affect significantly the successional changes of vegetation. Understanding the effects of fire impacts and the pathway of vegetation recovery is critical for community actions on resource management planning, land use decision, treatment procedure, habitat restoration, and on studies of ecological and economic complexities associated with wildfires.

Remote sensing data and geospatial modeling have been employed in monitoring of wildfires, assessing active fire characteristics, and analyzing post-fire effects (Fraser and Li 2002, Lentile et al. 2005, Ross et al. 2013, Yang 2014, Guo et al. 2017), in measuring forest structure and fuel loads (Skowronski et al. 2007) and in estimating fuel moisture for prediction of fire behavior (Dasgupta et al. 2007, Hao and Qu 2007). Multitemporal remote sensing data have been employed for assessing fire severity (Brewer et al. 2005, Wimberly and Reilly 2007, Miller and Thode 2007) and for simulating post-fire spectral response to burn severity (De Santis et al. 2009, Wang et al. 2011).

Increasing frequency and extent of wildfires demands significant amount of resources for wildfire management. A variety of programs have been established for wildfire research, management, and education. LANDFIRE, for example, produced consistent and comprehensive maps and data describing vegetation composition and structure, surface and canopy fuel characteristics, historical fire regimes, and ecosystem status across the United States (Rollins and Frame 2006).

Different models, such as Vegetation Dynamic Development Tool (VDDT), Tool for Exploratory Landscape Scenario Analyses (TELSA), Forest Vegetation Simulator (FVS), and LANDscape SUccession Model (LANDSUM), have been developed and employed for simulation of long-term forest succession and landscape dynamics (Merzenich et al. 2003, Beukema et al. 2003, Hemstrom et al. 2001, Hann and Bunnell 2001, Swanson et al. 2003, Klenner et al. 2000, Keane et al. 2004). LANDIS is a spatially explicit landscape model that is designed to simulate forest change over large spatial and temporal (10^1–10^3 years) scales with flexible spatial resolutions (He and Mladenoff 1999, Mladenoff and He 1999). LANDIS can simulate natural and anthropogenic disturbances and their interactions with adequate mechanistic realism and can simulate species-level forest succession in combination with disturbances and management practices. LANDIS modeling assumes that detailed, individual tree information and within-stand processes can be simplified, allowing large-scale questions about spatial pattern, species distribution, and disturbances to be adequately addressed. LANDIS can take remote sensing-derived thematic data as input, and the output is compatible with most GIS software for spatial analysis (Sturtevant et al. 2004, He et al. 2004). An improvement in LANDISv4.0a (Syphard et al. 2007) shortened the time interval of simulation from original design of 10 years to 1 year. Those features make LANDIS a candidate model for the simulation of short-term post-fire vegetation recovery. The LANDIS PRO version has been applied, e.g., in forest landscape modeling.

A critical challenge towards spatially explicit simulation of short-term post-fire vegetation recovery is data preparation for the establishment of initial conditions. Those include pre-fire vegetation status, impacted areas and burn severity, topographic locations, and biophysical conditions in an individual spatial unit. One of the critical LANDFIRE data products is the existing vegetation type (EVT). The EVT data represent the distribution of terrestrial ecological system classification developed by the NatureServe (Comer et al. 2003). A terrestrial ecological system is defined as a group of plant community types (associations) that tend to co-occur within landscapes with similar ecological processes, substrates, and/or environmental gradients. LANDFIRE project mapped EVTs using decision tree models, field reference data, Landsat imagery data, digital elevation model (DEM) data, and biophysical gradient data. The EVT products included EVT, vegetation canopy cover, and vegetation height. LANDFIRE Biogradients are the data layers that describe biophysical gradients that affect the distribution of ecosystem components across landscapes. The indirect gradients include slope, aspect, and elevation, and the direct gradients include temperature and humidity. Functional gradients describe the response of vegetation to direct and indirect gradients and include productivity, respiration, and transpiration. Therefore, the Biogradients data can be employed to define environmental parameters for post-fire vegetation recovery simulation.

Burn severity map is a key factor to quantify fire impacts on vegetation and soil (White et al. 1996, van Wagtendonk et al. 2004, De Santis and Chuvieco 2007, Keeley 2009) and to provide baseline information for monitoring restoration and recovery (Brewer et al. 2005). Differenced normalized burn ratio (DNBR) is a continuous index developed from pre- and post-fire Landsat imageries to measure burn severities. The Normalized Burn Ratio (NBR) and the DNBR are defined in the following equations:

$$\text{NBR} = (\text{TM Band 4} - \text{TM Band 7})/(\text{TM Band 4} + \text{TM Band 7}) \quad (46.1)$$

$$\text{DNBR} = \text{NBR}_{\text{pre-fire}} - \text{NBR}_{\text{post-fire}} \quad (46.2)$$

The DNBR dataset yields a burn severity with possible values ranging between −2,000 and +2,000. DNBR can team up with onsite estimation of composite burn index (CBI) for impact measurements.

The combination of LANDFIRE data products and the DNBR data can establish initial states and create parameters for spatially explicit and ecological process-based simulation of short-term post-fire vegetation recovery.

Methods

Data Preparation

This approach was tested in the Sanford Fire site, which is within the Dixie National Forest close to the Cedar City of the Utah state (Figure 46.1). The Sanford Fires were ignited in April and May 2002, for the purpose of reducing accumulated fuels, keeping pinyon/juniper from expanding further into sagebrush/grasslands, maintaining vegetation at different ages, returning fire to its natural role in the ecosystem,

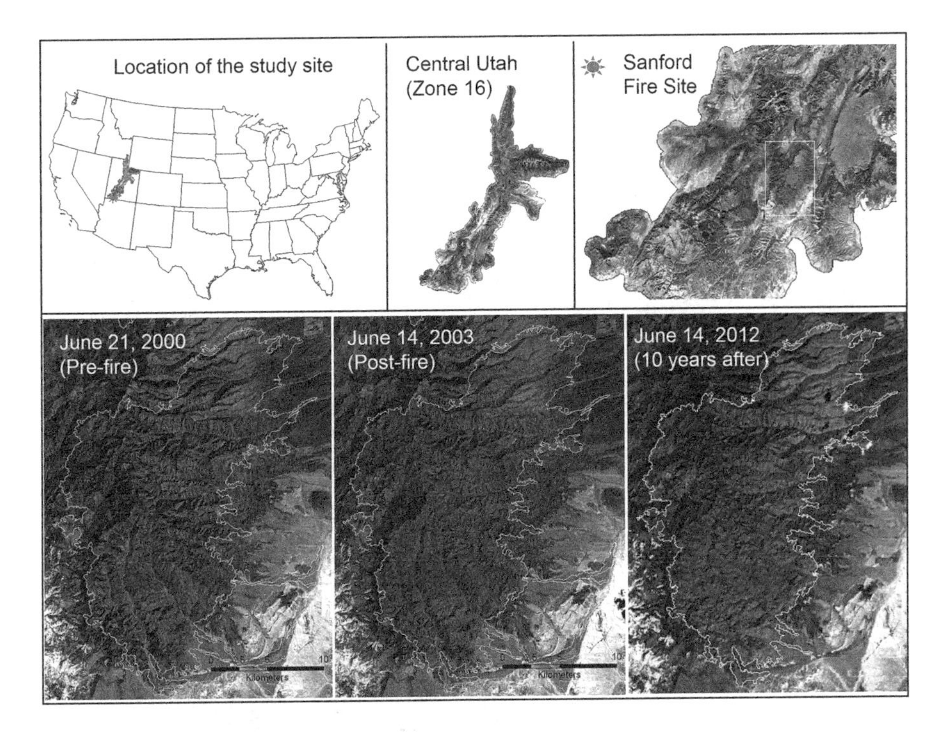

FIGURE 46.1 **(See color insert.)** The Sanford Fire site is located in southern section of the Central Utah Valley (Zone 16) of the LANDFIRE data products. A comparison of pre- and post-fire Landsat TM reflectance images and Landsat-7 ETM+ image 10 years after the fire illustrates the impacts of the affected areas and vegetation recovery.

and stimulating aspen suckering. Strong winds, low humidity, and increased temperatures at the time moved the prescribed burns outside their containment areas. On June 8, 2002, the two prescribed burns joined, fueled by strong winds, and the fire was then referred to as the Sanford Fire. The total amount of land within the perimeter of the Sanford Fire and the two prescribed burns was 31,579 ha. A comparison of pre- and post-fire Landsat images illustrates the affected areas and fire impacts on vegetation (Figure 46.1).

Pre-Fire EVT and DNBR Data

We referred the LANDFIRE data products as the pre-fire data as those were developed based on Landsat data 2 years prior to the Sanford Fire. We extracted spatial distribution of pre-fire EVT (Figure 46.2a), vegetation height, and percent canopy cover for fire impacted areas, and identified the ecosystem types (Table 46.1). We extracted the Biogradients data (e.g., Figure 46.2b) from the LANDFIRE prototype data to develop the land-type parameters that are required for LANDIS simulation modeling. We derived the DNBR data (Figure 46.2c) by the reflectance from pre-fire (2000) and post-fire (2003) Landsat TM data. The DNBR suggested that, in general, a threshold existed between about −100 and +150 DNBR units that marked an approximate breakpoint between burned and unburned areas. We considered the areas with DNBR values below this threshold as unburned. Within the burned area, increased DNBR values would correspond to increased burn severity on the ground.

Ground Verification

The project team visited the Sanford Fire site in 2003, 1 year after the fire, and conducted onsite estimations of the CBI for selected plots. CBI provided an index to represent the magnitude of fire effects. Through the numeric scale between 0.0 and 3.0, CBI described how much a fire had altered the biophysical conditions of a site. The CBI plots data provided reliable references to connect field measurements

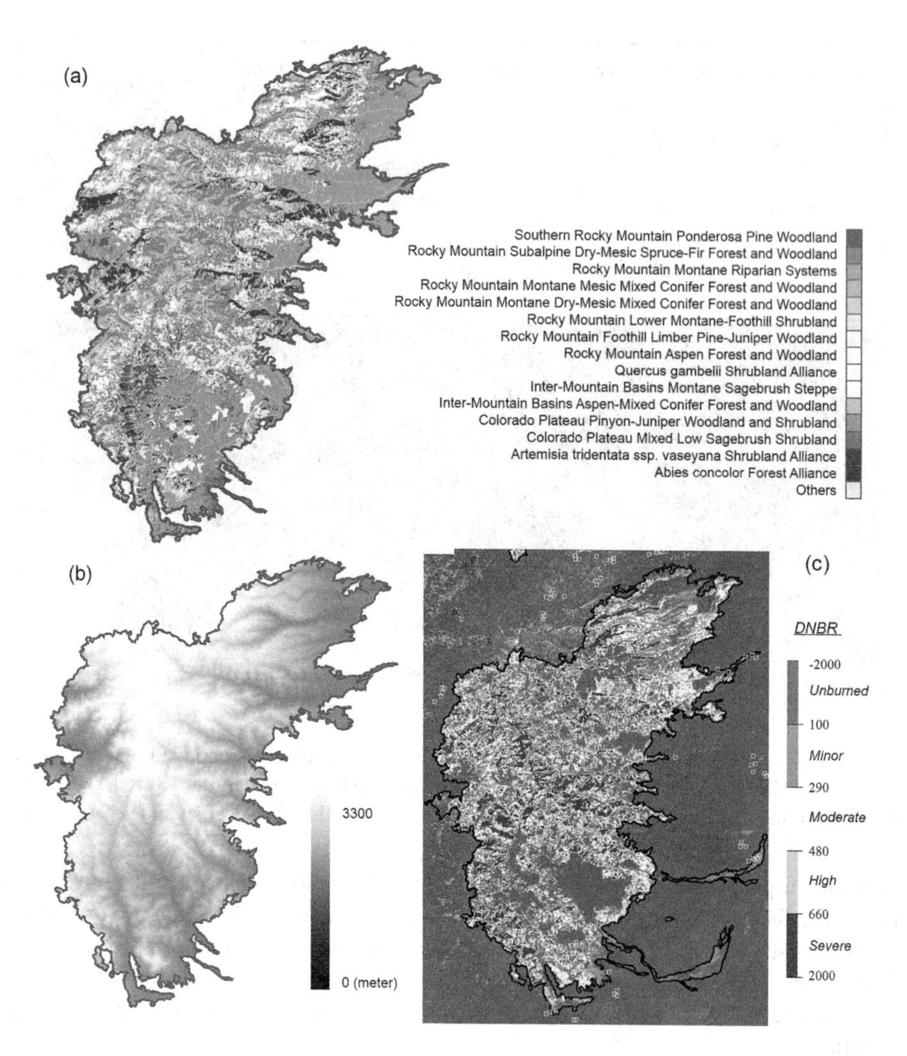

FIGURE 46.2 (See color insert.) The EVT data (**a**) and DEM (**b**) provided pre-fire ecosystem data and environmental setting for the Sanford Fire site. The DNBR data (c) provided scales of burn severity on each pixel location as estimated measurements of fire impacts. For Plot 66: Pre-fire vegetation was identified as Rocky Mountain Mesic Mixed Conifer Forest and Woodland, canopy >50% and <60%; canopy height (10, 25 m); DNBR = 996; CBI total = 2.79. The 2005 site visit observed 50% vegetation cover, predominantly herbaceous species, creeping barberry, with evident of dense resprout of aspen. There was no evidence of recovery of fir or coniferous. For Plot 32: Pre-fire vegetation was identified as Colorado Plateau Mixed Low Sagebrush Shrubland, canopy >30% and <40%; canopy height (0, 0.5 m); DNBR = 193; CBI total = 2.72. The 2005 site visit observed good recovery 3 years after the fire with 60%–70% vegetation cover, predominantly herbaceous species.

with remote sensing-derived DNBR data. The project team conducted field verification at the Sanford Fire site in September 2005 to observe the status of vegetation recovery 3 years after the fire and to examine the relationship between burn severity, fire age, and post-fire vegetation recovery (Figure 46.3). We employed a Trimble® GeoXT Global Positioning System unit for navigation to the selected CBI plot sites that were previously examined in 2003. We recorded locations of field transects and points of interests on both Landsat and DNBR data so that the burn severity and recovery status could be evaluated. We paid special attentions to locations with apparent signs of past burns for references. We documented the status of vegetation recovery in comparison with estimations from previous field observations (Table 46.2) by georeferenced field photographs.

TABLE 46.1 Pre-Fire EVT within the Sanford Fire Site

Code	Existing Vegetation Type from LANDFIRE Prototype Data	Abbreviations
2011	Rocky Mountain Aspen Forest and Woodland	RMA
2016	Colorado Plateau Pinyon-Juniper Woodland and Shrubland	CPP
2049	Rocky Mountain Foothill Limber Pine-Juniper Woodland	RMF
2051	Rocky Mountain Montane Dry-Mesic Mixed Conifer Forest and Woodland	RMD
2052	Rocky Mountain Montane Mesic Mixed Conifer Forest and Woodland	RMM
2054	Southern Rocky Mountain Ponderosa Pine Woodland	SRM
2055	Rocky Mountain Subalpine Dry-Mesic Spruce-Fir Forest and Woodland	RMS
2061	Inter-Mountain Basins Aspen-Mixed Conifer Forest and Woodland	IMA
2159	Rocky Mountain Montane Riparian Systems	RMR
2208	Abies concolor Forest Alliance	ACF
2064	Colorado Plateau Mixed Low Sagebrush Shrubland	CPM
2086	Rocky Mountain Lower Montane-Foothill Shrubland	RML
2125	Inter-Mountain Basins Montane Sagebrush Steppe	IMM
2217	*Quercus gambelii* Shrubland Alliance	QGS
2220	*Artemisia tridentata* ssp. *vaseyana* Shrubland Alliance	ATS

Source: Wang et al. (2011).

FIGURE 46.3 **(See color insert.)** An example of plot locations displayed on top of the post-fire Landsat TM image and DNBR map.

Source: Wang et al. 2011.

TABLE 46.2 Examples of Pre-Fire EVT, DNBR, CBI and Field Observations for Selected Referencing Plot Sites

Plot ID	Characteristics of Pre-Fire EVT	DNBR	CBI	Observed Recovery after 3 Years
T-B	Colorado Plateau Pinyon-Juniper Woodland and Shrubland. Forest canopy >20% and <30%; canopy height (10, 25 m)	369	2.61	65% vegetation cover predominantly herbaceous species
51	Colorado Plateau Pinyon-Juniper Woodland and Shrubland. Forest canopy >0% and <10%; canopy height (0, 5 m)	−18	0.74	50%–60% vegetation cover, predominantly herbaceous species. Pinyon Juniper survived; resprout of quicken aspens
90	Colorado Plateau Pinyon-Juniper Woodland and Shrubland. Forest canopy >40% and <50%; canopy height (10, 25 m)	497	2.75	<50% vegetation cover, 10% grasses, 30% mixed shrubs predominantly herbaceous species; no evidence of recovery of Mountain Mahogany that dominated plot area before fire
66	Rocky Mountain Montane Mesic Mixed Conifer Forest and Woodland. Forest canopy >50% and <60%; canopy height (10, 25 m)	996	2.79	50% vegetation cover, predominantly herbaceous species, dense resprout of aspen; no evidence of fir or coniferous recovery
32	Colorado Plateau Mixed Low Sagebrush Shrubland. Shrub canopy >30% AND <40%; shrub height (0, 0.5 m)	193	2.72	60%–70% vegetation cover, predominantly herbaceous species
86	Rocky Mountain Subalpine Dry-Mesic Spruce-Fir Forest and Woodland. Forest canopy >30% and ≤40%; forest height (10, 25 m)	1,099	2.92	40%–50% vegetation cover, predominantly herbaceous species, dense resprout of aspen
86B	Rocky Mountain Subalpine Dry-Mesic Spruce-Fir Forest and Woodland. Forest canopy >40% and ≤50%; forest height (10, 25 m)	401	1.42	40% vegetation cover, predominantly herbaceous species; sign of conifer seeding; resprout of aspen

Source: Wang et al. (2011).

Parameterization for LANDIS Simulation

Parameterization for LANDISv4.0a includes three major steps: (i) development of a land-type map and the species establishment coefficients for each land type; (ii) development of a species vital attribute table; and (iii) establishment of initial states for a simulation.

Land -Type Attribute and Establishment Coefficients

Land-type attributes encapsulate environmental variations and can be created from abiotic data sources such as climate, soil, geology, and topography. We employed LANDFIRE Biogradients data and adopted the method by Wimberly (2004) to express the topographic moisture gradient. We classified the landscape into three categories: *bottomlands*, *hillslopes*, and *ridges*. Bottomlands are distributed on flat terrain adjacent to major streams that include hydric and mesic sites in the study area. Hillslopes are areas with intermediate moist on hill sides. Ridges are the driest areas and on the gently sloping uplands. This study considered two slope aspects of northwest (NW) and southeast (SE) in defining the land type combined with the topographic moisture gradient. Elevation is another factor that defines a land type as spatial distributions of tree species are associated with variations of elevation ranges. We used the pre-fire EVT and the DEM data from the LANDFIRE data products to obtain distributions of vegetation species in different elevations. We calculated the areas in number of pixels represented by the EVT data within fire impacted areas. By referencing the calculated number of pixels from the EVT data, we divided the Sanford Fire site into four most influential elevation categories on vegetation distributions: <2,450 m, 2,450–2,700 m, 2,700–2,900 m, and >2,900 m. We then combined those features to create the land-type attribute file. As the slope and aspects affect mostly the areas on hillslopes rather than the bottomlands and ridges, we applied the slope and aspects to the hillslopes only. The final 16 land types are illustrated in Figure 46.4.

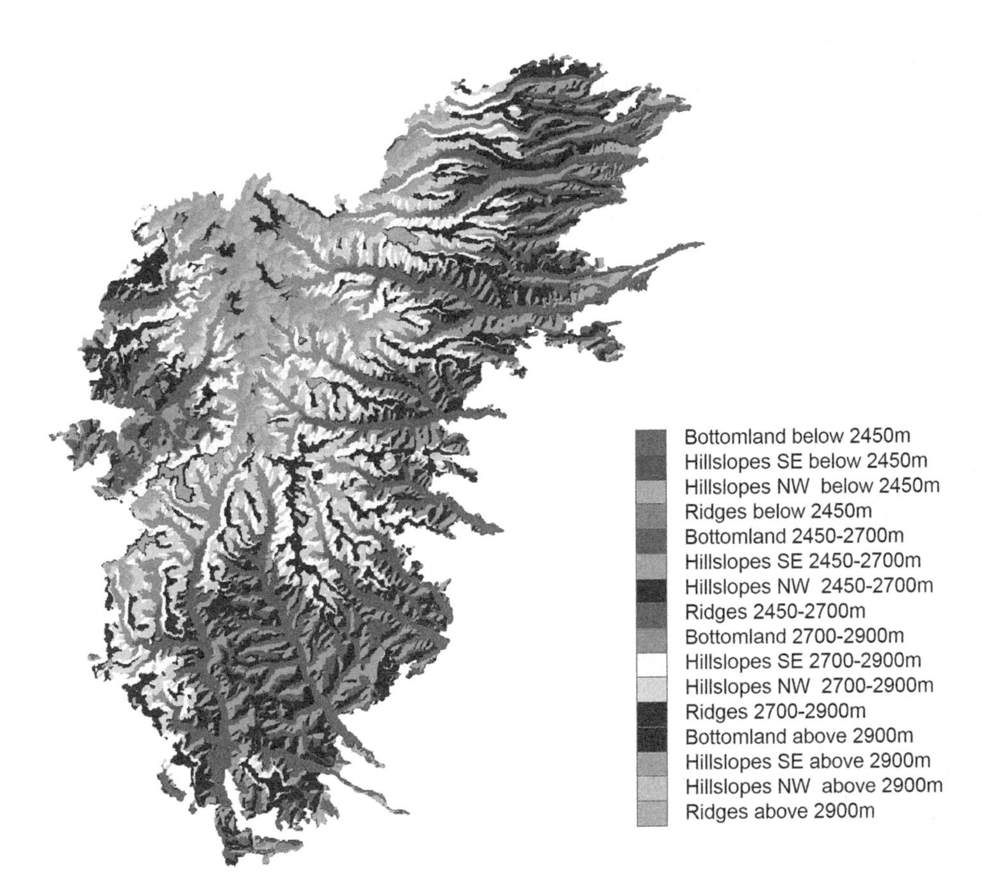

FIGURE 46.4 Land-type data developed by integrations of topographic moisture gradients, slope aspects, and elevations.
Source: Wang et al. 2011.

A species establishment coefficient is a number ranging from 0 to 1 that expresses the species' relative ability to grow on different land types. Coefficients are differentiated based on relative responses of species to soil moisture, climate, and nutrients. Reported studies estimated species establishment coefficients from published summaries of species characteristics (Burns and Honkala 1990, Sutherland et al. 2000) and from studies of community composition in present-day and pre-settlement forests (Cowell 1995, Wimberly 2004). As information about pre-fire vegetation composition within different land types could be extracted from EVT data, it would be efficient and effective to estimate establishment coefficients in each land type based on the existence of species from the pre-fire EVT. We obtained the percentage of vegetation species within defined land types based on EVT and calculated the occurrence frequencies by number of pixels for the main vegetation species within each of the 16 land types. We then obtained the ratio between the pixels of each vegetation species within each land type and the total pixels of that vegetation species, and the ratio between the pixels of each vegetation species within each land type and the total pixels of that land type. By multiplication of the two ratios, we derived the establishment coefficients for the 18 main vegetation species for each of the 16 land types.

Species Vital Attributes

Species vital attributes include longevity, mature age, shade tolerance, fire tolerance, effective seeding distance, maximum seeding distance, vegetation propagation probability, and maximum sprouting age. We derived the attributes for the simulated species from the Silvics of North America

(Burns and Honkala 1990) and Plant Database (NRCS 2009). We defined the effective seed dispersal ranges as 50 m for gravity-dispersed species, 100 m for large wind-dispersed winged seeds, 150 m for small wind-dispersed winged seeds, and 200 m for small plumed seeds (Sutherland et al. 2000, Wimberly 2004). For wind-dispersed seeds, we assumed the maximum dispersal distances as doubled the effective dispersal distance. For seeds with animal or bird dispersal vectors, we assumed the maximum dispersal distance to be 3,000 m. We added the annual grass category in final species attribute table (Table 46.3) to reflect the fact of significant recovery of annual grass species immediately after a wildfire.

Species Composition Map

The species composition map consists of species and their age classes. There are different ways to create species composition map (He et al. 1998). In this study, we referenced the canopy height from the LANDFIRE life-form data and the descriptions of the vegetation characteristics from the NRCS Plant Database to derive age information and generated the species composition map.

Fire Severity Classes and the Fire Regime

In LANDISv4.0a simulation, the fire effect module simulates which species age cohorts are killed on each burned pixel. We referenced the DNBR data to establish the initial state for the simulation. We considered six classes of fire severity as follows:

1. *No fire* (DNBR value <100)
2. *Severity class 1* (DNBR value 100–250)

TABLE 46.3 Species Vital Attribute Table for the Main Vegetation Species in the Study Area

Species Name	LONG	MATURE	SHADE	FIRE	EFFD	MAXD	VEG_P	SP_AG	RCLS_COEFF
Pinus edulis	600	25	1	2	50	3,000	0	0	0.5
Juniperus osteosperma	600	30	1	2	50	3,000	0	0	0.5
Pinus contorta	200	8	1	2	100	3,000	0.5	8	0.3
Picea engelmannii	600	40	4	2	100	200	0	0	0.5
Populus tremuloides	85	3	1	5	200	3,000	1	1	0.1
Pinus ponderosa	450	15	1	5	50	3,000	0	0	0.3
Pseudotsuga menziesii	500	14	2	3	100	3,000	0	0	0.3
Abies lasiocarpa	250	20	4	2	100	3,000	0	0	0.2
Abies concolor	350	40	4	5	100	200	0	0	0.3
Juniperus scopulorum	300	20	1	2	50	3,000	0	0	0.2
Quercus gambelii	120	6	1	5	50	3,000	1	1	0.1
Pinus flexilis	600	30	1	2	30	3,000	0	0	0.5
Cercocarpus montanus	54	10	3	5	100	200	1	1	0.1
Artemisia tridentata	45	2	1	5	30	60	0	0	0.1
Chrysothamnus nauseosus	35	4	1	2	150	300	1	1	0.1
Artemisia nova	45	2	1	2	50	100	0.2	1	0.1
Amelanchier utahensis	20	3	3	5	50	3,000	1	1	0.1
Annual grass	20	1	3	1	200	3,000	1	2	0.00005

Source: Wang et al. (2011).

LONG, maximum longevity (years); MATURE, age of reproductive maturity (years); SHADE, shade tolerance (1: least shade tolerant, 5: most shade tolerant); FIRE, fire tolerance (1: least fire tolerant, 5: most fire tolerant); EFFD, effective seeding distance (m); MAXD, maximum seedling distance (m); VEG_P, vegetation propagation coefficient; SP_AG, maximum age of vegetative propagation; RCLS_COEFF, reclassification coefficient (0–1).

3. *Severity class 2* (DNBR value 250–400)
4. *Severity class 3* (DNBR value 400–550)
5. *Severity class 4* (DNBR value 550–700)
6. *Severity class 5* (DNBR value >700).

Given that the DNBR data defined burn severity classes at a pixel level, we treated each pixel as a fire regime and set the mean fire size as the pixel size and the standard deviation of fire size as 0. In doing so, we were able to simulate the fire perimeter and severity as those defined by the DNBR data.

Simulation

To simplify the simulation, we combined two ecosystems of *Artemisia tridentata ssp. vaseyana Shrubland Alliance* (Table 1 Code 2220) and *Inter-Mountain Basins Montane Sagebrush Steppe* (Table 1 Code 2125) into Colorado Plateau Mixed Low Sagebrush Shrubland (Table 1 Code 2064) for the reason that these ecosystems share similar main sagebrush species. With added herbaceous category (HBR) to reflect the post-fire growth of annual grass species, the final simulation included 14 ecosystems as shown in the section "Result".

The simulation starts at year 0, i.e., the time right after the fire occurred. Then, the simulation proceeds at 1-year interval, as *year* (*i*) represents the simulated vegetation recovery at the *i*th year after the fire ($i = 1, 2, \ldots, 10$). We assumed that no new fire occurs in the 10-year simulation time period within any burned cell.

Result

A post-fire summary by the Dixie National Forest suggested that sagebrush would likely become re-established through seeds that are already present in the soil. Aspen was present throughout the burned area as "pure stands" or interspersed with conifers. Although the fire killed many aspen trees, it also enhanced aspen reproduction as the fire stimulated the growth of suckers from the aspen's extensive root system. In many instances, the fire left behind bare mineral soil and removed taller plants, which created a suitable condition for aspen seedlings to take root. Areas of mixed conifers, spruce fir, and ponderosa pine will take much longer to become after fires.

The simulation results illustrate post-fire change of vegetation species within the ecosystems 10 years after the fire impact (Table 46.4). For example, areas of HBR increased at different fire severity classes 10 years after the fire. The areas with a higher level of fire severity show more significant increase of herbaceous vegetation than the areas of "no-fire" category. The ecosystems that include the aspen species, e.g., the Rocky Mountain Aspen Forest and Woodland (RMA) and Abies concolor Forest Alliance (ACF), showed the similar recovery trends as the HBR, in particular for the higher burn severity classes. For the ecosystems of Colorado Plateau Pinyon-Juniper Woodland and Shrubland (CPP) and Rocky Mountain Subalpine Dry-Mesic Spruce-Fir Forest and Woodland (RMS), the recoveries are slower than those ecosystems with aspen species. The higher the fire severity, the slower the recovery would be in percentage of areas. Aspen would have a good recovery in the land types such as SE-facing hillslopes.

Species fire tolerance and species establishment are main drivers for post-fire species responses. The simulated results reflect the change of age cohorts in the ecosystems. For example, the simulated RMA shows increases in area after the fire, which reflects the resprout of aspen species in the RMA ecosystem. The higher level the fire severity classes, the more significant the increase in RMA with time. The increase in areas is gradually reduced through time with the age cohorts changing within the RMA areas. For the other ecosystems such as CPP and RMS, the changes are slow for the 10-year simulation. Changes of ecosystems are negligible for the no-fire areas in 10 years.

TABLE 46.4 The Simulated Ecosystems in Different Fire Severity Levels for the Pre-Fire, 0 Year and 10 Years in Percentage Areas

		No Fire (%)	Severity Class 1 (%)	Severity Class 2 (%)	Severity Class 3 (%)	Severity Class 4 (%)	Severity Class 5 (%)	Total (%)
RMA	Pre-fire	3.34	3.27	3.00	1.94	1.08	0.65	13.28
	0 year	3.34	3.27	6.16	4.95	3.26	0.00	20.97
	10 years	2.70	3.69	5.29	3.85	2.47	3.13	21.15
ACF	Pre-fire	0.45	0.63	0.90	1.10	1.42	2.08	6.57
	0 year	0.45	0.63	1.21	1.46	2.16	0.00	5.91
	10 years	1.16	2.60	3.71	3.48	3.79	3.64	18.38
HBR	Pre-fire	0.00	0.00	0.00	0.00	0.00	0.00	0.00
	0 year	0.00	0.00	0.00	0.00	0.00	0.00	0.00
	10 years	0.03	3.85	3.65	1.85	0.94	1.39	11.71
IMA	Pre-fire	2.32	2.79	3.16	3.02	2.18	1.67	15.13
	0 year	2.32	2.79	0.00	0.00	0.00	0.00	5.10
	10 years	1.79	2.91	1.25	0.98	0.71	0.93	8.58
RMS	Pre-fire	2.53	3.18	3.07	2.41	2.30	4.33	17.81
	0 year	2.53	0.00	0.00	0.00	0.00	0.00	2.53
	10 years	2.53	1.11	0.83	0.57	0.45	0.80	6.29
RMM	Pre-fire	0.19	0.24	0.31	0.36	0.58	1.18	2.87
	0 year	0.19	0.24	0.00	0.00	0.00	0.00	0.44
	10 years	0.76	1.19	0.39	0.32	0.27	0.59	3.52
CPP	Pre-fire	9.08	6.74	6.28	2.79	0.83	0.36	26.08
	0 year	9.08	0.00	0.00	0.00	0.00	0.00	9.08
	10 years	9.09	1.30	1.53	0.68	0.17	0.05	12.81
RMR	Pre-fire	0.83	0.74	0.69	0.42	0.16	0.07	2.91
	0 year	0.83	0.74	0.69	0.42	0.00	0.00	2.68
	10 years	0.86	1.08	0.93	0.52	0.06	0.04	3.48
RMD	Pre-fire	0.30	0.30	0.47	0.38	0.17	0.06	1.68
	0 year	0.30	0.30	0.47	0.38	0.00	0.00	1.45
	10 years	0.56	0.56	0.72	0.56	0.01	0.01	2.42
SRM	Pre-fire	1.37	0.79	0.60	0.35	0.17	0.14	3.42
	0 year	1.37	0.79	0.60	0.35	0.00	0.00	3.11
	10 years	1.11	0.56	0.39	0.20	0.01	0.00	2.26
CPM	Pre-fire	4.20	1.96	0.25	0.02	0.00	0.00	6.44
	0 year	4.20	1.96	0.25	0.02	0.16	0.00	6.60
	10 years	4.21	2.02	0.29	0.04	0.15	0.00	6.71
QGS	Pre-fire	0.66	0.52	0.08	0.01	0.00	0.00	1.27
	0 year	0.66	0.52	0.08	0.01	0.00	0.00	1.27
	10 years	0.66	0.53	0.09	0.01	0.00	0.00	1.29
RML	Pre-fire	0.68	0.34	0.09	0.01	0.00	0.00	1.12
	0 year	0.68	0.34	0.09	0.01	0.00	0.00	1.12
	10 years	0.59	0.37	0.12	0.03	0.01	0.02	1.12
RMF	Pre-fire	0.16	0.18	0.24	0.24	0.14	0.06	1.01
	0 years	0.16	0.00	0.00	0.00	0.00	0.00	0.16
	10 years	0.26	0.00	0.00	0.00	0.00	0.00	0.26
Other	Pre-fire	0.20	0.13	0.05	0.02	0.01	0.00	0.43
	0 year	0.20	10.23	9.64	5.46	3.45	10.61	39.59
	10 years	0.00	0.02	0.00	0.00	0.00	0.00	0.02
Total (%)		*26.30*	*21.79*	*19.19*	*13.08*	*9.03*	*10.61*	*100*

Source: Wang et al. (2011).

Conclusion

The combination of LANDFIRE data products, the DNBR data, and LANDISv4.0a modeling demonstrated an approach in the simulation of short-term post-fire vegetation recovery. LANDFIRE data products, such as EVT, life-form, and the Biogradients data, played a unique role in defining the land type, species establishment coefficients, and species attribute table. Pre-fire EVT provided an efficient and effective data source to estimate establishment coefficients for each land type based on the existence of the species. DNBR data made the measurements of fire severity classes possible. DNBR data, teamed up with LANDFIRE data products, defined fire regimes from identified fire impacts. The integration provided critical data to define and establish initial states for the simulation.

In simulation, each species responds differently to the levels of fire severity according to the fire tolerance of species and the level of fire severity. For the cells with moderate level of fire severity, LANDIS simulation was able to estimate changes in vegetation structure. Fire tolerance of tree species and fire severity determined the post-fire canopy gaps in different ecosystems. For example, as *Pinus ponderosa* could survive in a low level of fire severity, the fire severity would contribute to the simulation to create canopy gaps in ecosystems with *Pinus edulis*. The simulated spatial patterns for ecosystems and species demonstrated the trends of post-fire variation of vegetation, in particular for quick recovery species such as aspens. However, it's difficult to observe changes on slowly recovery species and ecosystems dominated by those species since this simulation was limited to 10 years only.

Climatic variations are among influencing factors for post-fire establishment of vegetation. Climatic conditions in early years after a wildfire should be critical for short-term post-fire vegetation recovery. Precipitations in summer months should increase survival rate of newly grown vegetation. Varying weather cycle may affect the establishment coefficient, given that seed germination depends on the patterns of precipitations. The simulated vegetation recovery would be more appropriate with climate-modified establishment coefficients, in particular for short-term simulations.

References

Beukema, S.J., Kurz, W.A., Klenner, W., Merzenich J. and Arbaugh M. Applying TELSA to assess alternative management scenarios. In *Systems Analysis in Forest Resources*, G. Arthaud and T. Barrett (Eds.) **2003**, pp. 145–154 (Amsterdam: Kluwer).

Brewer, C.K., Winne, J.C., Redmond, R.L., Opitz, D.W. and Mangrich, M.V. Classifying and mapping wildfire severity: A comparison of methods. *Photogrammetric Engineering and Remote Sensing* **2005**, 71, pp. 1311–1320.

Burns, R.M. and Honkala B.H. Silvics of North America: 1. Conifers; 2. Hardwoods. *Agriculture Handbook 654*, **1990**, vol. 2, p. 877 (Washington, DC: USDA Forest Service). Available online at: http://na.fs.fed.us/spfo/pubs/silvics_manual/table_of_contents.htm (accessed on August 12, 2019).

Comer, P., Faber-Langendoen, D., Evans, R., Gawler, S., Josse, C., Kittel, G., Menard, S., Pyne, M., Reid, M., Schulz, K., Snow, K. and Teague J. Ecological Systems of the United States: A Working Classification of U.S. Terrestrial Systems, **2003**. Available online at: http://na.fs.fed.us/spfo/pubs/silvics_manual/table_of_contents.htm (accessed on August 12, 2019).

Cowell, C.M. Presettlement Piedmont forests: Patterns of composition and disturbance in central Georgia. *Annals of the American Association of Geographers* **1995**, 85, pp. 65–83.

Dasgupta, S., Qu, J.J., Hao, X. and Bhoi, S. Evaluating remotely sensed live fuel moisture estimations for fire behavior predictions in Georgia, USA. *Remote Sensing of Environment* **2007**, 108, pp. 138–150.

De Santis, A. and Chuvieco, E. Burn severity estimation from remotely sensed data: Performance of simulation versus empirical models. *Remote Sensing of Environment* **2007**, 108, pp. 422–435.

De Santis, A., Chuvieco, E. and Vaughan, P.J. Short-term assessment of burn severity using the inversion of PROSPECT and GeoSail models. *Remote Sensing of Environment* **2009**, 113, pp. 126–136.

Fraser, R.H. and Li, Z. Estimating fire-related parameters in boreal forest using SPOT VEGETATION. *Remote Sensing of Environment* **2002**, 82, pp. 95–110.

Guo, M., Li, J., Xu, J., Wang, X., He, H. and Wu, L. CO_2 emissions from the 2010 Russian wildfires using gosat data. *Environmental Pollution* **2017**, 226, pp. 60–68.

Hann, W.J. and Bunnell, D.L. Fire and land management planning and implementation across multiple scales. *International Journal of Wildland Fire* **2001**, 10, pp. 389–403.

Hao, X. and Qu, J. Retrieval of real time live fuel moisture content using MODIS measurements. *Remote Sensing of Environment* **2007**, 108, pp. 130–137.

He, H.S., Mladenoff, D.J., Radeloff, V.C. and Crow, T.R. Integration of GIS data and classified satellite imagery for regional forest assessment. *Ecological Applications* **1998**, 8, pp. 1072–1083.

He, H.S. and Mladenoff, D.J. Spatially explicit and stochastic simulation of forest-landscape fire disturbance and succession. *Ecology* **1999**, 80, pp. 81–99.

He, H., Shang, B.Z., Crow, T.R., Gustafson, E.J. and Shifley, S.R. Simulating forest fuel and fire risk dynamics across landscapes: LANDIS fuel module design. *Ecological Modelling* **2004**, 180, pp. 135–151.

Hemstrom, M.A., Korol, J.J. and Hann, W.J. Trends in terrestrial plant communities and landscape health indicate the effects of alternative management strategies in the interior Columbia River basin. *Forest Ecology and Management* **2001**, 153, pp. 1–3.

Keane, R.E., Cary, G.J., Davies, I.D., Flannigan, M.D., Gardner, R.H., Lavorel, S., Lenihan, J.M., Li, C. and Rupp, T.S. A classification of landscape fire succession models: Spatial simulations of fire and vegetation dynamics. *Ecological Modelling* **2004**, 179, pp. 3–27.

Keeley, J.E. Fire intensity, fire severity and burn severity: A brief review and suggested usage. *International Journal of Wildland Fire* **2009**, 18(1), 116–126.

Klenner, W., Kurz, W.A. and Beukema, S.J. Habitat patterns in forested landscapes: Management practices and the uncertainty associated with natural disturbances. *Computers and Electronics in Agriculture* **2000**, 27, pp. 243–262.

Lentile, L.B., Holden, Z.A., Smith, A.M.S., Falkowski, M.J., Hudak, A.T., Morgan, P., Lewis, S.A., Gessler P.E. and Benson, N.C. Remote sensing techniques to assess active fire characteristics and post-fire effects. *International Journal of Wildland Fire* **2005**, 15, pp. 319–345.

Merzenich, J., Kurz, W.A., Beukema, S., Arbaugh, M. and Schilling, S. Determining forest fuel treatments for the Bitterroot front using VDDT. In *Systems Analysis in Forest Resources*, G. Arthaud and T. Barrett (Eds.) **2003**, pp. 47–59 (Amsterdam: Kluwer).

Miller, J.D. and Thode, A.E. Quantifying burn severity in a heterogeneous landscape with a relative version of the delta Normalized Burn Ratio (dNBR). *Remote Sensing of Environment* **2007**, 109, pp. 66–80.

Mladenoff, D.J. and He, H.S. Design and behavior of LANDIS, an object-oriented model of forest landscape disturbance and succession. In *Advances in Spatial Modeling of Forest Landscape Change: Approaches and Applications*, D. Mladenoff and W. Baker (Eds.) **1999**, pp. 163–185 (Cambridge: Cambridge University Press).

NRCS. Plant Database, **2009**. Available online at: http://plants.usda.gov (accessed on August 12, 2019).

Robinson, J.M. On uncertainty in the computation of global emissions from biomass burning. *Climatic Change* **1989**, 14, pp. 243–262.

Rollins, M.G. and Frame, C.K. The LANDFIRE prototype project: Nationally consistent and locally relevant geospatial data for wildland fire management. General Technical Report RMRS-GTR-175. Fort Collins: USDA Forest Service, Rocky Mountain Research Station, **2006**, p. 416. Available online at: www.treesearch.fs.fed.us/pubs/24484 (accessed on August 12, 2019).

Ross, A.N., Wooster, M.J., Boesch, H. and Parker, R. First satellite measurements of carbon dioxide and methane emission ratios in wildfire plumes. *Geophysical Research Letters* **2013**, 40, pp. 4098–4102.

Skowronski, N., Clark, K., Nelson, R., Hom, J. and Patterson, M. Remotely sensed measurements of forest structure and fuel loads in the Pinelands of New Jersey. *Remote Sensing of Environment* **2007**, 108, pp. 123–129.

Sturtevant, B.R., Gustafson, E.J. and He, H.S. Special issue: Modelling disturbance and succession in forest landscapes using LANDIS. *Ecological Modelling* **2004**, 180, pp. 1–232.

Sutherland, E.K., Hale, B.J. and Hix, D.M. Defining species guilds in the central hardwood forest, USA. *Plant Ecology* **2000**, 147, pp. 1–19.

Swanson, F.J., Cissel, J.H. and Reger, A. Chapter 9: Landscape management: Diversity of approaches and points of comparison. In *Compatible Forest Management*, R. Monserud, R. Haynes and A. Johnson (Eds.) **2003**, pp. 237–266 (Amsterdam: Kluwer).

Syphard, A.D., Yang, J., Franklin, J., He, H.S. and Keeley, J.E. Calibrating forest landscape model to simulate high fire frequency in Mediterranean-type shrublands. *Environmental Modelling and Software* **2007**, 12, pp. 1641–1653.

Turner, M.G. Disturbance and landscape dynamics in a changing. *World Ecology* **2010**, 91(10), pp. 2833–2849.

van Wagtendonk, J.W., Root, R.R. and Key, C. H. Comparison of AVIRIS and Landsat ETM+ detection capabilities for burn severity. *Remote Sensing of Environment* **2004**, 92, pp. 397–408.

Wang, Y., Zhou, Y., Yang, J. and He, H. Remote sensing assessments of wildfire impact and simulation modeling of short-term post-fire vegetation recovery within the Dixie National Forest (Chapter 14). In *Remote Sensing of Protected Lands*, Y. Wang (Ed.) **2011**, pp. 281–302 (Boca Raton, FL: CRC Press).

Westerling, A.L., Hidalgo, H.G., Cayan, D.R. and Swetnam, T.W. Warming and earlier spring increase western U.S. forest wildfire activity. *Science* **2006**, 313(5789), pp. 940–943.

White, J.D., Ryan, K.C., Key, C.C. and Running, S.W. Remote sensing of forest fire severity and vegetation recovery. *International Journal of Wildland Fire* **1996**, 6, pp. 125–136.

Wimberly, M.C. Fire and forest landscapes in the Georgia Piedmont: An assessment of spatial modeling assumptions. *Ecological Modelling* **2004**, 180, pp. 41–56.

Wimberly, M.C. and Reilly, M.J. Assessment of fire severity and species diversity in the southern Appalachians using landsat TM and ETM+ imagery. *Remote Sensing of Environment* **2007**, 108, pp. 189–197.

Wang, Y. (ed.) Yang, J. Fires: Wildland. In *Encyclopedia of Natural Resources: Land*, **2014**, pp. 210–213. (New York: Taylor and Francis).

47

Sustainable Agriculture: Social Aspects

Frederick H. Buttel
University of Wisconsin

The Definition of Sustainable Agriculture

It is widely agreed that sustainable agriculture is intrinsically a joint social and ecological construct. Thus, for example, Ikerd[1] stresses the "anthropocentric" as well as "ecocentric" nature of agricultural sustainability, noting that the essence of sustainable agriculture is that "we (in sustainable agriculture) are concerned about sustaining agriculture for the benefit of humans, both now and into the indefinite future." Sustainable agriculture is an agriculture that is "ecologically sound, economically viable, and socially responsible."[2] Allen similarly emphasizes that a conception of sustainable agriculture that fails to recognize the role of people, social actors, social institutions, and social movements is a limited one.[3]

Sustainable agriculture and organic farming are very closely related but are not exactly coterminous. Organic farming systems tend to be highly sustainable systems according to the ecological and social components of the definition of sustainable agriculture, and are often quite economically viable as well.[4,5] Many in the sustainable agriculture community, however, feel strongly that sustainability improvements in mainstream or conventional agriculture are as important as the expansion of organic farming and organic agro-food systems.

Structural Causes of the Lack of Sustainability

Agricultural sustainability is not only a vision of the long-term goals for agriculture, which imply measuring sticks for or indicators of its achievement.[6] Sustainability also implies a critique of or a set of concerns about "conventional" agriculture—that mainstream agriculture has shortcomings in terms of environmental soundness, economic soundness, and social justice—and suggests that there have been trends over time that jeopardize the achievement of these three ends.

The historical U.S. pattern of abundant land, limited and/or relatively expensive rural labor, and a strong tendency toward overproduction and low commodity prices has encouraged a capital-intensive, chemical-intensive monocultural system with a very high degree of enterprise and spatial homogeneity (or specialization). Commodity, trade, and public research policies have historically reinforced these tendencies to monoculture, specialization, and heavy reliance on chemicals, leading to an agriculture that has significant shortcomings on all of the criteria implied in the definition of sustainable agriculture.[7]

Sustainability and the Corporate-Industrial Agricultural Economy

Agricultural sustainability cannot be considered apart from the globalized and highly concentrated corporate economy within which sustainable agriculture, and agriculture as a whole, is enmeshed. Corporate concentration on input provision, food processing, and retailing limit the choices available to sustainable and other farmers. Corporate control over agriculture is increasing through processes such as the disappearance of "open markets" for agricultural products; the extension of vertical integration and contractual relationships; and the ever closer alignment of multinational chemical and seed companies on one hand, and food processors, manufacturers, and global trading companies on the other.[8] There is evidence that the increased corporate domination of agro-food systems has negative implications for achieving agricultural sustainability. Farmer contractees, for example, are typically directed contractually by "integrators" to use particular practices which, more often than not, are unsustainable.

It should be stressed, however, that corporations—often small ones, but also large firms—have been major agents in extending the scope of organic food production and organic marketing. There are two major nationwide organic chain grocery stores, and a number of corporations have been very successful in producing organic products such as potatoes, grapes, fresh fruits and vegetables, and so on. The corporate organization of the organic food industry represents a major dilemma for proponents of agricultural sustainability. Highly successful sustainable agriculture ideas will be attractive to private corporations, which will often be successful in diffusing products and practices. At the same time, corporate agricultural sustainability practices may undermine certain aspects of the sustainability agenda (e.g., large-scale monocultural production of organic potatoes in the Northern Intermountain states is associated with a considerable threat of soil erosion and loss of biodiversity).

There is considerable debate in the sustainable agriculture community as to whether the best indicator of the growth of sustainable agriculture is the national market share of organic food—much of which is due to the activities of very large corporations such as Whole Foods—or whether the extent of sustainable agriculture is most accurately measured by the prevalence of community-supported agriculture (CSA), farmers markets, local co-ops, community gardens, and direct marketing of food. Those with the strongest commitments to sustainable agriculture are also most likely to see food system localization and the re-creation of more local "foodsheds" as the heart and soul of a genuine and enduring sustainable agriculture.[9]

Sustainability as a Social Movement

Agricultural sustainability is as much a social movement as it is a set of technologies and production practices. Social definitions of sustainability—that is, the concerns, views, ideologies, and agendas of movement participants—are as or more important in shaping what sustainability is as ecological or biological indicators such as soil erosion rates, levels of soil organic matter, biodiversity, levels of runoff, and so on. What has put sustainability on the national and global agenda is primarily the organization and activities of three major types of groups: farmer-driven sustainable agriculture organizations and movements, consumer-driven or consumer-oriented sustainable agriculture initiatives (such as many CSA farms and most food co-ops and local food councils), and national and global environmental groups. Environmental groups have arguably done the most to increase the visibility and persuasiveness of the overall notion of sustainability, and of the particular concept of agricultural sustainability, though farmer-oriented groups do not always agree with environmentalists' agendas for agricultural sustainability.

Each of the three main types of sustainable agriculture organizations, however, agrees on the main components of the movement agenda: the need for more public research on sustainable practices, the need for public policy incentives for sustainability, the desirability of family farming, and the imperative to improve the environmental performance of agriculture. Increasingly, this consensus has extended to issues such as opposition to major trade liberalization agreements (the World Trade Organization and NAFTA) and opposition to genetic engineering of crops and foods.[10]

Public Policy and the Future of Agricultural Sustainability

Three necessary but insufficient conditions for the continued advance of sustainable agriculture are the development of improved sustainable technology; the presence of a vigorous and dynamic sustainability movement; and the development of a public policy environment that reduces the public policy disincentives to an environmentally sound, socially responsible, and economically viable agriculture. Of these three conditions, the public policy environment of agricultural sustainability is arguably the most important over the long term, even though the macro-public-policy environment is difficult to redirect, and there are areas of public policy disagreement among farmers, consumers, and environmental actors in the sustainable agriculture community. For example, some sustainability advocates, particularly those in the environmental community (and some farm groups in that community), believe that the federal (and global) levels of action are most important or efficacious, whereas others believe that at this time the local (community or regional) arena is where advocates can most easily make a difference. Nonetheless, it is apparent that over the long term sustainable agriculture cannot advance far in a public policy environment that involves the strong disincentives to sustainability that currently prevail. There is a need for more active federal, state, and local regulation of both the on-site and the off-site impacts of agriculture; for ending the commodity-driven pattern of federal agricultural policy that shovels the lion's share of subsidies in the direction of large, monocultural producers of overproduced commodities; and for redirection of the public agricultural research agenda. Some of the most innovative public policy ideas in sustainable agriculture are being developed in Europe. Green payments, taxes on pesticides and fertilizers, and the embracement of a "multifunctionality" approach to government agricultural policy are particularly promising policy instruments.[11] Each of these policy instruments would support sustainability and yield other benefits (reduced government outlays, rural development, reduced greenhouse gas emissions). Still, there are no technocratic shortcuts to sustainability. Sustainability is, and must remain, as much a social movement as it is a set of practices and measuring sticks of agro-food system performance.

References

1. Ikerd, J.E. Assessing the health of ecosystems: A socioeconomic perspective. Paper Presented at the First International Ecosystem Health and Medicine Symposium, Ottawa, Canada, 1994. (http://www.ssu.missouri.edu/faculty/jikerd/papers/Otta-ssp.htm), p. 2 (accessed October 2002).
2. Ikerd, J.E. New farmers for a new century. Paper Presented at the Youth in Agriculture Conference. Ulvik, Norway, February, 2000. (http://www.ssu.missouri.edu/faculty/jikerd/papers/Newfarmerl.htm), p. 3 (accessed October 2002).
3. Allen, P. Connecting the Social and Ecological in Sustainable Agriculture. In *Food for the Future: Conditions and Contradictions of Sustainability*; Allen, P., Ed.; Wiley: New York, 1993; 1–16.
4. Flora, C.B.; Francis, C.; King, L., Eds. *Sustainable Agriculture in Temperate Zones;* Wiley-Interscience: New York, 1990.
5. *Interactions between Agroecosystems and Rural Communities;* Flora, C.B., Ed.; CRC Press: Boca Raton, FL, 2001.
6. Airstars, G.A., Ed. *A Life Cycle Approach to Sustainable Agriculture Indicators: Proceedings. Center for Sustainable Systems;* University of Michigan: Ann Arbor, MI, 1999.
7. Buttel, F.H. The sociology of agricultural sustainability: Some observations on the future of sustainable agriculture. Agric. Ecosyst. Environ. **1993**, *46*, 175–186.

8. Heffernan, W.D. *Consolidation in the Food and Agriculture System*; National Farmers Union: Denver, CO, 1999. (http://www.nfu.org/images/heffernan_1999.pdf) (accessed October 2002).
9. Kloppenburg, J., Jr.; Hendrickson, J.; Stevenson, G.W. Coming into the foodshed. Agric. Human Values **1996**, *13*, 33–42.
10. Kirschenmann, F. Questioning Biotechnology's Claims and Imagining Alternatives. In *Of Frankenfoods and Golden Rice*; Buttel, F., Goodman, R., Eds.; Wisconsin Academy of Sciences, Arts, and Letters: Madison, WI, 2002; 35–61.
11. Organisation for Economic Cooperation and Development (OECD). *Multidimensionality: A Framework for Analysis*; OECD: Paris, 1998.

48

Sustainability and Sustainable Development

Alan D. Hecht
U.S. Environmental Protection Agency (EPA)

Joseph Fiksel
The Ohio State University

Introduction

Sustainable development was defined by the Brundtland Commission in a landmark 1987 United Nations report as "economic and social development that meets the needs of the present without compromising the ability of future generations to meet their own needs."[1] "Sustainability" has been widely characterized as a concept resting on three pillars of human well-being: environmental protection, economic prosperity, and social justice or equity (Figure 48.1).

The U.S. National Research Council defined sustainability as both a *goal* and a *process* that improves the economy, the environment, and society for the benefit of current and future generations.[2] As a *goal,* sustainability is aimed at meeting society's basic economic and social needs without undermining the natural resource base and environmental quality that are necessary for continuing to meet these needs in the future. Sustainability (or sustainable development) also describes the *process* whereby innovative tools, models, and approaches are applied in order to meet that goal, striving to advance economic prosperity while protecting natural resources.

FIGURE 48.1 The three pillars of sustainability.

In the United States, the concept of sustainable development has roots that are over a century old. President Theodore Roosevelt relied on Gifford Pinchot, Roosevelt's chief forester and the "father of American conservation," for advice to develop public land policies in the late 1890s. Pinchot believed that forests should be made to serve the nation's future as well as its present. He was an early advocate of sustainable development, writing in his 1910 book, *The Fight for Conservation,* "the central thing for which conservation stands is to make this country the best possible place to live in, both for us and for our descendants."[3]

Nearly 50 years later, in 1969, the National Policy and Environmental Act (NEPA) established a national goal of creating and maintaining "conditions under which [humans] and nature can exist in productive harmony, and fulfill the social, economic and other requirements of present and future generations of Americans."[4]

Sustainability has been investigated worldwide since the Club of Rome, composed of European economists and scientists, published "The Limits to Growth" in 1972.[5] This controversial report warned of the negative impact of economic growth on the environment and the need for strategic planning to balance economic growth and environmental protection. The concept of sustainability was further developed at the 1992 Earth Summit in Rio de Janeiro, where 179 nations endorsed the principles of Agenda 21, which declared, "The right to development must be fulfilled so as to equitably meet developmental and environmental needs of present and future generations."[6]

Today sustainability is incorporated into many government goals and policy objectives. The European Union has been promoting sustainability since 2000, when it formulated the Lisbon Strategy with the goal of achieving the most competitive and dynamic knowledge-based economy in the world. The European Council has stated that "Clear and stable objectives for sustainable development would present significant economic opportunities and the potential to unleash a new wave of technology innovation, generating growth and employment."[7]

Sustainability was advanced in the United States through Executive Orders issued by Presidents George W. Bush and Barack H. Obama, setting specific goals for how the U.S. government should manage Federal facilities. These were superseded by a comprehensive 2015 Executive Order, which included Federal sustainable procurement policies such as purchasing green and energy-efficient products.[8] At the same time, the U.S. government has sought to promote sustainability internationally through the first presidential "Policy Directive on Global Development," which aims to "place a premium on broad-based economic growth, democratic governance, game-changing innovations, and sustainable systems for meeting basic human needs."[9]

Around the world, sustainable business practices are advocated by numerous organizations such as the Global Reporting Initiative and the United Nations Environmental Programme (UNEP). At the state and local levels, an association of over 1,220 local governments from 70 different countries, known as ICLEI—Local Governments for Sustainability, offers technical consultation, training, and other resources to implement sustainable development.[10] Many cities in the United States have adopted sustainability goals, and numerous states have established governing bodies responsible for advancing sustainable practices.[11]

A Sense of Urgency

The Global Footprint Network estimates that humanity's *ecological footprint* has roughly doubled over the past 50 years.[12] This footprint represents the pressures on land and water area caused by the industrial and lifestyle patterns of human population. An urgent challenge for global society is to "decouple" the global ecological footprint from continued economic growth.

World population is expected to reach nine billion people by 2050.[13] Almost all the increase will take place in developing nations, where hundreds of millions of people are seeking more secure access to food, clothing, and shelter, as well as sanitation, education, health care, energy, communication, and

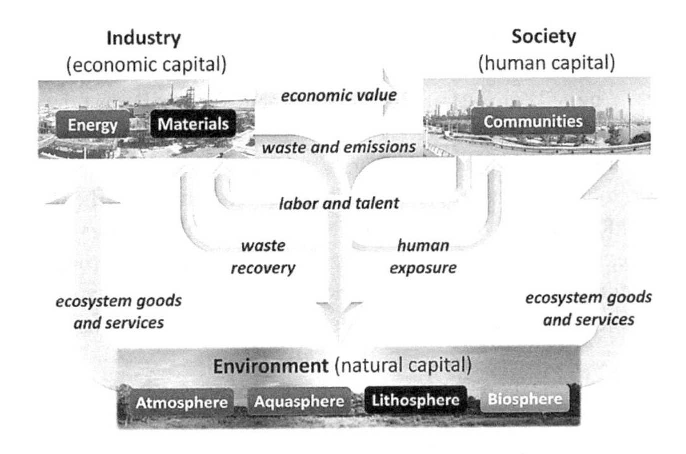

FIGURE 48.2 Systems view of sustainability, adapted from Fiksel's Triple Value Model.[15,23]

consumer goods. Coupled with increasing levels of resource consumption in developed nations, this trend will place significant stress on global ecosystems. Ironically, the pressures on natural resources in developing nations could impede industrialization and further widen existing income gaps between rich and poor.

In addition to its impact on global warming, the inexorable growth of the world economy is contributing to many disturbing trends: sea level is rising, available fresh water and arable land are growing scarce, forests are disappearing, and biodiversity is threatened by changing natural habitats. The impacts of population growth and economic development are detailed in the 2005 Millennium Ecosystem Assessment, which found that 15 of 24 important global ecosystem services are being degraded or used unsustainably.[14]

Accomplishing sustainable development in an increasingly interconnected global economy will require a holistic "systems approach" to problem solving. Figure 48.2 illustrates the flows of value from ecosystems to both industrial and social systems, as well as the flows of waste and emissions back into the environment. From the economic perspective, sustainability aims to create value for both shareholders and society while efficiently using resources and minimizing undesirable impacts. An urgent challenge for global sustainability is to enable continued economic growth while reducing the rates of both resource use and waste generation.[15] For example, companies that import components and feedstocks from overseas must be mindful of potential adverse consequences across their supply chains, including human rights violations and environmental pollution. Such indirect effects, known as *externalities*, are frequently ignored in cost–benefit calculations but may compromise supply chain viability over time.

Private Sector Opportunities

While governmental leadership is helpful in promoting sustainability, companies generally recognize that voluntary sustainable business practices can support corporate goals. Reducing resource demand typically leads to reduced purchasing and operating costs, so that sustainable development enhances a company's bottom line while reducing its environmental footprint. Moreover, sustainability enhances shareholder value by improving brand image, reputation, and stakeholder relationships. Consequently, sustainability strategies have increasingly been adopted by major companies around the world.

For example, hundreds of companies compete to be listed on the Dow Jones Sustainability Indexes, in which sustainability is defined as "a business approach that creates long-term shareholder value by embracing opportunities and managing risks that derive from economic, environmental and social developments."[16] A 2009 study published in the *Harvard Business Review* concluded that "sustainability is a mother lode of organizational and technological innovations that yield both bottom-line and top-line returns" and that "there is no alternative to sustainable development."[17]

The World Business Council for Sustainable Development (WBCSD) has developed an ambitious agenda to assist global industries in moving toward sustainable growth. The WBCSD Vision 2050 report coined the phrase "green race," and outlines a "pathway that will require fundamental changes in governance structures, economic frameworks, business and human behavior." The report argues that these changes are "necessary, feasible and offer tremendous business opportunities for companies that turn sustainability into strategy."[18] The non-profit organization CERES lays out a road map for sustainability based on assessing the company's baseline environmental and social performance, analyzing corporate management and accountability structures and systems, and conducting a materiality analysis of risks and opportunities.[19]

Financial industries, including banking, investment, and insurance have recognized the importance of sustainability and protection of natural resources. The Equator Principles, launched in Washington in June 2003, require that banks assess the social and environmental impacts of projects that they finance and have been adopted by 94 financial institutions in 37 countries. Influenced by advocacy groups such as the Rainforest Action Network, Citigroup went beyond the Equator Principles by refusing to fund projects that could result in illegal logging, other environmental damage, or harm to indigenous people.[20]

While these corporate and institutional commitments are valuable, scientific and technological innovations will be critical elements in achieving genuine progress in sustainability. The required scientific knowledge to apply a systems approach will come from many disciplines—including chemistry, biology, and physics as well as economic and social sciences. New approaches such as "transdisciplinary" research and "sustainability science" have been introduced to describe the fusion of multiple disciplines to generate new knowledge, which is important for advancing sustainable innovation.[21]

There are many excellent examples of innovations that dramatically improve the sustainability of products and processes. For example, traditional wood adhesives often contain formaldehyde, which is toxic to humans. Researchers have been able to develop new adhesives from soy flour, which are environmentally friendly, stronger, and cost competitive. These new adhesives have replaced more than 50% of formaldehyde-based adhesives and have earned a Presidential Green Chemistry Challenge Award.[22]

Finally, it is important to assure the continued resilience of environmental, economic, and social systems in the face of turbulent change. *Resilience* is defined as the capacity to not only absorb disruptions but also adapt and flourish in the face of turbulent change.[23] In order to cope with unpredictable events, whether caused by humans or nature, infrastructure and supply chain systems must be designed for long-term resilience. One example is harnessing ecosystem services to provide "green infrastructure" that helps to protect community-based water resource systems from storms, floods, and other climate disruptions.[24]

Eight Steps to Sustainability

Sustainability represents the next level of environmental protection. In previous centuries, the traditional approaches to environmental issues focused on land conservation and risk mitigation. However, the 21st century brings a new set of environmental challenges that are more complex and globally interconnected, including climate change, freshwater degradation, and biodiversity loss. These new challenges demand new, more integrated approaches that account for the linkages among environmental resources, national security, human well-being, and economic competitiveness.[25] In addition to regulatory approaches, a variety of strategies to achieve both sustainability and resilience, including public–private collaboration, integrated systems thinking, technological innovation, and adaptive management will be needed. The following suggests eight practical steps toward global sustainability.[26]

1. **Take the Long View.** Sustainability requires long-term thinking, not only in natural resource management and urban development but also in corporate strategic planning. Taking a long view leads us to consider emerging pressures on ecosystems at different scales of resolution and how the built environment and ecosystems can be managed in a synergistic way. For industrial innovators, the long view involves thinking about new business models in which material and energy resources can be used more effectively, while wastes and hazardous residuals can be reduced or eliminated.
2. **Understand the System Dynamics.** Humans are members of multiple complex, dynamic systems—in both the material world and cyberspace. As population growth and increased consumption place pressures on global ecosystems while the world becomes more tightly connected, small perturbations or changes in any one system can quickly cascade to other systems. Traditional industrial and engineered systems were not created with a systems view and may therefore be vulnerable to unexpected disruptions, such as natural disasters, industrial accidents, or deliberate sabotage. By understanding system vulnerabilities and leverage points, we can develop cost-effective, sustainable, and resilient management strategies.
3. **Define Ambitious Goals.** President Kennedy set a goal for the United States to land on the moon. This goal was achieved because it was fully supported with the needed resources and polices. Making sustainable development an explicit policy goal at all levels of government, as well as the private sector, can send a clear message about the need for breakthrough innovation and creative regulatory approaches that serve the collective interests of the public, business, and government. A commitment to sustainability can be a key component of technology and policy innovation that reinforces U.S. leadership and economic progress.
4. **Use Effective Tools.** In order to practice sustainable development, advanced tools and approaches are necessary. Science and technology are key underlying contributors, along with innovative environmental and economic policies. New instrumentation, data handling, and methodological capabilities have expanded our understanding of the environment and of how complex biological, chemical, and human systems interact. A growing body of economic and environmental assessment tools is available to support planners and decision makers at all levels of government and industry.
5. **Find the Right Collaborators.** Around the world, collaboration and partnerships among stakeholders are crucial to achieving solutions that are less polarized, more economically viable, and focused on balancing short- and long-term goals. Collaboration has begun among government, industry, and non-governmental organizations and has already yielded many sustainable innovations.
6. **Lead by Example.** The profound changes needed to achieve sustainability will require confidence and bold leadership. The most persuasive approach for overcoming uncertainty and hesitation is leading by example—demonstrating that these changes are both realistic and beneficial. Government can help by establishing framework conditions and incentives that encourage innovation and by initiating collaborative pilot projects that bridge knowledge gaps and overcome inertia.
7. **Measure and Track Progress.** The use of sustainability indicators is essential for an integrated systems approach to the multifaceted challenges of sustainability. Indicators can help managers and policy makers anticipate and assess key trends, provide early warning of potential disruptions, quantify progress toward sustainability goals, and support decision-making about complex trade-offs. A broad spectrum of indicators is available to capture the economic, environmental, and social attributes of the systems that we strive to manage.
8. **Learn and Adapt.** The path to sustainability cannot be planned precisely due to the enormous complexity and uncertainty inherent in global political, economic, social, and natural systems. To navigate this path successfully, we cannot cling to preconceived notions or prescriptive solutions. We must be prepared to learn from experience, rethink our assumptions, and continuously adapt to change. Taking a resilient approach in the short term will enable sustainability in the long term.

Conclusion

Current prevailing approaches to energy use and resource management are not sustainable. Moving from the status quo toward sustainability requires new approaches to problem solving that take an entire system into account, rather than isolated components. The challenge of achieving sustainability is urgent since unsustainable behavior—whether it is economic, social, or environmental—can lead to national and global crises. For example, when withdrawal rates from freshwater aquifers exceed recharge rates, water scarcity is the eventual result. Conversely, sustainable practices can preserve and protect natural capital and enhance economic growth. The challenge of realizing sustainability goals raises several practical questions: "What kind of government policies and business strategies can advance us toward sustainability?" "How can business and government work together to advance sustainable practices?" and "What can citizens do to advance the concept of sustainability?" Making sustainability operational is not something a government or business can do alone. A sustainable and thriving economy will require the convergence of four major elements: (i) regulations and policies, (ii) advances in science and technology, (iii) enlightened business practices, and (iv) public support and participation. A global commitment to accelerate movement toward sustainability is essential to safeguard social well-being, shared prosperity, and our natural environment. There is really no alternative.

Acknowledgments

The authors are grateful to Edward Fallon and Meadow Anderson for their helpful edits and comments to the original 2014 publication. The current version was updated by Dr. Joseph Fiksel, now an emeritus faculty member with The Ohio State University. Dr. Alan Hecht passed away in April 2019 after a distinguished career in public service.

Disclaimer

Views expressed in this entry are those of the authors and do not necessarily reflect the views or policies of the U.S. EPA. Mention of trade names or commercial products does not constitute EPA endorsement or recommendations for use.

References

1. World Commission on Environment and Development. *Our Common Future*; Oxford University Press: New York. 1987. www.un.org/documents/ga/res/42/ares42-187.htm (accessed March 2019).
2. National Research Council. 2011. Sustainability and the U.S. EPA. Washington, DC: The National Academies Press. https://doi.org/10.17226/13152.
3. Pinchot G. *The Fight for Conservation*; Doubleday, Page & Company: New York. 1910.
4. National Environmental Policy Act (NEPA). www.epa.gov/nepa (accessed March 2019).
5. Meadows D.H., Meadows D.K., Randers R., Behrens W.W. *The Limits to Growth*; Universe Books: New York. 1972.
6. United Nations. Division for Sustainable Development. *Agenda 21*. 1987. https://sustainabledevelopment.un.org/outcomedocuments/agenda21 (accessed March 2019).
7. Larsson A., Azar C., Sterner T.D., Strömberg D.B., Andersson B. *Technology and Policy for Sustainable Development*; Centre for Environment and Sustainability at Chalmers University of Technology and the Göteborg University: Goteborg. 2002.
8. Obama B.H., Executive Order 13693—Planning for Federal Sustainability in the Next Decade. 2015. www.govinfo.gov/content/pkg/FR-2015-03-25/pdf/2015-07016.pdf (accessed March 2019).
9. Obama B.H. *Fact Sheet: U.S. Global Development Policy*. 2010. www.whitehouse.gov/the-press-office/2010/09/22/fact-sheet-us-global-development-policy (accessed March 2019).

10. ICLEI - Local Governments for Sustainability. http://icleiusa.org/ (accessed March 2019).
11. Engel K.H., Miller M.L. State Governance: Leadership on Climate Change (Arizona Legal Studies Discussion Paper No. 07-37). In *Agenda for a Sustainable America*; Dernbach, J.C., Ed.; Environmental Law Institute: Washington, DC. 2009. http://papers.ssrn.com/sol3/papers.cfm?abstract_id=1081314 (accessed October 2011).
12. WWF and Zoological Society of London. *Living Planet Report 2018: Aiming Higher.* www.footprintnetwork.org/content/uploads/2018/10/LPR-2018-full-report.pdf (accessed March 2019).
13. United Nations. *World Population to 2300.* https://warwick.ac.uk/fac/soc/pais/research/researchcentres/csgr/green/foresight/demography/united_nations_world_population_to_2300.pdf (accessed March 2019).
14. Millennium Ecosystem Assessment. *Guide to the Millennium Assessment Reports*; Island Press: Washington, DC. 2005.
15. Fiksel J., *Design for Environment, Second Edition: A Guide to Sustainable Product Development*; McGraw-Hill: New York. 2011.
16. *Dow Jones Sustainability Indexes, in Collaboration with SAM. Corporate Sustainability.* www.sustainability-indices.com (accessed March 2019).
17. Nidumolu R., Prahalad C.K., Rangaswami M.R. Why sustainability is now the key driver of innovation. *Harvard Bus Rev.* 2009, *87*, 57–64.
18. World Business Council for Sustainable Development. *Vision 2050: The New Agenda for Business.* 2010. www.wbcsd.org/Overview/About-us/Vision2050 (accessed March 2019).
19. Ceres. *The 21st Century Corporation: The Ceres Roadmap for Sustainability.* www.ceres.org/ceres-roadmap (accessed March 2019).
20. Swiss Re Center for Global Dialogue. *Capitalizing on Natural Resources.* New Dynamics in Financial Markets. 9th International Sustainability Leadership Symposium; Swiss Re Centre for Global Dialogue: Rüschlikon. 2008.
21. Kates R.W., Parris T.M. Long-term trends and a sustainability transition. *Proc Nat Acad Sci.* 2003, *100* (14), 8062–8067.
22. U.S. EPA. *The Presidential Green Chemistry Challenge.* www.epa.gov/greenchemistry/green-chemistry-challenge-winners (accessed March 2019).
23. Fiksel J. *Resilient by Design: Creating Businesses That Adapt and Flourish in a Changing World*; Island Press: Washington, DC. 2015.
24. U.S. EPA. *Green Infrastructure.* www.epa.gov/green-infrastructure (accessed March 2019).
25. Hecht A.D., Fiksel J., Fulton S.C., Yosie T.F., Hawkins N.C., Leuenberger H., Golden J., Lovejoy T.E. Creating the future we want. *Sustain Sci Prac Pol.* 2012, *8*, (2), 62.
26. Hecht A.D. *Making America Green and Safe: A History of Sustainable Development and Climate Change.* Cambridge Scholars Publishing: Newcastle upon Tyne. 2018. p. 8.

49

Urban Environments: Remote Sensing

Xiaojun Yang
Florida State University

Introduction

The concentration of a large population in urban areas continues to be a major form of global change felt in both developing and developed counties alike (Foley et al., 2005). While urban growth and intensifying development are generally considered signs of the vitality of a regional economy, they have rarely been well planned, thus provoking the concerns over the degradation of our environment and ecosystem integrity (Grimm et al., 2008). Monitoring urban growth and landscape changes is critical to those who study urban dynamics and those who must manage resources and provide services in urban areas (Seto et al., 2012).

Assessment of urban growth and landscape changes involves the procedure of inventorying and mapping that can help derive a reliable information base (Yang, 2002). Urban and landscape patterns are observable and therefore can be mapped through ground surveys or remote sensing. While ground surveys are largely localized by nature, remote sensing makes direct observations across a large area of the land surface, thus allowing urban landscapes to be mapped in a timely and cost-effective mode. Nevertheless, the complexity of the urban environment challenges the applicability and robustness of remote sensing (Jensen and Cowen, 1999). Encouragingly, recent innovations in data, technologies, and theories in the wider arena of remote sensing and geospatial technologies have provided us with invaluable opportunities to advance the studies on the urban environment (Yang, 2011a).

This topical entry provides a brief overview on some major utilities of remote sensing for the urban environment. While an overwhelming share of the existing work has focused on using remote sensing to derive various urban physical features, we have witnessed a steady and strengthening expansion of the applicability beyond this domain and into other important and emergent areas to support various efforts toward urban sustainability (Seto et al., 2017). The following sections provide an overview on some major areas of research and discuss several issues needing further research.

Remote Sensing and Urban Analysis

Remote Sensing Systems for Urban Areas

Remote sensor data used for the urban environment should meet certain conditions in terms of spatial, spectral, radiometric, and temporal characteristics (Jensen and Cowen, 1999). There is a wide variety of passive and active remote sensing systems acquiring data with various resolutions that can be useful for urban analysis. Medium-resolution remote sensor data have been used to examine large-dimensional urban phenomena or processes since early 1970s when Landsat-1 was successfully launched (Yang, 2011b). Images acquired by the US Landsat program, French SPOT satellites, and the EU Copernicus Program, along with those by the Indian remote sensing satellites (IRS), the NASA Terra satellite, the China–Brazil Earth resources satellites (CBERS), are the principal sources of data under this category. With the launch of IKONOS, the world's first commercial, high-resolution imaging satellite, on September 24, 1999, very-high-spatial-resolution satellite imagery became available, which allow detailed work concerning the urban environment. Images under this category include those acquired by QuickBird (launched in 2001), FormoSat-2 (launched in 2004), WorldView-1 (launched in 2007), GeoEye-1 (launched in 2008), WorldView-2 (launched in 2009), Pleiades-1 (launched in 2011), SPOT-6 (launched in 2012), Gaofen-1 (launched in 2013), SPOT-7 (launched in 2014), WorldView-3 (launched in 2014), WorldView-4 (launched in 2016), FormoSat-5 (launched in 2017), among others. Independent of weather conditions, active remote sensing systems, such as airborne or space-borne radar, can be particularly useful for such applications as housing damage assessment or ground deformation estimation in connection to some disastrous events in urban areas (Dell'Acqua et al., 2011). Another active sensor system, similar in some respects to radar, is LIDAR (light detection and ranging), which can be used to derive height information useful for reconstructing three-dimensional city models (Li and Guan, 2011) or detailed land use mapping (Meng et al., 2012).

In addition to the above systems, data acquired from several new and emergent platforms or media, such as Unmanned Aerial Systems (UAS) (or drones), social sensing, and street views, have increasingly been used to support various urban applications (Jiang et al. 2016; Seiferling et al., 2017; Dodge, 2018).

Lastly, various image fusion techniques allow the combined use of images varying in spatial, spectral, temporal, or radiometric resolution, which can help improve urban mapping accuracy (Schneider, 2012; Huang et al., 2017; Arévalo et al., 2019).

Algorithms and Techniques for Urban Attribute Extraction

The urban environment is characterized by the presence of heterogeneous surface covers with large inter-pixel and intra-pixel spectral and spatial variations, thus challenging the applicability and robustness of conventional image processing algorithms and techniques. Largely built upon parametric statistics, conventional pattern classifiers generally work well for medium-resolution scenes covering spectrally homogeneous areas but not in heterogeneous regions such as urban areas or when scenes contain severe noises due to the increase of the image's spatial resolution. Developing improved image processing algorithms and techniques for working with different types of remote sensor data has therefore become a very active research area in urban remote sensing. For years, various strategies have been developed to improve urban mapping performance at the per-pixel, sub-pixel, or object level (Yang, 2011b; Verhoeye and Wulf, 2002; Walker and Briggs, 2007). While the pattern recognition approach at the per-pixel level has been widely adopted, the one at the sub-pixel level allows the decomposition of each pixel into independent endmembers or pure materials to map urban sub-pixel composition (Liu and Yang, 2013), which can be particularly useful for working with medium-resolution images. While the approach at the per-pixel or sub-pixel level heavily relies upon image spectral features, the object-oriented approach allows the combined use of spectral and spatial characteristics (Blaschke et al., 2014).

Several new and emergent areas of development are worth mentioning here. Firstly, for artificial intelligence, we have witnessed a steady expansion of moving beyond shallow learning algorithms and into deep learning models consisting of multiple processing layers to learn representations of data with hierarchical abstraction, which can help improve urban mapping from remote sensor data (Sharma et al., 2017, 2018). Secondly, multiple classifier systems (or classifier ensembles) have increasingly been used to generate better final outcomes for a classification task through combining a set of single classifiers (Shi and Yang, 2017). Lastly, cloud computing platforms, such as Google Earth Engine (GEE; https://earthengine.google.com/) and NASA Earth Exchange (NEX; https://c3.nasa.gov/nex/), are now available to execute large-scale remote sensor data analysis (Gorelick et al., 2017; Huang et al., 2017).

Urban Social and Environmental Analysis and Growth Modeling

Applying remote sensing to socioeconomic analysis has been an expanding research area in urban remote sensing. There are two major types of analysis. The first type centers on linking socioeconomic data to land change data derived from remote sensing in order to identify the drivers of landscape changes (Lo and Yang, 2002). This type of analysis usually needs to integrate image analysis with geographic information systems and spatial statistical analysis. The other type focuses on the development of indicators of urban socioeconomic status by combined use of remote sensing and census or field-survey data (Yu and Wu, 2006). For example, defining and measuring urban sprawl, which is considered as an important issue for urban and regional planning, involves not only urban spatial characteristics derived from remote sensing but also GIS-based socioeconomic conditions such as population density and transportation (Frenkel and Orenstein, 2011). Another example is the estimation of small area population; this type of information is critical for decision-making by both public and private sectors but is only available for one date per decade. Remote sensing can provide an alternative to derive reliable population estimation in a timely and cost-effective fashion (Wang and Cardenas, 2011).

Although urban areas are quite small relative to the global land cover, they significantly alter hydrology, biodiversity, biogeochemistry, and climate at local, regional, and global scales (Grimm et al., 2008). Understanding environmental consequences of urbanization is a critical concern to both the planning and global change science communities. Urban environmental analysis can help understand the status, trends, and threats in urban areas so that appropriate management actions can be planed and implemented. This is a research area in which remote sensing can play a critical role. For example, remote sensing can be used to derive the spatial distribution of impervious surfaces in urban areas that can be linked with water quality indicators (Yang, 2006). Remote sensing can be also used to estimate gross primary production in urban areas that can be associated with various settlement densities; such knowledge can help understand the net carbon exchange between land and the atmosphere due to urban development (Buyantuyev and Wu, 2009). Remote sensing can be used to characterize biodiversity in urban areas that can help assess how urbanization affects biodiversity (Hedblom and Söderström, 2010).

A group of important activities in urban studies is to understand urban dynamics and to assess future urban growth impacts on the environment. There are two major types of models that can be used to support such activities: analytical models that are useful to explain urban expansion and evolving patterns as well as dynamic models that can be used to predict future urban growth and landscape changes. The group of dynamic models can be particularly valuable because of their predictive power that can be used to imagine, test, and assess the spatial consequences of urban growth under specific socioeconomic and environmental conditions (Brown et al., 2013; van Vliet et al., 2016). The role of remote sensing is indispensable in the entire model development process from model conceptualization to implementation that includes input data preparation, model calibration, and model validation (Yang and Lo, 2003).

Conclusions and Further Research

This topical entry provides an overview on some major utilities of remote sensing applied in urban areas and discusses some major areas of research in urban remote sensing. While numerous exciting progresses have been made in urban remote sensing, there are several major conceptual or technical areas deserving further attention. Firstly, although the development of remote sensing has largely been technology driven, urban remote sensing professionals should be equipped with not only solid technical skills but also essentials of intellectual knowledge on the urban environment that can help better plan and implement an urban remote sensing project. Secondly, while existing digital image processing efforts overwhelmingly focus on detecting fast-paced changes, such as land use conversion, recent studies suggest the importance of monitoring continuous land use activities with lower change rates (such as land modifications) using satellite time series. Thirdly, more efforts are needed to balance the different needs by remote sensing and urban planning communities. Finally, more efforts are needed to develop innovative data models used for representing dynamic processes, to identify improved methods and techniques that can be used to deal with data incompatibility in terms of parameter measuring and sampling schemes, and to develop more realistic dynamic models that can be used to support various efforts toward urban sustainability.

References

Arévalo, P.; Olofsson, P.; Woodcock, C.E. Continuous monitoring of land change activities and post-disturbance dynamics from Landsat time series: A test methodology for REDD+ reporting. *Remote Sensing of Environment* 2019, doi:10.1016/j.rse.2019.01.013.

Blaschke, T.; Hay, G.J.; Kelly, M.; Lang, S.; Hofmann, P.; Addink, E.; Feitosa, R.Q.; Van der Meer, F.; Van der Werff, H.; Van Coillie, F.; Tiede, D. Geographic object-based image analysis–Towards a new paradigm. *ISPRS Journal of Photogrammetry and Remote Sensing* 2014, 87: 180–191.

Brown, D.G.; Verburg, P.H.; Pontius Jr., R.G.; Lange, M.D. Opportunities to improve impact, integration, and evaluation of land change models. *Current Opinion in Environmental Sustainability* 2013, 5(5): 452–457.

Buyantuyev, A.; Wu, J. Urbanization alters spatiotemporal patterns of ecosystem primary production: A case study of the Phoenix metropolitan region, USA. *Journal of Arid Environments* 2009, 73: 512–520.

Dell'Acqua, F.; Gamba, P.; Polli, D. Very-high-resolution spaceborne synthetic aperture radar and urban areas: Looking into details of a complex environment. In *Urban Remote Sensing: Modeling, Synthesis and Modeling in the Urban Environment*; Yang, X. Ed; Wiley-Blackwell: Chichester, UK, 2011; 63–73.

Dodge, M. Mapping II: News media mapping, new mediated geovisualities, mapping and verticality. *Progress in Human Geography* 2018, 42(6): 949–958.

Foley, J.A.; DeFries, R.; Asner, G.P., et al. Global consequences of land use. *Science* 2005, 309(5734): 570–574.

Frenkel, A.; Orenstein, D. A pluralistic approach to defining and measuring urban sprawl. In *Urban Remote Sensing: Modeling, Synthesis and Modeling in the Urban Environment*; Yang, X. Ed; Wiley-Blackwell: Chichester, UK, 2011; 165–181.

Gorelick, N.; Hancher, M.; Dixon, M.; Ilyushchenko, S.; Thau, D.; Moore, R. Google Earth Engine: Planetary-scale geospatial analysis for everyone. *Remote Sensing of Environment* 2017, 202: 18–27.

Grimm, N.B.; Faeth, S.H.; Golubiewski, N.E., et al. Global change and ecology of cities. *Science* 2008, 319(5864): 756–760.

Hedblom, M.; Söderström, B. Landscape effects on birds in urban woodlands: An analysis of 34 Swedish cities. *Journal of Biogeography* 2010, 37(7): 1302–1316.

Huang, H.; Chen, Y.; Clinton, N.; Wang, J.; Wang, X.; Liu, C.; Gong, P.; Yang, J.; Bai, Y.; Zheng, Y.; Zhu, Z. Mapping major land cover dynamics in Beijing using all Landsat images in Google Earth Engine. *Remote Sensing of Environment* 2017, 202: 166–176.

Jensen, J.R.; Cowen, D.C. Remote sensing of urban/suburban infrastructure and socio-economic attributes. *Photogrammetric Engineering and Remote Sensing* 1999, 65(5): 611–622.

Jiang, B.; Ma, D.; Yin, J.; Sandberg, M. Spatial distribution of city tweets and their densities. *Geographical Analysis* 2016, 48(3): 337–351.

Li, J.; Guan, H. 3D building reconstruction from airborne lidar point clouds fused with aerial imagery. In *Urban Remote Sensing: Modeling, Synthesis and Modeling in the Urban Environment*; Yang, X. Ed; Wiley-Blackwell: Chichester, UK, 2011; 75–91.

Liu, T.; Yang, X. Mapping vegetation in an urban area with stratified classification and multiple endmember spectral mixture analysis. *Remote Sensing of Environment* 2013, 133: 251–264

Lo, C.P.; Yang, X. Drivers of land-use/land-cover changes and dynamic modeling for the Atlanta, Georgia Metropolitan Area. *Photogrammetric Engineering and Remote Sensing* 2002, 68(10): 1073–1082.

Meng, X.; Currit, N.; Wang, L.; Yang, X. Detect residential buildings from lidar and aerial photographs through object-oriented land-use classification. *Photogrammetric Engineering and Remote Sensing* 2012, 78(1): 35–44.

Schneider, A. Monitoring land cover change in urban and peri-urban areas using dense time stacks of Landsat satellite data and a data mining approach. *Remote Sensing of Environment* 2012, 124: 689–704.

Seiferling, I.; Naik, N.; Ratti, C.; Proulx, R. Green streets - Quantifying and mapping urban trees with street-level imagery and computer vision. *Landscape and Urban Planning* 2017, 165: 93–101.

Seto, K.C.; Reenberg, A.; Boone, C.G.; Fragkias, M.; Haase, D.; Langanke, T.; Marcotullio, P.; Munroe, D.K.; Olah, B.; Simon, D. Urban land teleconnections and sustainability. *Proceedings of the National Academy of Sciences* 2012, 109(20): 7687–7692.

Seto, K.C.; Golden, J.S.; Alberti, M.; Turner, B.L. Sustainability in an urbanizing planet. *Proceedings of the National Academy of Sciences* 2017, 114(34): 8935–8938.

Sharma, A.; Liu, X.; Yang, X.; Shi, D.A. patch-based convolutional neural network for remote sensing image classification. *Neural Networks* 2017, 95: 19–28.

Sharma, A.; Liu, X.; Yang, X. Land cover classification from multi-temporal, multi-spectral remotely sensed imagery using patch-based recurrent neural networks. *Neural Networks* 2018, 105: 346–355.

Shi, D.; Yang, X. Mapping vegetation and land cover in a large urban area using a multiple classifier system. *International Journal of Remote Sensing* 2017, 38(16): 4700–4721.

van Vliet, J.; Bregt, A.K.; Brown, D.G.; van Delden, H.; Heckbert, S.; Verburg, P.H. A review of current calibration and validation practices in land-change modeling. *Environmental Modelling & Software* 2016, 82: 174–182.

Verhoeye, J.; Wulf, R.D. Land cover mapping at sub-pixel scales using linear optimization techniques. *Remote Sensing of Environment* 2002, 79(1): 96–104.

Walker, J.S.; Briggs, J.M. An object-oriented approach to urban forest mapping in Phoenix. *Photogrammetric Engineering and Remote Sensing* 2007, 73(5): 577–583.

Wang, L.; Cardenas, J.S.; Small area population estimation with high-resolution remote sensing and lidar. In *Urban Remote Sensing: Modeling, Synthesis and Modeling in the Urban Environment*; Yang, X. Ed; Wiley-Blackwell: Chichester, UK, 2011; 183–193.

Yang, X. Satellite monitoring of urban spatial growth in the Atlanta metropolitan region. *Photogrammetrical Engineering and Remote Sensing* 2002, 68(7): 725–734.

Yang, X. Estimating Landscape Imperviousness Index from Satellite Imagery. *IEEE Geosciences and Remote Sensing Letters* 2006, 3(1): 6–9.

Yang, X. What is urban remote sensing? In *Urban Remote Sensing: Modeling, Synthesis and Modeling in the Urban Environment*; Yang, X. Ed; Wiley-Blackwell: Chichester, UK, 2011a; 3–11.

Yang, X. Use of archival Landsat data to monitor urban spatial growth. In *Urban Remote Sensing: Modeling, Synthesis and Modeling in the Urban Environment*; Yang, X. Ed; Wiley-Blackwell: Chichester, UK, 2011b; 5–34.

Yang, X.; Lo, C.P. Modeling urban growth and landscape change for Atlanta metropolitan region. *International Journal of Geographical Information Science* 2003, 17(5): 463–488.
Yu, D.L.; Wu, C.S. Incorporating remote sensing information in modeling house values: A regression tree approach. *Photogrammetric Engineering and Remote Sensing* 2006, 72(2): 129–138.

Bibliography

Bhatta, B., *Analysis of Urban Growth and Sprawl from Remote Sensing Data*; Springer-Verlag: Berlin, Germany, 2010.
Donnay, J.P.; Barnsley, M.J.; Longley, P.A., Eds. *Remote Sensing and Urban Analysis*; Taylor & Francis: London, UK, 2001.
Gamba, P.; Herold, M., Eds. *Global mapping of Human Settlement: Experiences, Datasets, and Prospects*. CRC Press: Boca Raton, FL, 2009.
Jensen, R.R.; Gatrell, J.D.; McLean, D., Eds. *Geo-Spatial Technologies in Urban Environments: Policy, Practice, and Pixels* (2nd); Springer: Berlin, Germany, 2007.
Mesev, V., Ed. *Remotely-Sensed Cities*; CRC Press: Boca Raton, FL, 2003.
Netzband, M.; Stefanov, W.L.; Redman, C.L., Eds. *Applied Remote Sensing for Urban Planning, Governance and Sustainability*; Springer: Berlin, Germany, 2007.
Rashed, T.; Jürgens, C., Eds. *Remote Sensing of Urban and Suburban Areas*; Springer: Berlin, Germany, 2010.
Ridd, M.K.; Hipple, J.D., Eds. *Manual of Remote Sensing, Volume 5: Remote Sensing of Human Settlements*; ASPRS: Bethesda, MD, 2006.
Weng, Q., Ed. *Remote Sensing of Impervious Surfaces*; CRC Press: Boca Raton, FL, 2008.
Weng, Q., Ed. *Global Urban Monitoring and Assessment through Earth Observation*; CRC Press: Boca Raton, FL, 2014.
Weng, Q.; Quattrochi, D., Eds. *Urban Remote Sensing*; CRC Press: Boca Raton, FL, 2007.
Weng, Q.; Quattrochi, D.; Gamba, P.E., Eds. *Urban Remote Sensing* (2nd); CRC Press: Boca Raton, FL, 2018.
Xian, G.Z., *Remote Sensing Applications for the Urban Environment*; CRC Press: Boca Raton, FL, 2015.
Yang, X., Ed. *Urban Remote Sensing: Monitoring, Synthesis and Modeling in the Urban Environment*; Wiley-Blackwell: Chichester, UK, 2011.

Index

F

G

H

I

J

L

M

N

O

P

Q

R

S

T

U

V

W